QUANTUM GRAVITY 2

QUANTUM GRAVITY 2

A SECOND OXFORD SYMPOSIUM

EDITED BY

C. J. ISHAM
R. PENROSE
AND
D. W. SCIAMA

CLARENDON PRESS · OXFORD
1981

Oxford University Press, Walton Street, Oxford OX2 6DP
London Glasgow New York Toronto
Dehli Bombay Calcutta Madras Karachi
Kuala Lumpur Singapore Hong Kong Tokyo
Nairobi Dar es Salaam Cape Town
Melbourne Wellington
and associate companies in
Beirut Berlin Ibadan Mexico City

Published in the United States by
Oxford University Press, New York

British Library Cataloguing in Publication Data
Quantum gravity 2.
1. Gravitation—Congresses
2. Quantum theory—Congresses
I. Isham, C. J. II. Penrose, R.
III. Sciama, D. W.
531'.5 QC178

ISBN 0-19-851952-4

Printed in Great Britain by
Thomson Litho Ltd., East Kilbride, Scotland

PREFACE

The union of quantum theory with Einstein's general theory of relativity remains one of the major unattained goals of present-day theoretical physics. In February 1974, our first Oxford Symposium on Quantum Gravity was held, and we took stock of the ideas that were current at that time, publishing an account of them in our first Quantum Gravity volume.* In April 1980 a second conference on quantum gravity was held in Oxford, this time under the generous sponsorship of the Nuffield Foundation. Six years had passed since our first meeting and we sought to review and analyse the developments that had occurred in the intervening years. The second set of lectures is published in the present volume. Some of these provide review articles covering general or specific areas of interest whilst others are of a more technical nature, reflecting the increasing complexity and sophistication of current mathematical techniques.

Significant advances have been made in several branches of theoretical physics during the past decade. In particular, and of considerable relevance to quantum gravity, has been a renaissance in quantum field theory following the discovery that Yang—Mills theory is renormalizable if the masses of the particles are introduced via a spontaneous symmetry-breaking mechanism. The Salam—Ward, Weinberg, and Glashow theories of weak and electromagnetic interactions fall into this category and provide an important unification of two of the fundamental forces of nature. This has raised again the hope that all the basic interactions may be incorporated in one grand unified theory. The concept of unification has now permeated many approaches to quantum gravity, in contrast to much of the older work where attempts to quantize the gravitational field were usually made in isolation.

Perhaps the most striking example of this attitude is the theory of supergravity — a subject which has entirely arisen since the first Oxford meeting. This theory necessarily involves

*_Quantum gravity, an Oxford symposium_ (ed. C.J. Isham, R. Penrose, and D.W. Sciama). Oxford University Press, Oxford (1975).

fields other than the gravitational. Much effort is now being expended in an attempt to show that its most sophisticated version (n = 8 supergravity) can be employed as a genuine grand unified theory. A hopeful feature of supergravity is the disappearance of some of the infinities which have plagued conventional quantum field theory, by mutual cancellations between the various fields. The existence of infinitely many uncontrollable infinities had been a seemingly fatal flaw in conventional approaches to quantizing the gravitational field, so the hope that supergravity may resolve this problem is one of its main attractions.

Another significant modern development which has arisen out of the quantization of Yang–Mills theory is the study of non-perturbative phenomena. Many workers in quantum gravity have long held that the perturbative approach commonly employed – the expansion of the metric tensor around Minkowski space – is quite inappropriate in a theory that is as intrinsically non-linear as general relativity. Consequently the non-perturbative methods of Yang–Mills theory have been adapted to the gravitational case, providing one of the motivations for the study of gravitational instantons.

This has given a new impetus to the investigation of the role played by topology. The topological structure of the solutions of non-linear equations has received much attention of late. In the case of general relativity, this pertains to the topology of space-time itself. In the early formative years of quantum gravity John Wheeler consistently emphasized the significance of space-time topology. But it is only relatively recently that physicists have become aware of the mathematical tools that are necessary to implement these ideas; and many groups now actively study this subject from a wide variety of viewpoints.

At our 1974 meeting, we were greatly privileged to witness the announcement of Stephen Hawking's important discovery that, via quantum mechanical processes, black holes radiate particles with a thermal distribution. This has led, in the intervening years, to an intense study of quantum field theory in fixed background space-times, raising a number of fundamental conceptual and technical problems. Much physical insight has been gained, particularly in the close relation between the Hawking effect and the thermal state observed by a detector moving with a constant acceleration through a quantum vacuum. Interesting developments in the mathematical theory have also occurred, leading to the study of thermal Green's functions in space-times with a complex periodicity, particularly in relation to gravitational instantons, where the above topological ideas play a key role. The Hawking effect inevitably relates also to quantum gravity studies of the very early universe, and it has a number of astrophysical implications and interrelations with particle physics. These developments are in many ways exciting; yet mysteries remain, and some of the deeper issues are still unresolved, such as those which relate the Hawking effect to time-asymmetry questions in physics.

While it is clear that many of the techniques of 1980 are radically different from those of 1974, the prime objective remains the same — to attempt to obtain a unification of general relativity and quantum theory, two of the great intellectual achievements of this century. There is still the mystery of the Planck length, $\left(\frac{Gh}{c^3}\right)^{\frac{1}{2}} \sim 10^{-33}$ cm, at which the structure of space-time and quantum effects would become inextricably intertwined. The feeling that, at this distance, some profound change in our conceptual understanding of the physical world would be necessary, motivates and inspires much of the research into quantum gravity. But it may be that some completely new approach to the subject is required, our attempts to date representing primitive gropings in some not quite correct directions. Perhaps some modification of the ideas of quantum mechanics will of necessity accompany any eventual successful union of quantum theory with general relativity. Maybe such a union could lead to a resolution of some of the inherent interpretative difficulties of present-day quantum mechanics. Perhaps, instead (or as well), the ideas of space-time structure need overhauling, one possible approach to such questions being provided by the twistor programme. We do not know where the future will lead us. Yet we feel that much of the work to date has been exciting, and that some of it has yielded significant insights that are here to stay.

We hope very much that by publishing this volume we shall assist young researchers who wish to enter the field and take up the challenge of tackling one of the greatest problems of modern theoretical physics.

C.J. Isham
R. Penrose
D.W. Sciama

CONTRIBUTORS AND ADDRESSES

C.J. Isham — *The Blackett Laboratory, Imperial College, London SW7 2BZ*

T.W.B. Kibble — *The Blackett Laboratory, Imperial College, London SW7 2BZ*

M.J. Duff — *Physics Department, Imperial College, London SW7 2BZ*

G.T. Horowitz — *Institute of Advanced Study, Princeton.*

D. Deutsch — *Department of Astrophysics, South Parks Road, Oxford*

S.J. Avis — *Department of Applied Mathematics and Theoretical Physics, Silver Street, Cambridge*

N.D. Birrell — *Logica Limited, 64 Newman Street, London W1A 4SE*
formerly *Department of Mathematics, King's College, Strand, London WC2R 2LS*

P.C.W. Davies — *Department of Theoretical Physics, University of Newcastle-upon-Tyne*

D.W. Sciama — *Department of Astrophysics, South Parks Road, Oxford*
and *Center for Relativity, University of Texas, Austin, Texas 78712*

R.M. Wald — *Enrico Fermi Institute, University of Chicago, Chicago, Illinois 60637*

P. Penrose — *Mathematical Institute, University of Oxford, 24-29 St. Giles, Oxford.*

M.J. Rees — *Institute of Astronomy,*
University of Cambridge,
The Observatories,
Madingley Road,
Cambridge

G.G. Ross — *Theoretical Physics Department,*
University of Oxford,
1 Keble Road,
Oxford

J.B. Hartle — *Department of Physics,*
University of California,
Santa Barbara, CA 93106

K. Kuchař — *Department of Physics,*
University of Utah,
Salt Lake City,
Utah 8412

C.N. Pope — *Department of Applied Mathematics and Theoretical Physics,*
University of Cambridge,
Silver Street,
Cambridge

A. Ashtekar — *Physics Department,*
Syracuse University,
Syracuse, N.Y. 13210

M.R. Brown — *Department of Astrophysics,*
University of Oxford,
South Parks Road, Oxford

B.S. DeWitt — *Department of Physics,*
University of Texas,
Austin, Texas 78712

P. van Nieuwenhuizen — *Institute for Theoretical Physics,*
State of University of New York at Stony Brook,
Long Island, N.Y. 11794

S. Ferrara — *CERN,*
Geneva 23,
1211 Switzerland

and *Laboratori Nazionali di Frascati,*
INFN,
Frascati, Italy

K.S. Stelle — *CERN,*
Geneva 23,
1211 Switzerland

P.C. West — *Department of Mathematics,*
King's College,
Strand, London WC2R 2LS

R.S. Ward — *Department of Mathematics, Trinity College, Dublin*

J.S. Bell — *CERN, Geneva 23, 1211 Switzerland*

B. Mielnik — *Institute of Theoretical Physics, Warsaw University, Hoza 69, Warszawa, Poland*

R.C. Jennison — *Electronics Laboratories, The University, Canterbury, Kent CT2 7NT*

CONTENTS

PREFACE v

1. QUANTUM GRAVITY—AN OVERVIEW 1
C.J. Isham

2. IS A SEMI-CLASSICAL THEORY OF GRAVITY VIABLE? 63
T.W.B. Kibble

3. INCONSISTENCY OF QUANTUM FIELD THEORY IN CURVED SPACE-TIME 81
M.J. Duff

4. IS FLAT SPACE-TIME UNSTABLE? 106
G.T. Horowitz

5. PARTICLE AND NON-PARTICLE STATES IN QUANTUM FIELD THEORY 131
D. Deutsch

6. A REVIEW OF SOME ASPECTS OF FIELD THEORY ON TOPOLOGICALLY NON-TRIVIAL SPACE-TIMES 148
S.J. Avis

7. INTERACTING QUANTUM FIELD THEORY IN CURVED SPACE-TIME 164
N.D. Birrell

8. IS THERMODYNAMIC GRAVITY A ROUTE TO QUANTUM GRAVITY? 183
P.C.W. Davies

9. THE IRREVERSIBLE THERMODYNAMICS OF BLACK HOLES 210
D.W. Sciama

10. BLACK HOLES, THERMODYNAMICS, AND TIME-REVERSIBILITY 224
R.M. Wald

11. TIME-ASYMMETRY AND QUANTUM GRAVITY 245
R. Penrose

12. INHOMOGENEITIES FROM THE PLANCK LENGTH TO THE HUBBLE RADIUS 273
M.J. Rees

13. BARYON NUMBER ASYMMETRY IN GRAND UNIFIED THEORIES 304
G.G. Ross

14. PARTICLE PRODUCTION AND DYNAMICS IN THE EARLY UNIVERSE 313
J.B. Hartle

15. CANONICAL METHODS OF QUANTIZATION 329
K. Kuchař

16. THE ROLE OF INSTANTONS IN QUANTUM GRAVITY 377
C.N. Pope

17. ACAUSAL PROPAGATION IN QUANTUM GRAVITY 393
S.W. Hawking

18. QUANTIZATION OF THE RADIATIVE MODES OF THE GRAVITATIONAL FIELD 416
A. Ashtekar

19. IS QUANTUM GRAVITY FINITE? 439
M.R. Brown

20. A GAUGE INVARIANT EFFECTIVE ACTION 449
B.S. DeWitt

21. THE COSMOLOGICAL CONSTANT IN QUANTUM GRAVITY AND SUPER-GRAVITY 488
M.J. Duff

22. QUANTUM (SUPER)GRAVITY 501
P. van Nieuwenhuizen

23. A REVIEW OF BROKEN SUPERGRAVITY MODELS 520
S. Ferrara

24. REALIZING THE SUPERSYMMETRY ALGEBRA 549
K.S. Stelle and P.C. West

25. SOME REMARKS ON TWISTORS AND CURVED-SPACE QUANTIZATION 578
R. Penrose

26. THE TWISTOR APPROACH TO DIFFERENTIAL EQUATIONS 593
R.S. Ward

27. QUANTUM MECHANICS FOR COSMOLOGISTS 611
J.S. Bell

28. QUANTUM THEORY WITHOUT AXIOMS 638
B. Mielnik

29. THE QUANTIZED RESPONSE OF A STABLE PARTICLE OF TRAPPED ENERGY AND ITS BEHAVIOUR IN A GRAVITATIONAL FIELD 657
R.C. Jennison

1

QUANTUM GRAVITY—AN OVERVIEW

C.J. Isham

Blackett Laboratory, Imperial College, London SW7

1. INTRODUCTION

Six years have passed since the last Oxford Symposium on Quantum Gravity. During this time there have been radical changes in attitudes and methodology. Many of the theoretical physicists who have traditionally regarded quantum gravity as esoteric and rather pointless will now grudgingly admit that not only is it technically fascinating, but it may even impinge directly on their own, superficially remote, specialities. This significant change arises mainly from the way in which large portions of particle-physics based quantum field theory have been enthusiastically adapted to the gravitational situation. Somewhat unexpectedly, this has not been at the expense of the geometric or topological aspects of the classical theory of general relativity, but instead reflects the absorption of modern mathematical ideas into conventional quantum field theory.

The advances in theoretical physics that have been so instrumental in influencing research in quantum gravity are as follows:

1. There has been a general rapid evolution in quantum field theory following t'Hooft's discovery in the early 1970s that the Yang–Mills Higgs–Kibble theory could be successfully renormalized. This led to a dramatic shift in attitude of the particle physics community and the sterile S-matrix dominated 1960s gave way to the swinging second quantized 1970s to such an extent that the great majority of theoretical particle physicists now work in quantum field theory.

Of particular interest has been the emphasis on non-perturbative methods, as reflected in the studies of solitons, monopoles, and instantons. A striking feature is the way in which simple topological ideas have played an important role and indeed these days a basic knowledge of algebraic topology is almost indispensible. These developments have been avidly followed by workers in quantum gravity and the relevant techniques rapidly absorbed.

2. Related to this renaissance has been the evolution of unified field theories. There is the successful Salam/Weinberg unification of weak and electromagnetic interactions and much effort is being expended in the development of a grand unified theory (GUT) incorporating strong interactions. Indeed quantum chromodynamics (the Yang—Mills based theory of strong interactions) has been intensively investigated by itself and is responsible in part for the desire to develop non-perturbative methods. It is natural to speculate that gravity can be added to this list of forces to give a true unification of subatomic physics. It is notable that the current GUT mass scale is set at 10^{14}–10^{15} GeV; which is not that far removed from the 10^{18} GeV energy that characterises real quantum gravity effects (see below).

3. A totally new subject that has evolved since 1974 is supergravity. This arises naturally as the local gauge version of bose—fermion supersymmetry and possesses considerable attractions. For example many of the divergences that plague conventional quantum field theory mutually cancel, bose v. fermion, in a supersymmetric model. Also supergravity, when extended by a type of internal symmetry group, may lead in a natural way to a rigidly prescribed unified theory of gravitation and other forces.

4. Finally, but of the utmost importance, Hawking's announcement in 1974, of the quantum mechanically induced thermal radiation of black holes, led to an explosion of interest in quantum theory in various types of gravitational background and heralded the dawn of a new and exciting era in quantum gravity research.

In spite of this extensive and sometimes frantic activity, I can confidently state that there is as yet no viable quantum theory of gravity. Attitudes towards this regrettable situation vary widely. One school maintains that it is just a matter of time and that the problem will be cracked within the framework of 'conventional' quantum field theory, albeit perhaps with the aid of technical tools that are as yet undiscovered.

However a long held contrary belief is that general relativity and quantum theory are intrinsically incompatible and that, rather than merely developing technique, what is required is some fundamental breakthrough in our understanding of the relationships between space-time structure and quantum processes. Adherents of the former opinion tend to regard the second attitude as being rather defeatist and unproductive. However it should be admitted that, although the potentialities are still considerable, all conventional, or semiconventional methods *have* failed to work and it does no harm occasionally to reconsider precisely what is meant by quantizing gravity (Section 2).

Like many other branches of theoretical physics, quantum gravity is subject to fads and fancies and it would be a mistake to assume that all the earlier work is irrelevant or that the trends of the moment will continue into the future. A new student of the subject would be well advised to study some of the older ideas, as discussed for example in the various review articles contained in the proceedings of the first Oxford quantum gravity conference (Isham, Penrose, and Sciama 1975). Other useful references are Ashtekar and Geroch (1974); B.S. DeWitt (1970); Bergmann and Komar (1962); J.A. Wheeler (1963, 1968).

2. WHAT IS A 'QUANTUM THEORY OF GRAVITY'?

The question 'what exactly do you expect a quantum theory of gravity to look like?' if posed to a selection of physicists is likely to result in quite a spread of replies. Subdividing the question into

(a) What do you expect from such a theory? What will it do for us?

(b) How will it affect the rest of physics?

(c) What sort of technical apparatus should be used?

will not noticeably narrow the range of opinions. Indeed the past and present, sometimes heated, disagreements on methodology can often be traced to pronounced differences in the expectations of the protagonists, concerning precisely what in principle can be extracted from the theory they are trying so hard to construct.

There are however certain *a priori* questions which everyone will agree are worth asking. For example, in a quantum theory of gravity:

1. How much of the technical and conceptual structure of classical general relativity do we expect to retain? In particular, one thinks of the underlying smooth C^∞ manifold, the metric tensor, the local field equations $G_{\mu\nu}-\Lambda g_{\mu\nu} = T_{\mu\nu}$, the global topological properties of the manifold and global metric features such as lightcone structure and the existence of event horizons. It is very difficult to judge how many of these classical concepts should be present, in some form or another, in a quantized theory.

2. How much of the technical and conceptual structure of conventional quantum field theory do we expect to retain? For example, does the usual idea of a local quantum field $\hat{\phi}(x)$ make any sense at all or should we decide from the outset that, at a Planck scale, space-time structure is not that of a smooth manifold and therefore the local properties of fields become very unconventional? Similarly, what remains of the local commutativity of quantum fields in a theory where the lightcone is determined by the metric tensor, which is itself a dynamical variable? Perhaps the most crucial question of all concerns the role of perturbation theory. The notion of quanta (gravitons in our case) arises basically as a property of the Fock space quantization of a free field propagating in Minkowski space. It reappears in a conventional interacting theory through the Feynman–Dyson expansion of the Green's functions and S-matrix elements and the associated idea of the completeness of the asymptotic scattering states. This notion is not obviously appropriate in quantum gravity and indeed conventional perturbative methods of this type lead to a non-renormalizable theory. Ideally one wants to get away from perturbing in the Newtonian coupling constant since, even classically, this is known to be a dubious procedure. The situation is somewhat reminiscent of that in QCD, where the main struggle is to show why quarks and gluons do *not* appear as quanta, i.e. as asymptotic particles. It is perhaps attitudes to perturbation theory and to the highly nonlinear nature of Einstein's equation that most sharply distinguish the different schools of thought in quantum gravity.

Aside from these technical considerations, it has frequently been maintained that the conceptual structure of the Copenhagen interpretation of quantum mechanics is quite inappropriate when gravity is involved. Basically this is because the supposed 'classical background' has itself become part of the dynamical system. The most famous attempt to overcome this is the Everett–Wheeler 'many universe' theory. It has even been suggested that the Einstein equations are so highly non-linear, that the linear superposition principle of conventional quantum mechanics must be removed before a satisfactory quantum gravity theory can be constructed.

3. Do we expect to be able to quantize general relativity plus any collection of matter fields or must some special selection be made? This question is more important than it looks at first sight and again its answer sharply distinguishes different approaches. In the early work on quantum gravity most attempts were concerned with the gravitational field alone. On the other hand modern ideas on supergravity and unification suggest that only a special mixture of gravity and matter will work.

4. What sort of physical predictions can we expect to make from a quantum gravity theory? Indeed are there likely to be *any* experimental tests at all? If the answer to the second question is no, then we are engaged in a rather peculiar intellectual exercise which has no contact with the external world and whose ultimate aim is only internal consistency. This is more like pure mathematics than theoretical physics.

5. What is the significance of the Planck length $(G\hbar/C^3)^{\frac{1}{2}} \approx 10^{-33}$ cm ($= 10^{28}$ eV, 10^{-44} s, 10^{32} K) which must presumably arise in any theory which unifies gravity (G) with quantum mechanics ($\hbar$). The general opinion is that 'something odd' happens to space-time structure at this tiny distance or time, but opinions differ as to how this 'something' manifests itself. Indeed, should it come out of the theory (e.g. by cutting off divergent Feynman graph integrals), or by using an appropriate technical structure (e.g. a space-time lattice), should it in some way be fed in from the outset?

We shall return to these questions when discussing the various approaches that are currently being pursued with a view to considering tentative answers. Let me conclude on a cautionary note: it might be counter-productive to spend too much time pondering on the *a priori* structure of a quantum theory of gravity. To quote Abdus Salam (Isham *et al.* 1975, p. 532): 'In particle physics we have become conditioned never to ask what the theory can do for us; instead we humbly try to see what we can do for the theory.'

This pragmatic advice has been remarkably good in many branches of theoretical physics and in a subject that is as difficult as quantum gravity there is an arguable case for simply doing what is possible without worrying too much about the deeper conceptual problems.

3. THE REASONS FOR STUDYING QUANTUM GRAVITY

In view of the difficulty of the problem it seems sensible at this point to list the reasons why there is so much interest in quantizing the gravitational field. These are closely related to expectations of what the theory will give which in turn are intimately linked to the specific approach being employed. Many reasons have been cited since serious work began in the early 1950s but there have also been some important changes in attitude in recent years:

1. There is a purely intellectual desire to say *something* about the regime where both quantum and gravitational effects are important even if it is admitted that such regimes are experimentally inaccessible. This is often linked to the feeling that something rather mysterious happens at Planck energies 10^{28} eV (or equivalently at the Planck length 10^{-33} cm) and is accompanied by a suspicion that in order to construct a theory of quantum gravity some profound new development in theoretical physics is needed. In this sense the non-renormalizability of the conventional perturbative approach to quantum gravity is to be viewed as analogous to the 'ultraviolet catastrophe' that preceded the discovery of quantum mechanics.

2. The classical theory of general relativity admits gravitational collapse and possibly quantum theory can resolve the problem of the associated space-time singularities. Of specific interest is the existence of the event horizon, which, by the

cosmic censorship hypothesis, surrounds each singularity. As shown by Hawking in his famous black hole evaporation work (Hawking 1975*a*,*b*), an event horizon can have a significant influence on quantum field theory.

3. The most important space-time singularity for the human race was doubtless the 'big bang'. There is an intense interest in understanding quantum phenomena in the very early stages of the universe and quantum gravitational effects have been tentatively advanced to explain the isotropization and homogenization of the universe, the presence of the 3 K background, baryon asymmetry, and the number of neutrinos etc. There is currently a significant dialogue between elementary particle physicists with their grand unified field theories and early universe cosmologists. Much of this work does not involve quantum gravity proper but nevertheless it is of great interest and may ultimately play a significant part in the construction of a full quantum theory of gravity.

4. A related observation is that grand unified theories (GUT) almost seem to *need* a quantum theory of gravity. Indeed in a number of seminars recently, respectable high energy physicists talking about GUT have been heard to make remarks like 'of course we will assume that gravity gives us a cut off at the Planck length ...'! This is sometimes augmented by the belief that there is very little interesting physics between 10^3 GeV and 10^{15} GeV. The idea of a theory unifying electromagnetic, weak, strong and gravitational theories is an attractive one. In particular, in getting the gravitational sector correct, a specific structure may need to be chosen for the other three sectors and this may be experimentally testable. It is vital to emphasize that this might be the *only* experimental test of quantum gravity that we will ever have! It certainly does not seem conceivable that accelerators capable of reaching 10^{28} eV (and hence directly probing the gravity sector) will ever be built.

5. Many people have been deeply disturbed by the ultraviolet divergences that appear in conventional quantum field theory even if, as in the renormalizable case, they can be discreetly removed. An often expressed hope is that quantum gravity will remove these infinites by quantum smearing of the lightcone

(Pauli 1967; Deser 1970) and some calculations have been performed which support this (Khriplovich 1966; DeWitt 1964*a*; Isham, Abdus Salam, and Strathdee 1971). These all involve some type of non-perturbative method as reflected in the appearance of terms like $\log(Gm)$. On the other hand one of the principal hopes of extended supergravity theory is that it is actually *finite order by order* in perturbation theory because of the mutual cancellation of boson and fermion loops. This would be a perturbative result and as such would have a quite different status from the previous one although both lead to finite theories. In particular it has nothing to do with lightcone smearing; indeed the lightcone remains firmly fixed in the background Minkowski space about which the perturbations are made.

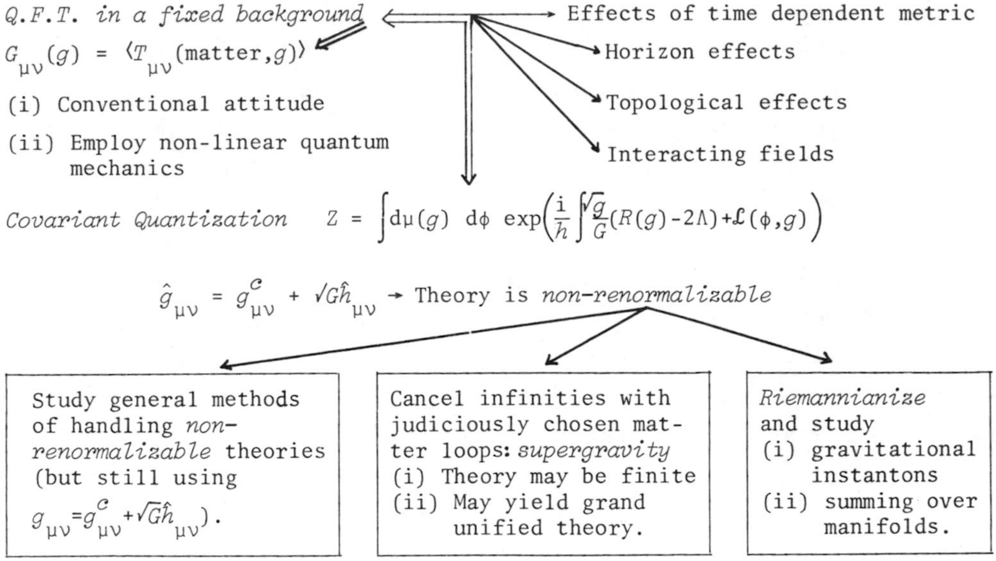

Questions:

1. Might we need all three boxes at once?
2. Will these conventional or semi-conventional methods ever work or should we try a more radical approach, e.g. *twistor theory*?
3. What happened to canonical quantization and the once popular study of 'quantum cosmology' models?

FIG. 1

4. THE CURRENT LINES OF ATTACK

Figure 1 summarizes the main approaches and subjects currently being studied in the attempt to construct a quantum theory of gravity. The logical (but not historical) starting point is perhaps the investigation of the construction and properties of a quantum field $\hat{\phi}$ propagating in a fixed background manifold M with metric tensor $g_{\mu\nu}$.

The early studies were of the particle production induced by a time-dependent metric in which the gravitational field energy is acquired by the quanta of the linear quantum field. The subject really caught fire only after Hawking's discovery that the event horizon surrounding a black hole leads to a transition between pure and mixed states. This may be viewed as an effect arising from the non-local metric structure. Complementary to this has been the recent interest in the quantum field theoretical rôle of the global topological properties of M. There has also been substantial investigation of interacting quantum field theories which possess properties not shared by their conventional Minkowski space counterparts. More details of these topics can be found in section 5.

The quantum field $\hat{\phi}$ will possess some sort of energy momentum tensor which might be expected to react back on the gravitational field via the Einstein equations. One way of realizing this is contained in the semi-classical approach to quantum gravity which may be summarized in the equations

$$G_{\mu\nu}(g) = \langle T_{\mu\nu}(\hat{\phi},g)\rangle \quad . \tag{4.1}$$

There has always been a school of thought maintaining that gravity should not be quantized at all and eqn. (4.1) is then proposed as the true (i.e. not merely an approximate) theory of gravity/matter interactions.

These equations are more subtle than may be apparent at first sight. The crucial point is that the construction of the quantum field $\hat{\phi}$ and its states (such as $|\rangle$) depends heavily on the properties of the background $g_{\mu\nu}$ which is itself determined via (4.1) by the properties of $\hat{\phi}$ and its states! Thus (4.1) needs to be solved within some type of self-consistent framework.

Apart from this general difficulty there are a number of specific technical problems. For example:

1. $T_{\mu\nu}(\hat{\phi},g)$ is formally divergent and needs to be renormalized and/or regularized. The resulting counter-terms involve the square of the Riemann curvature tensor and drastically modify the equations of motion. Quantum mechnically these extra terms lead to ghosts and this is reflected in the statement that when (4.1) is augmented by such terms, the solutions are intrinsically unstable when perturbed around flat space. (Horowitz and Wald 1978; G. Horowitz in this volume.)

2. The regularized $T_{\mu\nu}(\hat{\phi},g)$ possesses properties quite different from those of a classical energy momentum tensor (e.g. non-local functions of $g_{\mu\nu}$). This can be advantageous, as in the case of the violation of the Hawking–Penrose energy conditions which allows (4.1) to avoid a gravitational singularity (Hawking and Penrose 1970; Parker and Fulling 1973). In general however, it makes it difficult to prove that (4.1) possesses any sensible stable solutions. Additionally it is not obvious that the resulting object $g_{\mu\nu}$ has any interpretation as a metric tensor in the classical sense.

3. Aside from the self consistency problem mentioned above it is not physically clear how the state $|\rangle$ should be chosen. Indeed there may be situations where it would be natural to try and find a solution to (4.1) in which $|\rangle$ is an 'in' state and $\langle|$ is an 'out' state. Note that in this case $g_{\mu\nu}$ would not even be real.

4. If $\hat{\phi}$ is a linear field, the right hand side of (4.1) contains 'vacuum polarization' contributions from the single $\hat{\phi}$ loop in the background $g_{\mu\nu}$. However a single graviton loop would be of the same magnitude and produce the same type of counter-term. From this viewpoint it seems rather unnatural to keep $g_{\mu\nu}$ unquantized.

5. Theories which are equivalent at the fully quantized level become inequivalent in the semi-classical form because of the inapplicability of Borcher's type field redefinitions (M.J. Duff in this volume). Thus some, possibly *ad hoc*, selection procedure needs to be invoked.

The implicit dependence of the state $|\rangle$ on the metric tensor and hence on the solution to (4.1), has led to the observation that the states which solve (4.1) do not obey the superposition principle (Mielnik 1974; Kibble 1978, 1979; Smolin 1979). Thus it has been suggested that in constructing a quantum theory of gravity it is necessary to dispense with the linear structure of quantum mechanics. This may be a very deep remark which has applicability well beyond the semi-classical scheme epitomized by eqn. (4.1). However, it has only been implemented seriously in the latter approach in the recent work of Kibble and Randjbar-Daemi 1980; Kay, Kibble, and Randjbar-Daemi 1980. See also T. Kibble in this volume.

The quantum theory of gravity proper is mainly studied within the framework of a covariant formalism in which a four dimensional space-time manifold M is fixed once and for all and $g_{\mu\nu}$ is regarded as an operator defined on M. These days such investigations almost invariably employ a functional integral

$$Z[g_c,j] = \int d\mu(h)d\phi \, \exp\left(\frac{i}{\hbar}\int \frac{\sqrt{-g}}{G}(R(g)-\Lambda)+\mathcal{L}(\phi,g,j)\right) \qquad (4.2)$$

with a split

$$g_{\mu\nu} = g^c_{\mu\nu} + \sqrt{G}h_{\mu\nu} \qquad (4.3)$$

in which $g^c_{\mu\nu}$ is a classical solution to the Einstin equations and ϕ symbolically represents the matter fields. Thus Z is a functional of this background field $g^c_{\mu\nu}$ and the matter field sources j and variations with respect to $g^c_{\mu\nu}$ and j generate all the vertices and Green's functions of the theory.

It is clear that with this approach we are firmly embedded in the conventional particle physicist's language of Feynman diagrams, loop expansions, renormalizability, counter-terms etc. This aspect of the subject has been well understood for some time and there have been no startling developments during the last six years. Thus I shall not expand on this further, but will merely cite some useful references: Boulware and Deser 1975; De Witt 1964, 1967*b*,*c*, 1972, 1979; Duff 1973, 1975; Fadeev and Popov 1967, 1973; Fradkin and Tyutin 1970; Feynman 1963; Fradkin and Vilkovisky 1973, 1975*a*; t'Hooft 1973; Kallosh 1974; Mandlestam 1968*a*,*b*; van Nieuwenhuizen 1975; Popov and Fadeev 1972; Veltman 1975; Weinberg 1979.

The measure $d\mu(h)$ contains the well known Fadeev–Popov ghost factors necessary in any covariant gauge theory and may also possess additional terms which formally depend on $\delta^4(0)$ (arising from expressions like, for example, $\Pi\, dg_{\mu\nu}(x)$ (det $g(x)$)). The exact nature and status of these terms has always been the subject of some discussion. They arise naturally in making the transition from a canonical to a covariant path integral formalism (Fadeev 1969; Fradkin and Vilkovisky 1975*b*, 1977; Fradkin 1973; Leutwyler 1964). One attitude is that since, in dimensional regularization, $\delta^4(0)$ is formally zero these terms can simply be neglected. However this argument is only really valid in the coupling constant perturbation theory implicit in (4.3). In other, perhaps non-perturbative, schemes the precise form of the measure could be very important (see for example the remarks in Section 6 concerning Klauder's attitude to non-renormalizability).

It is worth noting that a cosmological term $\Lambda\sqrt{-g}$ has been included in the action in (4.2). Traditionally this term is neglected because its presence invalidates the expansion of the metric around a Minkowski space background. However, there has been considerable interest recently in studying the effects of such a term for a variety of reasons (Christensen and Duff 1980):

1. It occurs in the standard spontaneous symmetry breaking Lagrangian $\mathcal{L} = (-a^2\phi^2+\lambda\phi^4)\sqrt{-g}$, if $\langle\phi\rangle \neq 0$. Adding a counter-term to remove the effective $\Lambda\sqrt{-g}$ is not sufficient because at high temperatures when the symmetry is restored (so that $\langle\phi\rangle = 0$) the cosmological term will reappear. This is particularly relevant in the discussion of early universe phenomena.
2. Hawking's quantum foam picture of the structure of space-time at the Planck length scale requires a cosmological constant (Hawking 1978, 1979*a*).
3. A $\Lambda\sqrt{-g}$ counter term will usually be required if the matter field Lagrangian includes mass-like contributions. (This is not so however if massive supermultiplets are coupled to an external gravitational field since bose/fermion cancellation occurs (Zumino 1975).)
4. A cosmological constant appears if the SO(n) symmetry of extended supergravity is gauged (Freedman and Das 1977).

The major result of the quantum gravity programme based on (4.2) and (4.3) is that the theory is non-renormalizable. Or, to be more precise, it is known that an infinite number of possible counter-terms exist and there is no obvious reason why their coefficients should miraculously cancel, although it should perhaps be emphasized that, at the time of writing, even the two-loop contribution has not been explicitly calculated. If $g^c_{\mu\nu}$ satisfies the vacuum Einstein equations, the one-loop counter-terms in $Z[g^c]$ vanish apart from a topological contribution proportional to the Euler constant (t'Hooft and Veltman 1974). The one-loop counter-terms for matter plus gravity do not in general vanish apart from some rather special anomalous magnetic moment graphs. (Capper and Duff 1974; Capper, Duff, and Halpern 1974; Deser, Hung-Sheng Tsao, and van Niewenhuizen 1974; Deser and van Nieuwenhiuzen 1974*a*,*b*,*c*; Nouri-Moghadon and Taylor 1975; Van Proeyen 1977, 1979.)

Several attitudes can be adopted towards this somewhat catastrophic state of affairs. Various schemes (Section 6) have been proposed for handling non-renormalizable theories and it is possible that one of these may work in the gravitational context. However, most of them suffer from the 'defect' of being applicable to *any* non-renormalizable theory and it is arguable that gravity may be quantizable only by virtue of some special property of Einstein's equations.

A second possibility is to attempt to cancel the virulent effects of the graviton loops by introducing judiciously chosen matter loops. The success of such a scheme seems likely to require some sort of gravity/matter symmetry and this is precisely what is provided by supergravity (Section 8). Indeed this theory is two-loop finite (Grisaru 1977; Ferrara and van Nieuwenhuizen 1978) but unfortunately potential three-loop counter-terms are known to exist (Deser, Kay, and Stelle 1977). The current hope is that extended supergravity will be finite to all orders in perturbation theory and will in addition provide a grand unified theory of all the fundamental forces.

A third school of thought has arisen by considering the gravitational analogue of the non-perturbative instanton effects in Yang—Mills theory. This involves defining the functional integral (4.3) on a Riemannian rather than pseudo-Riemmanian manifold (Hawking 1979*a*,*b*). The effects of this are far reaching and are briefly discussed in Section 7. The way in which this approach might lead to a resolution of the non-renormalizability problem lies in the development of non-coupling constant based perturbative schemes plus the possibility of summing over all space-time manifolds in the functional integral. The idea that space-time should itself be subject to quantum fluctuations was vigorously advanced by Wheeler (1963, 1968). However, his ideas arose in the context of canonical quantization and no really plausible technique for implementing them was ever discovered. The idea has been revived by Hawking within his Riemannian programme and it is not inconceivable that some sort of sum over manifolds might one day be performed in a covariant functional integral. (See also Taylor 1979.)

It is noteworthy that one of the ambitions of the Riemannian programme is to free quantum gravity from perturbation theory based on the expansion $g_{\mu\nu} = g^{c}_{\mu\nu} + \sqrt{G}h_{\mu\nu}$. Expansions of this type are known to be bad in classical general relativity and they clearly misrepresent the global topological and lightcone structures of the pair $(M, g_{\mu\nu})$. From this point of view it is perhaps not surprising that the quantum theory fails. It should be observed however that the conventional interpretation of quantum field theory in terms of 'particles' (i.e. gravitons), S-matrices etc., is partly based on expansions of the form (4.3) and it is not at all obvious how one would attempt to interpret physically a quantum gravity scheme that was substantially different.

In relation to the reasons cited in Section 3 for studying quantum gravity, the particle picture is only immediately suitable for discussing the existence of a grand unified theory where of course the hadronic and leptonic sectors are described in the language of elementary particles. Perturbative schemes based on (4.3) are unlikely to be helpful in studying the space-time singularity problem or the possibility of the Planck length acting as a universal cut off in quantum field theory. Indeed a more careful analysis shows that the perturbation of, for example, an S-matrix element is really in terms of Gp^2 (p^2 is some momentum in the theory) which takes on the value 1 at

Planck energies. Thus the perturbative series would be divergent even if its individual terms were finite and hence any phenomena related to distances less than or equal to the Planck length cannot be usefully studied order by order in perturbation theory. At the very least one must attempt to sum infinite sets of Feynmann graphs and preferably a perturbation scheme not involving expansions in G should be employed.

The three boxes in Fig. 1 are not meant to be mutually exclusive. Perhaps the secret of quantum gravity lies in applying some prescription for handling non-renormalizable theories to extended supergravity in a Riemannian space based functional integral!

Neither should it be assumed that the schemes listed in Fig. 1 are the sum total of modern quantum gravity research. For example Penrose's twistor programme continues to flourish and possesses certain properties which, from the quantum gravity viewpoint, make it attractive. For example:

1. It concentrates on the dual concept (in the projective geometry sense) of null lines in space-time rather than space-time points. A point is rather defined as the intersection of two null lines and, if the lines are subject to quantum fluctuations (because $g_{\mu\nu}$ is), the concept of a point is smeared. (Penrose 1975; Penrose and MacCallum 1973.) Thus in the twistor programme the quantum fluctuations may set in at a level that is even more basic than that of the Wheeler—Hawking 'foaming' topologies.
2. Twistor techniques may be employed to find exact solutions to both the Yang—Mills and Einstein equations as reviewed by R. Ward in this volume. (Atiyah, Hitchin, Drinfeld, and Manin 1978; Atiyah and Ward 1977; Penrose 1976; Ward 1977.) Thus they may provide a technique for incorporating intrinsically the highly non-linear structure of general relativity.
3. Conventional Feynman graphs may be converted into twistorial form. However, 'twistor graphs' should never be infinite because they involve complex integrations over compact contours. Unfortunately, it is not clear if this really resolves the ultraviolet divergence problem since so far the twistor analogues of tree diagrams only have been computed.
4. It has been suggested that one twistor diagram may be in

some way equivalent to many (perhaps infinitely many) Feynman diagrams. This could lead to a new non-perturbative approach to quantum field theory.

5. Twistorial methods have been employed in classification schemes of elementary particles and as such might assist in the search for a truly grand unified theory (Hughston 1979).

Finally connoisseurs of the traditional approaches to quantum gravity will notice that no mention is made in Fig. 1 of canonical quantization. Indeed in the last six years there has been very little published on this approach. This seems a great pity as the canonical scheme is in many respects radically different from present day covariant quantization. It emphasizes different problems and is potentially applicable to a different class of situations. States are represented as wave functionals of three geometries and this is particularly appropriate for discussing the issue of quantum corrections to gravitational collapse. Current thinking on canonical quantization is reviewed by K. Kuchar in this volume.

Another topic that was very popular six years ago was quantum cosmology models (M. MacCallum 1975). These were models for canonical quantization in which all but a finite number n of degrees of freedom of the gravitational field were frozen out. Potentially this offered a totally different perturbative theory although no one ever constructed an inductive scheme which would allow n to be taken to infinity in some constructive sense. The idea of perturbing in the number of degrees of freedom has recently been resurrected in the context of Riemannian functional integral quantization (Hawking, Page, and Pope 1979, 1980).

5. QUANTUM FIELD THEORY IN A CURVED SPACE-TIME

Hawking's discovery (Hawking 1975*a,b*) that black holes radiate thermally via a quantum mechnical process, generated an intense interest in quantum field theory in a fixed, unquantized, space-time. The subject had been studied earlier by De Witt in a rather formalistic analysis of his background field methods for generating Green's functions (De Witt 1964, 1967*a,b*, 1972) and by various authors in a mainly cosmological context (Fulling

1973; Parker 1968, 1969, 1971; Sexl and Urbantke 1967; Zeldovich and Starobinsky 1972).

A typical example is afforded by a scalar field ϕ propagating in a four-dimensional space-time M equipped with a pseudo-Riemannian metric $g_{\mu\nu}$. Until recently attention focused on a linear system with the classical Lagrangian

$$\mathcal{L} = \surd(-\det g)\{g^{\mu\nu}\, \partial_\mu\phi\, \partial_\nu\phi - \mu^2\phi^2\} \tag{5.1}$$

In the 'massless' case ($\mu=0$) a coupling $1/6R\phi^2$ is frequently added. This renders the entire system conformally invariant under the transformations

$$g_{\mu\nu} \to \Omega^2 g_{\mu\nu}, \qquad \phi \to \Omega^{-1}\phi \tag{5.2}$$

The literature on the quantization of such a theory is quite considerable. I shall discuss only some general features and will cite a number of review articles for full details and extensive bibliographies. (De Witt 1975; Davies 1977, 1979; Gibbons 1977, 1979*a*; Isham 1977*a*; Parker 1976, 1979.)

Various quantization methods have been investigated. Probably the most widely adopted is a 'covariant' scheme in which one attempts to find field operators $\hat{\phi}(x)$ satisfying Heisenberg equations of motion

$$(g^{\mu\nu}\nabla_\mu\partial_\nu + \mu^2)\hat{\phi}(x) = 0 \tag{5.3}$$

and covariant commutation relations

$$[\hat{\phi}(x),\hat{\phi}(y)] = -i\hbar G(x,y). \tag{5.4}$$

Here $G(x,y)$ is the two-point distribution which generates solutions to the Cauchy problem:

$$\phi(y) = \int_\Sigma G(y,x)\overleftrightarrow{\partial}_\mu\, \phi(x)\, d\sigma^\mu \tag{5.5}$$

where Σ is a space-like hypersurface in M. In a globally hyperbolic manifold M a unique function $G(\ ,\ )$ exists (Leray 1952; Choquet-Bruhat 1969; Dimmock 1980) and has the support properties

$$\operatorname{supp}_x G(x,\psi) \subset J^+(\operatorname{supp}.\psi) \cup J^-(\operatorname{supp}.\psi) \tag{5.6}$$

$$\operatorname{supp}_y G(\psi,y) \subset J^+(\operatorname{supp}.\psi) \cup J^-(\operatorname{supp}.\psi) \tag{5.7}$$

where ψ is a smooth compact support test function and $J^+(A)$ (resp. $J^-(A)$) is the causal future (resp. causal past) of the set A defined as the set of points that can be connected to points in A by past directed (resp. future directed) time-like curves or null geodesics. (Penrose 1973; Hawking and Ellis 1973.)

Equations (5.3) and (5.4) can be made rigorous by introducing operator valued distributions and the solutions of (5.3) can be related to the classical theory via C^*-algebra techniques (Dimmock 1979, 1980; Hajicek 1977; Isham 1977*b*; Wald 1975). There exist many Fock space based representations of (5.3) and (5.4) and the major problem is selecting one that is appropriate for a given physical situation. The difficulty is that in a generic space-time there is no Poincaré group invariance and no canonical definition of positive frequency solutions with associated notions of vacuum and particle states. This problem has been rigorously solved in static or stationary space-times where the time-like Killing vector may be employed to give a natural definition of positive energy/frequency (Ashtekar and Magnon 1975; Kay 1978; Moreno 1977). Similarly in space-times with suitably asymptotically flat regions 'in' and 'out' states are well defined and the corresponding S-matrix may be computed from the Bogoliubov transformations connecting the 'in' and 'out' fields.

Various methods have been suggested for resolving this problem in a general space-time. For example the Hamiltonian diagonalization scheme has been subject to much critical appraisal. This technique employs a separate Hilbert space for each space-like hypersurface and choice of lapse and shift functions and the 'instantaneous' Hamiltonians are represented by positive, self-adjoint operators on the appropriate Hilbert spaces (Ashtekar and Magnon 1975; Kay 1977). This method was used in some discussions of cosmological models but was felt to possess physically undesirable characteristics and has now fallen into disuse. (Castagnino, Verbeure, and Weder 1975; Castagnino undated.) More recent suggestions for quantizing fields on general space-times are contained in the works of Ashtekar and Sen (1978); Castagnino, Foutssats, Laura, and Zandron (1980); Castagnino and Weder (1980); Rumpf (1976*a*,*b*); Rumpf and Urbantke (1977). Note that the problem is not alleviated by employing a functional integral representation for the Green's function generating functional

$$Z[j] = \int d\phi \exp\left(i \int_M (\mathcal{L}+\phi j)\right). \tag{5.8}$$

In order to compute (5.8) it is necessary to invert the differential operator $g^{\mu\nu}\nabla_\mu\partial_\nu + \mu^2$ and find the corresponding Feynman function $\Delta_F(x,y)$. However this latter problem is entirely equivalent to choosing a definition of positive frequency and hence a Fock space.

In the spirit of the times it is natural to contemplate starting with a quantum field on a Riemannian rather than pseudo-Riemannian manifold and indeed the construction of such 'Euclidean fields' has been extensively discussed by Wald (1980*a*). However, in general it is not possible to analytically continue the resulting Green's functions back to a Lorentzian space-time so this approach does not offer any resolution of the fundamental difficulty.

Another important problem lies in the intrinsically divergent nature of products of field operators. Thus in defining $\hat{T}_{\mu\nu}(\hat{\phi})$ it is necessary to renormalize and/or regularize and this subject alone has generated a substantial literature (see the review articles mentioned above). A number of regularization schemes have been developed amongst which the most important are probably zeta-function regularization, dimensional regularization, covariant point splitting, and adiabatic techniques.

These various methods have been applied to a number of cosmological situations with both isotropic and anisotropic backgrounds. The main object usually is to explore the quantum effects on the initial singularity and the removal of anisotropy. Aside from the general references cited above, some recent interesting papers include Fischetti, Hartle, and Hu (1979); Frolov and Vilkovisky (1979); Hartle and Hu (1979, 1980); Hu (1979*a*,*b*).

The presence of an event horizon has a profound effect on the quantum field theory. The intrinsic loss of information through the horizon induces a transition from an initially pure quantum state to a final mixed state. In the case of an eternal black hole with mass M, the vacuum states are defined by boundary conditions on the horizon (Fulling 1977; Unruh 1976) and the net result is the production of thermal radiation of field quanta with temperature

$$kT = \frac{\hbar C^3}{8\pi GM} \tag{5.9}$$

and an intrinsic entropy

$$S = \tfrac{1}{4}A \tag{5.10}$$

where A is the area of the event horizon (Hawking 1975*a*,*b*). These radiation results have been rederived in various ways. Notable examples are the derivations using thermal Green's functions (Gibbons and Perry 1978), analytic continuation in a path integral (Hartle and Hawking 1976), and functional integration (Gibbons and Hawking 1977*a*). The last mentioned is especially interesting in its relation with the Riemannian space approach to quantum gravity (Section 7).

The quantum field theoretical significance of the pure→mixed transition continues to be explored. In particular Page (1980) and Wald (1980*b*) have shown that it leads to a violation of the CPT theorem in the sense that there cannot exist an operator Θ such that

$$\$^{-1} = \Theta\$\Theta^{-1} \tag{5.11}$$

where $ is the superscattering matrix for pure and mixed states.

The idea that the CPT theorem should fail in quantum gravity has frequently been argued by Penrose (e.g. Penrose 1979).

Event horizons arise in other contexts in general relativity and radiation results have been obtained in cosmological situations (Gibbons and Hawking 1977*b*) and for an accelerated observer (Davies 1975; Fulling 1973). The latter case has been extensively investigated by Sanches (1979*a*,*b*) who relates the existence of thermal radiation to topological properties of coordinate transformations and singularities (see also Christensen and Duff 1978*b*).

Most of the earlier investigations of quantum field theory in a curved space-time were concerned with linear field equations. However in recent years there has been a fairly intensive investigation of interacting field theories. The LSZ reduction formula has been developed in globally hyperbolic space-times possessing 'in' and 'out' regions and hence asymptotic notions of particles and vacuum state (Birrell and Taylor 1978). Interaction picture perturbation theory has been applied in this context (Birrell and Ford 1979*a*) and, perhaps rather surprisingly, it has been shown that momentum space methods can be effectively applied in a curved space-time (Birrell 1979*a*,*b*).

The renormalization programme in a curved space-time contains subtleties that do not arise in Minkowski space (Birrell and Taylor 1978; Martelini, Sodano, and Vitiello 1978). Extra divergences can occur (possibly related to the global space-time topology) leading, within a dimensional regularization context, to a complicated pole structure (Bunch, Panangaden, and Parker 1980; Bunch and Panagaden 1978; Birrell 1979*b*; Ford and Yoshimura 1979). Another rather intriguing effect is the acausal propagation induced by the interactions (Ford 1980*a*; Drummond and Hathrell 1980). A fuller account of many of these features can be found in the review article by Birrell in this volume.

Another subject to which considerable attention has been devoted recently is the role played in quantum field theory by the global space-time topology of both Lorentzian and Riemannian manifolds. Different authors have different views concerning the length scale at which one might expect to see non-trivial topology, but the most prevalent is that one is really concerned with the structure of space-time at Planck lengths.

Unlike much of the previous material in this section the topological aspects have not been the subject of extensive reviews so I shall present a slightly more detailed analysis here. There are a number of ways in which space-time topology can influence quantum field theory:

1. Gauge theories

It is now appreciated that, even in Minkowski space, fibre bundle theory constitutes the appropriate mathematical language in which to discuss gauge theories. This is especially true in a general space-time M where there may be a delicate interplay between the topology of the internal symmetry group G and the topology of M. The classification of gauge sectors is equivalent to that of principal G-bundles and, in the case of manifolds M where dim $M \leqslant 4$, may be completely expressed in terms of certain characteristic cohomology classes (Avis and Isham 1979*a*; Dold and Whitney 1959). These are, for example, the first ($C_1 \in H^2(M;Z)$) and second ($C_2 \in H^4(M;Z)$) Chern classes for $U(n)$ bundles, the second Stieffel–Whitney class ($\omega_2 \in H^2(M;Z_2)$) and the first Pontryagin class ($p_1 \in H^4(M;Z)$) for $SO(n)$ bundles and bundles and additionally the Euler class ($e \in H^4(M;Z)$) for an $SO(4)$ bundle. The classes C_1 and C_2 may be freely specified whereas ω_2 and p_1 are subject to the constraint

$$\omega_2 \cup \omega_2 = p_1 \bmod 2 \tag{5.12}$$

The physical significance of these classes stems partly from their representability, via the Chern–Weil isomorphism. (See Kobayashi and Nomizu (1969) for a convenient summary) as functions of the Yang–Mills field strengths. Thus C_1 is represented by the Maxwell field two form $f_{\mu\nu}(G = U1)$ or $Tr\ F_{\mu\nu}(G = U(n))$ whilst p_1 and C_2 are proportional to the four form $Tr\ F \wedge F$.

A result of great mathematical and physical significance is the Atiyah–Singer index theorem. This relates the number of zero modes of various differential operators to characteristic classes. Physically, it arises in the statement that a theory that is formally invariant under chiral $U1$ transformations on spinor fields

$$\psi \to \exp(-i\alpha\gamma_5)\psi \tag{5.13}$$

possesses a quantum Noether current $j^{\mu 5}$ with an anomalous divergence.

For example in QCD with N_F flavours and N_C colours coupled to both the electromagnetic and gravitational fields, the index theorem states that the number of positive (n_+) and negative (n_-) chirality-zero eigenmodes is expressible in terms of the gauge fields as

$$n_+ - n_- = \frac{-N_F}{16\pi^2}\int Tr\, F_{\mu\nu}{}^*F^{\mu\nu} - \frac{N_F F_C}{384\pi^2}\int R_{\mu\nu ab}{}^*R^{\mu\nu ab}$$

$$+ \frac{N_F N_C}{16\pi^2}e^2\int f_{\mu\nu}{}^*f^{\mu\nu} \tag{5.14}$$

$$\text{where } {}^*f^{\mu\nu} = \frac{1}{2}\varepsilon^{\mu\nu\alpha\beta} f_{\alpha\beta} \text{ etc.} \tag{5.15}$$

This number $n_+ - n_-$ is directly proportional to $\nabla_\mu j^{\mu 5}$. (Christensen and Duff 1978*a*, 1979; Dowker 1978; Delbourgo and Salam 1972; Eguchi and Freund 1976; Hanson and Römer 1978; Nielson and Schroer 1977; Nielson, Römer, and Schroer 1978*a*,*b*.) Note that for simplicity the results are quoted for a compact Riemannian manifold. Non-compact manifolds require boundary term corrections (Atiyah, Patodi, and Singer 1973, 1975*a*,*b*,*c*, 1976; Gibbons, Pope, and Römer 1979).

Most of the well known properties of Minkowski space Yang–Mills theory (Jackiw and Rebbi (1976); Callan, Dashen, and Gross 1976; Jackiw 1979) have analogues in curved space-time. These include $|n\rangle$ vacua, $|\theta\rangle$ vacua, vacuum tunneling and CP violation (Deser, Duff, and Isham 1980). Tunneling can occur between $|n\rangle$ vacua defined on a space-like hypersurface Σ in a static space-time and the classification of the $|n\rangle$ vacua is in terms of the group $[\Sigma, G]$ of homotopy classes of maps from Σ into the internal symmetry group G. This depends directly on the topology of both Σ and G and $[\Sigma, G]$ may be completely characterized in terms of the cohomology groups $H^1(\Sigma;\pi_1(G))$ and $H^3(\Sigma;\pi_3(G))$ (Isham 1980).

Aside from these specific topological considerations there has been a substantial study of solutions to the coupled Einstein–Yang–Mills (or Maxwell) equations with particular emphasis on self-dual Yang–Mills fields which, in a Riemannian manifold, have vanishing energy momentum tensor (Boutaleb-Joutei and Chakrabarti 1979; Boutaleb-Joutei, Chakrabarti, and Comteb 1979*a*,*b*, 1980; Charap and Duff 1977*a*,*b*; Duff and Madore 1979; Pope and Yuille 1978).

Topological considerations also arise in the conformal anomaly – the non-vanishing of the regularized $\langle \hat{T}^{\mu}_{\mu} \rangle$ in a theory that is classically conformally invariant. This phenomenon has been widely studied and the reader is referred to the general review articles cited above for details and references. If $\langle \hat{T}^{\mu}_{\mu} \rangle$ is regularized using, for example, ζ-function methods the finite result may be expressed in terms of the eigenvalues of the linear field operators and has no topological significance. However if the background field is Ricci flat ($R_{\mu\nu} = 0$) or conformally flat ($C_{\mu\nu\rho\sigma} = 0$) then

$$\left\langle \int d^4x \sqrt{-g}\ \hat{T}^{\mu}_{\mu} \right\rangle^{\text{reg}} = -\frac{1}{180}\{2N_0 + 7N_{\frac{1}{2}} - 26N_1\}\chi \qquad (5.16)$$

where χ is the Euler number of the compact Riemannian space-time and N_0, $N_{\frac{1}{2}}$ and N_1 are respectively the number of spin 0, ½ and 1 particles (Duff 1977).

An explicit cancellation of the non-zero modes occurs in supergravity theories and the result is once again to express $\langle \hat{T}^{\mu}_{\mu} \rangle$ in terms of χ (Hawking and Pope 1978*b*; Christensen and Duff 1978*a*, 1979).

2. Spontaneous symmetry breaking

Spontaneous symmetry breaking plays an important part in the physical applications of local gauge theories. The problem of breaking an internal symmetry from the group G to a subgroup H is mathematically equivalent to reducing a principal G-bundle to an H-bundle. In a general space-time there may be topological obstructions to such breaking or, if it does occur, a given G-sector may break into one of a number of inequivalent

H-sectors. The obstructions to breaking may be expressed in terms of the characteristic classes of the original *G*-bundle and the inequivalent *H*-sectors may be related to the homotopical properties of the Higgs–Kibble scalar fields, which in the present context, may be viewed as cross sections of the associated vector bundle lying on G/H orbits.

For example consider the spontaneous breaking of the Salam–Weinberg weak/electromagnetic group $SU2 \times U1_Y$ down to $U1_{EM}$. This is possible if and only if the Chern classes C_1 and C_2 of the $SU2 \times U1_Y$ bundle are related in $H^4(M;Z)$ by $C_2 = C_1 \cup C_1$ in which case the first Chern class of the ensuing $U1_{EM}$ bundle is precisely C_1. Note that since $H^2(S^4;Z) = 0$ there can be no such breaking on the usual S^4 compactification of flat euclidean space unless $C_1 = C_2 = 0$. On the other hand $S^2 \times S^2$ does possess suitable cohomology elements and in this case breaking *is* possible from a non-trivial sector. Under these conditions the Salam–Weinberg model can admit non-trivial instanton solutions which could be important in early universe physics. Even if C_1 does not satisfy $C_2 = C_1 \cup C_1$ stable solutions of the field equations may still exist. Solutions, such as $\phi = 0$, that are conventionally unstable may be stabilized by the non-triviality of the vector bundle and by the space-time topology.

3. *Dimensional reduction*

The construction of theories in four dimensions by dimensionally reducing from a higher dimension has become a well established tool (Brink, Schwarz, and Scherk 1977; Gliozzi, Scherk, and Olive 1977; Cremmer and Scherk 1976*a,b,c*; Omero and Percaci 1980; Scherk and Schwarz 1979). It is interesting to enquire to what extent the topological properties of the higher dimensional manifold can affect the reduced theory. A rather elementary example is afforded by a scalar field Φ on $M \times S^1$ where M is four dimensional. This field is usually expanded as $\Phi(x,\theta) = \sum_n(\phi_n(x) \cos n\theta + \psi_n(x) \sin n\theta)$ which, when substituted into a five dimensional action (signature is -++++)

$$S = -\int d^4x \int_0^{2\pi} d\theta \sqrt{-g}\left\{\partial_{\hat\mu}\Phi\partial_{\hat\nu}\Phi g^{\hat\mu\hat\nu} + \mu_0^2\Phi^2\right\};\ \hat\mu,\hat\nu = 1 \ldots 5 \qquad (5.17)$$

gives, after integration over θ, an infinite set of four dimensional scalar fields with a mass spectrum $(\mu_0^2 + n^2/R^2)$. (I have assumed that the metric on $M \times S^1$ is in product form $ds = g_{\mu\nu}dx^\mu\, dx^\nu + R^2\, d\theta^2$.) Now suppose that Φ is regarded as a cross-section of a line bundle that is twisted over S^1. Then Φ may be represented by an anti-periodic function of θ and the spectrum becomes $\mu_0^2 + (n+\frac{1}{2})^2/R^2$. Thus a topological property of the five dimensional manifold (the existence of the non-trivial line bundle) is coded into the reduced theory.

There are many other examples of this type. A more subtle one is given by considering $SU2$ bundles over $M \times S^1$. The set $\mathscr{B}_{SU2}(M_5)$ of $SU2$-bundles over a 5-manifold M_5 appears in the exact sequence

$$\to H^3(M_5;Z) \to H^5(M_5;Z_2) \to \mathscr{B}_{SU2}(M_5) \to H^4(M_5;Z) \to 0 \qquad (5.18)$$

where the image of a given $SU2$ bundle ξ in $H^4(M_5;Z)$ is just the second Chern Class $C_2(\xi)$. It can be shown that the set of $SU(2)$ bundles with a given $C_2 = \alpha$ say, is

$$C_2^{-1}(\alpha) \approx H^5(M_5;Z_2)/\omega_2 \cup \{H^3(M_5;Z)\text{mod. } 2\} \qquad (5.19)$$

ω_2 is the second Stieffel–Whitney class of the manifold. If M_5 is compact and orientable then $H^5(M_5;Z_2) = Z_2$ and hence $C_2^{-1}(\alpha) = Z_2$ or 0. In particular if $M_5 = S^4 \times S^1$, $C_2^{-1}(\alpha) = Z_2$ and hence there are two $SU2$ bundles for each cohomology element $\alpha \in H^4(S^4{\times}S^1;Z) = Z$. It would be intriguing to see how this extra 'topological quantum number' is reflected in the reduced theory.

4. Non-linear σ-models

Non-linear σ-models are associated with non-linear realizations of a lie group G which are linear with respect to a subgroup H. The basic such realization is afforded by the transitive action of G on the coset space G/H:

$$\begin{aligned} G \times G/H &\to G/H \\ (g, g_0H) &\rightsquigarrow gg_0H \end{aligned} \qquad (5.20)$$

and the coordinates of G/H are viewed as a set of scalar fields Φ^i, $i = 1 \ldots \dim G/H$. Thus the fields may be regarded as maps from M into G/H and the homotopy classes $[M,G/H]$ can be expected to play some role in the classification of the solutions to the field equations. These are derived from the action

$$S = \int d^4x \sqrt{-g}\, h_{ij}\, g^{\mu\nu}\, \partial_\mu \phi^i\, \partial_\nu \phi^j \tag{5.21}$$

where h_{ij} is a G-invariant metric on G/H and the solutions are provided by harmonic maps from M into G/H. There is clearly a connection between these models and the spontaneous symmetry situation discussed above. Further details may be found in Misner (1978); Omero and Percacci (1980); Percacci (1980*a,b*).

5. *Non-simply connected space-times*

If the space-time manifold M is not simply connected (i.e. $\pi_1(M) \neq 0$) then many interesting phenomena can occur. These arise from the existence over M of non-trivial real line bundles whose cross sections are known as twisted scalar fields or automorphic scalar fields. Such bundles are classified by the cohomology group $H^1(M;Z_2)$ and a number of quantum field theory results (such as self-energies, vacuum polarizations, scattering matrices) depend on the choice of bundle.

Another related effect is the existence of inequivalent spinor structures which are again classified by $H^1(M;Z_2)$ (Milnor 1963). The topological and quantum field theoretic details plus a number of examples may be found in the following papers. Avis and Isham (1978, 1979*a,b*); Banach (1979*a,b*); Banach and Dowker (1978, 1979*a,b*); Birrell and Ford (1979); Chockalingham and Isham (1980); Denard and Spallucci (1980); DeWitt, Hart and Isham (1979); Ford (1978, 1980*a,b*); Isham (1978*a,b*); Kiskis (1978); Toms (1980*a,b,c,d*); Unwin (1979, 1980); Unwin and Critchley (1980). See also the article by S.J. Avis in this volume.

6. *Extended spin structures*

The global existence of spinor fields requires that at each point of M there exists an $SL(2,\mathbb{C})$ double covering of the local $SO(3,1)_0$ group of vierbein transformations. (If M is Riemannian, the corresponding groups are Spin 4 $\approx SU2 \times SU2$ and $SO4$). These coverings must patch together smoothly and this is possible if and only if the global topology is such that $\omega_2(M)$ – the second Stieffel–Whitney class – vanishes. An example of a space where this does not happen is $\mathbb{C}P^2$. Hawking and Pope suggested that in such cases the tangent space group should be extended by an internal $U1$ group and generalized spinors were defined as transforming under Spin $4_{Z_2} \times U1 \equiv$ Spin $^c4$ where the Z_2 equivalence relation is $(x,u) \equiv (-x,-u)$. (Hawking and Pope 1978.) In effect the undesirable twisting of the Spin 4 groups is undone in the $U1$ group.

The presence of the Z_2 group leads to a relation between the spin and $U1$ quantum numbers although unfortunately the result seems rather unphysical (Hawking and Pope 1978). The technique may be extended to a wide family of internal symmetry groups G by arranging that the non-vanishing $\omega_2(M)$ is cancelled by the appropriate characteristic class of the principal fibre bundle associated with the gauged G structure. (Avis and Isham 1980; Back, Freund, and Forger 1978; Forger and Hess 1979.)

7. Misner-Finkelstein kinks

The construction of a metric tensor on an oriented Riemannian (resp. Lorentzian) manifold M is equivalent to reducing the structure group from $GL(4,R)$ to $SO4$ (resp. $SO(3,1)$) which is in turn equivalent to constructing a cross-section of the associated $GL(4,R)/SO4$ (resp. $GL(4,R)/SO(3,1)$) bundle. Now $GL(4,R)/SO4 \approx R^{10}$ and hence there are no obstructions to finding such cross-sections and they are all homotopic. On the other hand $GL(4,R)/SO(3,1)$ is topologically non-trivial (Steenrod 1951) and as a result

(a) There may be no cross sections and hence no reduction of the structure group. (For example a necessary condition is that the Euler class of M vanish.)

(b) Homotopically inequivalent cross-sections may exist.

The latter property was first noticed by Finklestein and Misner (1959) who referred to the homotopically twisted metrics as 'kinks'. If the tangent bundle is trivial these metrics are classified by the set of homotopy classes $[M,P_3R]$. This topological aspect has been extensively studied by Shastri,

Williams, and Zvengrowski (1979).

It is not clear whether these kinks have any significant effects on the quantum field theory or if they are merely an intellectual curiosity. It is interesting that they only occur on Lorenzian space-times whereas all other topological properties that have been considered are applicable to both Riemannian and Lorentzian spaces or to Riemannian spaces alone.

8. Closed ≠ exact

A p-form ω that is exact ($\omega = d\phi$) is necessarily closed ($d\omega = 0$) since $d^2 = 0$ but the converse is not necessarily true. Indeed De Rham's theorem provides a direct connection between the set of closed but non-exact p-forms and the space-time topology:

$$\text{Closed } p\text{-forms}/\text{exact } p\text{-forms} = H^p(M;R). \qquad (4.22)$$

From one point of view the non-vanishing of these cohomology groups permits the existence of non-vanishing real characteristic classes and hence non-trivial gauge sectors. However the presence of closed but non-exact forms may also be related to phenomena of physical interest without invoking fibre bundle theory. For example if $H^2(M;R) \neq 0$ there exist Maxwell field tensors $F_{\mu\nu}$ which cannot be written as the curl of a vector potential (i.e. $F_{\mu\nu} \neq \partial_\mu A_\nu - \partial_\mu A_\nu$). This leads to the idea of 'charge without charge' which was extensively discussed in the past (Wheeler 1963, 1968). More recently the idea has been applied in a quantum field theoretic formalism by Ashtekar and Sen (1979) and Sorkin (1978) where it is shown to lead to superselection sectors. Another entertaining result is that of Duff and van Nieuwenhuizen (1980) who consider the theory of an antisymmetric tensor $f_{\mu\nu}$ invariant under the gauge transformations $f_{\mu\nu} \rightarrow f_{\mu\nu} + \partial_\mu\theta_\nu - \partial_\mu\theta_\nu$. Formally this is equivalent to a single spin 0 degree of freedom but the topological effects lead to a trace anomaly that is in fact different. Similar remarks also apply to closed, but not exact, field strengths of antisymmetric rank three potentials which in addition give rise to a gauge principle interpretation of the cosmological constant.

9. Gravitational instantons

The existence of non-trivial space-time topology is crucial in the study of gravitational instantons. It is important that a compact simply connected four dimensional manifold with spinors can be classified up to homotopy type by the Euler number and Hirzebruch signature. A related fact is the classification of simply connected four manifolds 'almost' as a topological sum of copies of $S^2 \times S^2$, K_3 and, if $\omega_2(M) \neq 0$, $\mathbb{C}P^2$ and its conjugate. I shall return to this feature in Section 7.

It should be clear from the preceding remarks that a substantial amount of time and effort has been expended in the study of quantum fields on a curved space-time. Indeed the subject is rather seductive, not least because it poses problems which, unlike those of quantum gravity proper, can be tackled with a justified hope of obtaining a solution in a reasonable amount of time. However, one should not lose sight of the ultimate goal which is the full quantization of the gravitational field itself.

Some of the features discussed above are likely to appear in some form in such a final theory. This seems particularly likely for those effects which are related to the global horizon and topological properties of a manifold. However, other results may be misleading. For example the difficulty in defining a unique vacuum state in a general space-time may disappear in a complete quantum gravity theory, especially if it is based on a perturbative scheme which does not involve expanding the metric about some background.

It is also necessary to be cautious when assessing the significance of the particle production work in cosmological settings. Much of this ignores completely the back reaction problem and even if it is incorporated in some type of $G_{\mu\nu} = \langle \hat{T}_{\mu\nu} \rangle$ scheme, the validity of the results as the first approximation to a theory that is non-renormalizable is rather dubious. This is especially true for those calculations which seek to explore and possibly to alleviate the initial space-time singularity.

6. APPROACHES TO NON-RENORMALIZABILITY

Various techniques have been suggested for handling non-renormalizable theories. Some are quite old and are well understood whilst others involve recent ideas that have not yet been fully assimilated. They are all, in principle, applicable to any non-renormalizable theory, which may of course be their weakness if gravity turns out to be quantizable only by virtue of some special properties of the Einstein equations. The problem of non-renormalizability is sufficiently serious to justify listing all the different approaches, even those that have lain dormant for a while.

1. *Summing infinite sets of graphs*

Even if quantum gravity were finite order by order in perturbation theory, there would still be a great incentive to sum infinite sets of Feynmann graphs. The effective expansion parameter is Gp^2 and hence the perturbation series would be divergent at the Planck energy $Gp^2 \sim 1$. Thus finite sets of graphs will never yield information on the Planck length domains of energy and distance.

However, in the present context the interesting possibility is that by first summing an infinite set of Feynmann integrands and then integrating, a finite answer may be obtained. For example $\int_0^1 \exp(-1/x^2)dx$ is finite but expanding the integrand and integrating term by term yields a series of infinities

$$\int_0^1 \left\{1 - \frac{1}{x^2} + \left\{\frac{1}{x^2}\right\}^2 \frac{1}{2!} - \dots\right\} dx = 1 - \infty + \infty \dots . \qquad (6.1)$$

The earliest summation attempts in quantum gravity were by Khriplovich (1966) and DeWitt (1964*a*) who summed overlap graphs of the form

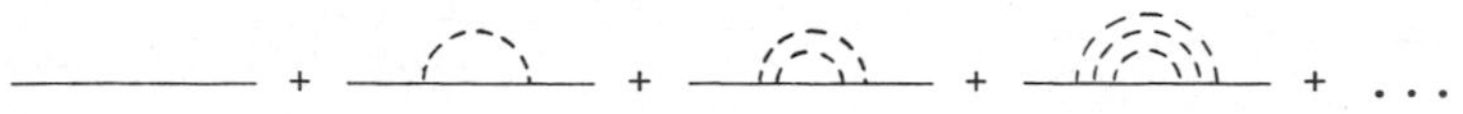

(6.2)

where the dotted line is the graviton interacting with a fermion. More sophisticated calculations were then performed using general methods for quantizing non-polynomial Lagrangians. (Isham *et al.* 1971, 1972.) These techniques were applied to quantum electrodynamics interacting with gravity and strings of graphs were computed of the form

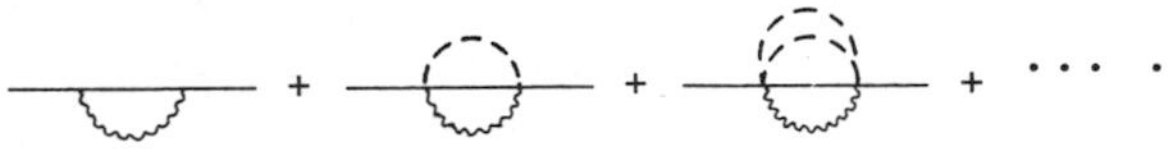

(6.3)

The first term in (6.3) is the usual photon contribution to the electron self energy and is of course logarithmically divergent. Nevertheless the whole series sums to a finite answer and gives

$$\frac{\delta m}{m} \simeq \frac{3\alpha}{4\pi} \log Gm_e \approx \frac{2}{11} . \tag{6.4}$$

The fine details of the results depend on the choice of interpolating field. The exponential $g_{\mu\nu} = \exp(\sqrt{G}h)_{\mu\nu}$ and rational $g_{\mu\nu} = \eta_{\mu\nu} + \sqrt{G}h_{\mu\nu}$ parametrizations give Green's functions of the form

$$\mathcal{G}(x,0) \sim \exp(-G\hbar/x^2) - 1 \tag{6.5}$$

and

$$\mathcal{G}(x,0) \sim \frac{x^2}{(G\hbar - x^2)} . \tag{6.6}$$

Clearly the infinite sum has radically changed the singularity structure. Thus (6.5) has an essential singularity on the Minkowskian light cone whereas (6.6) exhibits a shift ('quantum smearing') of the light cone to $x^2 = G\hbar$. It may also be shown that the exponential and rational parameterizations lead to theories that are local and non-local respectively. In fact non-polynomial theories split up into equivalence classes in the sense that a field transformation from one to the other changes the local structure.

Non-polynomial theories were shown to be unitary and gauge invariant but the techniques employed had drawbacks:

(a) Only a subset of graphs was summed and this subset did

not in itself possess properties that were physically acceptable. For example for large $p^2, \mathcal{G}(p^2) \sim \exp((p^2)^\alpha)$ for some $0 < \alpha < 1$, and unphysical ghosts in the Euclidean region frequently occurred.

(b) The results were not unambiguous.

(c) The techniques were computationally difficult and for this reason were only applied to gravity plus matter with simple vertices like $\psi\gamma^a A^\mu \psi e_{a\mu} \det e$. Pure quantum gravity was never tackled in this way.

Nevertheless these calculations remain the only full scale investigation of gravity as a universal cut off and regulator of nominally divergent Feynman integrals. Indeed the result in (6.4) is precisely what would be obtained if the original logarithmically divergent Feynman integral were cut off at the Planck length.

Interest in these methods has been dormant for a while but a Green's function with the behaviour of (6.5) has been employed by Brown in his recent work on the finiteness of gravity (M.R. Brown, in this volume).

2. Summation of complete theory

A related idea occurs in the recent work of Fradkin and Vilkovisky (1978). They introduce a cut off L in momentum space and conjecture that the divergent part of the effective action arising from multiloop contributions is, symbolically,

$$\text{Log } Z[g]_{\text{div}} \sim \left\{-\ln L^2 + G_{\text{ren}} L^2 - \frac{1}{2} G^2_{\text{ren}} L^4 \ldots\right\} \int R^2 \tag{6.7}$$

$$+ \left\{ (\ln L^2)^2 - 2G^2_{\text{ren}}(\ln L^2) + G^4_{\text{ren}} L^4 + \ldots\right\} \int R^3$$

where G_{ren} is the renormalized coupling constant and R^2 means both $R_{\mu\nu}R^{\mu\nu}$ and R^2 terms. Thus the coefficient of the Riemannian curvature squared may be summed as

$$- \ln L^2 + G_{\text{ren}} L^2 - \frac{1}{2} G^2_{\text{ren}} L^4 - \ldots = - \ln\left(\frac{1}{L^2} + G_{\text{ren}}\right) \tag{6.8}$$

whilst the contribution from the cubed terms is $[\ln (\frac{1}{L^2} + G_{\text{ren}})]^2$ and so on.

The crucial point is that as $L \to \infty$ these numbers converge to $\ln(G_{\text{ren}})$, $\ln(G_{\text{ren}})^2$ etc. In other words, the sum of the coefficients of any particular contribution to the effective action is finite although of course if an expansion such as (6.8) is made the series seems highly divergent.

This could be viewed as the most optimistic conjecture of the decade but nevertheless it would be well worth while to compute the first few terms in (6.7) to see whether their coefficients have the right values.

3. *Higher derivative interactions*

Suppose the Einstein Lagrangian is augmented with terms involving the square of the Riemann curvature. If these are regarded as extra perturbations then the ultraviolet divergence problem is even worse than before. However, if the metric is expanded as $g_{\mu\nu} = \eta_{\mu\nu} + \sqrt{G}h_{\mu\nu}$ there is a contribution $h_{\mu\nu}\Box^2 h_{\mu\nu}$ in R^2, and it is possible to view these fourth order derivative terms as part of the bilinear (free) Lagrangian. Under these circumstances the graviton propagator changes from (dropping indices) $\frac{1}{p^2}$ to $\frac{1}{p^2 - aG\ p^4}$.

The extra powers of momentum serve to regulate otherwise divergent integrals and the resulting theory may be shown rigorously to be renormalizable (Stelle 1977, 1978). Unfortunately writing

$$\frac{1}{p^2 - aG\ p^4} = \frac{1}{p^2(1-aGp^2)} = \frac{1}{p^2} + \frac{aG}{(1-aGp^2)} \tag{6.9}$$

we see that a pole at $p^2 = 1/aG$ is present with the wrong sign for unitarity; i.e. a 'ghost' has appeared.

A large number of variants of the basic $R + R^2$ idea have been published including some in which torsion appears as a propagating degree of freedom (e.g. Nevill 1980). These theories have recently been surveyed by Sezgin and van Nieuwenhuizen (1979) who show that they all possess ghosts. There are three possible reactions to the presence of these unphysical poles in the propagator:

(a) They are a disastrous feature and mean that $R + R^2$ type theories are not viable.

(b) The extra pole only appears at Planck energies which,

as mentioned previously are precisely the points at which the whole perturbation series diverges anyway. Thus it may be possible to remove the unphysical singularities by a selective summation of graphs. One very interesting example of this is due to Tomboulis (1977, 1980) who shows that the real poles vanish in a $1/N$ expansion when gravity is coupled to N massless fermions. Complex pairs of poles appear on the physical sheet but may be possibly interpreted in terms of the Lee–Wick mechanism.

(c) Since the theory is renormalizable it is subject to renormalization group analysis and attempts can be made to use these techniques to move the ghost mass off to infinity, where it does no harm. This is in effect a type of resummation and is discussed in Salam and Strathdee (1978) and Julve and Tonin (1978).

It should be noted that since R^2 terms arise as counter-terms in conventional quantum gravity it is necessary even there to decide whether to include $h_{\mu\nu}\Box^2 h_{\mu\nu}$ in the free Lagrangian or to regard it as part of the interaction.

4. Gravitons as bound states

It is not possible to write down local renormalizable interactions for high spin hadrons, but in practice this is not regarded as a matter for concern because such particles are viewed as resonances of an underlying (presumably renormalizable) quark field. Therefore the question naturally arises whether the graviton can also be regarded as a bound state, with the Einstein Lagrangian merely serving as an effective low energy description of the interpolating field. Unfortunately in such a picture it is difficult to explain why the graviton mass is exactly zero. What is needed is some analogue of the chiral invariance of low energy pion physics. A recent revival of these ideas is given by Adler (1976), Adler, Lieberman, Ng, and Hueng-Sheng Tsao (1976) and the subject is discussed at greater length in Weinberg (1979). (Weinberg's article contains a number of interesting observations on the whole non-renormalizability problem and should be regarded as mandatory reading.)

5. Klauder's ideas

For some time Klauder has been developing a new technique for handling non-renormalizable field theories. The idea is best approached through a simple example. Consider the Euclidean generating functional for the Green's functions of a single scalar field with a $\lambda\phi^p$ interaction ($p \geqslant 3$):

$$Z[j] = \int \exp\left(-\int (\partial_\mu\phi\partial_\mu\phi + \mu^2\phi^2 + \lambda\phi^p)\mathrm{d}^4x\right)\exp\left(\mathrm{i}\int \phi j \; \mathrm{d}^4x\right)\mathrm{d}\phi \tag{6.10}$$

Then if $p \leqslant 4$ (i.e. if the theory is renormalizable) field configurations which have a finite free action also give a finite contribution from the interaction term. This follows from the Sobolev inequalities:

$$\left\{\int |\phi|^p \; \mathrm{d}^n x\right\}^{2/p} \leqslant K\int (\partial_\mu\phi\partial_\mu\phi + \mu^2\phi^2)\mathrm{d}^n x \text{ for some } K \tag{6.11}$$

which holds if $p \leqslant (2n/n-2)$ (i.e. if $p \leqslant 4$ when $n = 4$). If $p \geqslant (2n/n-2)$ (i.e. if the theory is unrenormalizable) then there is no such K and there exist field configurations for which ($n = 4$)

$$\int (\partial_\mu\phi\partial_\mu\phi + \mu^2\phi^2)\mathrm{d}^4x < \infty \tag{6.11}$$

but

$$\int |\phi|^p \; \mathrm{d}^4x = \infty, \; p > 4. \tag{6.12}$$

Because of (6.12) these fields are formally suppressed in the functional integral and *they remain suppressed as the limit is taken* $\lambda \to 0_+$. Thus the $\lambda \to 0_+$ limit of (6.10) is *not* a conventional free field but rather a 'quasi-free' field in which the field configurations satisfying (6.12) are removed.

Klauder's basic contention is that in any non-renormalizable theory it is consequently *wrong* to adopt conventional perturbation theory around a free field but instead a perturbation around the quasi-free field should be employed. He substantiates his claims by illustrations drawn from quantum mechanical

Hamiltonians with singular potentials where precisely this type of behaviour occurs.

There are many other ideas arising from the above. For example, Klauder suggests that the measure in the functional integral should be 'prepared' for the interaction and shows that it is more appropriate to employ a measure that is invariant under the scale transformation $\phi(x) \to \Omega(x)\phi(x)$ rather than the usual translation invariant one. Formally, one uses

$$\prod_x \frac{d\phi(x)}{|\phi(x)|} \tag{6.13}$$

rather than the conventional

$$\prod_x d\phi(x). \tag{6.14}$$

Of course one might argue that (6.13) and (6.14) differ by $\delta^4(0)$ terms that can be dropped using dimensional regularization, but Klauder's point is that the type of perturbative framework in which this statement is phrased is *not* appropriate for a non-renormalizable theory. A measure of the type (6.13) can be obtained by adding auxiliary fields to a conventional theory in the form:

$$Z[j] = \int d\phi d\chi \exp\left(-\int\left\{\frac{1}{2}(\partial_\mu\phi\partial_\mu\phi+\mu^2\phi^2)+\lambda\phi^p-\frac{1}{2}\phi^p\chi^p+i\phi j\right\}d^4x\right) \tag{6.15}$$

It is perhaps interesting to note in passing that supergravity possesses auxiliary fields of this type (Stelle and West 1978; Ferrara and van Nieuwenhuizen 1978).

These considerations might also imply that perturbation theory of the type $\phi = \phi_0 + \lambda h$ should be replaced with some type of multiplicative transformation. In quantum gravity it suggests that the background field split $g_{\mu\nu} = g^c_{\mu\nu} + \sqrt{G}\, h_{\mu\nu}$ is inappropriate. However, a multiplicative split of the background metric seems unnatural. On the other hand, any tetrad of tangent vectors can be expressed as linear combinations of any other linearly independent set and a more appropriate split might therefore be

$$e^{\mu a} = f^{\mu b}\,\Lambda_b{}^a\,, \tag{6.16}$$

where $f^{\mu b}$ is a fixed background set of four independent tangent vectors and Λ is an arbitrary function from space-time into $GL(4,R)$. The perturbative expansion would arise by writing $\Lambda_a{}^b = (\delta_a{}^b + \sqrt{G}\, B_a{}^b)$ and would lead formally to the same on mass shell S-matrix elements as the usual theory. However (6.16) needs to be employed with a scale invariant measure and a quasi-free expansion before Klauder's ideas can be implemented.

In the language of canonical quantization the use of a scale covariant measure is equivalent to replacing the canonical commutation relations

$$[\phi(x),\pi(y)] = i\hbar\delta(x-y) \tag{6.17}$$

by the affine relations

$$[\phi(x),K(y)] = i\hbar\phi(x)\delta(x-y) \tag{6.18}$$

These had previously been introduced by Klauder (1972) into the canonical quantization of gravity as a means of implementing the positivity of g_{ij},

$$\det g_{ij} \geqslant 0.$$

Klauder's ideas are potentially of great importance and seem to have been overlooked by the theoretical physics community. A recent review article with a complete bibliography is Klauder (1979).

6. Asymptotic safety

In Weinberg (1979) it is suggested that in some circumstances the criterion of renormalizability may usefully be replaced by 'asymptotic safety' and, that if this is applied to a non-renormalizable theory, it may cut the infinite number of undetermined parameters down to a finite set. The basic idea is as follows.

Let $R = \mu^D f(E/\mu, X, \lambda)$ be a reaction rate for some process described by a renormalizable quantum field theory with a renormalized coupling constant λ and a renormalization point μ. E denotes some typical energy variable and X represents dimensionless parameters formed from ratios of momenta and μ. Then R satisfies the renormalization group equation

$$\left(\mu \frac{\partial}{\partial\mu} + \beta(\lambda) \frac{\partial}{\partial\lambda}\right) R(E,\mu,\lambda) = 0, \tag{6.19}$$

where

$$\beta(\lambda) = \mu \frac{\partial}{\partial\mu} \lambda(\lambda', {}^{\mu}/_{\mu'}) \Bigg|_{\substack{\lambda' = \lambda \\ \mu' = \mu}} \tag{6.20}$$

describes the dependence of λ on a shift in renormalization point.

If the energy is scaled by e^{t}, then

$$R(e^{t}E,\mu,\lambda) = e^{tD} R(E,\mu,\bar{\lambda}(t)) , \tag{6.21}$$

where $\bar{\lambda}(t)$ is the 'running coupling constant' satisfying

$$\beta(\bar{\lambda}(t)) = \frac{\partial\bar{\lambda}(t)}{\partial t} , \tag{6.22}$$

$$\bar{\lambda}(t = 0) = \lambda .$$

In particular if $\bar{\lambda}(t)$ approaches a fixed point λ^* as $t \to \infty$ the high energy behaviour is a simple scaling. Weinberg's conjecture is that in a non-renormalizable theory, most choices of the ambiguous coupling constants will lead to a momentum space behaviour with unphysical singularities at high energies. (Of course this cannot be observed order by order in perturbation theory.) This will be avoided however if the momenta has a simple scaling behaviour. Thus he proposes to generalize (6.19)–(6.22) to an infinite set of (essential) coupling constants and to look for fixed points. Such a theory is said to be asymptotically safe.

The crucial conjecture is that the renormalized coupling constants may all be chosen so that the running constants possess a fixed point and that the set of initial values with this property lies on a *finite* dimensional subspace of the infinite dimensional coupling constant space. Some justification is given for this hope by the observation that such behaviour arises if the ultraviolet fixed point can be related to an infrared induced second order phase transition.

Under these circumstances only a finite set of parameters needs to be adjusted to get a unique theory, and this is just as useful as the situation which arises in a conventional renormalizable theory. A technique much used in statistical physics in discussions of fixed points is analytic continuation in the space-time dimension. Using results of Christensen and Duff (1978*c*) and Gastmans *et al.* (1977), Weinberg observes that quantum gravity is asymptotically safe in $2 + \varepsilon$ dimensions (ε near 0_+) and the hope is that this behaviour is maintained as $\varepsilon \to 2$. A not unrelated recent paper is Smolin (1978) where the idea of putting quantum gravity on a lattice is resurrected.

7. Cancel the infinities

Boson and fermion loops appear with different signs and it is conceivable, therefore, that the graviton ultraviolet divergences may be cancelled by those of judiciously chosen matter fields. In general the addition of matter makes things worse (see the references cited in Section 4) so some care is obviously required. Clearly a resolution of the problem on these lines would depend critically on the details of the Einstein equations and, as such, does not fall within the category of the general techniques being discussed in this section. The only successful candidate to date is supergravity where the one loop contributions do indeed mutually cancel (Section 8).

Let me conclude this section by repeating that the techniques discussed above all start with a perturbative expansion $g_{\mu\nu} = g^c_{\mu\nu} + \sqrt{G}\, h_{\mu\nu}$ and then seek to extend or modify the quantum field theory in a way which sidesteps the infinite sets of ultraviolet divergent graphs and associated renormalization constants. Such perturbations lose some of the basic global structure of the classical theory of general relativity and may be fundamentally wrong. At the very least if one of these methods does work one would hope to recover this lost structure in some aspect of the quantum field theory.

7. THE RIEMANNIAN PROGRAMME

The functional integral has become an indispensable tool in modern quantum field theory. At a heuristic level the Green's function generating functional

$$Z[j] = \int d\phi \, \exp\left(\frac{i}{\hbar}\left\{ \int \{\partial_\mu\phi\partial_\mu\phi + \mu^2\phi + V(\phi)\}d^4x + \int j\phi d^4x\right\}\right) \quad (7.1)$$

provides, at the very least, an efficient way of handling the combinatorial aspects of the Feynman–Dyson expansion. Unlike the older operator methods, it also leads naturally to the correct Feynman rules for non-abelian gauge theories.

A significant advantage is obtained by computing the Green's functions on a Euclidean space (signature ++++) rather than the Lorentzian space (signature +++-) employed in (7.1). This has the effect of replacing in the action the time variable t by $-i\tau$ and (7.1) becomes

$$Z[j] = \int d\phi \, \exp\left(-\frac{1}{\hbar}\int\{\partial_\mu\phi\partial_\mu\phi + \mu^2\phi + V(\phi)\}\right)d^4x \, \exp\left(+\frac{i}{\hbar}\int j\phi\right). \quad (7.2)$$

Finite action solutions to (7.2) may be employed to discuss tunnelling through a potential barrier. Such tunnelling effects are non-perturbative and cannot be obtained from (7.1) to any finite order in perturbation theory. This aspect has been extensively developed in the Yang–Mills theory, and the classical solutions are the famous instantons.

At a rigorous level Euclideanization is widely used and (7.2) is interpreted as

$$Z[j] = \int d\mu(\phi) \, \exp\left(-\frac{1}{\hbar}\int V(\phi)d^4x\right)\exp\left(\frac{i}{\hbar}\int j\phi d^4x\right) \quad , \quad (7.3)$$

where $d\mu(\phi)$ is the Gaussian measure with covariance $(-\Delta + \mu^2)$ and may be written heuristically as

$$d\mu(\phi) \text{ "=" } d\phi \, \exp\left(-\frac{1}{\hbar}\int(\partial_\mu\phi\partial_\mu\phi + \mu^2\phi^2)d^4x\right) . \quad (7.4)$$

The term $\exp(-\frac{1}{\hbar}\int V(\phi)d^4x)$ is viewed as a multiplicative perturbation of this measure and is made rigorous by the introduction of suitable cut-offs and renormalization (Simon 1974).

The success of the Yang–Mills Euclidean programme led Hawking (1979*a*,*b*) to propose that a similar Riemannianization be applied to quantum gravity. (The word 'Euclideanization' frequently appears in the literature but this is potentially confusing in a curved space context and will not be used here.)

Thus, for example, the partition function becomes

$$Z = \int dg \, \exp\left(-\frac{1}{\hbar}S\right) , \tag{7.5}$$

$$S[g] = -\frac{1}{16\pi G}\int (R-2\Lambda)\sqrt{g}\, d^4x - \frac{1}{8\pi G}\int K\sqrt{k}\, d^3x , \tag{7.6}$$

where K is the trace of the second fundamental form of the boundary and k is the induced metric on the boundary. This term is necessary to cancel the second derivative parts of R and hence give a well posed boundary value problem. (Dirac 1958; Gibbons and Hawking 1977*a*). Green's functions are generated via the background field split $g_{\mu\nu} = g^c_{\mu\nu} + \sqrt{G}\, h_{\mu\nu}$ in which case Z becomes a generating functional $Z[g^c]$.

A number of reasons may be advanced for the use of Riemannian spaces:

1. The replacement of e^{iS} by e^{-S} should lead to a better defined functional integral, just as (7.3) is, via (7.4) more rigorous than (7.1).
2. Boundary value problems become elliptic rather than hyperbolic and are therefore more likely to be well posed.
3. In many situations it is both natural and desirable to compactify the Riemannian manifold. This generates interesting topologies and simplifies the mathematics needed to describe them.
4. The idea of microcausal quantum fields is problematical in quantum gravity where the light cone structure is itself dynamical. This is neatly side-stepped in a Euclidean formalism where the microcausal structure of the Lorentzian fields is related to the symmetry of the Euclidean Green's function. Thus in the path integral approach to quantum gravity it could be hoped that this difficult problem is resolved by the analytic continuation procedure by which the Riemannian manifold is related to a physical Lorentzian space-time (if it exists).

5. Similarly the existence of event horizons in a Lorentzian space-time is related to topological properties of a Riemannian continuation. In particular the Hawking radiation may be re-derived in this way (Gibbons and Hawking 1977).

6. By analogy with the Yang–Mills case, the use of a Riemannian formalism has lead to the study of 'gravitational instantons' defined as smooth complete solutions to the Riemannian Einstein equations (see below). Such instantons lead to non-perturbative effects which might provide insight into the renormalization problem.

7. The possibility arises of implementing the ideas of Wheeler (1963, 1968) and Hawking (1978) and performing sums over *different* manifolds. This could have profound effects on both the technical and conceptual structure of quantum gravity since space-time can no longer be regarded as a fixed background but is instead subject to quantum fluctuations.

These features all combine to make the Riemannian programme an attractive one. However a number of problems or questions also arise and should be properly considered. For example:

1. The action in (7.6) is unfortunately not positive definite and hence the functional integral is not as well defined as one might hope. The negative modes come from the conformal degrees of freedom and it has been suggested that they might be handled by analytic continuation method (Gibbons, Hawking, and Perry 1978). In effect this is based on the observation that

$$\int_{-\infty}^{\infty} \exp(-ax^2) = \sqrt{\left(\frac{\pi}{a}\right)} \quad a > 0 , \tag{7.7}$$

which is used to *define*

$$\int_{-\infty}^{\infty} \exp(bx^2) = \sqrt{\left(\frac{\pi}{-b}\right)} = \frac{1}{i}\sqrt{\left(\frac{\pi}{b}\right)}; \quad b > 0 \tag{7.8}$$

This is clearly a type of regularization scheme and its physical status in the quantum field theory is not absolutely clear (see Hawking 1979*a* for further comments).

2. In conventional quantum field theory the possibility of analytically continuing the Lorentzian Wightman functions to

the Euclidean Schwinger functions (and vice versa) is guaranteed by:

(a) The spectral condition which allows continuation into the forward tube.

(b) Lorentz covariance which enables the continuation to be taken into the extended forward tube.

(c) Locality of the quantum fields which allows continuation into the permuted extended forward tube in which the Euclidean points lie.

The first two conditions are not applicable in a curved space-time and in general it is not possible to embed a Lorentzian manifold in a complex manifold possessing a real Riemannian submanifold and vice versa. Thus the construction of the functional integral on a Riemannian manifold involves a conjectural leap and the implications for any given physical situation need to be carefully considered. This becomes especially difficult when a sum over manifolds is attempted. For example the n-point function (S_M is the action on the manifold M)

$$\tau(x_1 \ldots x_n) = \sum_M \int dg \, \exp(-S_M) \, g_{\mu_1\nu_1}(x_1) \, \ldots \, g_{\mu_n\nu_n}(\chi_n) \qquad (7.9)$$

cannot even be defined (let alone analytically continued) as there is no obvious way in which a common set of points $x_1 \ldots x_n$ can be identified on all of the manifolds M. It seems to be necessary to fix the asymptotic regions and only quantum fluctuate the manifolds in the 'middle' of the space-time. This leads rather naturally to an S-matrix picture and some specific illustrations of this may be found in Hawking *et al.* (1979, 1980); Hawking and Pope (1980).

3. In conventional quantum field theory the symmetry of the Euclidean Green's functions (and the associated interpretation of the Euclidean quantum fields as random variables) is related to local commutativity of the Lorentzian quantum fields. However, do we expect the fields in quantum gravity to be locally commutative or might there be Planck length corrections? In the latter case the applicability of the Riemannian functional integral would be doubtful.

4. The existence of non-perturbative aspects of instantons does not in itself relate directly to the problem of non-

renormalizability. Indeed these solutions are intended to be employed in a saddle point approximation to the functional ingegral via the background split $g_{\mu\nu} = g^{c}_{\mu\nu} + \sqrt{G}\, h_{\mu\nu}$ and the multiloop corrections will be as bad as usual. The resolution of renormalizability is more likely to come with some new approach to perturbation theory. For example Hawking (1979*a*) has suggested that tunnelling from one topology to another using Regge calculus to approximate the spaces may provide such a framework (Regge 1961; Wheeler 1963; Hartle and Sorkin 1979).

5. It is not clear what summing over manifolds means or indeed what manifolds should be included in such a sum. Stephen Hawking often advocates considering only manifolds M that are simply connected (i.e. $\pi_1(M) = 0$). Compact manifolds (admitting spinors) of this type may be classified up to homotopy by the Euler number χ and Hirzebruch signature τ. Both of these topological invariants may be expressed in terms of the curvature tensor:

$$\chi = \frac{1}{128\pi^2} \int R_{\mu\nu\alpha\beta} R_{\kappa\lambda\rho\sigma}\, \varepsilon^{\mu\nu\kappa\lambda} \varepsilon^{\alpha\beta\rho\sigma} \sqrt{g}\, d^4x$$

$$\tau = \frac{1}{96\pi^2} \int R_{\mu\nu\alpha\beta} R^{\mu\nu}{}_{\varkappa\delta}\, \varepsilon^{\alpha\beta\kappa\delta} \sqrt{g}\, d^4x$$

There is no obvious physical relation between manifolds that are homotopic but not diffeomorphic, so the implications of the classification being 'up to homotopy type' are not apparent.

Another approach is to build compact space M from the topological sums of copies of $S^2 \times S^2$, K_3 and $\mathbb{C}P_2$ all of which admit instanton solutions. Unfortunately it seems that only the topological sum of M and some finite set of $S^2 \times S^2$s can be so expressed, and therefore, it is unclear how useful this scheme will be.

In all this we have to assume $\pi_1(M) = 0$; but as remarked in Section 5 there are interesting phenomena associated with non-simply connected space-times and there is no real justification for making this restriction. Unfortunately if $\pi_1(M) \neq 0$ it is not possible even in principle to classify the space-times. A comprehensive recent account of the topological properties of four manifolds is Mandelbaum (1980).

Once one has admitted that 'something peculiar' happens to space-time topology at Planck distances the possibility arises that spaces should be considered that are not in any sense differentiable manifolds. For example Whitehead 1949 has shown that a simply connected four dimensional polyhedron is completely determined (and hence classified) by the structure of its various cohomology rings and maybe spaces of this type should also be included in a functional integral.

Many examples of gravitational instantons are now known and they may be partially classified by their boundary conditions in the following way.

1. Asymptotically Euclidean (AE). These would be the analogue of the Yang–Mills instanton vacuum states, but owing to the positive action theorem the only such solution is flat Euclidean space (Gibbons and Pope 1979).

2. Asymptotically locally Euclidean (ALE). These have a boundary S^3/Γ where Γ is some non-trivial subgroup of $SO4$ acting transitively on S^3. This identification of boundary points can lead to a violation of various discrete symmetries (e.g. TP, Perry 1979) in the S-matrix and the physical relevance of such solutions is still a matter for discussion.

3. Asymptotically flat (AF). This refers to metrics that are asymptotically flat in spatial directions and periodic in the time direction. Thus the boundary is $S^2 \times S^1$ and the periodic structure can be related to thermal radiation.

4. Asymptotically locally flat (ALF). This is a local version of 3. in which the boundary is some non-trivial S^1 bundle over S^2. As in the ALE case the physical significance of such solutions is not clear.

5. Compact without boundary. Examples are $\mathbb{C}P_2$, $S^2 \times S^2$ and K_3. Spaces of this type arise in Hawkings quantum foam considerations (Hawking 1978, 1979a) and as conformal compactifications of asymptotically Euclidean metrics.

A short but comprehensive review of this subject is Gibbons (1979) and a more leisurely discussion is contained in the article by C.N. Pope in this volume.

The Riemannian programme is an active one and has led to a substantial interaction between theoretical physicists and differential topologists. Its ultimate importance however will depend upon the invention of some viable non-coupling constant perturbation scheme.

8. SUPERGRAVITY

One of the more remarkable theoretical discoveries this decade is that bosons and fermions can be placed in the same multiplet of a 'supergroup' whose infinitesimal parameters contain anti-commuting elements (Gol'fand and Likhtmann 1971; Wess and Zumino 1974). Such a theory predicts a boson—fermion mass degeneracy that is not observed in nature and thus the super-symmetry must be broken. The Goldstone fermions associated with spontaneous breaking have the wrong properties to be neutrinos and hence the symmetry needs to be implemented as a local gauge invariance with the Higgs—Kibble mechanism in action.

This path leads naturally to supergravity as the gauge theory of the Poincaré group augmented by the anticommutation relations for spinorial charges:

$$\{Q^{\alpha},\bar{Q}^{\beta}\} = \gamma_{\mu}^{\alpha\beta}\, P^{\mu} \tag{8.1}$$

The local Lagrangian for such a theory was discovered in 1976 by Ferrara (1976*a*,*b*) and Deser and Zumino (1976) and since then considerable progress has been made. This subject is reviewed in the articles in this volume by S. Ferrera, P. van Nieuwen-hiuzen, and P. West so that I shall restrict myself to a few comments.

The supergravity Lagrangian is

$$\mathcal{L} = \det e\left\{\frac{1}{2G}\, R\Big(e,\omega(e,\psi)\Big) - \frac{i}{2}\,\psi_{\mu}\gamma_5\gamma_{\nu}\nabla_{\rho}\psi_{\kappa}\varepsilon^{\mu\nu\rho\kappa}\right\} \tag{8.2}$$

where $e_{a\mu}$ and ψ_{μ} are, respectively, the spin-two graviton (vier-bein) and the spin $3/_2$ Majorana spinor that together constitute the basic super multiplet. Torsional terms appear through the explicit form for $\omega_{\mu}(e,\psi)$ in terms of the usual spin connec-tion $\omega_{\mu ab}(e)$:

$$\omega_{\mu ab}(e,\psi) = \omega_{\mu ab}(e) + \frac{iG}{4}\left\{\bar{\psi}_{\mu}\gamma_{a}\psi_{b} + \bar{\psi}_{a}\gamma_{\mu}\psi_{b} - \bar{\psi}_{\mu}\gamma_{b}\psi_{\alpha}\right\} \tag{8.3}$$

The Lagrangian is invariant up to a total divergence under the transformations

$$\begin{aligned} \delta e_{a\mu} &= i\sqrt{G}\ \bar{\varepsilon}\gamma_a\ \psi_\mu \\ \delta\psi_\mu &= \frac{2}{\sqrt{G}}\ \nabla_\mu\Big(e,\omega(e,\psi)\Big)\varepsilon \end{aligned} \tag{8.4}$$

where ε is a Grassmann spinorial parameter.

The algebra generated by these transformations only closes if the field equations are used. This makes it impossible to construct invariants as simple functions of other invariants and also complicates the renormalization programme. This defect may be overcome by introducing auxiliary fields whose field equations are essentially trivial, but which enter non-trivially into the transformation laws (Breitenlohner 1977*a,b*). A minimal set containing two scalars and one vector was discovered by Ferrara and van Nieuwenhuizen (1978) and Stelle and West (1978*a*) and led quickly to a 'tensor calculus' for combining super multiplets (Ferrara and van Nieuwenhuizen 1978*c*; Stelle and West 1978*b,c*).

Supergravity theory has a better ultraviolet behaviour than conventional gravity plus matter and indeed the absence of superinvariant two-loop counter-terms means that the theory is two-loop finite (Grisaru 1977; Ferrara and van Nieuwenhuizen 1978). Unfortunately there exist possible three-loop counter-terms and there is no particular reason why their coefficients should vanish (Deser *et al.* 1977).

The desire for finiteness has led to the invention of extended supergravity theories which possess a larger symmetry group. The number of spinorial charges is increased and the anticommutation relations become

$$\{Q^{\alpha i},\bar{Q}^{\beta j}\} = \gamma_\mu^{\alpha\beta}P^\mu\delta^{ij} + \delta^{\alpha\beta}Z^{ij} + \gamma_5^{\alpha\beta}Z'^{ij} \qquad i,j = 1 \ldots N. \tag{8.6}$$

where Z^{ij} and Z'^{ij} are the central charges of Haag, Lopuszański and Sohnius (1975). The particle content of such a theory is rich and, if $2 \leqslant N \leqslant 8$, includes one spin 2 field and a number of spin 3/2 and lower spin particles (see the reviews in this volume for details). If $N > 8$ the top spin becomes at least 5/2 and many spin 2 particles appear. This is generally regarded as being physicably unacceptable and consequently attention has been concentrated mainly on theories with $N \leqslant 8$. Some of the important questions are:

1. Is supergravity renormalizable, or even finite, for some $N > 1$. A good omen is that the one loop β-function vanishes in the extended theory provided the $SO(N)$ group is gauged and $N \geqslant 5$ (Christensen, Duff, Gibbons, and Rocek 1980). There is some suggestion that this rather remarkable result is topological in origin and it may be no coincidence that $N \geqslant 5$ is precisely the condition for topological stability of $SO(N)$ bundles over a four dimensional manifold.

2. Is the particle spectrum physical? Unfortunately even for $N = 8$ it is not possible to fit all the known basic fields (quarks, leptons, W-mesons etc.) into the available multiplets. It may be necessary to fall back on the idea that the supergravity particles are the truly basic entities and physical particles arise as composite states.

3. How are the results to be interpreted? The relationship between quantum gravity and GUT is clear in this context but the techniques are less useful in connection with the other reasons for studying quantum gravity. For example if a 2 graviton-2 graviton S-matrix element is produced which is finite to all orders in perturbation theory, what use can be made of it? Essentially, one is faced here with the problem that plagues all conventional perturbative efforts to quantize gravity: even if the theory is finite it is still necessary to sum infinite sets of Feynman graphs before probing distances of the Planck length.

4. What is the resolution of the conceptual problems arising from the non-variance of the metric line element $ds^2 = g_{\mu\nu} dx^\mu dx^\nu$? These are related to the fundamental fact that fermion fields do not possess classical limits of the same types as bosons. Workers in supergravity usually ignore such conceptual difficulties, claiming that their interests are essentially pragmatic and their prime concern is to obtain a superinvariant Lagrangian. However it is possible that some truly deep connection between geometry and quantum physics is being overlooked in the process.

A number of techniques are being evolved to handle these sophisticated extended supergravity theories. One of the most useful seems to be dimensional reduction, in which a simple theory in a high dimension (11) reduces to *SO*8 theory in 4 dimensions: see for example, Cremmer and Scherk (1976); Cremmer, Julia, and Scherk (1978); Cremmer, Scherk, and Schwarz (1979); Cremmer and Julia (1979); Cremmer, Ferrara, Stelle, and West (1980); Sohnius, Stelle, and West (1980).

Considerable effort has also been directed towards superspace techniques in which space-time is augmented with anticommuting Grassmann parameters. Since the original ideas of Volkov and Akulov (1973) and Salam and Strathdee (1974) there have been significant developments from both the physical and the mathematical viewpoints. (Batchelor 1980; Berezin 1975; Kostant 1975; Rogers 1980). The idea of a real 'physical' superspace could provide the link between geometry and quantum theory referred to above.

Supergravity is one of the most rapidly evolving aspects of quantum gravity and the 'make or break' point must be fairly near. Extended supergravity theories are commendably rigid in their structure and basically will be either right or wrong: there is little room for manoeuvre.

If the *SO*8 theory does turn out to be finite, this will not be the end of the subject but the beginning; just as t'Hoofts's discovery of the renormalizability of Yang—Mills theory sparked off a new era in quantum field theory and particle physics, so a perturbatively finite theory of gravity would initiate a new and exciting chapter in the history of gravidynamics.

9. CONCLUSIONS

The construction of a successful quantum theory of gravity remains elusive after many generations of effort. (An early paper is Wereide (1923).) Two key questions that will be discussed in the next few years are:

1. Is extended supergravity finite and does it possess a bound state spectrum that fits in with the known elementary particles?
2. Can non-perturbative information be extracted from such

a finite or renormalizable theory to enable Planck distances to be probed?

If supergravity fails to work attention may fall again on one of the methods for handling non-renormalizable theories. Hopefully such a scheme will only render certain Lagrangians viable otherwise a Pandora's box of models will be opened up for both gravitational and non-gravitational forces. However, the prognosis in this direction is not good, partly because of technical difficulties and partly because of the deep-seated prejudice against non-renormalizable theories which will deter many potential workers.

If all conventional and semiconventional methods consistently fail then some difficult and deep questions are going to have to be asked. For example:

1. Does the problem hinge on the use of a metric perturbation $G_{\mu\nu} = \eta_{\mu\nu} + \sqrt{G}\, h_{\mu\nu}$? In other words, are Einstein's equations so intrinsically non-linear that *any* method based on a flat space expansion is bound to fail? This still leaves the door open for a successful resolution within the framework of conventional local quantum field theory provided that a suitable new perturbation scheme can be invented. This problem is similar in many respects to that of proving quark confinement in QCD and developments in that subject will be closely followed by those involved in quantum gravity.

2. Does the problem lie deeper in some profound breakdown of space-time continuum and quantum concepts at the Planck length? Some of the current topological work has this idea as its long term justification but probably does not go far enough in breaking up the space-time structure at these tiny distances. It is tempting for example to attempt to construct a quantum field theory on an arbitrary topological space that is *not* a smooth manifold. Unfortunately one loses one of the bedrocks of two centuries of successful physics — local differential field equations — and it is difficult to know with what it should be replaced.

Whatever the future may bring it is to be hoped that the recently forged links between quantum gravity, unified field theory and cosmology will not be lost. Many lessons have also been learnt by the classical relativists, especially in relation to perturbation theory, that are still largely unappreciated by the great majority of workers in quantum gravity. Hopefully the next ten years will see a collaboration between many branches of theoretical physics and mathematics in a combined attempt to solve what remains one of the outstanding physics problems of the century.

ACKNOWLEDGEMENTS

I am grateful to Mike Duff, Martin Sohnius, and Kelle Stelle for many discussions on quantum gravity.

REFERENCES

Adler, S.L. (1976). Linearized Hartree formulation of the Photon pairing problem. *Phys. Rev.* **D14**, 379.
——, Lieberman, J., Ng, Y.J., and Hung-Sheng Tsao (1976). Photon-pairing instabilities – a microscopic origin for gravitation. *Phys. Rev.* **D14**, 359.
Ashtekar, A. and Geroch, R. (1974). Quantum theory of gravitation. *Rep. Prog. Phys.* **37**, 1211.
—— and Magnon, A. (1974). Quantum fields in curved space-time. *Proc. Roy. Soc.* **A346**, 375.
—— and Sen, A. (1978). Quantum fields in curved space-times: selection and uniqueness of the vacuum state. Chicago preprint.
—— —— (1979). On the role of space-time topology in quantum phenomena: superselection of charge and emergence of non-trivial vacua. Chicago preprint.
Atiyah, M.F. and Ward, R. (1977). Instantons and algebraic geometry. *Comm. Math. Phys.* **55**, 111.
——, Hitchin, N.J., Drinfield, V.G., and Manin, Yu. I. (1978). Construction of instantons. *Phys. Lett.* **65A**, 185.
Atiyah, M.T., Patodi, W.K., and Singer, J.M. (1973). Spectral asymmetry and Riemannian geometry. *Bull. Lond. math. Soc.* **5**, 229.
—— —— —— (1975*a*). Spectral asymmetry and Riemannian geometry I. *Math. Proc. Camb. phil. Soc.* **77**, 43.
—— —— —— (1975*b*). Spectral asymmetry and Riemannian geometry II. *Math. Proc. Camb. phil. Soc.* **78**, 405.
—— —— —— (1976). Spectral asymmetry and Riemannian geometry III. *Math. Proc. Camb. phil. Soc.* **79**, 71.
Avis, S.J. and Isham, C.J. (1978). Vacuum solutions for a twisted scalar field. *Proc. Roy. Soc.* **A363**, 581.
—— —— (1979*a*). Quantum field theory and fibre bundles in a general space-time. In: *Recent developments in gravitation – Cargese 1978* (ed. S. Deser), Plenum Press, New York.
—— —— (1979*b*). Lorentz gauge invariant vacuum functionals for quantised spinor fields in non-simply connected space-times. *Nuc. Phys.* **B156**, 441.
—— —— (1980). Generalized spin structures on four dimensional space-times. *Comm. Math. Phys.* **72**, 103.
Back, A., Freund, P.G.O., and Forger, M. (1978). New gravitational instantons and universal spin structures. *Phys. Lett.* **77B**, 181.

Banach, R. (1980*a*). The quantum theory of free automorphic fields. *J. Phys.* **A13**, 2179.
—— (1980*b*). Topology and renormalizability. *J. Phys.* **A13**, L365.
—— and Dowker, J.S. (1978). Quantum field theory on Clifford–Klein space-times. The effective Lagrangian and vacuum stress-energy tensor. *J. Phys.* **A11**, 2255.
—— —— (1979*a*). Automorphic field theory – some mathematical issues. *J. Phys.* **A12**, 2527.
—— —— (1979*b*). The vacuum stress tensor for automorphic fields on some flat space-times. *J. Phys.* **A12**, 2545.
Batchelor, M. (1980). Two approaches to supermanifolds. *Trans. Am. math. Soc.* To appear.
Berezin, F.A. and Leĭtes (1975). Supermanifolds. *Soviet. Math.* **16**, 1218.
Bergmann, P.G. and Komar, A. (1962). *Recent developments in general relativity*. Pergamon, London.
Birrell, N.D. (1979*a*). Momentum space renormalization of $\lambda\phi^4$ in curved space-time. *J. Phys.* **A13**, 569.
—— (1979*b*). The application of quantum field theory to cosmology and astrophysics. Ph.D. thesis. London University.
—— and Ford, L.H. (1979*a*). Self interacting quantized fields and particle creation in Robertson Walker universes. *Ann. Phys.* (N.Y.). To appear.
—— —— (1980). Renormalization of self-interacting scalar field theories in a non-simply connected space-time. *Phys. Rev.* **D22**, 330.
—— and Taylor, J.G. (1978). Analysis of interacting quantum field theory in curved space-time. *J. math. Phys.* To appear.
Boulware, D.G. and Deser, S. (1975). Classical general relativity derived from quantum gravity. *Ann. Phys.* **89**, 193.
Boutaleb-Joutei, H. and Chakrabarti, A. (1979). Kerr–Schild geometry, spinors and instantons. *Phys. Rev.* **D19**, 457.
—— —— and Comteb, A. (1979*a*). Gauge field configurations in curved space-times I. *Phys. Rev.* **D20**, 1884.
—— —— —— (1979*b*). Gauge field configurations in curved space-times II. *Phys. Rev.* **D20**, 1898.
—— —— —— (1980). Gauge field configurations in curved space-times III. Self dual SU2 fields in Eguchi–Hanson space. *Phys. Rev.* **D21**, 979.
Breitenlohner, P. (1977*a*). A geometric interpretation of local supersymmetry. *Phys. Lett.* **B67**, 49.
—— (1977*b*). Some invariant Lagrangians for local supersymmetry. *Nucl. Phys.* **B124**, 500.
Brink, L., Schwarz, J. and Scherk, J. (1977). Supersymmetric Yang–Mills Theories. *Nucl. Phys.* **B121**, 77.
Bunch, T.S., Panangaden, P., and Parker, L. (1980). On renormalization of $\lambda\phi^4$ field theory in curved space-time I. *J. Phys.* **A13**, 901.
—— —— (1980). On renormalization of $\lambda\phi^4$ field theory in curved space-time II. *J. Phys.* **A13**, 919.
Callan, C.G., Dashen, R.F., and Gross, D.J. (1976). The structure of the gauge theory vacuum. *Phys. Lett.* **63B**, 334.
Capper, D.M. and Duff, M.J. (1974). The one-loop neutrino contribution to the graviton propagator. *Nucl. Phys.* **B62**, 147.
—— —— and Halpern, L. (1974). Photon corrections to the graviton propagator. *Phys. Rev.* **D10**, 461.
Castagnino, M. Second quantization of scalar fields in curved space-time. Undated preprint.
—— and Weder, R. (1980). The quantum equivalence principle and finite particle creation in expanding universes. Preprint.
—— Verbeure, A., and Weder, R.A. (1975). Catastrophies in the canonical quantization in an expanding universe. *Nuovo Cim.* **26B**, 396.

Castagnino, M., Foussats, A., Laura, R., and Zandron, O. (1980). The quantum equivalence principle and the particle model in curved space-time. Preprint.
Charap, J.M. and Duff, M.J. (1977*a*). Gravitational effects on Yang–Mills Topology. *Phys. Lett.* **69B**, 445.
—— —— (1977*b*). Space-time topology and a new class of Yang–Mills instanton. *Phys. Lett.* **71B**, 219.
Chockalingham, A. and Isham, C.J. (1980). Twisted supermultiplets. *J. Phys.* **A13**, 2723.
Choquet-Bruhat, Y. (1967). Hyperbolic partial differential equations on a manifold. In *Batelle Recontres* (ed. C. De Witt and J.A. Wheeler). W.A. Benjamin, New York.
Christensen, S.M. and Duff, M.J. (1978*a*). Axial and conformal anomalies for arbitrary spin in gravity and supergravity. *Phys. Lett.* **76B**, 571.
—— —— (1978*b*). Flat space as a gravitational instanton. *Nucl. Phys.* **B146**, 11.
—— —— (1978*c*). Quantum gravity in 2 + ε dimensions. *Phys. Lett.* **79B**, 213.
—— —— (1979). New gravitational index theorems and supertheorems. *Nucl. Phys.* **B154**, 301.
—— —— (1980). Quantizing gravity with a cosmological constant. *Nucl. Phys.* **B170**, 480.
—— ——, Gibbons, G., and Rocek, M. (1980). Vanishing one-loop β-function in n=4 supergravity. To appear, *Phys. Rev. Lett.*
Cremmer, E. and Julia, B. (1979). The *SO*8 supergravity. *Nucl. Phys.* **B159**, 141.
—— and Scherk, J. (1976*a*) Dual models in four dimensions with internal symmetries. *Nucl. Phys.* **B103**, 399.
—— —— (1976b). Spontaneous compactification of space in an Einstein–Yang–Mills–Higgs model. *Nucl. Phys.* **B108**, 409.
—— —— (1977). Spontaneous compactification of extra space dimensions. *Nucl. Phys.* **B118**, 61.
——, Julia, B., and Scherk, J. (1978). Supergravity theory in 11 dimensions. *Phys. Lett.* **76B**, 409.
——, Scherk, J., and Schwarz, J. (1979). Spontaneously broken N = 8 supergravity. *Phys. Lett.* **84B**, 83.
——, Ferrara, S., Stelle, K., and West, P. (1980). Off-shell N = 8 supergravity with central charges. LPTENS Preprint.
Davies, P.C.W. (1975). Scalar particle production in Schwarzschild and Rindler metrics. *J. Phys.* **A8**, 609.
—— (1977). Stress tensor calculation and conformal anomalies. In *Proc. Eighth Texas Symp. Relativistic Astrophys.* (ed. M. Papagiannis). New York Academy of Sciences.
—— (1979). Quantum fields in curved space. In *GRG Einstein centenary volume* (ed. A. Held).
Delbourgo, R. and Abdus Salam (1972). The gravitational corrections to PCAC. *Phys. Lett.* 40B, 381.
Denardo, G. and Spallucci, E. (1980). Dynamical mass generation in $S^1 \times R^3$. *Nucl. Phys.* **B169**, 514.
Deser, S. (1970). Talk given at the Austin Conference on Particle Physics. Unpublished.
—— and van Niewenhuizen, P. (1974*a*). Non-renormalizability of the quantized Einstein–Maxwell system. *Phys. Rev. Lett.* **32**, 245.
—— —— (1974*b*). One-loop divergences of quantized Einstein–Maxwell fields. *Phys. Rev.* **D10**, 401.
—— —— (1974*c*). Non-renormalizability of the quantized Dirac–Einstein system. *Phys. Rev.* **D10**, 411.
—— and Zumino, B. (1976). Consistent supergravity. *Phys. Lett.* **62B**, 335.

Deser, S., Duff, M.J., and Isham, C.J. (1980). CP violation in quantum gravity. *Phys. Lett.* **93B**, 419.
——, Kay, J., and Stelle, K. (1977). Renormalizability properties of supergravity. *Phys. Rev. Lett.* **38**, 527.
——, Hung-Sheng Tsao, and van Niewenhuizen, P. (1974). One-loop divergences of the Einstein–Yang–Mills system. *Phys. Rev.* **D10**, 3337.
De Witt, B.S. (1964*a*). Gravity – a universal regulator. *Phys. Rev. Lett.* **13**, 114.
—— (1964*b*). Dynamical theory of groups and fields. In *Relativity, Groups and Topology* (ed. C. De Witt and B.S. De Witt). Gordon and Breach, London.
—— (1967*a*). Quantum Theory of Gravitation I. The canonical theory. *Phys. Rev.* **160**, 1113.
—— (1967*b*). Quantum Theory of Gravitation II. The manifestly covariant theory. *Phys. Rev.* **162**, 1195.
—— (1967*c*). Quantum theory of gravitation III. Applications of the covariant theory. *Phys. Rev.* **162**, 1239.
—— (1970). Quantum theories of gravity. *Gen. Rel. Grav.* **I**, 181 (1970).
—— (1972). Covariant quantum geometrodynamics. In *Magic without magic* (ed. J. Klauder). W.H. Freeman, Reading, England.
—— (1975). Quantum field theory in a curved space-time. *Phys. Rev.* **19C**, 297.
—— (1979). Quantum gravity: the new synthesis. In *General relativity: an Einstein memorial volume* (ed. S.W. Hawking and W. Israel). Cambridge University Press.
——, Hart, C., and Isham, C.J. (1979). Topology and quantum field theory. In *Themes in contemporary physics* (ed. S. Deser). North-Holland, Amsterdam.
Dimock, J. (1979). Scalar quantum field in an external gravitational field. Princeton preprint.
—— (1980). Algebra of local observables on a manifold. Princeton preprint.
Dirac, P.A.M. (1958). The theory of gravitation in Hamiltonian form. *Proc. Roy. Soc.* **A246**, 333.
Dold, A. and Whitney, H. (1959). Classification of oriented sphere bundles over a four complex. *Ann. Maths.* **69**, 3.
Dowker, J.S. (1978). Another discussion of the axial vector anomaly and the index theorem. *J. Phys.* **A11**, 347.
Drummond, I.T. and Hathrell, S.J. (1980). Renormalization of the gravitational trace anomaly in QED. *Phys. Rev.* **D21**, 958.
Duff, M.J. (1973). A particle physicist's approach to the theory of gravitation. Trieste preprint IC/73/70.
—— (1975). Covariant quantization of gravity. In *Quantum gravity: an Oxford symposium* (ed. C.J. Isham, R. Penrose, and D. Sciama). Oxford University Press.
—— (1977). Observations on conformal anomalies. *Nuc. Phys.* **B125**, 334.
—— and Madore, J. (1979). Einstein–Yand–Mills pseudoparticles and electric charge quantization. *Phys. Rev.* **D18**, 2788.
—— and van Nieuwenhuizen, P. (1980). Quantum inequivalence of different field representations. Imperial College preprint.
Eguchi, T. and Freund, P. (1976). Quantum gravity and world topology. *Phys. Rev. Lett.* **37**, 1251.
Fadeev, L.D. (1969). The Feynman integral for singular Lagrangians. *Theor. math. Phys. SSSR.* **1**, 3.
—— and Popov, V.N. (1967). Feynman diagrams for the Yang–Mills field. *Phys. Lett.* **25B**, 29.
—— —— (1973). Covariant quantization of the gravitational field. *Sov. Phys. Usp.* **16**, 777.
Ferrara, S. and van Nieuwenhuizen, P. (1978*a*). The auxiliary fields of supergravity. *Phys. Lett.* **B74**, 333.
—— —— (1978*b*). Structure of supergravity. *Phys. Lett.* **78B**, 573.

Ferrara, S. and van Nieuwenhuizen, P. (1978*c*). Tensor calculus for supergravity. *Phys. Lett.* **76B**, 404.

Feynman, R.P. (1963). Lectures on gravitation. *Acta phys. pol.* **24**, 697.

Finkelstein, D. and Misner, C.W. (1959). Some new conservation laws. *Ann. Phys.* **6**, 230.

Fischetti, M.V., Hartle, J.B., and Hu, B.L. (1979). Quantum effects in the early universe I. Influence of trace anomalies on homogeneous, isotropic, classical geometries. *Phys. Rev.* D20, 1757.

Ford, L. (1979). Twisted spinor fields in flat space-times. In *Proc. Second Marcel Grossman Meet. Trieste* (ed. R. Ruffini).

—— (1980*a*). Vacuum polarization in a non-simply connected space-time. *Phys. Rev.* **D21**, 933.

—— (1980*b*). Twisted scalar and spinor strings in Minkowski space-time. *Phys. Rev.* **D21**, 949.

—— and Yoshimura, T. (1979). Mass generation by self interactions in non-Minkowskian space-times. *Phys. Lett.* **70A**, 89.

Forger, M. and Hess, H. (1979). Universal metaplectic structures and geometric quantization. *Comm. Math. Phys.* **64**, *269*.

Fradkin, E.S. (1973). Hamiltonian formalism in covariant gauge and the measure in quantum gravity. In *Proc. Xth Winter School Theor. Phys. Karpacz.*

—— and Tyutin, I.V. (1970). S-matrix for Yang–Mills and gravitational fields. *Phys. Rev.* **D2**, 2841.

—— and Vilkovisky, G. (1973). S-matrix for gravitational field. II Local measures, general relations; elements of renormalization theory. *Phys. Rev.* **D8**, 4241.

—— —— (1975*a*). Unitarity of S-matrix in gravidynamics and general covariance in quantum domain. *Lett. Nuovo cim.* **13**, 187.

—— —— (1975*b*). Quantization of relativistic systems with constraints. *Phys. Lett.* **55B**, 224.

—— —— (1977). Quantization of relativistic systems with constraints. Equivalence of canonical and covariant formalisms in quantum theory of gravitational field. CERN preprint.

—— —— (1978). Conformal invariance and asymptotic freedom in quantum gravity. *Phys. Lett.* **77B**, 262.

Freedman, D.Z. and Das, A. (1977). Gauge internal symmetry in extended supergravity. *Nucl. Phys.* **B120**, 221.

——, van Nieuwenhuizen, P., and Ferrara, S. (1976). Progress towards a theory of supergravity. *Phys. Rev.* **D13**, 3214.

—— —— (1976). Properties of supergravity theory. *Phys. Rev.* **D14**, 912.

Frolov, V.P. and Vilkovisky, G.A. (1979). Quantum gravity removes classical singularities and shortens the life of black holes. In *Proc. Second Marcel Grossman Meet. Trieste* (ed. R. Ruffine).

Fulling, S.A. (1973). Non-uniqueness of canonical field quantization in Riemannian space-time. *Phys. Rev.* **D7**, 2850.

—— (1977). Alternative vacuum states in static space-times with horizons. *J. Phys.* **A10**, 917.

Gibbons, G.W. (1977). Quantum processes near black holes. In *Proc. First Marcel Grossman Meet.* (ed. R. Ruffini). North Holland, Amsterdam.

—— (1979*a*). Quantum field theory in curved space-times. In *General relativity – an Einstein centenary survey* (ed. S.W. Hawking and W. Israel). Cambridge University Press.

—— (1979*b*). Gravitational instantons: a survey. Talk given at 1979 Lausanne Conference on Mathematical Physics.

—— and Hawking, S.W. (1977*a*). Action integrals and partition functions in quantum gravity. *Phys. Rev.* **D15**, 2752.

—— —— (1977*b*). Cosmological event horizons, thermodynamics and particle creation. *Phys. Rev.* **D15**, 2738.

Gibbons, G.W. and Perry, M.J. (1978). Black Holes and thermal Green's functions. *Proc. Roy. Soc.* **A358**, 467.
—— and Pope, C.N. (1979). The positive action principal and asymptotically Euclidean metrics in quantum gravity. *Comm. Math. Phys.* **66**, 267.
——, Hawking, S.W., and Perry, M.J. (1978). Path integral and the indefiniteness of the gravitational action. *Nucl. Phys.* **B138**, 141.
——, Pope, C.N., and Römer, H. (1979). Index theorem boundary terms for gravitational instantons. *Nucl. Phys.* **B157**, 377.
Gliozzi, F., Scherk, J., and Olive, D. (1977). Supersymmetry, supergravity theories and the dual spinor model. *Nucl. Phys.* **B122**, 253.
Golfand, Y.A. and Likhtman, E.P. (1971). Extension of the algebra of Poincaré group generators and violation of P invariance. *J.E.T.P. Lett.* **13**, 323.
Grisaru, M. (1977). Two-loop renormalizability of supergravity. *Phys. Lett.* **B66**, 75.
Haag, R., Lopuszański, J.T., and Sohnius, M. (1975). All possible generators of supersymmetries of the S-matrix. *Nucl. Phys.* **B88**, 275.
Hajicek, P. (1977). Observables for quantum fields on curved space-times. In *Proc. Bonn conf. diff. geom. methods math. phys. II* (ed. K. Bleuler, H.R. Petry, and A. Reetz). Springer-Verlag, Heidelberg, W. Germany.
Hanson, A. and Römer, H. (1978). Gravitational instanton contributions to Spin 3/2 axial anomaly. *Phys. Lett.* **80B**, 58.
Hartle, J.B. and Hawking, S.W. (1976). Path integral derivation of black hole radiance. *Phys. Rev.* **D13**, 2188.
—— and Hu, B.L. (1979). Quantum effects in the early universe II. Effective action for scalar fields in homogeneous cosmologies. *Phys. Rev.* **D20**, 1772.
—— —— (1980). *Phys. Rev.* D22. To appear.
—— and Sorkin, R. (1979). Boundary terms in the action for the Regge calculus. UCSB preprint.
Hawking, S.W. (1975*a*). Particle creation by black holes. *Comm. Math. Phys.* **43**, 199.
—— (1975*b*). Particle creation by black holes. In *Quantum gravity: an Oxford symposium* (ed. C.J. Isham, R. Penrose, and D.W. Sciama). Oxford University Press.
—— (1978). Space-time foam. *Nucl. Phys.* **B144**, 349.
—— (1979*a*). The path integral approach to quantum gravity. In *General relativity; an Einstein centenary survey* (ed. S.W. Hawking and W. Israel). Cambridge University Press.
—— (1979*b*). Euclidean quantum gravity. In *Recent developments in gravitation* (ed. M. Levy and S. Deser). Plenum, New York.
—— and Ellis, G. (1973). *The large-scale structure of space-time*. Cambridge University Press.
—— and Penrose, R. (1970). The singularities of gravitational collapse and cosmology. *Proc. Roy. Soc.* **A314**, 529.
—— and Pope, C.N. (1978*a*). Generalised spin structures in quantum gravity. *Phys. Lett.* **73B**, 42.
—— —— (1978*b*). Symmetry breaking by instantons. *Nucl. Phys.* **B146**, 381.
—— —— (1979). Yang–Mills instantons and the S-matrix. *Nucl. Phys.* **B161**, 93.
——, Page, D.N., and Pope, C.N. (1979). The propagation of particles in space-time foam. *Phys. Lett.* **86B**, 175.
—— —— —— (1980). Quantum gravitational bubbles. *Nuc. Phys.* **B170**, 283.
t'Hooft, G. (1973). An algorithm for the poles at dimension four in the dimensional regularization procedure. *Nucl. Phys.* **B62**, 444.
—— and Veltman, M. (1974). One-loop divergences in the theory of gravitation. *Ann. Inst. Henri Poincaré* **20**, 69.
Horowitz, G. and Wald, R. (1978). Dynamics of Einstein's equation modified by a higher order derivative term. *Phys. Rev.* **D17**, 414.

Hu, B.L. (1979*a*). Elementary particle physics and cosmology. In *Proc. first theor. particle phys. conf.* Academy of Science Press, Peking.

—— (1979*b*). Quantum field theories and relativistic cosmology. In *Proc. second Marcel Grossman conf. Trieste.*

Hughston, L.P. (1979). Twistors and Particles. *Lecture notes in physics* **97**. Springer-Verlag, New York.

Isham, C.J. (1977*a*). Quantum field theory in curved space-time — an overview. In *Proc. eighth Texas symp. relativistic astrophys.* (ed. M. Papagiannis). New York Academy of Sciences.

—— (1977*b*). Quantum field theory in a curved space-time — a general mathematical framework. In *Proc. Bonn Conf. diff. geom. methods math. phys.* (ed. K. Bleuler, H.R. Petry, and A. Reetz). Springer-Verlag, New York.

—— (1978*a*). Twisted quantum fields in a curved space-time. *Proc. Roy. Soc.* **A362**, 383.

—— (1978*b*). Spinor fields in four dimensional space-time. *Proc. Roy. Soc.* **A364**, 591.

—— (1981). Vacuum tunneling in static space-times. To be published in *Essays in Honour of Wolfgang Yourgrau* (ed. Alwyn van der Merwe). Plenum Press.

——, Abdus Salam, and Strathdee, J. (1971). Infinity suppression in gravity-modified quantum electrodynamics. *Phys. Rev.* **D3**, 1805.

—— —— —— (1972). Infinity suppression in gravity modified electrodynamics II. *Phys. Rev.* **D5**, 2548.

——, Penrose, R., and Sciama, D. (1975). *Quantum gravity: an Oxford symposium.* Oxford University Press.

Jackiw, R. and Rebbi, C. (1976). Vacuum periodicity in Yang–Mills quantum theory. *Phys. Rev. Lett.* **37**, 172.

Julve, J. and Tonin, M. (1978). Quantum gravity with higher derivative terms. *Nuovo Cim.* **46B**, 137.

Kallosh, R. (1974). Renormalization in non-abelian gauge theories. *Nucl. Phys.* **B78**, 293.

Kay, B.S. (1977). Quantum fields in time-dependent backgrounds and in curved space-times. Ph.D. thesis. University of London.

—— (1978). Linear spin-zero quantum fields in external gravitational fields I: a one particle structure for the stationary case. *Comm. Math. Phys.* **62**, 55.

——, Kibble, T.W.B., and Randjbar-Daemi, S. (1980). Renormalization of semi-classical theories. Imperial College preprint.

Khriplovich, I.B. (1966). Gravitation and finite renormalizations in quantum electrodynamics. *Sov. J. nucl. Phys.* **3**, 415.

Kibble, T.W.B. (1978). Relativistic models of non-linear quantum mechanics. *Comm. Math. Phys.* **64**, 73.

—— (1979). Geometrization of quantum mechanics. *Comm. Math. Phys.* **65**, 189.

—— and Randjbar-Daemi, S. (1980). Non-linear coupling of quantum theory and classical gravity. *J. Phys.* **A13**, 141.

Kiskis, J. (1978). Disconnected gauge groups and the global violations of charge conservation. *Phys. Rev.* **D17**, 3196.

Klauder, J. (1972). Soluble models of quantum gravitation. In *Magic without magic* (ed. J. Lauder). W.H. Freeman, Reading, England.

—— (1979). New measures for non-renormalizable quantum field theory. *Ann. Phys. N.Y.* **117**, 19.

Kobayashi, S. and Nomizu, K. (1969). *Foundations of differential geometry Vol. II.* Interscience, New York.

Kostant, B. (1975). Graded manifolds, graded Lie theory and prequantization. In *Differential Geometric Methods in Mathematical Physics Lecture Notes in Mathematics, Vol. 570.* Springer Verlag, New York.

Leutwyler, H. (1964). Gravitational field: equivalence of Feynman quantization and canonical quantization. *Phys. Rev.* **134**, 1155.
Leray, J. (1952). Hyperbolic partial differential equations. Mimeographed notes, Princeton.
MacCallum, M. (1975). Quantum cosmological models. In *Quantum gravity — an Oxford symposium* (ed. C.J. Isham, R. Penrose, and D.W. Sciama). Oxford University Press.
Mandelbaum, R. (1980). Four dimensional topology — an introduction. *Bull. Am. math. Soc.* **2**, 1.
Mandelstam, S. (1968*a*). Feynman rules for electromagnetic and Yang–Mills fields from the gauge-independent field theoretic formalism. *Phys. Rev.* **175**, 1580.
—— (1968*b*). Feynman rules for the gravitational field from the coordinate-independent field theoretic formalism. *Phys. Rev.* **175**, 1604.
Martellini, M., Sodano, P., and Vitiello, G. (1978). Vacuum structure for a field theory in curved spacetime. *Nuov. Cim.* **48A**, 341.
Mielnik, B. (1974). Generalised quantum mechanics. *Comm. Math. Phys.* **37**, 221.
Milnor, J. (1963). Spin structures on manifolds. *Enseign Math.* **9**, 198.
Misner, C.W. (1978). Harmonic maps as models for physical theories. *Phys. Rev.* **D18**, 4510.
Moreno, C. (1977). On quantization of free fields in stationary space-times. *Lett. math. Phys.* **1**, 407.
Neville, D.E. (1980). Gravity theories with propagating torsion. *Phys. Rev.* **D2**, 867.
Nielson, N.K. (1977). Axial anomaly and Atiyah–Singer theorem. *Nucl. Phys.* **B127**, 493.
——, Römer, H., and Schroer, B. (1978*a*). Anomalous currents in curved space. *Nucl. Phys.* **B136**, 475.
—— —— —— (1978*b*). Classical anomalies and local version of the Atiyah–Singer index theorem. *Phys. Lett.* **70B**, 445.
van Nieuwenhuizen, P. (1975). An introduction to covariant quantization of gravitation. In *Proc. first Marcel Grossman meet. Trieste* (ed. R. Ruffini).
Nouri-Moghadon, M. and Taylor, J.G. (1975). One-loop divergences for the Einstein charged meson system. *Proc. Roy. Soc.* **A344**, 87.
Omero, C. and Percacci, R. (1980). Generalized non-linear σ-models in curved space and spontaneous compactification. *Nucl. Phys.* **B165**, 351.
Page, D.N. (1980). Is black hole evaporation predictable? *Phys. Rev. Lett.* **44**, 301.
Parker, L. (1968). Particle creation in expanding universes. *Phys. Rev. Lett.* **21**, 562.
—— (1969). Quantized fields and particle creation in expanding universes I. *Phys. Rev.* **183**, 1057.
—— (1971). Quantized fields and particle creation in expanding universes II. *Phys. Rev.* **D3**, 346.
—— (1976). The production of elementary particles by strong gravitational fields. In *Proc. symp. asymptotic properties of space-time* (ed. F.P. Exposito and L. Witten). Plenum, New York.
—— (1979). Aspects of quantum field theory in curved space-time, effective action and energy momentum tensor. In *Recent developments in gravitation* (Cargese 1978) (ed. S. Deser and M. Levy). Plenum, New York.
—— and Fulling, S. (1973). Quantized matter fields and the avoidance of singularities in general relativity. *Phys. Rev.* **D7**, 2357.
Pauli, W. (1967). *Theory of relativity*. Pergamon, London.
Perry, M.J. (1979). TP inversion in quantum gravity. *Phys. Rev.* **D19**, 1720.
Penrose, R. (1973). Techniques of differential topology in relativity. SIAM Philadelphia.

Penrose, R. (1975). Twistor theory, its aims and achievements. In *Quantum gravity, an Oxford symposium* (ed. C.J. Isham, R. Penrose, and D. Sciama). Oxford University Press.

—— (1976). Non-linear gravitons and curved twistor theory. *Gen. Rel. Grav.* **7**, 31.

—— (1979). Singularities and time-asymmetry. In *General relativity — an Einstein centenary survey* (ed. S.W. Hawking and W. Israel). Cambridge University Press.

—— and MacCallum, M.A.H. (1973). Twistor theory: an approach to the quantization of fields and space-time. *Phys. Rep.* **6C**, 242.

Percacci, R. (1980*a*). Geometrical aspects of non-linear sigma models. Preprint from Institute Nazionale de Fisica Nucleare, Sezione di Trieste.

—— (1980*b*). Global definition of non-linear sigma model and some consequences. Preprint from Institute Nazionale di Fisica Nucleaire, Sezione di Trieste.

Pope, C.N. and Yuille, A.L. (1978). A Yang–Mills instanton in Taub NUT space. *Phys. Lett.* **B78**, 424.

Popov, V.N. and Fadeev, L.D. (1972). Perturbation theory for gauge-invariant fields. NAL-THY-57 Preprint. (Translation of Kiev report No: ITP67-36.)

van Proeyen, A. (1977). Quantum gravity corrections on the anomalous magnetic and quadrupole moments of a spin-1 particle. *Phys. Rev.* **D15**, 2144.

—— (1979). Gravitational divergences of the electromagnetic interactions of massive vector particles. *Nucl. Phys.* **B174**, 189.

Regge, T. (1961). General relativity without coordinates. *Nuov. Cim.* **19**, 558.

Rogers, A. (1980). A global theory of supermanifolds. *J. math. Phys.* To appear.

Rumpf, H. (1976*a*). Covariant description of particle creation in curved space-times. *Nuovo Cim.* **35B**, 321.

—— (1976*b*). Covariant treatment of particle creation in curved space-time. *Phys. Lett.* **61B**, 272.

—— and Urbantke, H.K. (1977). Covariant 'in–out' formalism for creation by external fields. Wien preprint.

Abdus Salam and Strathdee, J. (1974). Supersymmetry and non-Abelian gauges. *Phys. Lett.* **51B**, 353.

—— —— (1978). Remarks on high energy stability and renormalizability of gravity theory. *Phys. Rev.* **D18**, 4480.

Sanchez, N. (1979*a*). Analytic mappings: a new approach in particle production by accelerated observers. In *Proc. second Marcel Grossmann meet. Trieste* (ed. R. Ruffini).

—— (1979*b*). Topological invariants and thermal properties of analytic mappings. Meudon preprint.

Scherk, J. and Schwarz, J. (1979). Spontaneous breaking of supersymmetry through dimensional reduction. *Phys. Lett.* **82B**, 60.

Sexl, R.U. and Urbantke, H.K. (1967). Cosmic particle creation processes. *Acta phys. austriaca* **26**, 339.

Sezgin, E. and van Niewenhuizen, P. (1979). New ghost-free gravity Lagrangians with propagating torsion. *Phys. Rev.* **D21**, 3269.

Shastri, A.R., Williams, J.G., and Zvengrowski, P. (1979). Kinks in general relativity. Preprint, University of Alberta.

Simon, B. (1974). *The* $P(\phi)_2$ *Euclidean (quantum) field theory*. Princeton University Press.

Smolin, L. (1978). Quantum gravity on a lattice. Preprint.

—— (1979). What is the problem of quantum gravity? Santa Barbara preprint.

Sohnius, M.F., Stelle, K.S., and West, P.C. (1980). Dimensional reduction by Legendre transformations generates off-shell supersymmetric Yang–Mills theories. *Nucl. Phys.* **B173**, 127.

Sorkin, R. (1978). The quantum electromagnetic field in multiple connected space. Cardiff preprint.

Steenrod, N. (1951). *The topology of fibre bundles*. Princeton University Press.
Stelle, K.S. (1977). Renormalization of higher derivative quantum gravity. *Phys. Rev.* **D16**.
—— (1978). Classical gravity with higher derivatives. *Gen. rel. Grav.* **9**, 353.
—— and West, P.C. (1978*a*). Minimal auxiliary fields for supergravity. *Phys. Lett.* **B74**, 330.
—— —— (1978*b*). Tensor calculus for the vector multiplet coupled to supergravity. *Phys. Lett.* **77B**, 376.
—— —— (1978*c*). Relation between vector and scalar supermultiplets and gauge invariance in supergravity. *Nucl. Phys.* **B145**, 175.
Taylor, J.G. (1979). Quantizing space-time. *Phys. Rev.* **D19**, 2336.
Tomboulis, E. (1977). 1/*N* expansion and renormalization in quantum gravity. *Phys. Lett.* **70B**, 361.
—— (1980). Renormalizability and asymptotic freedom in quantum gravity.

Toms, D. (1980*a*). The Casimir effect and topological mass. *Phys. Rev.* **D21**, 928.
—— (1980*b*). Symmetry breaking and mass generation by space-time topology. *Phys. Rev.* **D21**, 2805.
—— (1980*c*). Interacting twisted and untwisted scalar fields in a non-simply connected space-time. *Annals of Physics*. To appear.
—— (1980*d*). Scalar electrodynamics in a non-simply connected space-time. *Phys. Lett.* **77a**, 303.
Unruh, W.G. (1976). Notes on black hole evaporation. *Phys. Rev.* **D14**, 870.
Unwin, S.D. (1979). Thermodynamics in multiply connected space-times. *J. Phys.* **A12**, L309.
—— (1980). Quantized spin-1 field in flat Clifford–Klein space-times. *J. Phys.* **A13**, 313.
—— and Critchley, R. (1979*b*). Atomic Ground state energy in multiply connected universes. Manchester University preprint.
Veltman, M. (1975). Quantum theory of gravitation. In *Methods in field theory - Les Houches 1975* (ed. R. Balian and J. Zinn-Justin). North-Holland, Amsterdam.
Volkov, D.V. and Akulov, V.P. (1973). Is the neutrino a Goldstone neutrino? *Phys. Lett.* **46B**, 109.
Wald, R. (1975). On particle creation by black holes. *Comm. math. Phys.* **45**, 9.
—— (1980*a*). On the Euclidean approach to quantum gravity. *Comm. math. Phys.* **70**, 221.
—— (1980*b*). Quantum gravity and time reversibility. *Phys. Rev.* **D21**, 2742.
Ward, R.S. (1977). On self-dual gauge fields. *Phys. Lett.* **61A**, 81.
Weinberg, S. (1979). Ultraviolet divergences in the quantum theories of gravitation. In *General relativity – an Einstein centenary survey* (ed. S.W. Hawking and W. Israel). Cambridge University Press.
Wereide, T. (1923). The general principle of relativity applied to the Rutherford–Bohr atom-model. *Phys. Rev.* **21**, 391.
Wess, J. and Zumino, B. (1974). Supergauge transformations in four dimensions. *Nucl. Phys.* **B70**, 39.
Wheeler, J.A. (1963). Geometrodynamics and the issue of the final state. In *Relativity groups and topology* (ed. C. DeWitt and B. DeWitt). Blackie, London.
—— (1968). Superspace and the nature of quantum geometrodynamics. In *Battelle rencontres 1967* (ed. C. DeWitt and J.A. Wheeler). W.A. Benjamin, New York.
Whitehead, J.H.C. (1949). On simply connected 4-dimensional polyhedra. *Comm. math. Helv.* **22**, 48.

Zeldovich, Ya. B. and Starobinsky, A.A. (1972). Particle production and vacuum polarization in an anisotropic gravitational field. *Sov. Phys. JEPT* **34**, 1159.
Zumino, B. (1975). Supersymmetry and the vacuum. *Nucl. Phys.* **B89**, 535.

2

IS A SEMI-CLASSICAL THEORY OF GRAVITY VIABLE?

T.W.B. Kibble

Blackett Laboratory, Imperial College, London SW7

1. INTRODUCTION

For the need to quantize the electromagnetic field there is good experimental evidence. In particular vacuum polarization is essential to obtain the excellent agreement between predicted and observed values of the electron magnetic moment and Lamb shift. However, there is no such direct evidence demanding the quantization of the gravitational field, and it is therefore pertinent to ask whether a semi-classical theory of gravity could be viable and, if so, whether its predictions would differ in any testable way from those of quantum gravity. Even the negative answer that most of this audience certainly expects would be interesting, as lending weight to the conventional view. On the other hand, no final answer can be given until at least one of the contending theories is shown to be consistent, which is certainly not yet the case. Until a quantum theory of gravity actually exists, its rival cannot be regarded as dead.

This debate has a long history (Møller 1962). Among the arguments that have often been advanced for the need to quantize the gravitational field are those originating with Bohr and Rosenfeld (1933) who showed that in a theory in which some fields are quantized and others not, some of the very basic principles of quantum mechanics — for example the uncertainty principle — could be violated. In a semi-classical theory of gravity (or anything else) the quantum-mechanical evolution is non-linear and does not respect the superposition principle. We really have no direct evidence for the validity of such principles in situations where the gravitational field is of importance, so this argument need not be regarded as conclusive (Rosenfeld 1963). There is, however, a genuine problem associated with the 'reduction of the wave packet' that occurs, according to the standard interpretation of quantum mechanics, in a measurement process. Even for ordinary quantum mechanics, as John Bell showed in a previous talk (see article in this volume), this phenomenon is problematic. It is considerably more so for a semi-classical, and therefore non-linear, theory. I shall return to this question in the concluding section of my talk.

Another objection to a semi-classical theory of gravity has been raised at this meeting by Mike Duff (see article in this volume), who showed that if only some fields are quantized then classically equivalent theories, related by transformation of variables, lead to inequivalent semi-classical theories. This is certainly correct, but is merely a reflection of the very special role played by the gravitational field in a semi-classical theory. If the metric is treated like any other field, it must of course be quantized. A semi-classical theory is necessarily based on the concept that the gravitational field has a distinguished place. In that context, it is not unreasonable that we must forbid transformations of the metric that are dependent on other fields.

In what follows I shall discuss from various aspects the viability of a semi-classical theory of gravity. The views expressed are my own, but the work on which they are based was done jointly with Seifallah Randjbar-Daemi and Bernard Kay (Randjbar-Daemi, Kay, and Kibble, 1980).

2. NON-LINEAR GENERALIZATIONS OF QUANTUM THEORY

Because a semi-classical theory necessarily introduces non-linearity into the quantum evolution, it will be useful to begin by discussing non-linear generalizations of quantum mechanics. The ordinary linear Schrödinger equation can of course be derived from an action principle, starting with the action integral

$$S[\psi] = \int dt\{\mathrm{Im}\,\langle\dot{\psi}(t)|\psi(t)\rangle - \langle\psi(t)|H|\psi(t)\rangle\}$$

in which the independent dynamical variables are the Schrödinger picture state $|\psi(t)\rangle$ and its conjugate. However this form is not very convenient for our purposes. This is because there is nothing to fix the scale of $|\psi(t)\rangle$, so that the corresponding 'Hamiltonian function', $\langle\psi(t)|H|\psi(t)\rangle$, is not the energy expectation value since it lacks the appropriate denominator. It is better to impose the normalization of $|\psi(t)\rangle$ as a constraint, using a Lagrange multiplier. Thus we take

$$\begin{aligned} S[\psi,\alpha] = \int dt\{\mathrm{Im}\langle\dot{\psi}(t)|\psi(t)\rangle - E(\psi(t)) \\ + \alpha(t)[\langle\psi(t)|\psi(t)\rangle - 1]\}, \end{aligned} \tag{2.1}$$

where E denotes the energy expectation value

$$E(\psi(t)) = \langle H\rangle_{\psi(t)} = \langle\psi(t)|H|\psi(t)\rangle . \tag{2.2}$$

Variation of α yields the constraint

$$\langle\psi(t)|\psi(t)\rangle = 1 , \tag{2.3}$$

while variation of $|\psi\rangle$ gives the Schrödinger equation in the form

$$\mathrm{i}|\dot{\psi}(t)\rangle = H|\psi(t)\rangle - \alpha(t)|\psi(t)\rangle . \tag{2.4}$$

The phase invariance of the action is reflected in the indeterminancy of α. It is possible to vary the phase of the state-vector $|\psi\rangle$ arbitrarily from each instant to the next. In practice it is of course convenient to remove this arbitrariness by some suitable convention.

It is quite easy to generalize the form (2.1) to accommodate non-linear models of various kinds, most simply by modifying the function E. Consider for example a simple relativistic quantum field theory, such as a self-interacting scalar field theory described by the Hamiltonian function

$$H = \int d^3x\left\{\frac{1}{2}\pi^2 + \frac{1}{2}(\nabla\phi)^2 + \frac{1}{2}m^2\phi^2 + \frac{1}{24}f\phi^4\right\}$$

where $\pi(x)$ is the canonical conjugate of the field $\phi(x)$, satisfying the canonical commutation relation

$$[\phi(x), \pi(y)] = i\delta_3(x-y).$$

For this theory $E = \langle H \rangle$ is the spatial integral of the expectation value of a local operator. The simplest type of generalization that will preserve the locality of the theory is to add terms which are spatial integrals of *products* of expectation values of local operators. For example, we might take (Kibble 1978)

$$E(\psi) = \langle H \rangle_\psi + \int d^3x \left[\frac{1}{8}\lambda \langle \phi^2 \rangle_\psi^{\;2} + \frac{1}{48}\mu \langle \phi^2 \rangle_\psi \langle \phi^4 \rangle_\psi \right].$$

By varying $|\psi\rangle$ we now obtain a non-linear generalization of the Schrödinger equation, of the form

$$i|\dot{\psi}(t)\rangle = H_{\psi(t)}|\psi(t)\rangle - \alpha(t)|\psi(t)\rangle , \tag{2.5}$$

where

$$H_\psi = H + \int d^3x \left\{ \frac{1}{4}\lambda \langle \phi^2 \rangle_\psi \phi^2 + \frac{1}{48}\mu \left[\langle \phi^4 \rangle_\psi \phi^2 + \langle \phi^2 \rangle_\psi \phi^4 \right] \right\}.$$

In other words, H_ψ is obtained from H by the replacements

$$m^2 \to m^2 + \frac{1}{2}\lambda \langle \phi^2 \rangle_\psi + \frac{1}{24}\mu \langle \phi^4 \rangle_\psi ,$$

$$f \to f + \frac{1}{2}\mu \langle \phi^2 \rangle_\psi .$$

The effective mass and coupling constant become functions of position and time, and depend on the quantum state $|\psi\rangle$ via expectation values of local operators. This is very similar to a Hartree–Fock model, though it is not intended to provide an approximate treatment of some many-particle interaction.

This model and other similar ones share many of the features of familiar relativistic quantum field theories, such as locality and Lorentz invariance. The proof of Lorentz invariance in the context of a Hamiltonian formalism is of course somewhat problematic, but the addition of non-linear interaction terms makes no essential difference.

Where they do differ essentially is in the non-linear character of the evolution equation (2.5). This means a violation of the superposition principle. Suppose that we start at time t_0 with a state expressed as a linear superposition of two others:

$$|\psi(t_0)\rangle = a_1|\psi_1(t_0)\rangle + a_2|\psi_2(t_0)\rangle \tag{2.6}$$

and follow the evolution to a later time t. Then in general,

$$|\psi(t)\rangle \neq a_1|\psi_1(t)\rangle + a_2|\psi_2(t)\rangle . \tag{2.7}$$

From the evolution of $|\psi_1\rangle$ and $|\psi_2\rangle$ separately we cannot deduce the evolution of $|\psi\rangle$.

Another consequence is the absence of any conserved measure of distance between states. Although equation (2.5) implies that the norm $\langle\psi|\psi\rangle$ is preserved in time, the same is not true of a scalar product $\langle\psi_1|\psi_2\rangle$. States which are initially close together can in principle move arbitrarily far apart (subject of course to remaining normalized). Thus it is possible at least in principle to violate the uncertainty principle.

These non-linear models are not intended to be realistic. Their role is merely illustrative of one direction in which it is possible to generalize quantum mechanics.

3. THE SEMI-CLASSICAL THEORY OF GRAVITY

Let me now turn to the incorporation of gravity. The classical Einstein equations are of course derivable from the action integral

$$S[g] = \frac{1}{16\pi G_N}\int d^4x(-g)^{\frac{1}{2}}R \tag{3.1}$$

where R is the curvature scalar and G_N is Newton's constant.

It is a remarkable fact that the equations of semi-classical gravitation theory may be derived from an action principle merely by combining (2.1) and (3.1). We take (Kibble and Randjbar-Daemi 1980)

$$S[g,\psi,\alpha] = S[g] + S[\psi,\alpha] \tag{3.2}$$

where the independent dynamical variables are the metric $g_{\mu\nu}(x)$, the quantum state-vector $|\psi(t)\rangle$ (with its conjugate) and the Lagrange multiplier $\alpha(t)$. Of course in $S[\psi,\alpha]$ the Hamiltonian operator H must be that appropriate to a quantum field in the curved background described by $g_{\mu\nu}$.

Variation of α and $|\psi\rangle$ yield as before the equations (2.3) and (2.4), with this metric-dependent Hamiltonian, while variation of the metric yields Einstein's equations in the form

$$G_{\mu\nu}(x) = -8\pi G_{\mathrm{N}} \langle T_{\mu\nu}(x)\rangle_\psi . \tag{3.3}$$

This may not be immediately obvious. To prove it, consider the action integral for the quantum field ϕ namely

$$W_\phi = \int \mathrm{d}^4x\, \mathscr{L}(\phi)$$

where $\mathscr{L}$ is the Langrangian density, for example

$$\mathscr{L} = (-g)^{\frac{1}{2}}\left[\frac{1}{2}g^{\mu\nu}\partial_\mu\phi\partial_\nu\phi - \frac{1}{2}m^2\phi^2 - \frac{1}{24}f\phi^4\right] .$$

The canonically conjugate fields π are defined as usual by

$$\pi = \frac{\partial\mathscr{L}}{\partial\dot{\phi}} . \tag{3.4}$$

The field equations in Hamiltonian form may be obtained from the action

$$W_{\phi\pi} = \int \mathrm{d}t\left\{\int \mathrm{d}^3x\,\pi\dot{\phi} - H(t)\right\} .$$

The Euler–Lagrange equations $\delta W_{\phi\pi}/\delta\pi = 0$ are equivalent to (3.4). Thus when these equations are satisfied,

$$-\frac{1}{2}(-g)^{\frac{1}{2}}T^{\mu\nu} = \frac{\delta W_{\phi}}{\delta g_{\mu\nu}} = \frac{\delta W_{\phi\pi}}{\delta g_{\mu\nu}} = -\frac{\delta}{\delta g_{\mu\nu}}\int \mathrm{d}t\, H(t) \ ,$$

from which (3.3) follows at once.

It is important to realize that although the Schrödinger equation (2.4) is still formally linear, the quantum evolution is non-linear. This is because the metric $g_{\mu\nu}$ depends on the quantum state $|\psi\rangle$ via the Einstein equation (3.3). Thus the superposition principle is again violated. Just as in the case of an explicitly non-linear theory, (2.6) still leads in general to (2.7).

In addition to this implicit non-linearity it would also be possible – perhaps even useful – to add extra explicit non-linearities similar to those discussed in the preceding section, but involving direct interaction between the non-linearity and the gravitational field. For example, we could add a term such as

$$\int \mathrm{d}^4x\,(-g)^{\frac{1}{2}}RF(\langle\phi^2\rangle_\psi) \tag{3.5}$$

where F is some polynomial. This yields an explicitly non-linear Schrödinger equation of the form (2.5) with

$$H_\psi = H + H_\psi^{n\ell}$$

where the non-linear part of the Hamiltonian is

$$H_\psi^{n\ell} = -\int \mathrm{d}^3x\,(-g)^{\frac{1}{2}}RF'(\langle\phi^2\rangle_\psi)\,\phi^2 \ . \tag{3.6}$$

The Einstein equation then takes the form

$$G_{\mu\nu} = -\frac{8\pi G_N}{1 + 16\pi G_N F}\left[\langle T_{\mu\nu}\rangle - 2F_{;\mu\nu} + 2g_{\mu\nu}F_{;\lambda}{}^{\lambda}\right]$$

which has some resemblance to the equations of the Brans–Dicke theory. However I shall not pursue the implications of this idea here.

Up to now we have used the Schrödinger picture, but the equations derived therein have at best a formal significance. To obtain equations that can be given a more exact meaning, it is better to transform (formally) to the Heisenberg picture. In the present context this means removing at least the linear part of the time-dependence from the states. Making the usual formal unitary transformation using $H(t) - \alpha(t)$ as Hamiltonian we arrive at a formalism in which the quantized field ϕ obeys the covariant field equation

$$(-g)^{-\frac{1}{2}}\partial_{\mu}\left[(-g)^{\frac{1}{2}}g^{\mu\nu}\partial_{\nu}\phi\right] + m^2\phi + \frac{1}{6}f\phi^3 = 0 . \tag{3.7}$$

If extra non-linear terms such as (3.5) are present in the action, the time dependence they generate cannot be transferred to the operators. In that event the states retain some time-dependence, described by the equation

$$\mathrm{i}|\dot{\psi}(t)\rangle = H_{\psi}^{n\ell}(t)|\psi(t)\rangle , \tag{3.8}$$

where $H^{n\ell}$ is given by (3.6), but in which the operators now obey (3.7). Unlike the Schrödinger picture, this formalism exhibits the covariance of the theory in a rather direct way. Indeed, one can write (3.8) in a form very similar to the Tomonaga version of the interaction picture, namely

$$\mathrm{i}\frac{\delta}{\delta\sigma(x)}|\psi(\sigma)\rangle = -(-g)^{\frac{1}{2}}RF'(\langle\phi^2\rangle_{\psi})\,\phi^2|\psi(\sigma)\rangle$$

where σ denotes a space-like surface.

For simplicity, however, I shall assume that no explicitly non-linear terms such as (3.5) are present. In this case we have a true Heisenberg picture, in which the quantum state is time-independent. Moreover I shall set the scalar-field self-coupling constant f to zero. Then ϕ obeys the covariant Klein–Gordon equation, and the metric is determined in terms of the state by Einstein's equation (3.3).

The major problem that has to be faced is of course how to handle the infinities that arise in the theory, particularly in evaluating the expectation value of the energy–momentum tensor. To examine this problem in a relatively simpler context, I shall discuss a model that shares some of the features of the semi-classical theory of gravity.

4. RENORMALIZATION OF A SIMPLE SEMI-CLASSICAL MODEL

Let us consider a classical scalar field $v(x)$ (which plays a role similar to that of the metric) interacting with a quantized scalar field $\phi(x)$ (Randjbar-Daemi, Kay, and Kibble 1980). We take the action integral to be

$$S[v,\phi,\alpha] = \int d^4x \frac{1}{2}[\partial_\mu v(x)\partial^\mu v(x) - \mu^2 v^2(x))] + \int dt\left\{\text{Im}\langle\dot{\psi}|\psi\rangle - E(\psi) + \alpha[\langle\psi|\psi\rangle - 1]\right\} \tag{4.1}$$

where

$$E(\psi) = \langle\psi|H_v|\psi\rangle ,$$

with

$$H_v = H + \frac{1}{2}\lambda \int d^3x \; v(x)\phi^2(x) . \tag{4.2}$$

As before H is the Hamiltonian for a free Klein–Gordon field ϕ of mass m.

Variation of the action integral yields the Schrödinger equation

$$i|\dot{\psi}\rangle = H_v|\psi\rangle - \alpha|\psi\rangle ,$$

together with the constraint

$$\langle\psi|\psi\rangle = 1 ,$$

and the field equation for v,

$$(\partial^2 + \mu^2)v = -\frac{1}{2}\lambda\langle\psi|\phi^2|\psi\rangle . \tag{4.3}$$

If we transform to the Heisenberg picture, the v-field equation (4.3) is formally unaltered, while for ϕ we obtain

$$(\partial^2 + m^2)\phi = -\lambda v\phi \ . \tag{4.4}$$

It is easy to develop a perturbation theory by expanding both ϕ and v in powers of λ. In zeroth order, they satisfy the free Klein–Gordon equations

$$(\partial^2 + m^2)\phi^{(0)} = 0, \qquad (\partial^2 + \mu^2)v^{(0)} = 0,$$

while $|\psi\rangle$ is taken to be some chosen state in the $\phi^{(0)}$ Fock space. Then (4.3) and (4.4) can be written as integral equations,

$$v(x) = v^{(0)}(x) + \tfrac{1}{2}\lambda \int d^4y\ \Delta_r(x-y,\mu)\,\langle\psi|\phi^2(y)|\psi\rangle \ , \tag{4.5}$$

$$\phi(x) = \phi^{(0)}(x) + \lambda \int d^4y\ \Delta_r(x-y,m)v(y)\phi(y) \ . \tag{4.6}$$

We could equally well have written Δ_F in place of Δ_r, incorporating different boundary conditions.

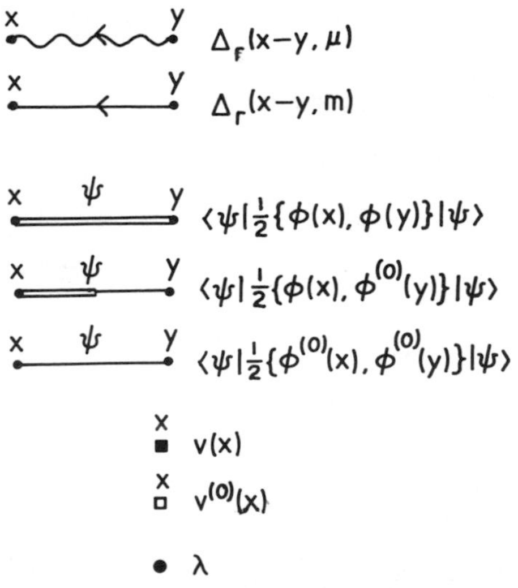

FIG. 1.

FIG. 2.

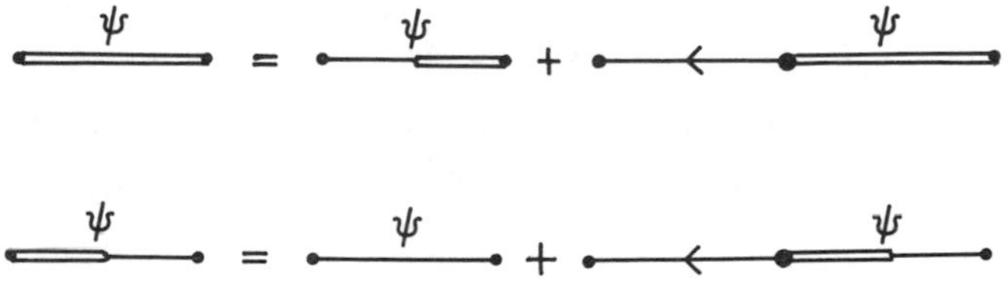

FIG. 3.

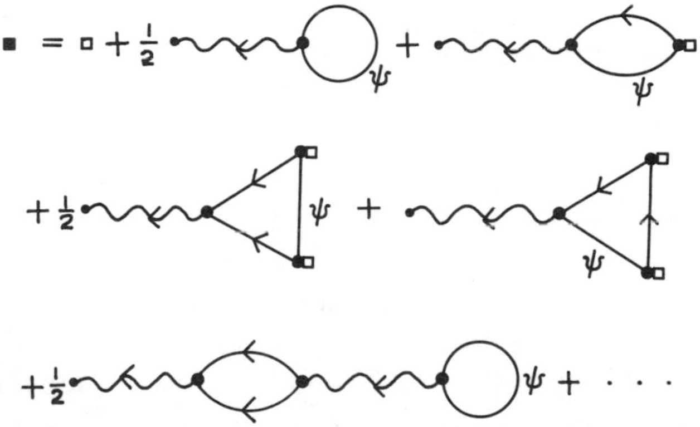

FIG. 4.

It is convenient to introduce the diagrammatic notations shown in Fig. 1. Note than when $|\psi\rangle$ is the vacuum state $|\Omega\rangle$ the single-line function shown is simply $\frac{1}{2}\Delta_1(x-y,m)$. In this notation the equation (4.5) for v may be written in the form shown in Fig. 2, while (4.6) yields for the two-point functions the pair of equations of Fig. 3. Straightforward iteration yields for example for v the series illustrated in Fig. 4.

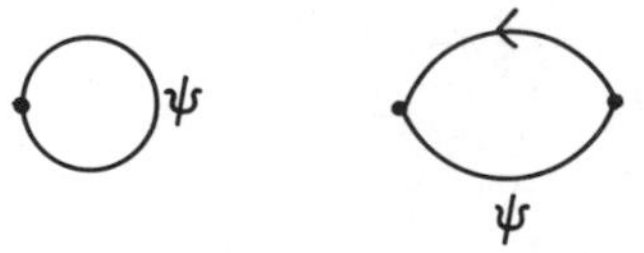

FIG. 5.

The only diagrams in this series that represent divergent contributions are the one- and two-point bubbles of Fig. 5. These can be cancelled by adding two appropriate counter terms to the action namely

$$\Delta S[v] = -\int d^4x\{\delta\Phi v(x) + \tfrac{1}{2}\delta\mu^2 v^2(x)\} \tag{4.7}$$

where $\delta\Phi$ and $\delta\mu^2$ are divergent constants. These constants must be fixed in the usual way by imposing suitable renormalization conditions.

Firstly, we require that if quantum state is the vacuum, $|\psi\rangle = |\Omega\rangle$, and if the zeroth order v field vanishes, $v^{(0)} = 0$, then so does the full v field. This fixes $\delta\Phi$ to cancel precisely the first diagram of Fig. 5 for the case $|\psi\rangle = |\Omega\rangle$.

Second is what might be thought of as a mass renormalization for v. We again take $|\psi\rangle = |\Omega\rangle$ but suppose that $v^{(0)}$ is different from zero. Then we require that in the Fourier transform $\tilde{v}(p)$ of $v(x)$ the only singularity of mass-shell delta-function type is at $p^2 = \mu^2$. This condition has the effect of making $\delta\mu^2$ precisely cancel the second diagram of Fig. 5, again for the case $|\psi\rangle = |\Omega\rangle$.

Since the divergent parts of these diagrams are easily seen to be independent of the state $|\psi\rangle$ it follows that these conditions serve to remove all divergences from the series of Fig. 4.

5. EXTENSION TO SEMI-CLASSICAL GRAVITY

Up to a point it is straight-forward to extend these considerations to the case of the semi-classical theory of gravity.

The divergences in this case arise in evaluating the expectation value of the energy-momentum tensor which appears on the right-hand side of equation (3.3). As in the case of the simple $v - \phi^2$ model we can develop a perturbation theory starting from the flat-space Minkowski metric, $g^{(0)}_{\mu\nu} = \eta_{\mu\nu}$, and a free Klein–Gordon field $\phi^{(0)}$ satisfying $(\partial^2 + m^2)\phi^{(0)} = 0$.

Since there is no self-interaction of the scalar field, only single scalar-loop diagrams contribute, exactly as in Fig. 4. The resulting divergences have a well-known structure. They can be cancelled by at most four counter terms which may be written (de Witt 1975)

$$\delta S = -\int d^4x(-g)^{\frac{1}{2}}\{\delta\Lambda + \delta KR(x) + \delta AR^2(x) + \delta BR_{\mu\nu}(x)R^{\mu\nu}(x)\} \tag{5.1}$$

where K stands for $1/16\pi\, G_N$. The first two counter terms may be regarded as renormalizing the cosmological constant and Newton's constant, respectively. The other two have the effect of introducing unwanted higher-derivative terms, if regarded as contributions to the classical action. I shall discuss them later.

Although the expression (5.1) is familiar, its interpretation here is not entirely standard. In conventional treatments of quantization in a curved space-time the divergences can be assembled into terms of the form (5.1). But if the space-time is represented by a given background metric, there is no term in the action to which they can be regarded as corrections. Therefore they cannot be interpreted in the usual sense as renormalization terms; indeed, their status is somewhat obscure. It is only when the dynamics of the gravitational field itself is taken into account that we can interpret (5.1) as a set of counter-terms.

The most important consequence of this change of interpretation is concerned with the removal of the ambiguities associated with finite changes in the divergent constants in (5.1). Such ambiguities are of course inherent in any renormalization theory, but in treatments of quantization in a gravitational background field they are especially problematic. Since the divergences in that case cannot be regarded as corrections to parameters in the theory such as masses and coupling constants, finite ambiguities cannot be removed by imposing renormalization conditions on these parameters. Instead, they are usually handled by a variety of *ad hoc* procedures, such as minimal subtraction in the dimensional regularization scheme, for which there is no real justification.

By contrast, in the treatment presented here, where the dynamics of the classical gravitational field is itself described by a term in the same action integral that yields the quantum dynamics, we can straightforwardly regard the terms in (5.1) as corrections to various parameters in the theory, though of course to do this for the two terms quadratic in the Riemann tensor requires an extension of the theory that brings with it other problems.

The cosmological term $\delta\Lambda$ may be fixed by requiring that flat empty space be a possible solution to the equations. Specifically there is a solution in which the metric is the Minkowski metric $\eta_{\mu\nu}$ and the quantum state is the Fock-space vacuum $|\Omega\rangle$ of the free quantum field ϕ. To fix δK we may consider for example a single-particle state of near-zero momentum and demand that there be a near-static solution in which asymptotically at large spatial distances the metric has the correct Schwarzschild form.

It is not hard in principle to invent similar conditions to fix the arbitrary constants in δA and δB. One might hope to be able to do this in such a way as to ensure the stability of the vacuum and single-particle states of the theory. However, the work of Horowitz and Wald suggests that this is in fact impossible. They conclude that in a semi-classical theory of gravity the vacuum state is necessarily unstable against very short wavelength perturbations, on the scale of the Planck length.

Several attitudes are possible to this result. It rests on apparently very plausible assumptions, though its applicability is restricted to massless theories. There is, moreover, one technical assumption that might be questioned, namely that the quantum stress energy cannot depend on sixth or higher-order derivatives of the metric (Horowitz 1980, p.447). However the result may well hold without this assumption.

One conclusion might be that a semi-classical theory of gravity cannot ever be a complete self-consistent theory. At best, it can only be an approximation to quantum gravity, though it is not at all clear that there is any regime where such an approximation would be valid. Even if it were, in solving the theory, we should have in some way artificially to suppress the unstable short wavelength modes which the theory is incapable of describing correctly. On the other hand quantum gravity as it presently exists is no more capable of providing a complete self-consistent description. Indeed, the necessary counter-terms quadratic in the Riemann tensor render the quantum gravity vacuum equally unstable.

An alternative conclusion would be that if a semi-classical theory of gravity does exist it does not possess a vacuum state with the assumed properties. Possibly its symmetries are broken. For example, the true ground state might involve a spontaneous breakdown of translational invariance, leading to 'crystallization' of the vacuum on a scale of the Planck length. It is hardly likely that such a state would represent physical reality, perhaps. My point is only that demonstrating the non-existence of a true vacuum state does not of itself prove the inconsistency of the theory.

6. EXPERIMENTAL TESTS

So far I have been discussing the self-consistency of the semi-classical theory of gravity, but it is also worth asking whether, if both semi-classical and quantum theories of gravity could be made consistent, there would be any way of deciding between them. At first sight the case might seem hopeless because specifically quantum-gravity effects would not be expected to show up except at extreme energies. Nevertheless there are distinctions that can be tested, arising from the inherent non-linearity of the quantum evolution equation in the case of a semi-classical theory.

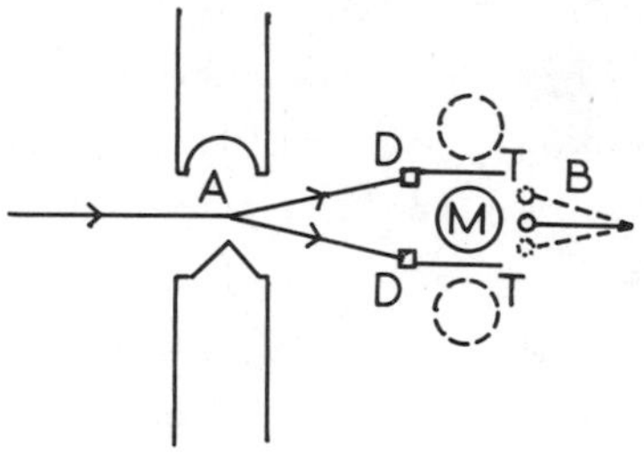

FIG. 6.

To focus ideas, let us consider a specific experimental arrangement (Fig.6). Take a beam of neutral spin-½ atoms polarized in, say, the x direction. Allow it to pass through a Stern–Gerlach apparatus (A) oriented so as to separate the beam into two, according to the value of the y component of spin. Arrange for each of these two beams to strike a detector (D) coupled to a trap-door mechanism (T) that allows a large mass (M) to move in one direction or the other. Near the large mass suspend a test body (B) which can move in response to the gravitational field of M.

Now consider what happens when a single atom enters this apparatus. According to the standard theory, which of course implicitly incorporates quantum gravity, the answer is quite clear. The mass M moves in one direction or the other according to which detector captures the atom. This movement is faithfully followed by a movement of the test body in the corresponding direction. However, according to semi-classical theory this apparently obvious conclusion is wrong. For in that theory the source of the gravitational field is the expectation value of the energy–momentum tensor. Now the quantum state is a superposition of two states in which the mass M is moved in opposite directions. Hence the gravitational field will in fact be one-half the sum of the fields that would be produced classically by masses in the two positions. Consequently the test-body will be pulled equally in both directions and will not move!

I do not suppose many physicists will believe this analysis, but it might nonetheless be interesting to perform such an experiment, because it is apparently, in this rather indirect way, a test of quantum gravity.

This proposed experiment is of course very similar to ones that have played an important part in discussions of quantum measurement theory. Indeed it illustrates very clearly the fact that semi-classical theory (whether of gravity or anything else) is inconsistent with conventional measurement theory involving 'reduction of the wave packet'. For evidently the test-body does not move only so long as we do not know in which direction the large mass has moved. As soon as we look at the system and determine the result of the measurement, it must suddenly jump one way or the other. This is clearly nonsense. (Similar problems of observer dependence have been discussed in a rather different context by Gibbons and Hawking 1977). It is not hard to locate the origin of this inconsistency. One commonly adopted way of avoiding at least some of the paradoxes of measurement theory is to say that the wave function does not represent the state of the system but only our knowledge of the state of the system. It is then entirely reasonable that the wave function should change suddenly when we make a measurement and acquire new information. It is also reasonable that different observers should use different wave functions to describe the same system, depending on their relative knowledge of it. But in this interpretation the wave function has only a passive role and cannot play an active part in the dynamics. If the source of the gravitational field is to be the expectation value of the energy—momentum tensor, it must be the expectation value in a state which is independent of the observer, and moreover one which is not subject to the sudden changes involved in reduction of the wave packet. Thus a consistent semi-classical theory certainly demands a new version of measurement theory. However, this is not necessarily a conclusive argument against semi-classical theory, since there are in any case other reasons for seeking a radical change in quantum measurement theory.

We must conclude that anyone who seriously wishes to propose a semi-classical theory of gravity has formidable obstacles to overcome. But the attempt to see how far one can go with such a theory is still worthwhile, if only because one cannot really judge quantum gravity without having some idea of what the alternatives might be.

ACKNOWLEDGEMENTS

I am indebted to my collaborators Seifallah Randjbar-Daemi and Bernard Kay for many fruitful discussions, and to Mike Duff and Chris Isham for criticizing an earlier draft of this talk.

REFERENCES

Bohr, N. and Rosenfeld, S. (1933). *Kgl. Dansk Vidensk. Selsk, Math.-fys. Medd.* **12**, 8.
De Witt, B.S. (1975) *Phys. Rep.* **C19**, 295.
Gibbons, G.W. and Hawking, S.W. (1977). *Phys. Rev.* **D15**, 2738.
Horowitz, G.T. (1980). *Phys. Rev.* **D21**, 1445.
Horowitz, G.T. and Wald, R.M. (1980). *Phys. Rev.* **D21**, 1462.
Kibble, T.W.B. (1978). *Comm. Math. Phys.* **64**, 73.
Kibble, T.W.B. and Randjbar-Daemi, S. (1980). *J. Phys.* **A13**, 141.
Møller, C. (1962). *Les theories relativistes de la gravitation.* CNRS, Paris.
Randjbar-Daemi, S., Kay, B.S., and Kibble, T.W.B. (1980). *Phys. Lett.* **91B**, 417.
Rosenfeld, S. (1963). *Nucl. Phys.* **7B**, 353.

3

INCONSISTENCY OF QUANTUM FIELD THEORY IN CURVED SPACE-TIME

M.J. Duff

Physics Department, Imperial College, London

1. INTRODUCTION

This Second Oxford Quantum Gravity Conference affords us the opportunity not only to learn of the many new and exciting results in the subject, but also to pause and reflect on some of the old questions in quantum gravity that are baffling us even today. And the oldest question surely is 'Should we quantize the gravitational field at all?' After all, the severe problems involved in quantization have prompted many physicists to argue that perhaps all the fields in nature should be governed by the laws of quantum field theory, with the exception of gravity itself. One is thus led to the idea of 'Quantum field theory in a curved space-time', the title of this present session.

Such investigations fall roughly into two categories: those where the gravitational field acts as a prescribed external source with no dynamics of its own; and those where the 'back reaction' of gravitational effects on quantum matter contribute to the gravitational dynamics via the semi-classical Einstein equations.

$$G_{\mu\nu} = \langle\psi'|\hat{T}_{\mu\nu}|\psi\rangle$$

where the ψs denotes some appropriately chosen state(s).

Let me declare at the outset my own prejudices against taking seriously any such semi-classical approach. The idea of not quantizing gravity seems to be the very antithesis of the economy of thought which is surely the basis of theoretical physics. Our ultimate goal should be the unification of the strong, electromagnetic, weak, and gravitational forces within an all-embracing quantum framework. QFT in curved space-time, however, is not a unification. It is a mongrel, and as such deserves to be put down.

From the point of view of supergravity (Freedman, van Nieuwenhuizen, and Ferrara 1976; Deser and Zumino 1976), of course, leaving the gravitational field classical would be absurd. Since in supergravity the graviton and the other particles of lower spin are simply different components of one and the same multiplet, not quantizing gravity would be as silly as quantizing the proton but not the neutron.

Indeed, one of the beauties of supergravity is that ultraviolet divergences which arise from quantum matter in the background gravitational field might be cancelled against those due to quantum gravity itself. I mention this because it highlights a feature of quantum gravity which ought to be well-known: gravitational vacuum polarization effects due to quantum matter are of exactly the same order of magnitude as those due to gravitons (gravity couples to everything with equal strength, including itself). There was a while ago a myth circulating in the literature according to which there exists a regime in nature where quantizing photons, neutrons etc, in a curved space-time was, in some sense, an 'approximation'; and that the quantization of gravity *per se* (if indeed it were to be done at all) could be postponed to a later date. It is to be hoped that this particular mythology, at least, has now disappeared.

In supergravity, in fact, not quantizing gravity would be worse than silly; it would be positively inconsistent. The consistency problems of coupling spin -3/2 fields to gravity (or to electromagnetic and Yang–Mills gauge fields) which are cured by supergravity would of course reappear if one were perverse enough to quantize one field and not the other. Already with spin -3/2 fields therefore, there is an obvious sense in which QFT in curved space-time is inconsistent. Though this is not the kind of inconsistency I wish to discuss here since there is as yet no empirical evidence for spin -3/2 fields. Moreover, the correctness of supergravity, or any other quantum theory of gravity, has not yet been established.

In a subject like quantum gravity, therefore, which consists of a whole mountain of prejudice resting upon the flimsiest of empirical foundations, it would be much more desirable if we could establish some more objective criterion for the correctness or falsehood of QFT in a curved space-time. Although we cannot as yet appeal to experiment (see, however, Kibble in this volume) what we can appeal to is inner consistency. It is the purpose of this article to establish that, in the sense to be described below, QFT in a curved space-time is internally inconsistent.

The crux of the argument will be the idea of 'field redefinitions'.

2. FIELD REDEFINITIONS

Let us denote the gravitational field by $g_{\mu\nu}(x)$ and the matter fields with which it interacts by the generic symbol $\phi^i(x)$. Next consider a classical theory described by an action functional

$$S[g,\phi]$$

The equations of motion are given by

$$\frac{\delta S}{\delta g_{\mu\nu}} = 0, \qquad \frac{\delta S}{\delta \phi^i} = 0 \ . \tag{2.1}$$

Now consider another action $S'[g',\phi']$ obtained from the first by field redefinitions, i.e.

$$S'[g',\phi'] = S[g,\phi] \tag{2.2}$$

where

$$g'_{\mu\nu} = g'_{\mu\nu}(g,\phi) = g_{\mu\nu} + \ldots$$

$$\phi'^i = \phi'^i(g,\phi) = \phi^i + \ldots . \tag{2.3}$$

To keep things simple, let us assume that these changes of variable do not involve derivatives of the fields. We shall further assume that they are not too pathological in the sense that the primed and the unprimed fields coincide in the weak-field limit (thus excluding $\phi' \sim \phi^2, \phi^{-1}$ or ln ϕ etc). At this classical level, therefore, the actions of $S[g,\phi]$ and $S'[g',\phi']$ do not prescribe two different physical theories (boundary conditions on the primed fields being determined from those on the unprimed). They describe one and the same theory, simply called by different names.

We now pose the question 'Do S and S' describe the same quantum theory?' I will maintain that the answer is YES provided we quantize *all* the fields, both gravity and matter, but NO if we quantize only matter. So by handing a 'quantum field theorist in curved space-time' the same classical theory written in infinitely many different ways, he will come up with infinitely many inequivalent quantum theories and has no way of deciding which one is correct. By contrast, in quantum gravity the action S and S' will lead to different off-mass-shell Green's functions, but provided we ask only physical questions e.g. on mass-shell S-matrix elements, the results will be the same. The invariance of the S-matrix under field redefinitions is a long-established rule of thumb in quantum field theory. In axiomatic QFT, it may even be proved rigorously under certain special assumptions (Borchers' Theorem). I shall not be concerned with any rigorous arguments, but will attempt to illustrate the point with some simple examples. In what follows, the statement that the theories 'are the same' will mean that they are the same order by order in perturbation theory. The question of non-perturbative effects will be postponed till Section 5.

Within the conventional perturbative framework one may, if so desired, make the point pictorially by drawing all Feynman diagrams contributing to a given process (e.g. ϕ—ϕ or ϕ—g or g—g scattering). The set of diagrams derived from S will be different, graph by graph, from that derived from S' because graviton loops and matter loops are mixed up by the transformations (2.3). Taken in their entirety, however, both sets of diagrams will yield the same on-shell amplitude for any given process. (Note, incidentally, that external lines do not get mixed up, by virtue of the requirement that the primed and unprimed fields coincide asymptotically.) If, on the other hand, one takes the view that gravity should not be quantized, then those graphs involving internal graviton lines (which were necessary to maintain the equivalence in the fully quantized theory) are now absent. The remaining graphs will lead to inequivalence results.

We might also take note that these arguments in no way

contradict the commonly used and perfectly respectable device of introducing an external c-number source $J(x)$ into the quantum theory and probing the response of the system to variations in that source. The introduction of a generating functional $Z[J]$, which yields the n-point functions upon differentiating with respect to J, is a far cry from introducing a classical gravitational field if we believe, as we surely must, that the gravitational field is itself dynamical. Not to include the dynamics of the gravitational field would be even more inconsistent, of course, inconsistent with the earth's orbit around the sun, for example! In keeping with this we shall always include the Einstein $\sqrt{g}R$ term in the Lagrangians to be considered. (For simplicity we eschew the alternative possibilities, of the kind discussed by Adler, Lieberman, Ng, and Tsao (1976), in which the graviton emerges as a bound-state.)

At this stage of the discussion on field redefinitions, some readers might well consider that I am unduly labouring an almost obvious point. Yet, to the best of my knowledge, it is a point which has never been discussed in the, by now vast, literature on QFT in a curved space-time. I shall proceed therefore to labour it further with some examples and an explicit calculation. This done, I shall return to its implications in Section 5, where I hope to show that the point is perhaps not quite so obvious after all.

3. AN EXAMPLE

Consider the interaction of gravity with N real scalars ϕ'^{i} ($i=1,\ldots N$) described by the action (with $K^2 = 4\pi G/3$)

$$S'[g',\phi'] = \int d^4x\sqrt{g'}\left[-\frac{1}{12K^2}R' + \frac{1}{2}g'^{\mu\nu}\partial_\mu\phi'^{i}\partial_\nu\phi'^{i} + \frac{1}{12}R'\phi'^{i}\phi'^{i}\right] \tag{3.1}$$

i.e. the scalars are massless with a conformal coupling and do not self-interact. Now make the change of variables

$$g'_{\mu\nu} = \Omega^2 g_{\mu\nu}, \quad \phi'^{i} = \Omega^{-1}\phi^{i}; \qquad \Omega^2 \equiv 1 + K^2\phi^{i}\phi^{i} \tag{3.2}$$

The new action is given by

$$S[g,\phi] = \int d^4x\sqrt{g}\left[-\frac{1}{12K^2}R + \frac{1}{2}\gamma_{ij}(\phi)g^{\mu\nu}\partial_\mu\phi^i\partial_\nu\phi^j\right] \qquad (3.3)$$

where

$$\gamma_{ij}(\phi) \equiv \delta_{ij} - \frac{\phi^i\phi^i}{K^{-2} + \phi^i\phi^i} \qquad (3.4)$$

Thus the conformal term has disappeared and the scalars now undergo a quite complicated self-interaction akin to that of the non-linear σ-model.

[The fields ϕ^i still provide a linear representation of SO(N), i.e.

$$\delta\phi^i = f^{ijk}\phi^j\delta\beta^k; \qquad \beta^k \text{ arbitrary} \qquad (3.5)$$

but also a non-linear realization of SO(N,1) i.e.

$$\delta\phi^i = (1 + K^2\phi^j\phi^j)^{\frac{1}{2}}\delta\alpha^i; \qquad \alpha^i \text{ arbitrary} \qquad (3.6)$$

as opposed to the SO(N+1) of the conventional σ-model. Thus the topology of the space in which the fields take their value is left unchanged by the transformations (3.2). Note, incidentally, that (3.6) reveals a hidden SO(N,1) symmetry of the action (3.1): S' is invariant under the transformations

$$\delta\phi'^i = \delta\alpha^i - K^2\phi'^i\phi'^j\delta\alpha^j \qquad (3.7)$$

$$\delta g'_{\mu\nu} = 2g'_{\mu\nu}(1 - K^2\phi'^j\phi'^j)^{-1}K^2\phi'^i\delta\alpha^i \qquad (3.8)$$

This follows from the field redefinition (3.2), but is far from obvious in the original formulation. Indeed, the action (3.1) has been much studied in the literature but, to the best of my knowledge, this symmetry has hitherto gone unnoticed.]

I think it is clear that faced with two such actions (3.1) and (3.3), a proponent of QFT in a curved space-time would reach very different conclusions by quantizing matter and not gravity. For example,

	S'	S
Is One-loop Exact?	Yes	No
Is The Theory Renormalizable?	Yes	No

In the semi-classical approach, the primed theory would be exact to one-loop (to be pedantic, one particle irreducible vertices would be exact to one-loop) since the scalars suffer no self-interaction. There would, consequently, be only a finite number of distinct counter-terms in contrast to the unprimed theory where, as is well-known, the σ-model requires an infinite number. On the other hand, the fully quantized theory would yield, consistently, the answers NO and NO to both questions.

Note that in the literature the question is sometimes posed: 'Will a theory renormalizable in flat-space remain renormalizable in the presence of gravity'? But, as this example illustrates, the question as it stands is really meaningless since one can obtain either answer depending on which form one chooses to write the classical theory. The question of renormalizability can be formulated consistently only if one adopts

Either (A) A fully quantized theory of gravity and matter.

Or (B) The point of view that one must single out one choice of field variable for preferential treatment in the quantum theory, from among the infinitely many equivalent choices available in the classical theory.

Before enlarging upon these remarks, let us examine some concrete calculations which demonstrate the inequivalence of different choices of field variable within the semi-classical approach.

4. A CONCRETE CALCULATION

For the purposes of these explicit calculations, let us consider the action of (3.1) with N=1.

$$S'[g',\phi'] = \int d^4x\sqrt{g'}\left[-\frac{1}{12K^2}R' + \frac{1}{2}g'^{\mu\nu}\partial_\mu\phi'\partial_\nu\phi' + \frac{R'}{12}\phi'^2\right]. \quad (4.1)$$

The classical equations of motion are

$$-\Box'\phi' + \frac{R'}{6}\phi' = 0 \quad (4.2)$$

and

$$G'_{\mu\nu} = 6K^2T'_{\mu\nu} \quad (4.3)$$

where

$$T'_{\mu\nu} = \nabla'_\mu\phi'\nabla'_\nu\phi' - \frac{1}{2}g'_{\mu\nu}\nabla'^\rho\phi'\nabla'_\rho\phi' + \frac{1}{6}\left[G'_{\mu\nu}\phi'^2 + g'_{\mu\nu}\Box'\phi'^2 - \nabla'_\mu\nabla'_\nu\phi'^2\right]. \quad (4.4)$$

By virtue of the conformal invariance of the matter action

$$R' = -6K^2T'^\mu_\mu = 0\ . \quad (4.5)$$

Under the transformations (3.2), the new matter action would again take on the σ-model form with $N=1$. However, since the N=1 σ-model may, by means of a further field redefinition, be rendered non-self-interacting, it follows that there exist transformations which eliminate the conformal coupling at no expense. They are

$$g'_{\mu\nu} = g_{\mu\nu}\cosh^2 K\psi, \qquad \phi' = K^{-1}\tanh K\psi\ , \quad (4.6)$$

under which the S' action of (4.1) becomes

$$S[g,\psi] = \int d^4x\sqrt{g}\left[-\frac{1}{12K^2}R + \frac{1}{2}g^{\mu\nu}\partial_\mu\psi\partial_\nu\psi\right]. \quad (4.7)$$

Thus, for a single massless scalar, the $R'\phi'^2$ coupling may be transformed away completely (another little known fact, amusing in own right). The new equations of motion are

$$-\Box\phi = 0 \tag{4.8}$$

and

$$G_{\mu\nu} = 6K^2 T_{\mu\nu} \tag{4.9}$$

where

$$T_{\mu\nu} = \nabla_\mu\psi\nabla_\nu\psi - \frac{1}{2}g_{\mu\nu}\nabla^\rho\psi\nabla_\rho\psi \ . \tag{4.10}$$

The stress-tensor is no longer traceless and

$$R = -\ 6K^2 T^\mu_\mu = 6K^2\nabla^\rho\psi\nabla_\rho\psi \tag{4.11}$$

Next we turn to the quantum theory. We wish to demonstrate the equivalence of the primed and unprimed systems in the fully quantized approach and their inequivalence in the semi-classical approach. Rather than compute on-shell scattering amplitudes, which would be quite complicated, we shall confine our attention to on-shell divergences. In particular, we shall consider the one-loop counterterms ΔS and $\Delta S'$ which must be added to S and S' respectively in order to remove the one-loop ultraviolet divergences in the quantum theory. These calculations were carried out in collaboration with Ward Goldthorpe (Duff and Goldthorpe 1980).

1. *Quantum Gravity and Matter*

Using the background field method and dimensional regularization (with $\varepsilon \equiv n-4$) we found, up to terms vanishing with field equations (4.2), (4.3) and (4.4), the primed counter-term

$$\Delta S' = \frac{1}{\varepsilon} \cdot \frac{1827}{160\pi^2}K^4 \int d^4x\sqrt{g'}(1 - K^2\phi'^2)^{-4}(\nabla'^\rho\phi'\nabla'_\rho\phi')^2 \tag{4.12}$$

and, up to terms vanishing with the field equations (4.8), (4.9) and (4.10), the unprimed counter-term

$$\Delta S = \frac{1}{\varepsilon} \cdot \frac{203}{640\pi^2} \int d^4x\sqrt{g}\ R^2 \ . \tag{4.13}$$

(This latter result was first obtained by t'Hooft and Veltman (1964).) But according to the transformation rule (4.6)

$$36K^4\sqrt{g'}(1 - K^2\phi'^2)^{-4}(\nabla'^{\rho}\phi'\nabla_{\rho}\phi')^2$$

$$= \sqrt{g}(6K^2\nabla^{\rho}\psi\nabla_{\rho}\psi)^2 = \sqrt{g}\ R^2 \tag{4.14}$$

on using (4.11). A comparison of (4.12) and (4.13) therefore yields

$$\Delta S' = \Delta S \ . \tag{4.15}$$

2. *Quantum Field Theory in Curved Space-time*

When only the scalar field is quantized, the counter-terms depend only on the gravitational field. Since the scalars suffer no self-interaction moreover, the one-loop divergences are exact. Because of the conformal coupling, the primed counter-term is conformally invariant. One finds

$$\Delta S' = \frac{1}{\varepsilon} \cdot \frac{1}{1920\pi^2} \int d^4x\sqrt{g'}\ C'_{\mu\nu\rho\sigma}C'^{\mu\nu\rho\sigma} \tag{4.16}$$

where $C'_{\mu\nu\rho\sigma}$ is the Weyl tensor. The unprimed counter-term, on the other hand, is given by

$$\Delta S = \frac{1}{\varepsilon} \cdot \frac{1}{1920\pi^2} \int d^4x\sqrt{g}\left[C_{\mu\nu\rho\sigma}C^{\mu\nu\rho\sigma} + \tfrac{5}{3}R^2\right] . \tag{4.17}$$

But according to the transformation rule (4.6)

$$\sqrt{g'}\ C'_{\mu\nu\rho\sigma}C'^{\mu\nu\rho\sigma} = \sqrt{g}\ C_{\mu\nu\rho\sigma}C^{\mu\nu\rho\sigma} \neq \sqrt{g}\left[C_{\mu\nu\rho\sigma}C^{\mu\nu\rho\sigma} + \tfrac{5}{3}R^2\right] \tag{4.18}$$

on using (4.11).

Note that $R \neq 0$ even classically. In the curved space-time approach, the classical $T_{\mu\nu}$ is replaced by some matrix-element of the operator-valued $T_{\mu\nu}$ between appropriate states. We have deliberately avoided a detailed discussion of this 'back reaction problem' since there are almost as many versions of this as there are papers on QFT in curved space-time. In any event, $R \neq 0$.

A comparison of (4.16) and (4.17) now yields

$$\Delta S' \neq \Delta S \;.$$

Thus this simple example of on-shell counter-terms serves to illustrate the equivalence of the primed and unprimed systems when we quantize gravity and matter, but their inequivalence when only matter is quantized. Similar remarks apply to the conformal anomalies in the trace of the energy momentum tensor (Capper and Duff 1974; Duff 1975; Deser, Duff, and Isham 1976*b*). Although for purposes of illustration we have exhibited only the N=1 case these one-loop calculations have also been performed for arbitrary N starting from the actions (3.1) and (3.3) (Duff and Goldthorpe 1980). The details are more complicated than for N=1, but the conclusions are the same.

There are many other extensions and generalizations one might contemplate; allowing $m^2\phi^2$ mass terms and $\lambda\phi^4$ interactions, for example. Moreover, such terms would in any case be forced upon us by the transformations (3.2) if we allowed for a cosmological $\Lambda\sqrt{g}$ term in the Einstein Lagrangian, since

$$\sqrt{g}' = \sqrt{g}\Omega^4 = \sqrt{g}(1 + 2K^2\phi^2 + K^4\phi^4) \;. \qquad (4.20)$$

Note that the equivalence check in the fully quantized theory would now require the calculation of quantum gravity effects with a cosmological constant. Fortunately, the evaluation of counterterms and anomalies in the presence of a cosmological constant has now been carried out (see the article by the author in this volume), and are consistent with the on-shell invariance of the full quantum theory under field redefinitions.

Indeed, there is one closely-related context where the point of view put forward in the present article might have some immediate practical significance, namely spontaneous symmetry breakdown and cosmology. There is a trivial, but vitally important, field redefinition of a kind not discussed so far:

$$\phi' = \phi - \eta \qquad \eta = \text{constant} \tag{4.21}$$

i.e. the shift encountered in the Higgs—Kibble effect from the field ϕ with non-vanishing vacuum expectation value η to the field ϕ' with vanishing expectation value. Although this transformation does not involve the metric directly, it leads to a huge cosmological constant in the Einstein equations (Linde 1974; Dreitlein 1974; Veltman 1975). This may be cancelled by the addition to the Lagrangian of another cosmological term with the opposite sign, but this enormous additional term will then reappear at high temperatures when the symmetry is restored and ϕ recovers a vanishing expectation value i.e. in the very early stages of the Universe. (Bludman (1979) has recently argued that the effects of such a term may even be drastic enough to avoid the Big Bang singularity.) The point I wish to make is that even if one adopts the semi-classical approach, and opts for alternative 2 of Section 3, one cannot avoid the transfer of dynamical information from the matter sector to the gravitational sector which spontaneous symmetry breaking inevitably involves. Quantizing one field and not the other is therefore susceptible, in principle, to all the attendant consistency problems. This is presently being investigated.

5. QUESTIONS AND ANSWERS

At the first Oxford Quantum Gravity Conference in 1974, I remarked that quantum gravity was a subject of two cultures (Duff 1975). Six years later, in spite of considerable progress, this state of affairs has not much changed. Someone once said that no physicist ever changes his ideas: it is simply that those holding the old ideas eventually fade away! If previous experience in quantum gravity is any guide, I anticipate that the inconsistency of QFT in a curved space-time under field redefinitions will elicit two different responses: (A) Tell us something new, (B) We do not believe you.

Those of persuasion (A) need read no further and so in this section I will try to anticipate questions from the rest.

Q. If matter is quantum and gravity classical, what right have you to mix up q-numbers with c-numbers?
A. At no time have I mixed up q-numbers and c-numbers. I simply present you with the same *classical* theory written in two different ways, then ask you to go away and quantize it. You will return with two different quantum theories.
Q. I am still not satisfied. In Section 3 you listed two alternatives. What is wrong with alternative 2? After all, it is axiomatic in the semi-classical approach that gravity is different from all other fields, why not insist further that field transformations which mix it with matter fields be forbidden?
A. The first response is purely pragmatic. The problem of finding a Lagrangian which describes the grand unified theory of everything is already difficult enough. This is a problem we both share. QFT in curved space-time, however, is making things infinitely more difficult. Even if presented with such a Lagrangian, you would then have to decide which of the infinitely many ways you are going to write it! More seriously though, in the search for the correct theory of nature one is traditionally guided by various principles, like symmetry requirements, unitary, causality, renormalizability etc. What is lacking in the semi-classical approach is some principle which will select one choice of variables as 'correct' and reject all the others.
Q. You mention renormalizability (a point to which I shall return), but in Section 3 we saw that one choice of variable gave a renormalizable theory and the other not. Why not invoke renormalizability as the principle on which to base one's choice?
A. Unfortunately, as the example in Section 4 demonstrated, one still encounters the inconsistency problems even under the subclass of field redefinitions which respect the semi-classical renormalizability.
Q. All right, but one choice involved a non-minimal $R\phi^2$ coupling. What about Minimal Substitution? Suppose someone

handed me the correct theory of strong, weak, and electromagnetic interactions in flat-space, why not simply apend gravity by the rules of minimal coupling?

A. You mean, for example, that

$$\frac{1}{2}\partial_\mu\phi\partial^\mu\phi + \lambda\phi^4$$

would become

$$\frac{1}{2}\sqrt{g}\; g^{\mu\nu}\partial_\mu\phi\partial_\nu\phi + \lambda\sqrt{g}\phi^4$$

Q. Yes, what is wrong with that?

A. But even treated semi-classically, this theory yields counter-terms like $\lambda R\phi^2$ (Freedman *et al.* 1974) which would then have to be included at the tree level to maintain renormalizability thus contradicting your original assumption of minimal substitution.

Q. So to maintain renormalizability, I would have to abandon minimal coupling as the required principle, but at least renormalizability has narrowed things down a bit.

A. True, you can arrange for only a finite number of distinct counter-terms but these involve R^2 and $R_{\mu\nu}R^{\mu\nu}$ which, being fourth order in derivatives, give rise to the usual ghost problems (De Witt 1965; Stelle 1977, 1978; Sezgin and van Nieuwenhuizen 1980). In this respect you are no better off than in quantum gravity (indeed, you may be worse off; see Section 6), and you are still left with the problem of which field variables to pick.

Q. Let me try a change of tack. Suppose I concede that QFT in curved space-time leads to inequivalent quantum results starting from the same classical theory written in different ways, and suppose I concede further that there is no obvious principle which singles out one particular choice; is quantum gravity really any better? In Section 4, for example, you showed the equivalence of the counter-terms which govern the divergent part of the on-shell amplitudes, should you not also check that the finite parts are equivalent?

A. Yes indeed, that omission was only through laziness. There is, however, a formal proof of this equivalence. Let us, following De Witt (1965) denote all the fields in the theory, both gravity and matter, by ϕ^i and functional derivatives with respect to ϕ^i by a comma. Again for simplicity, let us consider the functional integral at one-loop only. It is obtained by making the background field split

$$\phi^i(x) \to \phi^i(x) + h^i(x) \tag{5.1}$$

and integrating over the quantum fields h^i:

$$\exp(-W[\phi]) \sim \int Dh \, \exp(-h^i S,_{ij}[\phi]h^j) \sim (\det S,_{ij})^{-\frac{1}{2}} \tag{5.2}$$

$\Gamma[\phi] = S[\phi] + W[\phi]$ in the generating functional for one-particle-irreducible vertices from which we compute the S-matrix. Under a change of variable

$$\phi^i \to \phi'^i = \phi'^i(\phi) \,, \tag{5.3}$$

the action is a scalar

$$S'[\phi'] = S[\phi]$$

and

$$S',_{i'} = \frac{\partial\phi^i}{\partial\phi'^{i'}} S,_i \tag{5.4}$$

Hence

$$S',_{i',j'} = \frac{\partial\phi^i}{\partial\phi'^{i'}} S,_{ij} \frac{\partial\phi^j}{\partial\phi'^{j'}} + \frac{\partial^2\phi^i}{\partial\phi'^{i'}\partial\phi'^{j'}} S,_i \,. \tag{5.5}$$

Going on-shell, i.e. using the background field equations $S,_i = 0$, leads to

$$(\det S',_{i'j'}) = (\det S,_{ij})\left[\det \frac{\partial\phi}{\partial\phi'}\right]^2 . \tag{5.6}$$

Since $S,_{ij}$ is a differential operator, the ordering of the factors in (5.5) is important, but of no consequence in the result (5.6) which could have been obtained in an equivalent way by substituting (5.5) in (5.2) and changing integration variables to

$$h'^{i'} = h^j \frac{\partial \phi'^{i'}}{\partial \phi^j} . \tag{5.7}$$

The Jacobian factor would then emerge in the transformed measure.

The factor $\det(\partial\phi'/\partial\phi)$ is formally infinite. It may be converted, using det = exp tr ln, into the exponential of a $\delta^4(0)$ term which is set equal to zero in dimensional regularization (Capper and Liebbrandt 1973; t'Hooft 1978). After regularization therefore

$$W'[\phi'] = W[\phi] \tag{5.8}$$

and the finite parts of the amplitudes will also be equivalent.

Q. What about non-perturbative effects for which the Jacobian factor may no longer be quite so innocuous as it is in perturbation theory? (Klauder 1979).

A. This involves delicate questions about the definition of the measure in the functional integral which, as you point out, are ignored in the standard perturbative framework. Closely related to this is the change in the range of field variables. In (4.6) for example, if ψ ranges from $-\infty$ to ∞, then ϕ is constrained to lie between -1 and 1; a feature which is also ignored. Pending a rigorous definition of the functional integral, I do not know a complete answer to such questions, except to say that a change of integration variable in the functional integral ought not to change its value provided, of course, one is sufficiently careful in one's manipulations and provided one integrates over *all* the fields in the theory. (I presume this would also hold true, even if one considered field transformations of the pathological type not considered in (2.3). How to perform actual calculations in this case is another question.) The problem with QFT in a curved space-time is that one is choosing to integrate over some of the fields and not others. Different choices would then certainly make a difference.

Unfortunately, concrete examples of non-perturbative effects are thin on the ground, especially in quantum gravity. Let us consider instead a known example in flat space-time where one can begin to get to grips with the problem of field redefinitions in a non-perturbative context, namely instantons in the non-linear σ-model:

$$\mathscr{L} = \frac{1}{2} g_{ij}(\phi)\partial_\mu\phi^i\partial^\mu\phi^i \ . \tag{5.9}$$

This time, we are considering the conventional σ-model so that $g_{ij}(\phi)$ is the metric on a group manifold with non-trivial topology, typically the sphere (Isham 1969). The explicit form of the Lagrangian (5.9) may be written in many different ways by making a change of coordinate basis,

$$\phi^i \to \phi'^i(\phi) = \phi^i + \ldots \tag{5.10}$$

which in perturbation theory does not affect on-shell amplitudes. Now when topological instanton effects are taken into account, failure to recognize that manifolds with non-trivial topology require more than one coordinate patch to cover them, can lead to errors. However, provided such subtleties are taken into account, then consistency is maintained (Deser, Duff and Isham 1976*a*).

Although taken from flat space-time, this example, if anything, only re-enforces the case. The ability to make field redefinitions (i.e. to move from one coordinate patch to another) is positively necessary! This is contrary to the spirit of QFT in a curved space-time where one is forced into the corner of singling out one choice of variables for preferential treatment.

Q. I would prefer to return to quantum gravity and consider topological effects there. In comparing counter-terms and trace anomalies in Section 4, you ignored terms proportional to

$$\chi = \frac{1}{32\pi^2}\int d^4x\sqrt{g}\,(R_{\mu\nu\rho\sigma}R^{\mu\nu\rho\sigma} - 4R_{\mu\nu}R^{\mu\nu} + R^2) \tag{5.11}$$

presumably on the grounds that the integrand is a total divergence. But in space-times with non-trivial topology χ does not vanish; it is a topological invariant, namely the Euler number. And as you yourself have pointed out *ad nauseam*

(Duff 1977; Christensen and Duff 1978*a,b*; 1979, 1980; and again in this volume), χ must then be included in the counter-terms and integrated trace anomaly.

A. Thank you for reminding me. My only reason for not including these terms was their irrelevance to the argument at hand. The coefficient of χ is the same in the fully quantized theory whichever choice of variable one picks. A different coefficient is obtained if gravity is not quantized, of course, but it again turns out to be insensitive to the choice of field variable.

Q. Ah! Now you are hoist with your own petard. In a recent paper with van Nieuwenhuizen (Duff and van Nieuwenhuizen 1980; see also this volume) you claim that the gauge theory of a rank two asymmetric tensor potential $A_{[\mu\nu]}$, although related by field redefinitions to the theory of a single scalar field A, nevertheless gives rise to gravitational counterterms and anomalies which differ precisely by the Euler number χ. Is this not in stark contradiction to your principal argument that theories which are equivalent at the classical level should remain equivalent at the quantum level?

A. Not so. It is true that classical equivalence should imply quantum equivalence. The reason $A_{[\mu\nu]}$ and A yield inequivalent quantum results is because they were never equivalent even at the classical level. As explained in the quoted references, the two theories are *not* related by field redefinitions when space-time has a non-trivial topology. This topological inequivalence at the classical level then reappears at the quantum level in the guise of different anomalies and counter-terms.

Another example requiring extreme caution concerns chiral transformations

$$\psi \rightarrow \psi' = \psi + i\theta\gamma_5\psi \tag{5.12}$$

(θ infinitesimal) of massive fermion fields in the Einstein–Dirac action $S_{\text{E-D}}$. This induces a change in the mass term

$$m\bar{\psi}'\psi' = m\bar{\psi}\psi + 2i\theta m\bar{\psi}\gamma_5\psi \ . \tag{5.13}$$

In the quantum theory, the fermion measure in the functional integral is not invariant under (5.11) owing to the presence of zero-modes (Fujikawa 1979), and the Atiyah–Singer theorem then implies that (5.12) must be compensated by an effective topological, CP violating, addition to $S_{E\text{-}D}$ proportional to $\int d^4x\sqrt{g}\,{}^*R_{\mu\nu\rho\sigma}R^{\mu\nu\rho\sigma}$ (Deser, Duff, and Isham 1980). Quantum mechanically, therefore

$$S_{E\text{-}D} + 2i\theta m\int d^4x\sqrt{g}\,\psi\gamma_5\psi \equiv S_{E\text{-}D} + \frac{\theta}{192\pi^2}\int d^4x\sqrt{g}\,{}^*R_{\mu\nu\rho\sigma}R^{\mu\nu\rho\sigma}\,. \tag{5.14}$$

The equivalence may also be seen by integrating the axial-current anomaly

$$\nabla^\mu J^5_\mu = 2im\bar{\psi}\gamma_5\psi - \frac{1}{192\pi^2}\,{}^*R_{\mu\nu\rho\sigma}R^{\mu\nu\rho\sigma}\,. \tag{5.15}$$

(N.B. The m = o case must be handled separately.) Does this contradict the assertion that equivalent classical theories lead to equivalent quantum theories? I think this is more a question of semantics than of physics. Within the framework of functional integrals, one may often trade off effects in the measure by additions to the argument of the exponential i.e. by additions to the classical Lagrangian. Since the addition in question is a total divergence, this will not change the classical field equations (inasmuch as the epithet 'classical' is, in any event, meaningful when dealing with anti-commuting fermions), even though it is necessary to maintain equivalence when one sums over different topological sectors in the quantum theory. (N.B. If gravity is not quantized, this addition is an irrelevant phase factor in the functional integral.) In so far as the two theories which are equivalent at the quantum level have equivalent classical field equations, therefore, this example is in harmony with our previous remarks.

Q. This is all very well and good but you have until now avoided the crucial question, the one to which I promised to return. You maintain that the semi-classical approach is inconsistent because it yields ambiguous results and one has no criterion for deciding which is correct. Yet the principal objection to quantizing gravity itself is that Einstein's theory is non-renormalizable! The avoidance of this disaster was a primary, albeit unfulfilled, hope of QFT in curved space-time. Non-renormalizability implies an infinite number of distinct counter-terms and hence an infinite number of undetermined parameters thus robbing the quantum theory of any predictive power. In this respect quantum gravity is no less ambiguous than quantum field theory in a curved space-time. Unless this problem is solved, is not all this talk of quantizing gravity just so much hot air?

A. That is a very good question!

6. RENORMALIZABILITY: CONFRONTATION WITH EXPERIMENT

The aim of any theory must surely be to make predictions which can be compared with experiment (in principle if not yet in practice). And the non-renormalizability of quantum gravity has been, and remains, the single most important barrier to this aim. Nor can QFT in a curved space-time claim to have solved the problem. Even if one forbids the kind of field redefinitions of Section 3, which anyway spoil the semi-classical renormalizability, the addition to the Lagrangian of R^2 and $R_{\mu\nu}R^{\mu\nu}$ terms necessary to absorb the one-loop divergences leads to the aforementioned troubles of ghosts and lack of unitarity.

This unitarity problem is common to almost all theories of gravity plus matter. There are, however, two known examples of theories for which R^2 and $R_{\mu\nu}R^{\mu\nu}$ terms are not required: pure quantum gravity which is on-shell finite at one loop (t'Hooft and Veltman 1975) and supergravity (simple and extended) which is on-shell finite to two-loop order (Grisaru 1977). Note that counter-terms which can never vanish with the field equations like $C^3_{\mu\nu\rho\sigma}$, required (presumably) at two-loops in pure gravity, and $C^4_{\mu\nu\rho\sigma}$ required (perhaps) at three-loops in supergravity (Deser, Kay, and Stelle 1977) pose no extra problems as far as ghosts are concerned. Since they are at least $O(h^3)$ in the quantum perturbations $h_{\mu\nu}$, they affect only the vertices but not the propagators. *A priori*, one might also worry about terms quadratic in the curvature but containing higher numbers of derivatives since these are superficially $O(h^2)$. Remarkably, however, the Bianchi identities always come to the rescue. For example $\nabla^{\mu *}C_{\mu\nu\rho\sigma}$ vanishes on shell, and hence $C \,\Box\, C$ which is superficially $O(\hbar^2)$is converted to C^3 which is in fact $O(\hbar^3)$. (This means, incidentally, that within the background field method, the *exact* two-point function requires no counter-terms which do not vanish with the field equations even though it contains infinitely many sub-divergencies!) In the light of the above, one might be able to make out a case for demanding that, whatever the final resolution of the renormalizability problem, finiteness at one-loop is a necessary prerequisite for avoiding the ghost problem. Alternative ways out have been suggested. See, for example Salam and Strathdee (1978), and also Weinberg (1979) who points out that the location of the ghost poles correspond to energies at which perturbation theory would in any case have already broken down.

What is the moral of the tale so far? In the author's opinion, the inadequacy of QFT in a curved space-time has served to spotlight the imperative of solving the renormalizability problem in quantum gravity. What is required is a renormalizable theory of quantum gravity plus matter or, failing that, an explanation of why renormalizability is not necessary (e.g. 'asymptotic safety' (Weinberg 1979)). I shall not list here all the ingenious attempts in the latter direction but simply remark that, if correct, they would create yet another puzzle. Why has the criterion of renormalizability proved so successful in leading to the Weinberg–Salam theory of weak and electromagnetic interactions, the QCD theory of the strong interactions, and the Grand Unified Theories of all three? The renormalizability of these models would then appear to be an unnecessary luxury. Another possibility which does not fall neatly into either category, is that the correct theory of gravity and matter may be ostensibly non-renormalizable by power counting but finite order by order in perturbation theory. Our only hope in this direction, so far, is supergravity. The idea that this theory may be literally finite because of mutual cancellation of divergences between bosons and fermions is a beautiful one, and one of which Dirac, who never warmed to the idea of divergences in physics, ought to approve.

Ultimately, experiment will be the final arbiter. In this connection, it is remarkable how attitudes to the empirical testing of quantum gravity have mellowed since the last Oxford conference. Many enthusiasts were then content to concede that their efforts were of no practical value until Planck energies of 10^{19} GeV were considered. The more optimistic maintained that it might first be necessary to understand physics at 10^{19} GeV in order to understand physics at present energies. Since hard-nosed particle physicists are now claiming the necessity of first understanding physics on the other side of the grand unification 'desert' at 10^{15} GeV, such optimism no longer seems quite so outrageous. There seem to me several interconnected ways in which quantum gravity may be important at energies below the Planck scale.

1. Newton's constant G may not be the only relevant gravitational parameter. As discussed in Section 4 and elsewhere in this volume, there is also the cosmological constant which, though (apparently) vanishingly small at the present epoch, is predicted to be huge above the electro-weak unification temperature and even huger above the grand unification temperature. On a cosmological scale, therefore, quantum effects of gravity with a cosmological constant may be felt long after the Planck time ('long' in the sense that a desert is long, albeit that in cosmological terms the 'desert' lasted only for a fraction of a second, and we have been living in the oasis ever since).

2. There exist topological and instanton effects in gravity that are independent of the strength of the gravitational interaction but determined rather by the global topology of space-time. Under what circumstances such topologies might be important (i.e. above or below Planck energies), however, remains somewhat obscure. These are in addition to, and distinct from, the astrophysical phenomena associated with event horizons and exploding black holes (Hawking 1974) whereby small effects accumulate into large ones over the age of the universe.

3. Related to both the above are situations where G is again inconsequential but what is important is the number of elementary particle species. This is relevant not only for

cosmological considerations, where the number of massless degrees of freedom (like gravitons) is important, but also when vacuum energies come into play as in the Casimir effect. (For a review see De Witt 1975.)

In much the same vein, and perhaps most important of all, the universality of the gravitational interaction means that 'gravity counts the number of elementary particles' (Duff 1975). Thus, the problem of determining the correct elementary particle spectrum, and the problem of constructing a self-consistent renormalizable/finite theory of quantum gravity may have one and the same solution (Capper, Duff, and Halpern 1975).

In this sense, supergravity, which makes very definite and rigid predictions on the kinds of particle with which gravity may interact, represents a quantum theory of gravitation which is already falsifiable at present energies! Indeed, some cynics might argue that it goes one better: it is not only falsifiable but false! The $O(8)$ symmetry of $N = 8$ supergravity cannot accommodate the standard $SU(3) \times SU(3) \times U(1)$ as a subgroup. Thus there is room for the electron but not the muon, the Z^0 boson but not the $W^{\pm}$ etc. In this respect, the present state of supergravity calls to mind Dr Johnson's description of woman preachers: 'Like a dog walking on his hind legs, it is not done well; the remarkable thing is that it is done at all'. That a theory of gravitation should predict any particle spectrum is surely the remarkable thing.

The future of supergravity, in fact, is far from bleak. As noted by Julia and Cremmer (1978), the hidden local $SU(8)$ invariance of $N = 8$ supergravity is large enough to allow the possibility that particles which seem 'elementary' at present energies may be bound states of the fundamental fields of supergravity. (Curtright and Freund 1979; Ellis, Gaillard, Maiani, and Zumino 1980). Although a gamble, the rewards of this ambitious endeavour are potentially so great that the understanding of $N = 8$ supergravity is a project to which all of us in quantum gravity ought perhaps to be turning our attention. Up until now, it is the straight particle physics aspects of supergravity which have received the closest scrutiny but my feeling is that there is also a wealth of geometrical and topological information contained in extended supergravity that is only just beginning to be tapped.

None of this would be possible, of course, with QFT in a curved space-time. There is of course a much less extreme attitude to the semi-classical approach: one may simply regard it as a theoretical laboratory in which to test out one or two ideas which will come to fruition only in the fully quantized theory. Even here, however, it is vital to be aware of the limitations of such an approach; limitations which I hope to have shown to be much more severe than is generally appreciated.

In conclusion therefore, 'Quantum field theory in curved space-time' is guilty until proven innocent. The burden of proof should now shift, if indeed it has not already done so, from those who would quantize gravity to those who would not. I have presented here the case for the prosecution; the case for the defence may be found elsewhere in this volume.

ACKNOWLEDGEMENTS

I would like to thank C.J. Isham and W. Goldthorpe for their invaluable help. Conversations with B. Kay, T.W.B. Kibble, S. Randjbar-Daemi, M. Rocek, and K.S. Stelle are also gratefully acknowledged.

REFERENCES

Adler, S.L., Lieberman, J., Ng, Y.J., and Tsao, H.S. (1976). *Phys. Rev.* **D14**, 359.
Bludman, S.A. (1979). University of Pennsylvania Preprint.
Capper, D.M. and Duff, M.J. (1974). *Nuovo Cim.* **23A**, 173.
—— and Liebbrandt, G. (1973). *Nuovo Cim. Lett.* **6**, 117.
——, Duff, M.J., and Halpern, L. (1974). *Phys. Rev.* **D10**, 461.
Christensen, S. and Duff, M.J. (1978*a*). *Phys. Lett.* **76B**, 571.
—— —— (1978*b*). *Phys. Lett.* **79B**, 213.
—— —— (1979). *Nucl. Phys.* **B154**, 301.
—— —— (1980). *Nucl. Phys.* **B170**, 480.
Cremmer, E. and Julia, B. (1978). *Phys. Lett.* **80B**, 48.
Curtright, T.L. and Freund, P.G.O. (1979). *Supergravity* (ed van Nieuwenhuizen and Freedman).
Deser, S. and Zumino, B. (1976). *Phys. Lett.* **62B**, 335.
——, Duff, M.J., and Isham, C.J. (1976*a*). *Nucl. Phys.* **B111**, 45.
—— —— —— (1976*b*). *Nucl. Phys.* **B114**, 29.
—— —— —— (1980). *Phys. Lett.* **93B**, 419.
——, Kay, J., and Stelle, K.S. (1977). *Phys. Rev. Lett.* **38**, 527.

De Witt, B.S. (1965). *Dynamical theory of groups and fields*. Gordon and Breach, New York.
——(1975). *Phys. Reports*. **196**, 295.
Dreitlein, J. (1974). *Phys. Rev. Lett.* **33**, 1243.
Duff, M.J. (1975). *Quantum Gravity: An Oxford Symposium*. (ed Isham, Penrose, and Sciama) Oxford University Press.
—— (1977). *Nucl. Phys.* **B125**, 334 and Abstracts of contributed papers GR8 Conference, Waterloo.
—— and Goldthorpe, W. (1980). To appear.
—— and van Nieuwenhuizen, P. (1980). *Phys. Lett.* **94B**, 179.

Ellis, J., Gaillard, M.K., Maiani, L., and Zumino, B. (1980). CERN Preprint.
Freedman, D.Z., Murzinich, I.J., and Weinberg, E.J. (1974). *Ann. Phys. (N.Y.)* **87**, 95.
——, van Nieuwenhuizen, P., and Ferrara, S. (1976). *Phys. Rev.* **D13**, 3214.
Fujikawa, K. (1979). *Phys. Rev. Lett.* **12**, 1195.
Grisaru, M. (1977). *Phys. Lett.* **66B**, 75.
Hawking, S. (1974). *Quantum Gravity: An Oxford Symposium* (ed Isham, Penrose, and Sciama) Oxford University Press.
t'Hooft, G. and Veltman, M. (1975). *Ann. Inst. Henri Poincaré*. **A20**, 69.
Isham, C.J. (1969). *Nuovo Cim.* **61A**, 188.
Klauder, J.R. (1979). *Ann. Phys. (N.Y.)*. **111**, 19.
Linde, A.D. (1974). *Pis'ma zh. eksp. teor. Fiz.* **19**, 32.
Salam, A. and Strathdee, J. (1978). *Phys. Rev.* **D18**, 4480.
Sezgin, E. and van Nieuwenhuizen, P. (1980). *Phys. Rev.* D22, 301.
Stelle, K.S. (1977). *Phys. Rev.* **D16**, 953.
—— (1978). *GRG J.* **9**, 353.
Veltman, M. (1975). *Phys. Rev. Lett.* **34**, 777.
Weinberg, S. (1979). *General Relativity: An Einstein Centenary Survey* (ed Hawking and Israel) Cambridge University Press.

4

IS FLAT SPACE-TIME UNSTABLE?

G.T. Horowitz

Department of Physics, University of California, Santa Barbara, CA 93106

1. INTRODUCTION

Although one does not yet have a quantum theory of gravity, one might hope to gain insight into some of the features of this theory by investigating various models and approximations. An important feature of any quantum theory is the nature of its ground state. It is therefore of interest to investigate what can be learned about the ground state of quantum gravity.*

Rather than attempt to describe all aspects of this ground state, we focus on one aspect which is simple to describe, yet physically important; the expectation value of the metric in this state $\langle 0|g_{ab}|0\rangle$.† In particular we consider the question: Is it true that this expectation value will be flat, i.e.,

$$\langle 0|g_{ab}|0\rangle = \eta_{ab} \ ? \tag{1.1}$$

If (1.1) is not satisfied, then a state for which the expectation value of the metric *is* flat would presumably be unstable, and decay into the true ground state. In this sense, flat space-time would be unstable in quantum gravity.

---

* It must be remarked that there is a possibility that quantum gravity will not have any state resembling a 'ground state'. In fact, slight evidence for this possibility comes from the fact that in the canonical approach to quantum gravity, one obtains a Hamiltonian which depends on the three-geometry. This three-geometry can be interpreted as carrying the time information. Thus, one obtains a 'time dependent Hamiltonian'. (I thank K. Kuchar for bringing this to my attention.) However, for the present discussion, we shall assume that the concept of a ground state is meaningful in quantum gravity.

† To simplify the discussion, we ignore the possibility of fluctuations in the manifold topology, and concentrate on fluctuations in the space-time geometry. In the approximations we will soon consider, the manifold will be fixed to be R^4.

There are at least two classical reasons why one might expect (1.1) to be satisfied. The first is that flat space-time is a 'preferred' metric in the classical theory. For example, it is the only metric invariant under the Poincaré group. The second reason is flat space-time is the 'ground state' of classical relativity, in the following sense. Consider the space of all asymptotically flat vacuum solutions to Einstein's equation on R^4. Let E denote the function on this space which assigns to each solution its ADM energy. (Arnowitt, Deser, and Misner 1961; Geroch 1977.) Then it is known (Schoen and Yau 1979) that flat space-time is the global minimum of this function.

However several model field theories (e.g. massless scalar electrodynamics) are now known which have the property that the quantum ground state does not correspond to the classical state of minimum energy. (Coleman and Weinberg 1973; Gross and Neveu 1974.) This property has been given the name 'dynamical symmetry breaking'. It is possible that a similar phenomena occurs in quantum gravity.

There is, however, at least one important difference between dynamical symmetry breaking in quantum field theories in flat space-time, and dynamical symmetry breaking (if it exists) in quantum gravity. This difference concerns the role of Poincaré invariance in defining the ground state. In the original discussions of dynamical symmetry breaking, the quantum field which developed a non-zero ground state expectation value was a scalar field (or a composite of two Fermi fields), and $\langle\phi\rangle$ (or $\langle\bar{\psi}\psi\rangle$) was required to be a constant in space-time, consistent with a Poincaré invariant ground state. More recently, the possibility of dynamical symmetry breaking has been investigated in Yang–Mills' theories.* In this case the Poincaré group is combined with the gauge group to obtain a more general notion of invariance (Bergmann and Flaherty 1977; Jackiw 1978), but still the ground state expectation values are required to be invariant under these generalized transformations. In quantum gravity, on the other hand, the requirement that the ground state be Poincaré invariant is simply not applicable. (Indeed, it's hard to even find a meaning for this requirement.) Thus the ground state expectation value of the metric is not

* See, e.g., Pagels and Tomboulis 1978; Ovrut 1979.

a priori constrained, and could be anything from a simple space-time of constant (positive or negative) curvature, to a complicated space-time with large curvature on small scales.

Our discussion of the ground state of quantum gravity will consist of investigating three (rather severe) approximations to this theory. Thus our results will, at best, be merely suggestive. We begin by reviewing the quantum theory of linearized gravity. As is well known, $\langle g_{ab}\rangle = \eta_{ab}$ in this theory. Next we consider semi-classical relativity, which is a theory of gravity coupled to matter in which gravity is treated classically and matter is treated quantum mechanically. We find that flat space-time is unstable in this theory, which might be interpreted as evidence that $\langle g_{ab}\rangle \neq \eta_{ab}$. Strictly speaking, however, semi-classical relativity is not an approximation to quantum gravity. In the fourth section, we consider an approximation to a full theory of quantum gravity, which is the closest analogue to semi-classical relativity. It turns out that this approximation is equivalent to the leading order in a $1/N$ expansion (where N is the number of matter fields coupled to gravity). In this approximation, we find that for a certain range of the parameters of the theory, there is no evidence for $\langle g_{ab}\rangle \neq \eta_{ab}$, but if the parameters are outside this range, there is evidence for a non-flat ground state expectation value of the metric. In the final section, we compare our results for quantum gravity, with a similar analysis of massless Q.E.D.

2. LINEARIZED THEORY

We begin with a review of the quantum theory of linearized gravity. Since this is simply the theory of a free spin two field h_{ab}, the ground state is the standard vacuum state and one has immediately

$$\langle 0|h_{ab}|0\rangle = 0 \tag{2.1}$$

Since $g_{ab} = \eta_{ab}I+h_{ab}$ (where I is the identity operator), equation (2.1) implies

$$\langle 0|g_{ab}|0\rangle = \eta_{ab} \tag{2.2}$$

It is instructive to pass from this 'particle representation' of the quantum theory, to the 'field representation'. The ground state wave function for linearized gravity was found by Kuchar (1970). The configuration space for this theory can be obtained as follows. Fix a space-like three plane S in Minkowski space-time. The configuration space C consists of all linearized, three dimensional Ricci tensors $\dot{R}_{\alpha\beta}$ on S which are tracefree. (This is just a gauge invariant formulation of the usual definition of C involving transverse, tracefree perturbations. Notice that this formulation yields the correct number of degrees of freedom for linearized gravity: there are six independent components of $\dot{R}_{\alpha\beta}$ which are restricted by three Bianchi identities and one trace condition leaving two independent degrees of freedom.)

The ground state wave function on C turns out to be*:

$$\psi_0(\dot{R}) = N \exp\left\{-\frac{1}{4\pi^2\hbar}\int d^3x \int d^3x'\dot{R}_\alpha{}^\gamma(x)\dot{R}_{\gamma\beta}(x')n^\alpha n^\beta\right\} \tag{2.3}$$

where $n^\alpha = \dfrac{x^\alpha - x'^\alpha}{\|x-x'\|}$, and N is a normalization constant. By taking the Fourier transform of the integrand, one can show that it is non-negative and vanishes only when $\dot{R}_{\alpha\beta} = 0$. Therefore ψ_0 is peaked about flat space. In fact, by symmetry, it's clear that the expectation value of the Ricci tensor in this state vanishes:

$$\langle\psi_0|\dot{R}_{\alpha\beta}|\psi_0\rangle = 0 \tag{2.4}$$

Of course, equation (2.3) also contains information about fluctuations about this expected value. Suppose $\dot{R}_{\alpha\beta}$ has support in a region of characteristic size L, and has typical magnitude (say, in an orthonormal frame) of R. Then the probability of finding the curvature to be $\dot{R}_{\alpha\beta}$ is roughly

$$N \exp\left(-\frac{R^2L^6}{\hbar}\right) . \tag{2.5}$$

---

* We use geometrical units: $c = 8\pi G = 1$, but $\hbar \neq 1$.

Therefore large amplitude fluctuations can be extremely likely, provided they exist on sufficiently small scales.

If all large amplitude fluctuations in the ground state of the linearized theory were suppressed, then one might argue that the non-linearities of the full theory would not significantly alter this ground state. However, we have just seen that this is not the case. It is therefore possible that the ground state of the full theory of quantum gravity will be very different from this linearized result.

3. SEMI-CLASSICAL RELATIVITY

We now wish to consider a non-linear theory and investigate the effects of interactions on the ground state. Of course classical relativity is non-linear, but we already know that the ground state of this theory is flat space-time. Perhaps the simplest generalization of classical relativity which includes some quantum effects is semi-classical relativity. This is a theory of gravity interacting with matter in which gravity is treated classically (described by general relativity) and matter is treated quantum mechanically (described by quantum field theory). We now investigate whether the ground state of semi-classical relativity is flat space-time.

Let g_{ab} be a space-time metric, ϕ a quantum matter field on this space-time, and $|\xi\rangle$ an in-state of the quantum field. The field equations for semi-classical relativity consist of the usual operator field equation for ϕ in the curved space-time, together with the semi-classical Einstein equation

$$G_{ab} = \langle\xi|T_{ab}|\xi\rangle \tag{3.1}$$

where T_{ab} is an effective stress energy of the quantum field. As is well known, since T_{ab} is the product of field operators, all expectation values of T_{ab} formally diverge and must be regularized. One can show (Wald 1977, 1978) that there exists a regularized T_{ab} which satisfies a short list of reasonable axioms. Unfortunately, this T_{ab} is only unique up to the addition of conserved local curvature tensors. There are two conserved local curvature tensors which have the dimension

cm^{-4}. These can be obtained by varying the Lagrangians R^2 and $C_{abcd}C^{abcd}$. Since there is no way at present to fix the coefficients of these two tensors, semi-classical relativity should be regarded as a theory with two free parameters.

One solution to the semi-classical Einstein equation is flat space-time $g_{ab} = \eta_{ab}$ and the in-vacuum state of the quantum field $|\xi\rangle = |0_-\rangle$. We are interested in the question of whether this solution represents the ground state of semi-classical relativity, i.e., is it the lowest energy configuration? Unfortunately, we cannot answer this question directly. The proof that flat space-time is the lowest energy configuration in classical relativity, is not applicable to semi-classical relativity because the quantum stress energy does not satisfy the energy condition. Therefore, the existence of negative energy solutions cannot yet be ruled out. Another way to test whether the ground state of semi-classical relativity is flat space-time is to see whether this solution is stable. Since the ground state of a theory must be stable, any evidence that this solution is not stable, can be interpreted as evidence that this solution is not the ground state. We now investigate the stability of flat space-time in semi-classical relativity. For simplicity, we restrict consideration to massless, non-self interacting quantum fields.

To test stability, we linearize the semi-classical Einstein equation about the solution $g_{ab} = \eta_{ab}$, $|\xi\rangle = |0_-\rangle$:

$$\dot{G}_{ab} = \langle 0_-|\dot{T}_{ab}|0_-\rangle \tag{3.2}$$

where a dot above a tensor denotes the first order change in that tensor. If there exist perturbations satisfying (3.2) that grow exponentially in time, then flat space is unstable (at least to 1st order).

Before one can solve this equation one needs an expression for the linearized quantum stress energy $\langle 0_-|\dot{T}_{ab}|0_-\rangle$ in terms of the gravitational perturbation h_{ab} that created it. Capper, Duff, and Halpern (1974) obtained such an expression in momentum space using dimensional regularization*. But since

---

* More precisely, they obtained an expression for the Fourier transform of $\langle 0_+|\dot{T}_{ab}|0_-\rangle$ (see Section 4).

we are looking for exponentially growing solutions which don't have a Fourier transform, we need a position space expression for $\langle 0_-|\dot{T}_{ab}|0_-\rangle$. Such an expression can be obtained (up to numerical constants) in a manner which is independent of any regularization procedure. We now summarize how this is accomplished. (For details see Horowitz 1980).

Since $\langle 0_-|\dot{T}_{ab}|0_-\rangle$ is linear and continuous in h_{ab}, there must exist a Green's function $H_{ab}{}^{m'n'}(x,x')$ such that:

$$\langle 0_-|\dot{T}_{ab}(x)|0_-\rangle = \hbar \int H_{ab}{}^{m'n'}(x,x')h_{m'n'}(x')\mathrm{d}^4x' \quad . \qquad (3.3)$$

where the indices ab denote tensors at the point x and $m'n'$ denote tensors at x'.

Thus the problem of determining the linearized stress energy reduces to the problem of determining $H_{ab}{}^{m'n'}$. This Green's function must satisfy the following five properties:

G1. Poincaré invariance
G2. Symmetry in ab, and $m'n'$
G3 Divergence free on any index
G4. Support on the past lightcone of x
G5. Dimensions cm^{-8}.

The first property follows from the fact that H is independent of the perturbation h_{ab}. The second follows from the symmetry of $\langle 0_-|\dot{T}_{ab}|0_-\rangle$ and h_{ab} respectively. The third is a result of two properties of $\langle 0_-|\dot{T}_{ab}|0_-\rangle$: conservation and gauge invariance. The fourth follows from causality and the fact that we are considering massless fields. Finally, the last property follows from dimensional analysis (plus an assumption that there exist no unnecessary length scales in the problem).

Surprisingly, it turns out that these five properties determine $H_{ab}{}^{m'n'}$ up to numerical constants. It should perhaps be emphasized that we are using a Green's function here in a manner which is different from most applications in physics. Typically, one introduces a Green's function to help solve a certain linear differential equation, and this differential equation is an essential ingredient in determining this Green's function. In the present context, however, one does not have a simple differential equation relating $\langle 0_-|\dot{T}_{ab}|0_-\rangle$ to h_{ab}. One is determining the Green's function solely from the properties it must satisfy.

The first three properties above imply that the linearized stress energy can be expressed in terms of a scalar Green's function acting on the linearized form of the two conserved local curvature tensors of dimension cm^{-4}. Both of these tensors can be simply expressed in terms of the linearized Einstein tensor. Varying the Lagrangian $C_{abcd}C^{abcd}$ and linearizing off flat space-time yields

$$\dot{A}_{ab} = -2\nabla^2\dot{G}_{ab} - \frac{2}{3}\nabla_a\nabla_b\dot{G}^m{}_m + \frac{2}{3}\eta_{ab}\nabla^2\dot{G}^m{}_m . \tag{3.4}$$

Varying the Lagrangian R^2 and linearizing yields

$$\dot{B}_{ab} = 2\eta_{ab}\nabla^2\dot{G}^m{}_m - 2\nabla_a\nabla_b\dot{G}^m{}_m . \tag{3.5}$$

The remaining properties above can be used to determine the scalar Green's function. The result is (Horowitz 1980):

$$\langle 0_-|\dot{T}_{ab}(x)|0_-\rangle = \hbar\left\{\int H_\lambda(x-x')[a\dot{A}_{ab}(x') + b\dot{B}_{ab}(x')]\mathrm{d}^4x' + \alpha\dot{B}_{ab}(x)\right\} . \tag{3.6}$$

The scalar Green's function H depends on a dimensional parameter λ, and is defined as follows. Fix a point x in Minkowski space-time, and let u,v be the standard retarded and advanced null co-ordinates with origin x. For any test function f, the action of H_λ on f is defined to be:

$$\int H_\lambda(x-x')f(x')\mathrm{d}^4x' = \int_{-\infty}^{0}\mathrm{d}u\int_0^{4\pi}\mathrm{d}\Omega\left[\left.\frac{\partial f}{\partial u}\right|_{v=0}\ln(-u/\lambda) + \frac{1}{2}\left.\frac{\partial f}{\partial v}\right|_{v=0}\right] . \tag{3.7}$$

Roughly speaking, H_λ integrates the 'derivative of f' over the past light cone of x. The action of H_λ on a tensor is defined by applying H_λ to each Cartesian component of the tensor.

The formula for the linearized stress energy (3.6) involves three dimensionless parameters, a, b, and α, and one parameter λ with dimensions of cm^1. The parameters a and b have unique values which depend on the particular massless quantum field being considered. These values are independent of the regularization procedure used to define the stress energy and can be obtained from calculations by e.g. Christensen 1978. For a conformally invariant scalar field, for example, one finds that $a = \frac{1}{3840\pi^3}$, $b = 0$. The values of the parameters α and λ, on the other hand, are not independent of the regularization procedure. They are thus the two free parameters in semi-classical relativity. From (3.6) it is clear that $\langle 0_-|\dot{T}_{ab}|0_-\rangle$ has both a non-local and a local contribution. This decomposition, however, depends on the value of λ. From (3.7) we find:

$$\int H_{\lambda'}(x-x')f(x') - \int H_{\lambda}(x-x')f(x') = 4\pi \ln(\lambda/\lambda')f(x) \quad .(3.8)$$

Thus changing λ changes H_λ by a multiple of the δ-function.

We now wish to solve the linearized semi-classical Einstein equation. For definiteness we consider a massless conformally invariant scalar field. (The results for other massless fields are similar.) From equations (3.4)-(3.6) it is clear that $\dot{G}_{ab} = \langle 0_-|\dot{T}_{ab}|0_-\rangle$ is simply an equation on a conserved second rank tensor $\dot{G}_{ab}$. (If one wants to obtain the metric perturbation then one can always impose, say, the Lorentz gauge and solve the appropriate wave equation with source $\dot{G}_{ab}$.)

We begin by considering conformally flat perturbations. In this case:

$$\dot{G}_{ab} = \nabla_a\nabla_b\phi - \nabla^2\phi\eta_{ab} \tag{3.9}$$

where ϕ is a positive function on Minkowski space-time. Substituting this into (3.6) we find that the non-local term in the expression for the linearized stress energy vanishes. (This is simply a reflection of the fact that conformally flat space-times do not create conformally invariant particles (Parker 1969).) The semi-classical Einstein equation thus reduces to:

$$\nabla_a\nabla_b\phi - \nabla^2\phi\eta_{ab} = 6\alpha\hbar(\nabla_a\nabla_b\nabla^2\phi - \nabla^4\phi\eta_{ab}) \quad . \tag{3.10}$$

To test the stability of flat space-time we try a solution of the form $\phi = e^{\omega t}$. Substituting into (3.10) we find that there exist exponentially growing solutions if $\alpha < 0$. One can show (Horowitz and Wald 1978) that if $\alpha \geqslant 0$, there exist no exponentially growing conformally flat perturbations.

We now look for solutions in which $\dot{G}_{ab}$ is trace free. In this case tensor $\dot{B}_{ab}$ vanishes and the semi-classical Einstein equation becomes:

$$\dot{G}_{ab}(x) = -\frac{\hbar}{1920\pi^3}\int H_\lambda(x-x')\nabla^2\dot{G}_{ab}(x')\mathrm{d}^4x'. \tag{3.11}$$

Since we are looking for exponentially growing solutions, we try

$$\dot{G}_{ab} = L_{ab}\mathrm{e}^{\omega t} \tag{3.12}$$

where L_{ab} is constant, tracefree, and orthogonal to $\nabla^a t$ (so $\dot{G}_{ab}$ is conserved). Substituting (3.12) into (3.11) and using (3.7) to evaluate the integral yields

$$-1 = \frac{\hbar}{960\pi^2}\omega^2(\ln\lambda^2\omega^2 + 2\gamma - 1) \tag{3.13}$$

where γ is Euler's constant. There exist solutions to this equation for real ω only when $\lambda \leqslant \lambda_{\text{crit}} \equiv \mathrm{e}^{-\gamma}\sqrt{\frac{\hbar}{960\pi^2}}$. However, when λ is greater than this critical value (which is slightly less than the Planck length) one can show that there exists solutions to (3.13) for complex ω where Re $\omega > 0$. Perturbations with these frequencies would therefore oscillate with exponentially growing amplitude and are still evidence of instability. In addition to these spatially homogeneous perturbations, one can show that there even exist exponentially growing perturbations which vanish near spatial infinity.

Our conclusion, therefore, is that for any values of the free parameters α and λ, there exist exponentially growing perturbations off flat space-time in semi-classical relativity. Flat space-time is thus unstable in this theory and cannot be the ground state.

4. BEYOND SEMI-CLASSICAL RELATIVITY

The semi-classical theory we have been discussing in the previous section is really quite classical, in the sense that the objects of interest are the same as the objects of classical relativity, i.e., real four dimensional space-times, and the questions of interest are the same as in the classical theory, i.e., do there exist horizons, singularities, etc.? In order to investigate the possible significance (for quantum gravity) of the instability of flat space-time in this semi-classical theory, one must pass to a more complete quantum theory. Functional methods in general, and the effective action in particular, have been very successful in increasing our understanding of quantum field theories in flat space-time (e.g. Faddeev 1975). Recently these techniques have been applied to gravity (De Witt 1979; Fischetti, Hartle, and Hu 1979). We now use the effective action to investigate the instability of flat space-time in semi-classical relativity. We will consider a certain approximation to the effective action (which corresponds to the closest analogue in the full theory to semi-classical relativity), and test the conjecture that $\langle g_{ab}\rangle = \eta_{ab}$ in this approximation.

Clearly, the approximation to the effective action which is closest to semi-classical relativity, is one in which the entire matter contribution to the effective action is included ('quantum matter'), but only the lowest order in $\hbar$ gravitational contribution is included ('classical gravity'). In this semi-classical approximation, the effective action is:

$$\Gamma_s[g_{ab}] = S[g_{ab}] + Y[g_{ab}] \tag{4.1}$$

where g_{ab} is an asymptotically flat space-time, S is the classical gravitational action:

$$S[g_{ab}] = \frac{1}{2}\int R\sqrt{-g}\ \mathrm{d}^4x + \text{boundary terms} \tag{4.2}$$

and Y is the log of the vacuum persistence amplitude for the matter field, i.e., if $|0_-\rangle$ and $|0_+\rangle$ denote the in and out vacuum states of the matter field in this space-time, then

$$\exp(iY[g_{ab}]) = \langle 0_+ | 0_- \rangle \quad . \tag{4.3}$$

The real part of Y is formally divergent and must be regularized. In analogy with the quantum stress energy, one finds that the regularization procedure is not unique, and the resulting finite value of Y depends on two free parameters. In fact, one can show that

$$\frac{\delta Y}{\delta g^{ab}} = -\frac{1}{2}\frac{\langle 0_+ | T_{ab} | 0_- \rangle}{\langle 0_+ | 0_- \rangle} \tag{4.4}$$

so the two free parameters in the definition of Y are precisely the same as the parameters that arise in the definition of the quantum stress energy (λ and α).

The approximation we have made to the effective action was motivated by the fact that it is the closest analogue to the semi-classical theory discussed in the previous section. Unfortunately, however, it does not appear to correspond to any systematic approximation of the effective action. In particular it is not clear under what circumstances (if any) this approximation might be expected to resemble the exact result. Fortunately, it turns out that with only slight modification, our approximation can be seen to be the leading order in an expansion in terms of a small parameter. The idea is simply that the quantum effects of gravity itself become less important as one increases the number of other fields in the problem. Thus, the modification that is needed is to couple gravity to N matter fields, and the expansion is in terms of the parameter $1/N$*. After appropriately rescaling the gravitational coupling constant G (so that the $N \to \infty$ limit exists) one finds that $\Gamma_s[g_{ab}]$ (defined by (4.1)) is indeed the leading term in a $1/N$ expansion of the effective action.

To test the conjecture that $\langle g_{ab} \rangle = \eta_{ab}$, we now investigate

---

* I thank R. Wald for bringing to my attention the existence of a connection between the semi-classical approximation and the $1/N$ expansion. I also thank B. Kay for several discussions on this topic.

the poles of the lowest order graviton propagator computed off flat space-time in this approximation. In particular, we are interested in the possible existence of tachyonic poles (space-like k^2). The connection between the ground state expectation value of the field, and tachyonic poles in the propagator of the field, has been studied in model field theories. It is known, for example, that in the $O(N)$ symmetric $\lambda\phi^4$ theory (Coleman, Jackiw, and Politzer 1974; Abbott, Kang, and Schnitzer 1976) and Gross–Neveu model (Gross and Neveu 1974) (at least to leading order in $1/N$) the propagators of the theory contain tachyonic poles when computed off the 'false' ground state, but have no tachyonic poles when computed off the 'true' ground state. It is possible that a similar connection exists in quantum gravity.

Tomboulis (1977) calculated the graviton propagator off flat space-time to leading order in the $1/N$ expansion. We now present an alternative calculation which emphasizes the connection between the $1/N$ expansion and semi-classical relativity.

Since our approximation to the effective action Γ_s contains only the lowest order gravitational contribution, we can view Γ_s as the 'classical' action for a theory of gravity. The lowest order graviton propagator is then simply a Green's function for the linearized field equation which follows from this action. The field equation is:

$$\frac{\delta\Gamma_s}{\delta g^{ab}} = 0 \tag{4.5}$$

which, from equations (4.1) and (4.4) is

$$G_{ab} - \frac{\langle 0_+|T_{ab}|0_-\rangle}{\langle 0_+|0_-\rangle} = 0 \, . \tag{4.6}$$

Therefore, the field equation obtained from the leading order term in a $1/N$ expansion of the effective action differs from the semi-classical Einstein equation (3.1) only in which matrix element of the quantum stress energy is being considered. Since we want to compute the graviton propagator off flat space-time, we linearize (4.6) off the solution $g_{ab} = \eta_{ab}$:

$$\dot{G}_{ab} - \langle 0_+|\dot{T}_{ab}|0_-\rangle = 0 \tag{4.7}$$

The lowest order graviton propagator is a solution to (4.7) with a δ-function source on the right hand side. The poles of this propagator are therefore determined by solutions to the homogeneous equation. We now investigate these solutions. We again restrict consideration to a conformally invariant scalar field.

Equation (4.7) is, of course, very similar to the linearized semi-classical Einstein equation (3.2). As in that case, we need an expression for $\langle 0_+|\dot{T}_{ab}|0_-\rangle$ in terms of the metric perturbation h_{ab} (or linearized Einstein tensor $\dot{G}_{ab}$) before we can find solutions. Unfortunately we cannot obtain such an expression in the same manner as we did in the previous section. The reason is that the Green's function which yields $\langle 0_+|\dot{T}_{ab}(x)|0_-\rangle$ (unlike the one which yields $\langle 0_-|\dot{T}_{ab}(x)|0_-\rangle$) does not have support on the past lightcone of the point x, i.e., it does not satisfy property G4. Using the properties that $\langle 0_+|\dot{T}_{ab}|0_-\rangle$ does satisfy, and proceeding as far as we can, we find that:

$$\langle 0_+|\dot{T}_{ab}(x)|0_-\rangle = \hbar \int G(x-x')\alpha\dot{A}_{ab}(x')\,d^4x' + \alpha\hbar\dot{B}_{ab}(x) \tag{4.8}$$

when G is an undetermined scalar Green's function. We cannot determine which Green's function it is without a calculation.

Fortunately, we can use the fact that $\langle 0_-|\dot{T}_{ab}|0_-\rangle$ is already known to simplify the calculation required to determine $\langle 0_+|\dot{T}_{ab}|0_-\rangle$. We begin with the fact that:

$$\langle 0_+|T_{ab}|0_-\rangle = \sum_{n_-} \langle 0_+|S|n_-\rangle\langle n_-|T_{ab}|0_-\rangle \tag{4.9}$$

where S is the S-matrix and $|n_-\rangle$ is a complete set of in-states for the matter field. To first order off flat space-time:

$$\langle 0_+|\dot{T}_{ab}|0_-\rangle = \langle 0_-|\dot{T}_{ab}|0_-\rangle + \sum_{|n_-\rangle} \langle 0_+|\dot{S}|n_-\rangle\langle n_-|\mathring{T}_{ab}|0_-\rangle \tag{4.10}$$

where $\mathring{T}_{ab}$ is the quantum stress energy in flat space-time. The second term on the right is readily calculated in momentum space (Hartle and Horowitz 1981). Recall that our goal is to find the poles in the leading order graviton propagator. These poles are given by values of k^2 such that $h_{ab} = \mathring{h}_{ab}\exp(ik^m x_m)$ ($\mathring{h}_{ab}$ constant) is a solution to Eq. (4.7). Therefore, we only need to find solutions to (4.7) which have a Fourier transform, and hence a momentum space expression for $\langle 0_+|\dot{T}_{ab}|0_-\rangle$ is sufficient. (In fact, we will soon discover that every solution to (4.7) has a Fourier transform, so no solutions will be missed by going to momentum space.) From (3.6) the Fourier transform of the in-vacuum expectation value of the stress energy is:

$$\langle 0_-|\hat{\dot{T}}_{ab}|0_-\rangle = \frac{\hbar}{3840\pi^3}\hat{H}_\lambda\hat{\dot{A}}_{ab} + \alpha\hbar\hat{\dot{B}}_{ab} \tag{4.11}$$

where a hat above a tensor denotes the Fourier transform of that tensor and $\hat{H}_\lambda$ is the Fourier transform of the Green's function H_λ. One can show that

$$\hat{H}_\lambda = -2\pi[\ln\lambda^2|k^2| + 2\gamma - 1 + i\pi\theta_-(k)] \tag{4.12}$$

where γ is Euler's constant and θ_- is a step function which takes the value -1 inside the future lightcone, +1 inside the past lightcone, and zero elsewhere. From (4.8), the Fourier transform of the in-out matrix element of the stress energy is of the form:

$$\langle 0_+|\hat{\dot{T}}_{ab}|0_-\rangle = \frac{\hbar}{3840\pi^3}\hat{G}\hat{\dot{A}}_{ab} + \alpha\hbar\hat{\dot{B}}_{ab} \tag{4.13}$$

where $\hat{G}$ can be determined from (4.10). One finds that

$$\hat{G}_\lambda = -2\pi[\ln\lambda^2|k^2| + 2\gamma - 1 - i\pi\theta(k^2)] \tag{4.14}$$

where $\theta(k^2)$ is the step function that takes the value +1 inside both the future and past lightcones and zero elsewhere (and we've added a subscript λ to indicate the dependence of $\hat{G}$ on the parameter λ).

Comparing equation (4.12) and (4.14) one finds that the Green's function yielding the in-vacuum expectation value of the stress energy $\hat{H}_\lambda$ differs from the one yielding the in-out matrix element of the stress energy G_λ only in the sign of their imaginary part. This could have been expected. The sign of the imaginary part can be viewed as determining the boundary conditions for the Green's function by specifying the contour taken in the complex frequency plane to avoid the branch cut in $\ln\lambda^2 k^2$. The imaginary part $i\pi\theta_-(k)$ corresponds to the contour appropriate for the 'retarded' Green's function: $\langle 0_-|\dot{T}_{ab}|0_-\rangle$ vanishes to the past of the support of the perturbation. The imaginary part $-i\pi\theta_-(k)$ (not shown above) corresponds to the contour appropriate for the 'advanced' Green's function. This Green's function yields the out-vacuum expectation value of the stress energy $\langle 0_+|\dot{T}_{ab}|0_+\rangle$ which vanishes to the future of the perturbation. Finally $-i\pi\theta(k^2)$ corresponds to the contour appropriate for the 'Feynman' Green's function for the linearized stress energy: $\langle 0_+|\dot{T}_{ab}|0_-\rangle$ does not vanish anywhere in space-time. (This follows, for example, from the fact that the Fourier transform of $\langle 0_+|\dot{T}_{ab}|0_-\rangle$ differs from that of $\langle 0_-|\dot{T}_{ab}|0_-\rangle$ by the addition of a term which has support confined to the past lightcone of momentum space. The Fourier transform of this addition term will thus be non-zero everywhere in position space.)

The fact that the Green's function yielding $\langle 0_+|\dot{T}_{ab}|0_-\rangle$ has support everywhere in Minkowski space-time, has the following interesting consequence: there are no exponentially growing perturbations of flat space-time in this theory.

(If h_{ab} is an exponentially growing perturbation, then $\langle 0_+|\dot{T}_{ab}|0_-\rangle$ is infinite at every point in space-time and thus cannot be equal to $\dot{G}_{ab}$.) Therefore if one takes the view that (4.6) is the equation for a classical gravitational field, then flat space-time satisfies the classical tests of stability. However, this is not the view adopted here. Instead we take the view that (4.6) arises from an approximation to a full quantum theory, and therefore one should investigate quantum mechanical tests of stability, e.g., the existence of poles in the graviton propagator computed off flat space-time.

Since all solutions to (4.7) have a Fourier transform, a position space expression for the Green's function G_λ is not needed to find solutions. This is fortunate since it is unlikely that there exists a simple expression for G_λ, analogous to (3.7) for H_λ. Recall that the Feynman Green's function for even the massless wave equation is rather complicated.

We are finally in a position to answer the question of whether there exist tachyonic solutions to (4.7). Taking the Fourier transform of this equation (and using (4.8) and (4.14)) we obtain:

$$\hat{\dot{G}}_{ab} + \frac{\hbar}{1920\pi^2}[\ln\lambda^2|k^2| + 2\gamma - 1 - i\pi\theta(k^2)]\hat{\dot{A}}_{ab} - \alpha\hbar\hat{\dot{B}}_{ab} = 0 \, . \tag{4.15}$$

We again solve (4.15) for $\hat{\dot{G}}_{ab}$ (rather than $\hat{h}_{ab}$) which eliminates any question about the gauge dependence of solutions. Since $\dot{G}_{ab}$ is conserved, $\hat{\dot{G}}_{ab}$ is orthogonal to k^a. Any solution $\hat{\dot{G}}_{ab}$ can be decomposed into a pure trace and a tracefree part. We first consider pure trace solutions:

$$\hat{\dot{G}}_{ab} = \frac{1}{3}\,\hat{\dot{G}}^m{}_m\left(\eta_{ab} - \frac{k_a k_b}{k^2}\right) \, . \tag{4.16}$$

(These are just the conformally flat perturbations.) Substituting (4.16) into (4.15) and using (3.4) and (3.5) we obtain:

$$\left(1 + 6\alpha\hbar k^2\right)\left(\eta_{ab} - \frac{k_a k_b}{k^2}\right)\hat{\dot{G}}^m{}_m = 0 \, . \tag{4.17}$$

Therefore, if the free parameter α is less than zero, then there exist solutions with space-like k^a, i.e., for $\alpha < 0$ there exist tachyonic poles in the graviton propagator. If $\alpha \geqslant 0$. then there exist no conformally flat tachyonic solutions.

We now consider the tracefree solutions to (4.15). Substituting

$$\hat{\dot{G}}^m{}_m = 0 \tag{4.18}$$

into (4.15) yields

$$\left\{1 + \frac{\hbar k^2}{960\pi^2}[(\ln\lambda^2|k^2| + 2\gamma - 1 - i\pi\theta(k^2))]\right\}\hat{\dot{G}}_{ab} = 0 \ . \quad (4.19)$$

There exist solutions whenever the quantity in brackets vanishes. The presence of the step function $\theta(k^2)$ implies that there cannot exist solutions for time-like k^a. For space-like k^a (4.19) is equivalent to (3.13) governing the frequency for exponentially growing solutions to the linearized semi-classical Einstein equation. Thus, there exist tachyonic solutions if and only if the free parameter λ is less than or equal to the critical value $\lambda_{crit} \equiv e^{-\gamma}\sqrt{\frac{\hbar}{960\pi^2}}$. Of course, in analogy with the complex frequency solutions to (3.13) we find that when $\lambda > \lambda_{crit}$ there exist complex values of k^2 at which the quantity in brackets vanishes. It is not clear, however, what the significance of these complex values of k^2 are. On the one hand, one knows that $\dot{h}_{ab}\exp(ik^m x_m)$ for complex k^m *cannot* be a solution to (4.7), since this perturbation grows exponentially in some direction and thus $\langle 0_+|\dot{T}_{ab}|0_-\rangle$ will be infinite. On the other hand, one can always consider the expression for the graviton propagator in momentum space and investigate its analyticity properties for complex k^2. One then finds poles for complex k^2 (whenever $\lambda > \lambda_{crit}$).

To summarize, we have found that if $\alpha < 0$ or $\lambda \leqslant \lambda_{crit}$, then there exist tachyonic poles in the leading order $1/N$ expansion of the graviton propagator computed off flat space-time. This can be interpreted as further evidence that, for this range of the parameters, $\langle g_{ab}\rangle \neq \eta_{ab}$. If, however, $\alpha \geqslant 0$ and $\lambda > \lambda_{crit}$, then there are no tachyonic poles. There are, however, poles for complex k^2. The significance of these poles is not yet clear. There are indications, however, that a meaningful quantum theory with $\langle g_{ab}\rangle = \eta_{ab}$ and a unitary S matrix can exist in this case (Lee and Wick 1970).

5. A COMPARISON: MASSLESS Q.E.D.

We conclude by comparing our previous analysis of the ground state of quantum gravity with a similar analysis of quantum electrodynamics for massless fermions.

We begin this comparison by recalling the ground state of

the free electromagnetic field. The configuration space consists of all divergence-free vector fields B^α on a space-like plane in Minkowski space-time. The ground state wavefunction on this configuration space is given by:

$$\psi_0(B) = N\left\{\exp - \frac{1}{4\pi^2\hbar} \int d^3x \int d^3x' \frac{B^\alpha(x)B_\alpha(x')}{|x^\alpha - x'^\alpha|^2}\right\} \tag{5.1}$$

where N is a normalization constant. Notice that this formula is very similar to the ground state wave function for the linearized theory of gravity (2.3). In particular, the ground state expectation value of B^α vanishes.

We next consider the massless Q.E.D.-analogue of semi-classical relativity. This theory, semi-classical electrodynamics, consists of a classical Maxwell field F_{ab} interacting with a (massless) quantum Dirac field ψ via

$$\nabla_a F^{ab} = \langle\xi|J^b|\xi\rangle \tag{5.2}$$

where J^a is the charge current operator for ψ, and $|\xi\rangle$ is an in-state for ψ. (One also has the usual field equation for ψ in the presence of the external field F_{ab}, but this equation will not be needed in the following discussion.)

One solution to (5.2) is $F_{ab} = 0$, $|\xi\rangle = |0_-\rangle$. This solution can be viewed as the 'vacuum state' of semi-classical electrodynamics. We now ask whether this 'vacuum state' is the ground state. As in the case of semi-classical relativity, we attempt to answer this question by testing the stability of this state. Linearizing (5.2) about its vacuum solution yields

$$\nabla_a \dot{F}^{ab} = \langle 0_-|\dot{J}^b|0_-\rangle \ . \tag{5.3}$$

We now look for exponentially growing solutions to this equation.

We are in the (by now familiar) position of needing an expression for the in-vacuum expectation value of the linearized current $\langle 0_-|\dot{J}^b|0_-\rangle$ in terms of the classical Maxwell perturbation $\dot{F}_{ab}$, in order to solve (5.3). Once again, we use the technique of introducing a Green's function for the linearized current, and determining this Green's function by its general properties. Since $\langle 0_-|\dot{J}^a|0_-\rangle$ depends linearly and continuously on $\dot{F}^{ab}$, there exists a Green's function $G^a{}_{b'c'}(x,x')$ such that:

$$\langle 0_-|\dot{J}^a(x)|0_-\rangle = \frac{e^2}{\hbar}\int G^a{}_{b'c'}(x,x')\dot{F}^{b'c'}(x')\mathrm{d}^4x' . \qquad (5.4)$$

The Green's function $G^a{}_{b'c'}$ must satisfy the following five properties (cf. G1—G5):

E1. Poincaré invariance

E2. Anti-symmetry in b' and c'

E3. Divergence free on the index 'a'

E4. Support on the past lightcone of x

E5. Dimensions cm^{-5}

In close analogy to the gravitational case, one can show that these five properties uniquely determine $G^a{}_{b'c'}$ up to a numerical constant K. One finds:

$$\langle 0_-|\dot{J}^b(x)|0_-\rangle = \frac{e^2K}{\hbar}\int H_\lambda(x - x')\nabla_a\dot{F}^{ab}(x')\mathrm{d}^4x' \qquad (5.5)$$

where H_λ is precisely the *same* Green's function which appeared in the formula for the in-vacuum expectation value of the quantum stress energy (3.7). The value of the constant K can be determined from calculations by, e.g. Schwinger (1951) and turns out to be $K = -\frac{1}{48\pi^3}$. Substituting (5.5) into (5.3) we obtain:

$$\nabla_a\dot{F}^{ab}(x) = -\frac{e^2}{48\pi^3\hbar}\int H_\lambda(x - x')\nabla_a\dot{F}^{ab}(x')\mathrm{d}^4x' . \qquad (5.6)$$

Recall (from (3.8)) that changing the dimensional constant λ changes the Green's function H_λ by a multiple of the δ-function. Therefore, unlike the gravitational case, λ is not a free parameter in semi-classical electrodynamics. A change in λ is equivalent to a change in the coupling constant e^2. Thus, already at this semi-classical level, one sees evidence for the renormalizability of Q.E.D. In fact, from (3.8) and (5.6) one easily derives the 'renormalization group' equation for how e^2 changes when the 'renormalization point' λ is varied:

$$e^2(\lambda') = \frac{e^2(\lambda)}{1 + \frac{e^2(\lambda)}{12\pi^2}\ln(\lambda'/\lambda)} . \tag{5.7}$$

In other words, (5.6) is unchanged when its parameters $(e^2(\lambda),\lambda)$ are replaced with $(e^2(\lambda'),\lambda')$ provided that $e^2(\lambda')$ is related to $e^2(\lambda)$ by (5.7) (and $\lambda' > \lambda \exp\left\{-\frac{12\pi^2}{e^2(\lambda)}\right\}$, so that e^2 remains positive). The fact that $e^2 \to 0$ as $\lambda' \to \infty$ can be interpreted as a reflection of the well known screening of electric charge.

We now solve (5.6). Since we are testing for instability, we try a solution of the form

$$\dot{F}_{ab} = V_{[a}t_{b]}e^{\omega t} \tag{5.8}$$

where V_a is a constant vector field and $t_b = \nabla_b t$. (Notice that $\dot{F}_{ab}$ is curl-free.) Substituting (5.8) into (5.6) yields the following algebraic equation on ω:

$$1 = \frac{e^2}{24\pi^2\hbar}(\ln\lambda^2\omega^2 + 2\gamma - 1) \tag{5.9}$$

where γ is Euler's constant. This equation has the following real solution for ω:

$$\omega_0 = \frac{1}{\lambda}\exp\left(\frac{12\pi^2\hbar}{e^2} + \frac{1}{2} - \gamma\right) \tag{5.10}$$

Therefore, in analogy with semi-classical relativity, the vacuum solution $F_{ab} = 0$, $|\xi\rangle = |0_-\rangle$ of semi-classical electrodynamics is unstable.* Notice that the frequency for these unstable modes ω_0 is invariant under simultaneously changing λ and $e^2(\lambda)$ according to (5.7) (i.e., ω_0 is a 'renormalization group invariant').

The instability of the vacuum solution in semi-classical electrodynamics helps to clarify the analogous instability of

---

* This instability has also been noticed by R. Wald (private communication).

flat space-time in semi-classical relativity. The view has sometimes been expressed that the instability of flat space-time in semi-classical relativity is a result of the presence of fourth order derivatives in the semi-classical Einstein equation. Notice, however, that the equation for semi-classical electrodynamics does not contain fourth order derivatives and yet its vacuum solution is still unstable. It appears that the instabilities in these semi-classical theories are more a result of the non-local nature of the equations of motion (leading to small effects building up over a long time) rather than the presence of higher order derivatives.

It is clear from (5.10) that the exponentially growing solutions we have found are not analytic in the coupling constant e . Similarly, one can show that the exponentially growing solutions to the linearized semi-classical Einstein equation are not analytic in the gravitational coupling constant G. These instabilities would therefore not have been found by solving the semi-classical equations order by order in powers of the coupling constant. It is only when the effects of the coupling constant to all orders is included that the instability appears.

Recall that in the case of gravity, we found that the semi-classical approximation to the effective action was equivalent to the first term in a $1/N$ expansion of this effective action where N is the number of matter fields coupled to gravity. Similarly, one can show that the semi-classical approximation to the effective action for massless Q.E.D. is equivalent to the first term in a $1/N$ expansion where N is the number of fermions coupled to the electromagnetic field. We now ask whether there exist tachyonic poles in the photon propagator computed to lowest order in this $1/N$ expansion.

In analogy with the gravitational case, the poles of this photon propagator are determined by the solutions to (*cf.*(4.7))

$$\nabla_a \dot{F}^{ab} - \langle 0_+ | \dot{J}^b | 0_- \rangle = 0 . \tag{5.11}$$

Taking the Fourier transform of this equation and using equations (4.14) and (5.5) one finds (*cf.* (4.15)):

$$k_a \hat{F}^{ab}\left\{1 - \frac{e^2}{24\pi^2\hbar}[\ln\lambda^2|k^2| + 2\gamma - 1 - i\pi\theta(k^2)]\right\} = 0 \ . \quad (5.12)$$

The poles of the photon propagator are the values of k^2 for which the expression in brackets vanishes. For any value of e^2, there exists a positive value of k^2 (space-like k^a) for which this expression vanishes. Therefore, there exists a tachyonic pole in the photon propagator computed to leading order in $1/N$.

In conclusion, we have found that massless Q.E.D. is similar to gravity in that the 'vacuum' solution to the semi-classical equation is unstable, for all values of the parameters in the theory. It is different from gravity, however, with regard to the existence of tachyonic poles in the propagator (computed off the vacuum solution to leading order in the $1/N$ expansion). For gravity, there exist tachyonic poles only for a certain range of the parameters of the theory ($\lambda \leqslant \lambda_{\text{crit}}$ or $\alpha < 0$) whereas for massless Q.E.D., there exists a tachyonic pole for every value of the parameter in the theory. Landau (1955) has shown that this tachyonic pole remains even if the fermions are given a mass. Furthermore, in Q.E.D. one does not have the possibility that this tachyonic pole is a result of computing the propagator off the 'wrong' ground state. The requirement that the ground state be Poincaré invariant implies $\langle F_{ab}\rangle = 0$. Therefore it appears that either the lowest order term in a $1/N$ expansion of Q.E.D. with N fermions is not reliable (higher order corrections remove the tachyonic pole) or Q.E.D. is not a well-defined quantum theory.

Somewhat surprisingly, quantum gravity is slightly better behaved in this approximation, since there exists no tachyonic poles in the graviton propagator for certain values of the parameters λ and α. Of course one must keep in mind the fact that the parameters λ and α only arise because quantum gravity is not renormalizable. Thus one might expect that the inclusion of terms which are higher order in powers of $1/N$ will introduce new free parameters and this theory will lose all predictive power.

There are indications, however, that this is not the case

(Tomboulis 1977). Thus, one has the possibility of obtaining a quantum theory of gravity coupled to matter, defined perturbatively in $1/N$, with only two free parameters (in addition to Newton's constant G).

In terms of the ground state of quantum gravity, the results discussed here (instability of flat space-time in semi-classical relativity, tachyonic poles in the graviton propagator for certain λ, α, etc) are, at best, merely suggestive. Further investigation is clearly needed before a more complete understanding of the situation is achieved.

ACKNOWLEDGEMENTS

It is a pleasure to thank James Hartle, Roman Jackiw, Bernard Kay, and Robert Wald for interesting discussions.

The author would also like to acknowledge the support of the National Science Foundation for part of this work.

REFERENCES

Abbott, L.F., Kang, J.S., and Schnitzer, H.J. (1976). *Phys. Rev.* **D13**, 2212.
Arnowitt, R., Deser, S., and Misner, C.W. (1961). *Phys. Rev.* **122**. 997.
Bergmann, P.G. and Flaherty, E.J. (1977). *J. Math. Phys.* **19**, 212.
Capper, D.M., Duff, M.J., and Halpern, L. (1974). *Phys. Rev.* **D10**, 461.
Christensen, S.M. (1978). *Phys. Rev.* **D17**, 946.
Coleman, S. and Weinberg, E. (1973). *Phys. Rev.* **D7**, 1888.
——, Jackiw, R., and Politzer, H.D. (1974). *Phys. Rev.* **D10**, 2491.
De Witt, B. (1979). In *General Relativity: An Einstein Centenary Survey* (ed S.W. Hawking and W. Israel). Cambridge.
Faddeev, L.D. (1975). In *Methods in Field Theory*. Les Houches session 28.
Fischetti, M.V., Hartle, J.B., and Hu, B.L. (1979). *Phys. Rev.* **D20**, 1757.
Geroch, R. (1977). In *Asymptotic Structure of Space-time* (ed F.P. Esposito and L. Witten). New York, Plenum.
Gross, D.J. and Neveu, A. (1974). *Phys. Rev.* **D10**, 3235.
Hartle, J.B. and Horowitz, G.T. (1981). *Phys. Rev.* **D**, to appear.
Horowitz, G.T. (1980). *Phys. Rev.* **D21**, 1445.
——and Wald, R.M. (1978). *Phys. Rev.* **D17**, 414.
Jackiw, R. (1978). *Phys. Rev. Lett.* **41**, 1635.
Kuchař, K. (1970). *J. Math. Phys.* **11**, 3322.
Landau, L.D. (1955). In *Niels Bohr and the Development of Physics* (ed W. Pauli). New York, McGraw-Hill.
Lee, T.D. and Wick, G.C. (1970). *Phys. Rev.* **D2**, 1033.
Ovrut, B.A. (1979). *Phys. Rev.* **20**, 1446.
Pagels, H. and Tomboulis, E. (1978). *Nucl. Phys.* **B143**, 485.
Parker, L. (1969). *Phys. Rev.* **183**, 1057.
Schoen, R. and Yau, S.-T. (1979). *Phys. Rev. Lett.* **43**, 1457.
Schwinger, J. (1951). *Phys. Rev.* **82**, 664.

Tomboulis, E. (1977). *Phys. Lett.* **70B**, 361.
Wald, R.M. (1977). *Commun. Math. Phys.* **54**, 1.
——(1978). *Phys. Rev.* **D17**, 1477.

5

PARTICLE AND NON-PARTICLE STATES IN QUANTUM FIELD THEORY

D. Deutsch

Department of Astrophysics, Oxford University

1. PARTICLES FROM FIELDS

The most elementary, but in a way the most surprising success of quantum field theory is its ability to describe *particles*. The particulate properties of matter in quantum field theory are neither an approximation (like 'geometrical optics' in classical field theory) nor an external constraint imposed for empirical reasons, but seem to be consequent upon the process of quantization itself. The non-commuting algebra of the field operators certainly *allows* the construction of a space of physical state vectors, spanned by a basis of 'many-particle states'. Their properties entirely justify the name. In particular, an observable $\hat{N}$, the 'particle number', then exists which has only non-negative integer eigenvalues, and is diagonal in this basis. There is a unique no-particle state, the 'vacuum'. Moreover, important quantum numbers often correspond to operators approximately diagonal in this same basis, in which case the particles may justifiably be said to 'possess values' of those numbers. This is all true in cases where interactions between the (putative) particles may be neglected or treated in perturbation theory. Where interactions are too strong, or where the topology of space-time is not trivial (see Ashtekhar 1978) or in space-times with horizons (Fulling 1979) it is known that the ordinary notion of 'particle' may break down, but these are separate effects from those discussed in this paper, which may even occur in free field theory in flat space-time. Specifically, we shall find that field theory generically permits other states in addition to the many-particle states, and I shall argue that they exist in reality.

2. QUANTIZATION

Let us now go back to first principles and see exactly how the nature and properties of states are determined during the construction of a quantum theory.

For most physical systems, no other way of obtaining the quantum theory is known than the 'quantization' of an associated classical theory — usually, but not necessarily, *the* classical theory of that system. A set of 'quantization rules' provides the mapping

CLASSICAL THEORY	—— quantization rule ——→	QUANTUM THEORY

Logically, a quantization rule has a status similar to that of a physical theory, and may be tested by testing the quantum theories derived from it. Thus, general quantization rules are always chosen so as to give the 'right' answer for systems with finite numbers of degrees of freedom; that is, they reproduce the well known and highly successful quantum theories of such systems.

This constraint is not, unfortunately, sufficient for quantum field theories. For definiteness, let us consider a non-interacting real scalar field $\phi(x)$ propagating on a fixed classical space-time background $\mathcal{M}$. Suppose for the moment that $\mathcal{M}$ is globally hyperbolic. There are several classical theories for $\phi(x)$, useful in different contexts, the most important of which are all described by the stationary action principle

$$\frac{\delta S[\phi(x)]}{\delta\phi} = 0 \qquad (2.1)$$

where

$$S[\phi(x)] = -\frac{1}{2}\int_{\mathcal{M}} d^4x\, g^{\frac{1}{2}}[\phi^{;\mu}(x)\phi_{;\mu}(x) + (\xi R + m^2)\phi^2(x)] \qquad (2.2)$$

Here m is the mass of a field quantum. When $m = 0$ and $\xi = \frac{1}{6}$, $\phi(x)$ transforms as a density under conformal transformations of the background metric. When $\xi = 0$, ϕ is said to be 'minimally coupled' to the space-time geometry. The quantization of the theory (2.2) would not be fundamentally different if ϕ were a higher-spin, weakly interacting field. For simplicity of exposition it seems best to use the following abbreviated 'canonical' scheme to quantize the theory of $\phi(x)$. Let $\mathscr{M}$ be foliated by a family of space-like hypersurfaces $\mathscr{S}_t$, whose unit future-pointing normal vector field is

$$n^{\mu}(x) \equiv n^{\mu}(t,\underline{x}) \equiv \left\|\frac{\partial}{\partial t}\right\|^{-1}\frac{\partial}{\partial t} \,. \tag{2.3}$$

t is a parameter labelling the hypersurfaces and the $\underline{x}$ label points in $\mathscr{S}_t$. Then the quantum operator version of ϕ satisfies the following 'equal time' commutation relations

$$\begin{aligned} [\hat{\phi}(t,\underline{x}),\ n'.\nabla'\hat{\phi}(t',\underline{x}')]_{t'=t} &= i\delta(\underline{x},\underline{x}')\hat{1} \\ [\hat{\phi}(t,\underline{x}),\ \hat{\phi}(t,\underline{x}')] &= 0 \\ [n.\nabla\hat{\phi}(t,\underline{x}),\ n'.\nabla'\hat{\phi}(t',\underline{x}')]_{t'=t} &= 0 \,. \end{aligned} \tag{2.4}$$

These relations constitute one of the three legs upon which quantum theories conventionally rest, namely the algebra of the $\hat{\phi}$ operators.

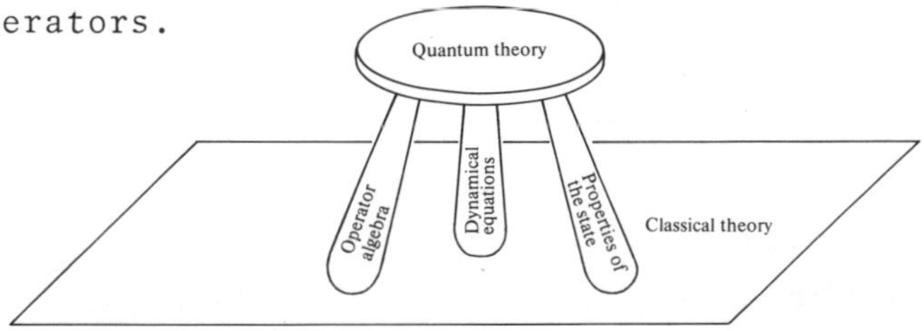

It does not matter much whether we obtained (2.4) by reciting the mystic incantations of canonical quantization (see e.g. Lurié 1968).

$$[\hat{\phi}(t,\underline{x}),\ \hat{\pi}(t,\underline{x})] = i\delta(\underline{x},\underline{x}')\hat{1}$$

or covariant quantization (see De Witt 1965)

$$[\hat{\phi}(x),\ \hat{\phi}(x')] = i\tilde{G}(x,x')$$

or otherwise. The second 'leg', the dynamical equations for $\hat{\phi}(x)$, ensures that (2.4) holds on every space-like hypersurface if it holds on any one, and this is a sufficient condition for (2.4) to be a consistent quantization rule. The dynamical equations themselves result from the quantum analogue of the stationary action principle (2.1)

$$\frac{\delta\hat{S}}{\delta\phi(\underline{x})} = [\Box - (\xi R + m^2)]\hat{\phi}(x) = 0 \tag{2.5}$$

For a system with a finite number of degrees of freedom (e.g. if there were finitely many points $\underline{x}$) these first two 'legs' would in fact express the entire quantum theory. The third leg, the physical states, could be constructed more or less unambiguously as rays in the vector space in which the $\hat{\phi}$ are operators. Let us try to do this for our field theory.

3. STATES; CONVENTIONAL CONSTRUCTION OF

First we must solve (formally!) (2.4) and (2.5). Because of the Cauchy completeness of $\mathscr{M}$ it is possible, and very convenient, to identify any solution $\Phi(x)$ of the classical dynamical equation (2.1) with its Cauchy data on some $\mathscr{S}_t$

$$\Phi(x) \Longleftrightarrow \{\Phi(t,\underline{x}),\ n.\nabla\Phi(t,\underline{x})\} \tag{3.1}$$

Let $\{U_k(x)\}$ be any complete orthonormal basis for real functions on $\mathscr{S}_t$, in the sense

$$\sum_k U_k(\underline{x})U_k(\underline{x}') = \delta(\underline{x},\underline{x}')$$

$$\int d\Sigma\ U_k(\underline{x})U_{k_1}(\underline{x}) = \delta_{kk_1} \tag{3.2}$$

where $d\Sigma$ is the covariant volume element on $\mathscr{S}_t$, k is a generalized index (discrete and/or continuous) and $\sum_k$ the corresponding summation/integration. A complete set of solutions, $\Phi_k{}^{(+)}(x)$ and $\Phi_k{}^{(-)}(x)$ of (2.1) on $\mathscr{M}$ corresponds, via Cauchy development, to the union of two sets of such Cauchy data on some $\mathscr{S}_t$, for example

$$\Phi_k^{(+)}(x) \Longleftrightarrow \{U_k(\underline{x}),\ \sum_{k_1} M_{kk_1}^{(+)}U_{k_1}(\underline{x})\}$$

$$\Phi_k^{(-)}(x) \Longleftrightarrow \{U_k(\underline{x}),\ \sum_{k_1} M_{kk_1}^{(-)}U_{k_1}(\underline{x})\} \tag{3.3}$$

with the $M^{(\pm)}_{kk_1}$ real, non-singular and distinct but otherwise arbitrary. The quantum dynamical equations (2.5) differ from (2.1) only in that their solutions are to be complex operators rather than real c-numbers. Accordingly, (3.3) may be used to construct the general solution of (2.5). Allowing M_{kk_1} now to be complex, with $M^{(-)}_{kk_1} = M^{(+)*}_{kk_1}$ and performing an orthonormalization on (3.3) we obtain a set of complex solutions of (2.1), $u_k(x)$ and $u_k{}^*(x)$,

$$u_k(x) \Longleftrightarrow \sum_{k_2} [\mathrm{i}(M^\dagger - M)]^{-\frac{1}{2}}_{kk_2}\{U_{k_2}(\underline{x}), \sum_{k_1} M_{k_2k_1}U_{k_1}(\underline{x})\} \tag{3.4}$$

which are complete and orthonormal according to

$$\mathrm{i}\sum_k [u^*_k(x)n'.\nabla' u_k(x') - n.\nabla u^*_k(x)u_k(x')]_{x,x'\in\mathscr{S}_t} = \delta(\underline{x},\underline{x}')$$

$$\mathrm{i}\sum_k [u_k(x)n'.\nabla' u_k(x') - n.\nabla u_k(x)u_k(x')]_{x,x'\in\mathscr{S}_t} = 0$$

$$\mathrm{i}\int_{\mathscr{S}_t} \mathrm{d}\Sigma\,[u_k(x)n.\nabla u_{k_1}(x) - u_{k_1}(x)n.\nabla u^*_k(x)] = \delta_{kk_1}$$

$$\mathrm{i}\int_{\mathscr{S}_t} \mathrm{d}\Sigma\,[u_k(x)n.\nabla u_{k_1}(x) - u_{k_1}(x)n.\nabla u_k(x)] = 0$$

Then the general Hermitian solution of (2.5) must be

$$\hat{\phi}(x) = \sum_k [\hat{a}_k u_k(x) + \hat{a}^*_k u^*_k(x)] \tag{3.6}$$

where the $\hat{a}_k$ are fixed operators chosen so as to satisfy the commutation relations (2.4). These amount to

$$[\hat{a}_k, \hat{a}^*_{k_1}] = \delta_{kk_1}\hat{1}$$

$$[\hat{a}_k, \hat{a}_{k_1}] = 0 \tag{3.7}$$

The quantity $M^{(\pm)}_{kk_1}$ determines a so-called 'complex structure' on the space of real solutions of (2.1). It is to be thought of as a geometrical object on that space. It is convenient to consider it both in the 'k-representation', as M_{kk_1} and in the '$\underline{x}$-representation', as $\vec{M}$, a linear integral operator on $\mathscr{S}_t$

$$M_{kk_1} = \int U_{k_1}(\underline{x})\vec{M}U_k(\underline{x})\,\mathrm{d}\Sigma$$

$$\vec{M}\mathrm{f}(x) = \int \mathrm{d}\Sigma \sum_{kk_1} U_k(\underline{x})M_{kk_1}U_{k_1}(\underline{x}_1)\mathrm{f}(\underline{x}_1) \ . \tag{3.8}$$

I should like to stress, since the reader may have lost track of the fact in the above welter of formalism, that the choice of 'complex structure' is of no significance in (3.6). No generality is lost by choosing a particular $\vec{M}$. The expansion of $\hat{\phi}(x)$ in the form (3.6) is thus valid regardless of the choice of representations for the $\hat{a}_k$ and $\hat{a}_k^*$, to which we now turn. The construction of these representations is tantamount to the construction of all the physical states and their properties, the 'third leg' of the quantum theory. The conventional procedure may be expressed as follows:
Consider the operator

$$\hat{N} = \sum_k \hat{a}_k^*\hat{a}_k \ . \tag{3.9}$$

(3.7) implies that if $|\psi\rangle$ is an eigenstate of $\hat{N}$, then $\hat{a}_k^*|\psi\rangle$ and $\hat{a}_k|\psi\rangle$ either vanish or are also eigenstates of $\hat{N}$ with eigenvalues respectively one greater and one less than for $|\psi\rangle$. Thus the $\hat{a}_k$ and $\hat{a}_k^*$ are ladder operators for the eigenstates of $\hat{N}$, and there exists a state $|\mathrm{vac}\rangle$, the vacuum, annihilated by all the $\hat{a}_k$ (since $\hat{N}$ is non-negative definite). The

$$|n_1,\ldots,n_k,\ldots\rangle \propto \prod_k \left(\hat{a}_k^*\right)^{n_k}|\mathrm{vac}\rangle \tag{3.10}$$

have all the properties of the 'many-particle states' mentioned above, with $\hat{N}$ as their particle-number operator. If there is only one vacuum state (let us in fact assume this from now on) then the states (3.10) must be precisely the eigenstates of $\hat{N}$ and hence form a complete basis for a space of physical states, *Fock space* $\mathcal{F}_a$. In this basis the a_k^* are represented as

$$\langle n_1\ldots n_k\ldots|\hat{a}_k^*|m_1\ldots m_k\ldots\rangle = \delta_{n_1m_1+1}\ldots\delta_{n_km_k+1}\ldots \tag{3.11}$$

Physically significant quantities are, of course, covariant under arbitrary changes of basis in $\mathcal{F}_a$; that is, they are all tensorial objects on $\mathcal{F}_a$ *qua* Hibert space. However the nomenclature 'vacuum state', 'one-particle state' etc, is not. Since this nomenclature begs some of the questions we shall wish to answer, it will be convenient to distinguish between $\mathcal{F}_a$ *including* some given many-particle structure, which we call 'the Fock space $\mathcal{F}_a$', and $\mathcal{F}_a$ just as a Hibert space of undistinguished states, which we shall call 'the Fock space *sector* $\mathcal{F}_a$'. A sector may also be thought of as an equivalence class of Fock spaces related by arbitrary unitary transformations.

Thus the many-particle states, and only they, appear to emerge without further postulates from the operator algebra, as determined by the quantization rules. But the above argument includes a hidden assumption, whose falsity would allow the existence of further states — our non-particle states. The assumption is an apparent technicality: the convergence of the sum (3.9) in the definition of $\hat{N}$. Note for the moment only that if it diverged and $\hat{N}$ did not exist, the above formal argument demonstrating the existence and exhaustiveness of many-particle states would be invalid.

4. NON-PARTICLE STATES

In order to motivate more physically the postulate that $\hat{N}$ should not exist for real-life field theories, and to make contact with other work in this area, let us creep up on the problem of the existence of (3.9) from another direction.

I stated earlier that the choice of complex structure $\vec{M}$ was without physical content in the field expansion (3.6). The *only* property acquired from the mode functions by the quantum field $\hat{\phi}(x)$ is its obedience to the dynamical equations, which does not depend on $\vec{M}$. This desirable covariance is unfortunately not preserved by the conventional construction of states (3.9) (3.10). For if we had chosen a different basis, say

$$v_k(x) = \sum_{k_1} [\alpha_{kk_1} u_{k_1}(x) + \beta^*_{kk_1} u^*_{k_1}(x)] \tag{4.1}$$

where for orthonormality the Bogoluibov (De Witt 1975)

coefficients α_{kk_1} and β_{kk_1} obey

$$\sum_{k_1} [\alpha_{kk_1}\alpha^*_{k_2k_1} - \beta_{kk_1}\beta^*_{k_2k_1}] = \delta_{kk_2}$$

$$\sum_{k_1} [\alpha_{kk_1}\beta_{k_2k_1} - \beta_{kk_1}\alpha_{k_2k_1}] = 0 \tag{4.2}$$

and had made the equally valid expansion

$$\hat{\Phi}(x) = \sum_k [\hat{b}_k v_k(x) + \hat{b}^*_k v^*_k(x)] , \tag{4.3}$$

then the conventional procedure would have given us many-particle states based on a different state as vacuum, say $|b,\text{vac}\rangle$ (call the other one $|a,\text{vac}\rangle$), and related by the $\hat{b}_k$ and $\hat{b}^*_k$ as ladder operators. Covariance of the quantum theory with respect to the choice of modes would imply that each of the b-states could be identified as a linear superposition of a-states in such a way that the vector space structure of the sector $\mathscr{F}_a$ (but not necessarily the particle nomenclature) be preserved. Then a change in the complex structure would give rise to a mere change in nomenclature for the physical states, but leave their properties unaffected. The two Fock spaces $\mathscr{F}_a$ and $\mathscr{F}_b$ would then be related by a unitary transformation, and the a- and b-quantum theories '*unitarily equivalent*'. With respect to the b-nomenclature, $|a,\text{vac}\rangle$ would just be some linear superposition of many-b-particle states. It is easy to calculate just 'how many' (on average):

$$\langle a;\text{vac}|\hat{N}_b|a;\text{vac}\rangle = \langle a,\text{vac}|\sum_k \hat{b}^*_k\hat{b}_k|a,\text{vac}\rangle \tag{4.4}$$

To evaluate this, note that the identity of (3.6) and (4.3) yields the following connecting relation between the $\hat{a}$s and the $\hat{b}$s

$$\hat{b}_k = \sum_{k_1} [\alpha^*_{kk_1}\hat{a}_{k_1} - \beta_{kk_1}\hat{a}^*_{k_1}] , \tag{4.5}$$

whence

$$\langle a,\text{vac}|\hat{N}_b|a,\text{vac}\rangle = \sum_{kk_1} \beta_{kk_1}\beta^*_{kk_1} , \tag{4.6}$$

which brings us back to convergence; for there is no reason

why the Bogoliubov coefficients defined by an arbitrary pair of sets of modes should obey the rather stringent spectral condition necessary for the convergence of (4.6). And if (4.6) diverges, then necessarily the state $|a,\text{vac}\rangle$ can be found nowhere in $\mathscr{F}_b$; moreover if $\hat{N}_b$ existed then, from (4.5), $|a,\text{vac}\rangle$ could not be a physical state at all. Thus if the formalism of quantum theory is to be covariant with respect to the choice of mode functions — and surely it was morally wrong to have made the expansion (3.6) if we thought it would not be — then the space of physical states cannot be a Fock space — in other words, not all the states are particle states. The simplest structure it could have is that of a union of all the possible unitarily inequivalent Fock spaces (i.e. the direct sum of all possible Fock space sectors). That this is a disjoint union (i.e. that separate sectors are distinguishable) follows from the fact that 'unitary equivalence' is an equivalence relation.

What does it mean, physically, to say that the particle number $\hat{N}_a$ is 'not an observable'? It means that any apparatus which under the single-Fock-space hypothesis would measure $\hat{N}_a$ will, according to a multiple-sector theory, measure something else. For example, it might measure

$$\hat{P}_a \hat{N}_a \hat{P}_a \ , \tag{4.7}$$

the restriction of $\hat{N}_a$ to the 'a'-sector, where

$$\hat{P}_a = \sum_{n_1, n_2 \dots} |a; n_1, n_2 \dots\rangle \langle a; n_1, n_2 \dots|$$

In fact, the physical significance of globally constructed formal operators such as $\hat{N}$ is suspect on grounds other than these (see e.g. Deutsch and Candelas 1979 for a criticism of the 'total energy operator'). To be safe, when making appeals to 'physical regularity' as a criterion for the acceptability of states, one should consider only the n-point functions

$$\hat{\phi}(x_1)\hat{\phi}(x_2)\hat{\phi}(x_3) \ \dots \ \hat{\phi}(x_n) \ . \tag{4.9}$$

Only these (or, if you like, suitable smeared out values of

these) can really be measured in a laboratory. Even then, there are problems of renormalization when two or more of the x_i coincide.

A simple example

Here is a simple example, in Minkowski space-time, of a viable free field theory with *two* Fock space sectors (generalizing it to an infinite number is left as an exercise for the reader).

The usual mode functions for $\phi(x)$ in Minkowski space-time are

$$v_{\underline{k}}(x) = v_{\underline{k}}(t,\underline{x}) = (2\pi)^{-3/2}(2k)^{-\frac{1}{2}}\exp(-ikt + i\underline{k}.x) \quad (k \equiv |\underline{k}|) \tag{4.10}$$

By the conventional method of constructing states, already described, these give rise to a vacuum state, say $|v;\text{vac}\rangle$ and particle states related by ladder operators $\hat{a}_k^{(v)}$ and $\hat{a}_k^{(v)*}$. Let us append to this v-Fock space $\mathscr{F}_v$ an unconventional u-sector, $\mathscr{F}_u$, based on the modes

$$u_{\underline{k}}(x) = \alpha_k v_{\underline{k}}(x) + \beta_k^* v_{-\underline{k}}^*(x) \tag{4.11}$$

(not summed over $\underline{k}$), where α_k and β_k obey the simplified Bogoliubov relation

$$|\alpha_k|^2 - |\beta_k|^2 = 1 \tag{4.12}$$

and where

$$\sum_k |\beta_k|^2 \tag{4.13}$$

diverges (at the upper end only, so that the issue shall not be obscured with infrared divergences).

Now (4.13) would be the 'expectation value of the number of u-particles in the v-vacuum state' and vice versa – if it existed – but we have agreed that this cannot be measured. More worrying, at first, is that the total energy difference per unit volume between $|u,\text{vac}\rangle$ and $|v,\text{vac}\rangle$, namely

$$\langle u;\text{vac}|\sum_k k\hat{a}_k^{*(v)}\hat{a}_k^{(v)}|u;\text{vac}\rangle - \langle v;\text{vac}|\sum_k k\hat{a}_k^{*(v)}\hat{a}_k^{(v)}|v;\text{vac}\rangle = \int d^3\underline{k}k|\beta_k|^2 \tag{4.14}$$

is even more divergent than (4.13). Notice that (4.14), being an energy *difference* between two states does not need to be renormalized, so its divergence cannot be blamed on the usual infinities of field theory. It is a consequence of the 'wild' choice (4.11) of basis functions. An infinite energy difference would, I suppose, be too much to tolerate between two states of the same theory. It would be a powerful reason for abandoning the u-sector. But let us remember the warning of the previous section against globally defined operators. The relevant physically measurable quantity is the true *energy density* operator

$$\hat{\rho}(x) \equiv n^{\mu}n^{\nu}\hat{T}_{\mu\nu}(x) = \frac{1}{2}[(n.\nabla\hat{\phi}(x))^2 + \hat{\phi}_{;\mu}(x)\hat{\phi}^{;\mu}(x)] \qquad (4.15)$$

(for the $\xi = 0$, $m = 0$ field).

There is no guarantee that the volume integral of this is a well defined operator, even though one can formally evaluate it as

$$\hat{E} = \int_{\mathscr{S}_t} d\Sigma\ \hat{\rho}(t,\underline{x}) = \sum_k k\hat{a}_k^*\hat{a}_k \times \int_{\mathscr{S}_t} d\Sigma\ .$$

Now (4.15) is really the coincidence limit of derivatives of an n-point function (4.9) and if we regard it as such, it is a simple matter to evaluate the difference between its expectation values in $|u,\text{vac}\rangle$ and $|v,\text{vac}\rangle$ (again, this obviates the necessity of renormalization).

$$\langle\hat{\rho}(t,\underline{x})\rangle_{\text{ren}} = \langle u;\text{vac}|\hat{\rho}(t,\underline{x})|u,\text{vac}\rangle - \langle v;\text{vac}|\hat{\rho}(t,\underline{x})|v;\text{vac}\rangle$$

$$= \lim_{t'\to t} \int d^3\underline{k}k|\beta_k|^2\cos[k(t-t')]\ . \qquad (4.17)$$

If we interchanged the order of the integral and the limit, we should recover (4.14), and the formal identity (4.16) between 'energy density' and 'total energy per unit volume'. But such an interchange is invalid, and the latter quantity has no physical significance.

There is an infinity of choices of β_k for which (4.17) exists but (4.13) and (4.14) do not. As a simple example, take

$$\beta_k = (k_0/k)^{\frac{1}{2}} \qquad (4.18)$$

for some constant k_0. Then

$$\langle\hat{\rho}(t,\underline{x})\rangle_{\text{ren}} = 4\pi k_0 \lim_{t'\to t} \int_0^\infty dk k^2 \cos[k(t-t')]. \tag{4.19}$$

This oscillating integral, like all n-point functions, must be regarded as a distribution in $t-t'$ (or, if you like, notice that it arises essentially from the completeness relation (3.2) which converges only in a generalized sense). As such it exists and has the value zero. Notice that there exist choices of β_k which give (4.17) any desired value, including negative values.

From the 'experimental' point of view, the finiteness of $\langle u;\text{vac}|\,\hat{\rho}\,|u;\text{vac}\rangle_{\text{ren}}$ is in itself strong evidence of the admissibility of the u-states. The experimentalist should surely be reluctant to regard a state as unphysical if its apparent pathologies leave the energy density finite, since no paradoxical result can then be predicted for any observation. Even if some other n-point function, say $\langle u;\text{vac}|\hat{X}|u,\text{vac}\rangle$ were infinite, no 'X-meter' could detect the fact unless it could itself supply the infinite energy required to display the paradoxical reading. In fact in the above example, the n-point functions are all finite.

In summary, there are no physical grounds on which the u-states can be excluded from the theory of the free scalar field in Minkowski space. Thus, there might be *non-particulate matter* in this very room — though the success of the one-sector hypothesis in field theory to date presumably indicates that the state of the world has at most a very small component in the u-sector(s).

5. FURTHER COMMENTS ON THE SINGLE FOCK SPACE CONJECTURE

What I have been calling the 'one-sector' or 'single Fock space' conjecture is of course the currently accepted one. Much work (see especially Ashtekhar and Magnon 1975) has been done attempting to find a general criterion for preferring one Fock space sector over all the others.

When $\mathscr{M}$ is Minkowski space-time, of course, the conventional Fock space $\mathscr{F}_v$ is physically unique in many ways, but no one has yet found a satisfactory generalization of one of these ways to curved manifolds. For example $\mathscr{F}_v$ is the only sector which contains a Poincaré invariant state, viz. the 'Minkowski vacuum' $|v,\text{vac}\rangle$. But Poincaré invariance has no analogue in curved manifolds. Likewise, the complex structure which generates the conventional modes (4.10) and $\mathscr{F}_v$ is evidently

$$\vec{M} = i(-\underline{\nabla})^{\frac{1}{2}} \qquad (5.1)$$

where $\mathscr{S}_t$ is a t=constant hypersurface, and $\underline{\nabla}^2$ the Laplacian on $\mathscr{S}_t$.

But in a curved manifold one could add any scalar invariant of the intrinsic or extrinsic curvature of $\mathscr{S}_t$ to (5.1) without altering (4.10) in the flat space limit. Another peculiar property of the Minkowski vacuum state is that the expectation value in it of the Hamiltonian operator is formally at a minimum with respect to changes of state. Unfortunately, the Hamiltonian is an infinitely ambiguous concept: according to Fulling (1979) there exists for each sector $\mathscr{F}$ a Hamiltonian which correctly generates the classical theory of $\phi(x)$ and whose expectation value finds its minimum in one of the states of $\mathscr{F}$.

Nevertheless some Hamiltonians are more natural than others, in particular, the 'total energy' '$\hat{E}$'. Still trying to retain the single-Fock-space construction, one is led to propose (cf. Ashtekhar and Magnon 1975) that at an instant $\mathscr{S}_t$ the vacuum state $|\vec{M};\text{vac}\rangle$ of the preferred physical Fock space is the state which minimizes

$$\langle\vec{M};\text{vac}|\hat{E}|\vec{M};\text{vac}\rangle \qquad (5.2)$$

with respect to infinitesimal changes in $\vec{M}$. This identification of the 'true vacuum' as the 'state of least energy' has a pleasing physical interpretation.

Of course to make this interpretation, and indeed for the sake of mathematical decency, we should really be minimizing the renormalized value of (5.2),

$$\langle\vec{M};\text{vac}|\hat{E}|\vec{M};\text{vac}\rangle_{\text{ren}}\,. \qquad (5.3)$$

Fortunately, the difference between (5.2) and (5.3), though

infinite, is a c-number. Thus, provided that only state-independent manipulations are made of the divergent quantity (5.2), it will attain its formal minimum at the same value of $\vec{M}$ as would (5.3) its real minimum.

Likewise, our earlier warning about the unphysical nature of $\hat{E}$ need not concern us here.

Evaluating the '*formal*' minimum of (5.2) involves precisely postponing the divergent sum or integral in the definition of $\hat{E}$ until after the expectation value has been taken. In other words, by

$$\frac{\delta}{\delta\vec{M}}\langle\vec{M},\mathrm{vac}|\int_{\mathscr{S}_t}\hat{\rho}(\underline{M})\,d\Sigma|\vec{M};\mathrm{vac}\rangle = 0 \tag{5.4}$$

we can only mean

$$\frac{\delta}{\delta\vec{M}}\int_{\mathscr{S}_t}\langle\vec{M};\mathrm{vac}|\hat{\rho}(\underline{x})|\vec{M};\mathrm{vac}\rangle\,d\Sigma = 0\ . \tag{5.5}$$

Now it turns out (Deutsch and Najmi 1981) that (4.0) has a unique solution for $\vec{M}$, viz.

$$\vec{M} = \vec{M}_{\min} = -2\xi\chi + i(-\underline{\nabla}^2)^{\frac{1}{2}} \tag{5.6}$$

where χ is the trace of the second fundamental form on $\mathscr{S}_t$. If there *is* to be a preferred sector, this leads to a powerful argument in favour of its being $\mathscr{F}_{\min}$, the sector containing $|\vec{M}_{\min};\mathrm{vac}\rangle$; no other sector contains a state of 'least energy' (with respect to infinitesimal changes in the complex structure).

In fact, notwithstanding my persuasive plea for the covariance of quantum field theory with respect to the choice of mode functions, there is little doubt that $\mathscr{F}_{\min}$ would be generally accepted as the only physical sector, were it not for a certain catastrophic defect: in generic space-time backgrounds, the sector $\mathscr{F}_{\min}$ depends on the hypersurface $\mathscr{S}_t$ on which (5.3) is minimized. That is, not only is the vacuum state generically a function of time (that would merely constitute particle production), but the preferred Fock spaces at successive instants $\mathscr{S}_t$ are not unitarily equivalent. (See Fulling 1979). In the Schrödinger picture, one would say that the time evolution of the state vector does not respect the boundaries of the supposedly preferred sector. States regular on $\mathscr{S}_t$ suffer an 'infinite amount of particle production' after an infinitesimal time. Unfortunately this pathology, unlike the one in our little 2-sector example, is *not* confined to the expectation values of quantities, like $\hat{N}$, which cease to be observable in a multi-sector theory, so it would seem that introducing the additional sectors is not sufficient to solve the problem. No doubt some alternative, weaker and more physical regularity criterion is called for. Nevertheless it seems to me that these two properties of $\mathscr{F}_{\min}$, namely its unique possession of a 'least-energy state' and its susceptibility to Fulling's pathology under the single-Fock-space conjecture, together make the conjecture very unattractive. Before abandoning it, though, I must in all fairness make three more remarks in its defence.

Firstly, there is a trivial way of constructing a preferred sector without the Fulling pathology. Take one preferred instant $\mathscr{S}_{t_0}$ and construct $\mathscr{F}_{t_0} = \mathscr{F}_{\min}$ for it. Then choose the preferred sector $\mathscr{F}_t$ at every other instant to be that into which $\mathscr{F}_{t_0}$ evolves. It is known (Wald 1978) that if $\mathscr{M}$ is flat in a neighbourhood of $\mathscr{S}_{t_0}$ and of $\mathscr{S}_{t_1}$ (say), then $\mathscr{F}_{t_0}$ and $\mathscr{F}_{t_1}$ are unitarily equivalent. From our present point of view this is a sort of 'superselection rule' (though no conserved quantity is apparently involved). But it does mean that the single-Fock space conjecture is perfectly viable for space times containing flat 'in'- and 'out'-regions (in practice, asymptotically flat will probably do). Of course, the trouble is that the real universe contains no flat 'in'-region, and quite possibly no flat 'out'-region either. There seems to be no natural generalization of this construction to space-times without static in/out regions.

Secondly it should be pointed out that (4.15) is not the only possible definition of 'total energy' based on $\hat{T}_{\mu\nu}$. It might be that

$$\int \hat{T}_{\mu\nu} n^{\mu} \xi^{\nu} d\Sigma \tag{5.7}$$

with some geometrical specification of the time-like vector field ξ^{ν} such that it is a time-like Killing vector field whenever $\mathcal{M}$ admits one, will turn out to be more appropriate. The expectation value of (5.7) can be minimized with respect to choices of complex structure, but the result (Deutsch and Najmi 1981) seems to suffer from the same above-mentioned defects as $\vec{M}_{\min}$.

Finally, the minimization of an energy with respect to the *complex structure* (as opposed to the states within a given allowed set) has never been given a good physical interpretation. So the above arguments are not as 'physical' as they look.

6. THEORY AND EXPERIMENT

In laboratory physics, one does not ask of a physical theory that it predict the state of the system being studied. The system is usually *prepared* in some state before measurements are made, and all trace of its state prior to preparation is consequently lost. What is tested in such experiments is essentially the system's law of motion, but not its initial conditions. In astrophysics, we make conjectures about not only the laws of motion, but also the actual states of systems of interest, such as stars and galaxies. Nevertheless, we do not expect these states to be determined by fundamental physical principles but to be more or less contingent. The less contingent, of course, the better, but all explanations of apparently contingent properties of systems must eventually invoke assumptions about the state of the universe itself — e.g. its homogeneity, isotropy, etc. — which are not to be 'explained' (except perhaps on a deeper level than physics has yet attained, a level consequently beyond the scope of this article) but just tested experimentally.

The proposal of this article, that all Fock sectors should appear on an equal footing in quantum field theory, can be thought of as the partial implementation of a suggestion by Dennis Sciama (private communication), questioning the necessity of finding the 'true' physical Fock space. He has pointed out that the location of the cosmological state vector *within* this space is in any case 'left to experiment' — so why should we expect there to exist an *a priori* principle restricting it to a certain sector?

Once postulated, however, the existence of new sectors must (in principle) have locally testable consequences also. Linear superpositions of states of different sectors are allowed. But the design of experiments directly to test the postulate (e.g. by attempting to isolate some non-particulate matter) must be severely hampered by Wald's (1979) 'superselection rule' that a system which is initially in a state of $\mathscr{F}_{min}$ in a static space-time background will return to $\mathscr{F}_{min}$ whenever the background becomes static again. Indirect effects, say on scattering amplitudes in virtue of the presence of extra channels for intermediate processes, may be more accessible.

ACKNOWLEDGEMENT

Work supported in part by the Science Research Council and NSF grant PMY 7826592.

REFERENCES

Ashtekhar, A. (1978). University of Chicago Preprint.
—— and Magnon, A. (1975). *Proc. Roy. Soc.* **A.346**.
Deutsch, D. and Candelas, P. (1979). *Phys. Rev.* **D**.
—— and Najmi, A. (1981). Paper in preparation.
De Witt, B.S. (1965). *Dynamical Theory of Groups and Fields*. Gordon and Breach.
—— (1975). *Phys. Rep.* **19C**, 6.
Fulling, S.A. (1979). *J. Rel. Grav.* **10**, 10.
Lurié, D. (1968). *Particles and Fields*. Wiley.
Wald, R.M. (1979). *Ann. Phys.* **118**.

6

A REVIEW OF SOME ASPECTS OF FIELD THEORY ON TOPOLOGICALLY NON-TRIVIAL SPACE-TIMES

S.J. Avis

Department of Applied Mathematics and Theoretical Physics, and Trinity Hall, Cambridge

1. INTRODUCTION

In this seminar I hope to indicate to you some of the points of contact between the topology of a four dimensional space-time manifold, and the description of fields that propagate on it. For the most part the discussion will be limited to classical fields, although the quantum nature of spinor fields will be an essential ingredient at one point.

Models of the physical world that involve classical general relativity usually begin with a four dimensional pseudo-Riemannian manifold. Such a manifold possesses a metric of Lorentzian type, and if – as I shall assume – the manifold is separately both space and time orientable then it has a local $SO_{\uparrow}(1,3)$ structure. In order to avoid closed time-like loops the space-time manifold is assumed to be non-compact.

Although most of the present discussion deals with pseudo-Riemannian manifolds, four dimensional Riemannian manifolds – that is to say manifolds which have a metric with positive definite signature – will also be mentioned briefly since their use is an essential part of the space-time 'foam' approach to quantizing gravity (Hawking 1978, 1979*a*,*b*; Wheeler 1964). In this picture not only does one sum over all metrics on a *given* four dimensional manifold of fixed topology in the Feynman path integral, but one also sums over all the possible topologies that a four dimensional manifold can have: although the manifolds are usually taken to be compact. This approach, together with the 'classical' observation that the topology of space-time may be important on the cosmological length scale leads one to look at field theories that are defined on topologically non-trivial manifolds; be they Riemannian or pseudo-Riemannian.

For simplicity first consider a multiplet of spin zero fields

$\{\varphi_a\}_{a=1}^{n}$ that is defined on Minkowski space-time and which transforms under the action of some Lie group G — the 'internal' symmetry group — via the representation D of G i.e. $\varphi_a(x) \to D(g)_a^b\varphi_b(x)$ where $g\in G$ and x varies over all points of $\mathbb{R}^4$. Here φ is simply a function from the space-time, which is here $\mathbb{R}^4$ into the vector space $\mathbb{R}^n$ (or $\mathbb{C}^n$) which carries the representation D. To translate this definition into one involving an arbitrary four dimensional manifold $\mathcal{M}$ is trivial. However, there is an equally valid construction which is a generalization of this simple view of φ being a function from $\mathcal{M}$ into $\mathbb{R}^n$ (or $\mathbb{C}^n$) (Isham 1978*a*). Since it involves the idea of a fibre bundle I shall first give the relevant definitions.

If $\mathcal{M}$ and G are respectively the manifold and Lie group under consideration then a principal G-bundle (Husemoller 1966), ξ, over $\mathcal{M}$ consists of a manifold of dimension dim G × dim $\mathcal{M}$, usually written as $E(\xi)$ and called the total space of ξ, together with a projection map $p:E(\xi)\to\mathcal{M}$. The fibres, that is the pre-images $p^{-1}(x)\subset E(\xi)$ of the points $x\in\mathcal{M}$, are each required to have a free G-action under which an element of any fibre of $E(\xi)$ is mapped into the same fibre. In this, and in the following, all manifolds, maps, and group actions etc. are all taken to be smooth (C^∞). An alternative and more intuitive definition of such a G-bundle is the following (Isham 1978*a*; Steenrod 1951). Again one starts with a total space $E(\xi)$ and a projection p, but one then introduces a collection of open sets $\{U_\alpha\}_{\alpha\in A}$ — here A is just some index set e.g. {1,2,3,..} — isomorphisms $\{h_\alpha : p^{-1}(U_\alpha)\to U_\alpha\times G\}_{\alpha\in A}$, and transition functions $g_{\alpha\beta}:U_\alpha\cap U_\beta\to G$ defined whenever $U_\alpha\cap U_\beta\neq\emptyset$. These are required to satisfy

$$\left.\begin{array}{ll}\text{(i)} & \bigcup_{\alpha\in A} U_\alpha = \mathcal{M} \\ \text{(ii)} & \text{if } h_\alpha(y) = (x,g) \text{ then } x = p(y) \\ \text{(iii)} & \text{if } U_\alpha\cap U_\beta \neq \emptyset \text{ then } (h_\alpha\circ h_\beta^{-1})(x,g) = (x, g_{\alpha\beta}(x)\cdot g) \\ & \text{where } x\in U_\alpha\cap U_\beta,\ g\in G, \text{ and } y\in p^{-1}(U_\alpha)\end{array}\right\} \qquad (1.1)$$

If ξ and $\bar{\xi}$ are both G-bundles over the manifold $\mathcal{M}$ then, without loss of generality, their respective transition functions — $\{g_{\alpha\beta}\}$ and $\{\bar{g}_{\alpha\beta}\}$ — may both be assumed to be based on the same collection of open sets $\{U_\alpha\}$. The two bundles ξ and $\bar{\xi}$ are said to be equivalent (Husemoller 1966; Steenrod 1951) if and only if there are (smooth) functions $h_\alpha : U_\alpha \to G$ for each $\alpha \in A$ which satisfy

$$\bar{g}_{\alpha\beta}(x) = h_\alpha(x)\cdot g_{\alpha\beta}(x)\cdot h_\beta(x)^{-1} \text{ where } x \in U_\alpha \cap U_\beta \,. \tag{1.2}$$

On any given manifold one can always construct the 'trivial' G-bundle whose total space is $\mathcal{M} \times G$ and whose transition functions take the simple form $g_{\alpha\beta}(x) = e$, where e is the identity element of G. Furthermore, any G-bundle that is equivalent to the trivial one will also be called trivial.

Given a G-bundle ξ and representation D of G on the vector space V — which is here either $\mathbb{R}^n$ or $\mathbb{C}^n$ — an associated vector bundle $\xi[V]$ can be constructed and which consists of a total space $E(\xi[V])$, a projection $\bar{p}$ from $E(\xi[V])$ onto $\mathcal{M}$, and a set of isomorphisms $\{\bar{h}_\alpha : \bar{p}^{-1}(U_\alpha) \to U_\alpha \times V\}_{\alpha\in A}$ such that whenever $U_\alpha \cap U_\beta$ is not empty

$$(\bar{h}_\beta \circ \bar{h}_\alpha^{-1})(x,v) = (x, g_{\beta\alpha}(x)\cdot v) \tag{1.3}$$

for all $x \in U_\alpha \cap U_\beta$ and $v \in V$. It should be realized that the total spaces of these associated vector bundles are not necessarily of the form $\mathcal{M} \times V$. A very simple example is provided by considering the possible real line bundles over the one dimentional circle S^1. There are precisely two: the trivial one whose total space is just the open cylinder i.e. $\mathbb{R} \times S^1$, and the non-trivial one whose total space is identical to the (open) Möbious band, which is not the same as $\mathbb{R} \times S^1$.

A multiplet of scalar fields φ can now be introduced as a cross-section of $\xi[V]$. That is φ is a function from $\mathcal{M}$ into $E(\xi[V])$ which satisfies $\bar{p}\varphi(x) = x$ for all $x \in \mathcal{M}$. For each open set U_α there is a local representative $\varphi_{(\alpha)}$ of φ which are required to satisfy

$$\left.\begin{aligned} &\text{(i) } \varphi_{(\alpha)} : U_\alpha \to V \\ &\text{(ii) for all } x \in U_\alpha \cap U_\beta,\ \varphi_{(\alpha)}(x) = D(g_{\alpha\beta}(x))\cdot\varphi_{(\beta)} \end{aligned}\right\} \tag{1.4}$$

In order to end up with an action which is invariant under local G transformations (to be defined later) one needs to introduce, as usual, gauge fields. Thus define the covariant derivatives on each U_α by

$$\mathscr{D}_\mu = \partial_\mu + ieD(A_{(\alpha)\mu}) \tag{1.5}$$

where e is a coupling constant, and U_α has been taken to be both a co-ordinate chart for the manifold $\mathscr{M}$ — the index μ referring to these co-ordinates — and a trivializing chart for the G-bundle ξ: clearly this can always be done. The $A_{(\alpha)\mu}$ in equation (1.5) are the gauge fields and take their values in the Lie algebra of G — the factor 'i' occurs in (1.5) because the generators of the compact Lie group G can be taken to be Hermitean. The $A_{(\alpha)\mu}$ for different U_α are related by the equation (whenever $x \in U_\alpha \cap U_\beta$)

$$A_{(\beta)\mu}(x) = g_{\beta\alpha}(x)\cdot A_{(\alpha)\mu}(x)\cdot g_{\beta\alpha}(x)^{-1} + ie(\partial_\mu g_{\beta\alpha}(x))\cdot g_{\beta\alpha}(x)^{-1} \tag{1.6}$$

which comes from the requirement that

$$(\mathscr{D}_\mu\varphi)_{(\beta)}(x) = D(g_{\beta\alpha}(x))\cdot(\mathscr{D}_\mu\varphi)_{(\alpha)}(x) \ . \tag{1.7}$$

To conclude this discussion of scalar fields the notion of gauge equivalence will be described. The field configurations specified by $\{\varphi_{(\alpha)}\}$, $\{A_{(\alpha)\mu}\}$ and $\{\bar{\varphi}_{(\alpha)}\}$, $\{\bar{A}_{(\alpha)\mu}\}$ are said to be gauge equivalent if there are functions $h_\alpha : U_\alpha \to G$ satisfying

$$\left.\begin{aligned} \bar{\varphi}_{(\alpha)}(x) &= D(h_\alpha(x))\cdot\varphi_{(\alpha)}(x) \\ \text{and} \\ \bar{A}_{(\alpha)\mu}(x) &= h_\alpha(x)\cdot A_{(\alpha)\mu}(x)\cdot h_\alpha(x)^{-1} + ie(\partial_\mu h_\alpha(x))\cdot h_\alpha(x)^{-1} \end{aligned}\right\} \tag{1.8}$$

Clearly, gauge equivalent field configurations are physically equivalent. (A more extended, though still fairly short, introduction to the formalism can be found in Daniel and Viallet (1980).)

To summarize then, in order to define on a space-time

manifold $\mathcal{M}$ a multiplet of scalar fields φ transforming under some representation of G, one can take an arbitrary G-bundle ξ defined over $\mathcal{M}$ and φ then becomes a cross-section of a, possibly non-trivial, associated vector bundle $\xi[V]$ (Isham 1978*a*). The gauge fields A_μ that are needed to form the covariant derivatives are then connections on ξ. Such a procedure is of course well known in the study of instantons on the Riemannian four sphere S^4 (there is a vast literature on this subject and a good introduction can be found in Coleman (1977)). Indeed, unless one has very good reasons why only the trivial G-bundles need be taken into account, one must consider the possible effects of the other, non-trivial G-bundles, whenever these exist.

There are two questions that are immediately suggested. First, given an arbitrary four dimensional manifold $\mathcal{M}$ and some compact Lie group G as internal symmetry group, how many distinct equivalence classes of G-bundles do there exist on $\mathcal{M}$, and how may they be distinguished from one another. Second, what are the main physical differences between field theories based on the trivial G-bundle to those based on a non-trivial bundle? An almost complete solution to the first stated problem can be given in terms of the cohomology groups of the four dimensional manifold $\mathcal{M}$, and I refer the interested reader to Avis and Isham (1979*a*) for the details. It is to the second stated problem that the remainder of this seminar is addressed.

2. SPONTANEOUS SYMMETRY BREAKING

In present day particle physics one of the key mechanisms used in the unified description of the weak and electromagnetic interactions is that of spontaneous symmetry breaking. In this mechanism the Higgs fields – which form a multiplet φ of scalar fields forming some representation of a Lie group G – are 'forced' to lie in the set of absolute minima of the Higgs potential – usually a polynomial in φ – over the entire space-time manifold. Whilst no problems are encountered on Minkowski space-time there may be potential topological obstructions to this mechanism when the Higgs field is a cross-section of a vector bundle associated with a non-trivial G-bundle on a topologically non-trivial space-time manifold. Some simple models will illustrate this and suggest what might happen in general.

Consider first the $U(1)$ gauge theory defined on Minkowski space-time (with signature +1,-1,-1,-1) and consisting of a complex scalar field φ and a (real) gauge field A_μ with Lagrangian density

$$\mathscr{L} = -\frac{1}{4}F_{\mu\nu}F^{\mu\nu} + \frac{1}{2}(\mathscr{D}_\mu\varphi)^+(\mathscr{D}^\mu\varphi) - \frac{\lambda^2}{2}(\varphi^+\varphi - a^2)^2 \tag{2.1}$$

where

$$F_{\mu\nu} = \partial_\mu A_\nu - \partial_\nu A_\mu$$

$$\mathscr{D}_\mu = \partial_\mu + ieA_\mu$$

and e, λ and a are real constants with $a^2 > 0$. Then the energy density for such a theory is

$$\mathscr{E} = \frac{1}{2}(F_{0i}F_{0i}+F_{12}F_{12}+F_{13}F_{13}+F_{23}F_{23}) + \frac{1}{2}(\mathscr{D}_0\varphi^+\mathscr{D}_0\varphi+(\mathscr{D}_i\varphi)^+(\mathscr{D}_i\varphi)) + \frac{\lambda^2}{2}(\varphi^+\varphi-a^2)^2 \tag{2.2}$$

where the index i runs over the spatial indices 1,2,3. Hence the classical state of lowest energy (i.e. zero energy) must satisfy the equations

$$F_{\mu\nu} \equiv 0, \qquad \mathscr{D}_\mu\varphi \equiv 0, \qquad \varphi^+\varphi \equiv a^2 \tag{2.3}$$

These equations are easily solved to give the classical ground state as any field configuration which is gauge equivalent to

$$A_\mu(x) \equiv 0 \quad \text{and} \quad \varphi(x) \equiv a\ . \tag{2.4}$$

Now consider the same theory but defined on a general space-time $\mathscr{M}$ instead of Minkowski space-time. Then with the usual tensor conventions the Lagrangian density (on each co-ordinate chart) is again given by (2.1). Furthermore, if $\mathscr{M}$ is assumed static (so that the concept of energy makes sense) then the field configurations, if any, that yield the absolute minimum energy must again satisfy the equations (2.3). If the theory is based on the trivial $U(1)$ bundle on $\mathscr{M}$ then these equations might well have a solution. However, if $\mathcal{M}$ admits a non-trivial $U(1)$ bundle ζ, and φ were taken to be a cross-section of the complex line bundle $\zeta[\mathbb{C}]$ associated with ζ then $\varphi^+\varphi \equiv a^2$ could not be satisfied everywhere on $\mathcal{M}$. Because if such a φ were to exist then it would be a nowhere-vanishing cross-section of $\zeta[\mathbb{C}]$ which would thus be a trivial bundle (Husemoller 1966): but $\zeta[C]$ is trivial if and only if ζ is trivial (Husemoller 1966), and so ζ must be trivial, which is a contradiction. (The proofs of these statements are completely analogous to proving that the existence of a globally defined verbein field on a four dimensional Riemannian manifold implies that both the tangent bundle and the underlying SO(4) bundle are trivial.) In other words, either there is no lowest energy cross-section of $\zeta[\mathbb{C}]$ or if there is then it certainly does not satisfy equations (2.3), and its energy is strictly greater than zero. An example of a manifold which exhibits the above phenomena is provided by $\mathbb{R}$(time) $\times$ ($\mathbb{R} \times \mathbb{S}^2$)(space) endowed with any static metric, since the topology of this space-time admits non-trivial $U(1)$ bundles.

In order to see what the quantitative effects of this phenomenon might be it is necessary to simplify the model still further. Although this simplified model could be given in four dimensions it turns out that precisely the same effects occur in its restriction to two dimensions and so I shall here only describe the two dimensional model; the four dimensional one is given in Avis and Isham (1978).

Thus consider a real scalar field φ defined on the flat two dimensional space-time with topology $\mathbb{R}$(time) $\times$ $\mathbb{S}^1$(space) (i.e. the flat two dimensional cylinder) and with a Lagrangian invariant under $\varphi \to -\varphi$ given by

$$\mathscr{L} = \frac{1}{2}(\partial_\mu\varphi)(\partial^\mu\varphi) - \frac{\lambda^2}{2}(\varphi^2 - a^2)^2 \ . \tag{2.5}$$

This is the $\mathbb{Z}_2 := \{1,-1\}$ version of the $U(1)$ model given above. Note that since the gauge group is discrete there are no gauge fields.

It is easy to convince oneself (and easy to prove rigorously) that there are only two equivalence classes of real-line bundles over $\mathbb{R}^1 \times \mathbb{S}^1$. By 'unwrapping' the spatial circle, cross-sections of the trivial $\mathbb{R}$-bundle can be represented as functions defined on the entire (t,θ) plane, $\mathbb{R}^2$, satisfying the periodic condition

$$\varphi(t,\theta) = \varphi(t,\theta + 2\pi) \quad \text{for all } (t,\theta) \in \mathbb{R}^2 \, . \tag{2.6}$$

In these co-ordinates the metric induced on $\mathbb{R}^2$ from the original flat $\mathbb{R}^1 \times S^1$ is given by

$$ds^2 = dt^2 - R^2 d\theta^2 \tag{2.7}$$

where R is the radius of the original spatial circle. In this case, of course, the lowest energy field configuration is either $\varphi = +a$ or $\varphi = -a$: neither of these is symmetric under multiplication by -1 and the usual spontaneous symmetry breaking has occurred.

However, cross-sections of the non-trivial $\mathbb{R}$ bundle on $\mathbb{R}^1 \times S^1$ can be represented as real valued functions on $\mathbb{R}^2$ satisfying, in the same co-ordinate system as before, the antiperiodic condition

$$\varphi(t,\theta) = -\varphi(t,\theta + 2\pi) \quad \text{for all } (t,\theta) \in \mathbb{R}^2 \tag{2.8}$$

Then, it can be shown (Avis and Isham 1978) that in this case the states of lowest energy are given by

$$\left.\begin{array}{l} \text{(i) } \varphi \equiv 0 \quad \text{for } 2\lambda aR \leqslant 1 \\ \\ \text{and} \\ \\ \text{(ii) } \varphi(t,\theta) = \surd(a^2-w)\, sn\left[\lambda k\Big/\left(\dfrac{a^2+w}{2}\right)(\theta+\theta_0);\gamma\right] \text{ for } 2\lambda aR > 1 \end{array}\right\} \tag{2.9}$$

where θ_0 is a constant, $\gamma = (a^2-w)/(a^2+w)$ and w is the solution to the transcendental equation

$$\pi\lambda R\Big/\left(\frac{a^2+w}{2}\right) = K(\gamma)$$

where

$$K(\gamma) = \int_0^{\frac{\pi}{2}} \frac{dy}{\sqrt{(1- \quad \sin^2 y)}}$$

is the quarter period of the Jacobi elliptic function $sn(x;\gamma)$ (Abramowitz and Stegun 1964). The behaviour of the ground state for $2\lambda Ra > 1$ is understandable as one would expect that as $R \to \infty$ the ground state would tend pointwise to the usual two dimensional flat space-time kink, $a\cdot\tanh\left(\frac{Ra\lambda\theta}{\sqrt{2}}\right)$: this is precisely what happens. However, the fact that the ground state is zero for $2\lambda Ra \leqslant 1$ is somewhat surprising at first sight, but it is easily verified to be so by considering small oscillations about $\varphi \equiv 0$ and noting that the condition $2\lambda Ra \leqslant 1$ is the condition for the stability of the zero solution.

It would be interesting to see which properties of this $\mathbb{Z}_2$ model on $\mathbb{R}^1 \times S^1$ (and its four dimensional analogue defined on $\mathbb{R}^1 \times S^1 \times S^1 \times S^1$ (Avis and Isham 1978)) generalize to four dimensional manifolds admitting non-trivial G-bundles, where G is a compact Lie group, and what the quantum implications of this phenomenon might be. Further work is needed here.

3. SPIN-STRUCTURES AND QUANTUM FIELD THEORY

In this section the effects of space-time topology on spinor fields will be discussed (Avis and Isham 1979*b*; Isham 1978*b*). First, however, it is necessary to show how spinor fields are defined on general space-time manifolds.

Let $\mathcal{M}$ be a separately both space and time orientable four dimensional pseudoriemannian manifold, then for each coordinate chart U_α of $\mathcal{M}$ there is a set of four linear independent vector fields $\{e_{(\alpha)a}\}_{a=0}^{3}$ defined on U_α, and by taking suitable linear combinations they can be assumed to be orthonormal in the sense that

$$\left.\begin{aligned} \eta^{ab} e_{(\alpha)a\mu} e_{(\alpha)b\nu} &= g_{\mu\nu} \\ \text{and} \qquad g^{\mu\nu} e_{(\alpha)a\mu} e_{(\alpha)b\nu} &= \eta_{ab} \end{aligned}\right\} \tag{3.1}$$

where $(\eta^{ab}) = \mathrm{diag}(1,-1,-1,-1)$, $g_{\mu\nu}$ are the components of the space-time metric in the chart U_α, and $e_{(\alpha)a\mu}$ are the components of $e_{(\alpha)a}$ in the chart U_α. Now suppose that U_α and U_β are two charts on $\mathcal{M}$ with a non-empty intersection, then the second equation of (3.1) shows that $e_{(\alpha)a} = k_{\alpha\beta a}{}^{b} e_{(\beta)b}$ where $k_{\alpha\beta a}{}^{b} k_{\alpha\beta c}{}^{d} \eta_{bd} = \eta_{ac}$ i.e. $k_{\alpha\beta}$ is an O(1,3) matrix valued function on $U_\alpha \cap U_\beta$. However, $\mathcal{M}$ is both space and time orientable and so the vector fields $e_{(\alpha)a}$ can be chosen so that $k_{\alpha\beta}(x)$ is an $SO_\uparrow(1,3)$ matrix whenever $x \in U_\alpha \cap U_\beta$. In other words the $k_{\alpha\beta}$ form the transition functions of an $SO_\uparrow(1,3)$ bundle over $\mathcal{M}$ which will be denoted by τ. (This follows because $k_{\alpha\gamma} = k_{\alpha\beta} \cdot k_{\beta\gamma}$ whenever $U_\alpha \cap U_\beta \cap U_\gamma \neq \emptyset$ and $k_{\alpha\alpha} \equiv \mathbb{1}$ (Husemoller 1966; Steenrod 1951). In fact it is easily seen that the tangent bundle of $\mathcal{M}$ is the associated bundle formed from τ by using the usual matrix action of $SO_\uparrow(1,3)$ on $\mathbb{R}^4$.

From τ it is necessary to construct an SL(2,$\mathbb{C}$) bundle $\bar{\tau}$ which is a double covering of τ in the same way that SL(2,$\mathbb{C}$) is a double covering of $SO_\uparrow(1,3)$. More precisely, one must try to construct $\bar{\tau}$ so that there is a two to one map f from $E(\bar{\tau})$ onto $E(\tau)$ satisfying

$$f(s \cdot A) = f(s) \cdot \Lambda(A) \quad \text{for all } s \in E(\bar{\tau}) \text{ and } A \in SL(2,\mathbb{C}) \qquad (3.2)$$

where Λ denotes the usual two to one map from SL(2,$\mathbb{C}$) onto $SO_\uparrow(1.3)$ and the '$\cdot$' denotes the (right) action of either SL(2,$\mathbb{C}$) on $\bar{\tau}$ or $SO_\uparrow(1,3)$ on τ. As well as (3.2) it is necessary that a point of $E(\bar{\tau})$ which is in the fibre over $x \in \mathcal{M}$ be mapped into the $SO_\uparrow(1,3)$ fibre over x contained in $E(\tau)$ i.e. $\bar{p}(s) = p(f(s))$ for all $s \in E(\bar{\tau})$ where $\bar{p}$ is the projection from $E(\bar{\tau})$ onto $\mathcal{M}$. The pair $(\bar{\tau}, f)$ is called a spin-structure and if one exists, Dirac four component spinor fields can be defined as cross-sections of an associated vector bundle $\bar{\tau}[\mathbb{C}^4]$, where the SL(2,$\mathbb{C}$) action on $\mathbb{C}^4$ is given by the Dirac representation of SL(2,$\mathbb{C}$). Two component Weyl spinors can be defined in an analogous manner.

Unfortunately, however, not all manifolds admit a spin-structure since it can be shown that a necessary and sufficient condition for the existence of a spin-structure is the vanishing of the second Stiefel–Whitney class $w_2(\mathcal{M})$ of the manifold (Milnor 1963): there do exist manifolds for which this class is not zero. Fortunately, a result due to Geroch (1968, 1970) states that the tangent bundle to any globally hyperbolic (space and time orientable) manifold is necessarily trivial and so automatically has a vanishing second Stiefel–Whitney class. For those manifolds with $w_2(\mathcal{M}) \neq 0$ there are constructions available which enable generalized spin-structures to be defined (Avis and Isham 1980; Back, Freund, and Forger 1978; Hawking 1978; Hitchin 1974; Petry to appear; Whiston 1975) and I refer the interested reader to the references for the details.

From now on I shall assume that $\mathcal{M}$ is globally hyperbolic and so has a spin-structure. As a consequence of the tangent bundle being trivial $E(\tau)$ is isomorphic to $\mathcal{M} \times SO_{\uparrow}(1,3)$ and the projection p is given by $p(x,g) = x$ for all $x \in \mathcal{M}$ and $g \in G$. It then follows that the total space of any spin-structure for $\mathcal{M}$ must be of the form $\mathcal{M} \times SL(2,\mathbb{C})$ with projection given by $\overline{p}(x,A) = x$ for all $x \in \mathcal{M}$ and $A \in SL(2,\mathbb{C})$. Thus any spin-structure map $f: E(\overline{\tau}) \to E(\tau)$ can be expressed as

$$f(x,A) = (x, \Omega(x)\cdot\Lambda(A)) \quad \text{for } x \in \mathcal{M} \text{ and } A \in SL(2,\mathbb{C})$$

where Ω is some function from $\mathcal{M}$ into $SO_{\uparrow}(1,3)$ (Isham 1978*b*). For reasons that will be given later, it is convenient to say that the two spin-structures specified by Ω_1 and Ω_2 are equivalent if and only if there is a (smooth) function S from $\mathcal{M}$ into $SL(2,\mathbb{C})$ such that $\Lambda(S(x)) = \Omega_1(x)\Omega_2(x)^{-1}$. Such a function, S, is said to be a 'lift' of the function $\Omega_1(x)\cdot\Omega_2(x)^{-1}$ and those functions from $\mathcal{M}$ into $SO_{\uparrow}(1.3)$ which lift will be called 'small' whilst those that do not will be called 'large'.

Let $[\Omega]$ denote the set of all those (smooth) functions from $\mathcal{M}$ into $SO_{\uparrow}(1,3)$ which are equivalent to Ω, and let H be the set of all such equivalence classes. Then it can be shown (Avis 1980) that H has the structure of an abelian group with multiplication defined by

$$[\Omega_1]\cdot[\Omega_2] = [\Omega_1\cdot\Omega_2] \tag{3.3}$$

where $(\Omega_1\cdot\Omega_2)(x) = \Omega_1(x)\cdot\Omega_2(x)$, and identity defined as the class $[1]$ where the function 1 is identically equal to the unit matrix in $SO_{\uparrow}(1,3)$. In fact H is isomorphic to the cohomology group $H^1(\mathscr{M}; \mathbb{Z}_2)$ and so the order of this group gives the number of inequivalent spin-structures on $\mathscr{M}$.

Since $\mathscr{M}$ is parallelizable there is a globally defined vierbein field $\{e_a\}^3_{a=0}$ whose components satisfy in each co-ordinate chart of $\mathscr{M}$

$$g^{\mu\nu}e_{a\mu}e_{b\nu} = \eta_{ab}, \text{ and } \eta^{ab}e_{a\mu}e_{b\nu} = g_{\mu\nu} \ . \tag{3.4}$$

It should be pointed out, however, that there is no natural choice of vierbein field e_a for any given metric $g_{\mu\nu}$. In certain special cases — e.g. on metrically flat space-time — there may be an 'obvious' choice of vierbein: but this has no intrinsic significance.

Given a globally defined vierbein field $\{e_a\}^3_{a=0}$ a Lagrangian density for a spin half Dirac field can be constructed as follows

$$\mathscr{L}(e,\psi) = \left\{\frac{i}{2}(\bar{\psi}\gamma_a(\mathscr{D}_\mu\psi) - (\mathscr{D}_\mu\bar{\psi})\gamma_a\psi)e^{a\mu} - m\bar{\psi}\psi\right\}\det e \tag{3.5}$$

where the spin half covariant derivative is defined as (with $\bar{\psi} = \psi^+\gamma_0$)

$$\left.\begin{aligned} &\mathscr{D}_\mu\psi = \partial_\mu\psi + iB_\mu\psi, \quad \mathscr{D}_\mu\bar{\psi} = \partial_\mu\bar{\psi} - i\bar{\psi}B_\mu \\ &B_\mu = \frac{1}{4}\Gamma_{\mu ab}\sigma^{ab} \quad \text{with} \quad \sigma^{ab} = \frac{i}{2}[\gamma^a,\gamma^b] \\ &\Gamma_{\mu ab} = e_{b\nu}e_a{}^\nu{}_{,\mu} + e_{b\alpha}e_a{}^\nu\left\{{\alpha \atop \nu\mu}\right\} \end{aligned}\right\} \tag{3.6}$$

and $\left\{{\alpha \atop \nu\mu}\right\}$ is the usual (metric) Christoffel symbol of the second kind. The Dirac matrices $\gamma^1,\gamma^2,\gamma^3$ are anti-Hermitean, γ^0 is Hermitean and they satisfy the defining relation $\{\gamma^a,\gamma^b\} = 2\eta^{ab}$. Using the Lagrangian the vacuum generating functional of a quantized spin half field in the presence of a background metric (that is specified by the e_a) can be written as

$$Z[e] = \int \mathscr{D}\psi\cdot\mathscr{D}\bar{\psi}\ \exp\left[i\int_{\mathscr{M}} \mathscr{L}(e,\psi)\right] \tag{3.7}$$

where $\mathscr{D}\psi$ denotes the functional measure for the spinor field ψ.

Suppose $\Omega:\mathscr{M}\to SO_{\uparrow}(1,3)$ is small so that there exists a lift S_Ω with $\Lambda(S_\Omega(x)) = \Omega(x)$ then

$$Z[e\Omega] = \int \mathscr{D}\psi\cdot\mathscr{D}\bar{\psi}\ \exp i\int_{\mathscr{M}} \mathscr{L}(e\Omega,\psi) \tag{3.8}$$

$$= \int \mathscr{D}\psi\cdot\mathscr{D}\bar{\psi}\ \exp i\int_{\mathscr{M}} \mathscr{L}(e,S_\Omega^{-1}\psi) \tag{3.9}$$

$$= \int (\mathscr{D}S_\Omega\psi)\cdot(\mathscr{D}\overline{S_\Omega\psi})\ \exp i\int_{\mathscr{M}} \mathscr{L}(e,\psi)$$

thus

$$Z[e\Omega] = Z[e] \tag{3.10}$$

where the functional measure has been assumed — as is normally done — to be SL(2,$\mathbb{C}$) invariant so that $\mathscr{D}(S_\Omega\psi) = \mathscr{D}\psi$. The crucial step in deriving (3.10) is the assumption that Ω is small, since without this one cannot pass from (3.8) to (3.9). In a similar manner it can be shown that $Z[e\Omega_1] = Z[e\Omega_2]$ whenever Ω_1 and Ω_2 are equivalent. If, however, Ω_1 and Ω_2 are not equivalent then the above argument says nothing about the relationship between $Z[e\Omega_1]$ and $Z[e\Omega_2]$. Indeed, calculation shows that on the flat space-time with topology $\mathbb{R}^3 \times S^1$ the resulting quantum theories are different (De Witt, Hart, and Isham to be published; Ford 1979). Thus it seems likely that on those space-times which have more than one spin-structure the quantum field theory depends on the equivalence class of the vierbein field chosen to represent the metric (where the two vierbein fields e_a and $\bar{e}_a$ are said to be equivalent if and only if the map Ω from $\mathscr{M}$ into $SO_{\uparrow}(1,3)$ determined by $\bar{e}_a = e_b\Omega^b{}_a$ is small).

It is natural to try and see if there are any ways of finding generating functionals that do not depend on the equivalence class of the vierbein field chosen to represent the metric.

For each $h\in H$ let Ω_h be any (smooth) map from $\mathscr{M}$ into $SO_{\uparrow}(1,3)$ with $[\Omega_h] = h$ then equation (3.10) shows that $Z_h[e] := Z[e\Omega_h]$ does not depend on the particular Ω_h chosen to represent h. Thus consider the generating function defined by

$$Z^{\chi}[e] = \sum_{h \in H} \chi(h) Z_h[e] \tag{3.11}$$

where I have assumed that the group H is finite so that the sum makes sense, and that the χ is a group homomorphism from H into $\mathbb{Z}_2 = \{+1,-1\}$. The generating functional (3.11) is analogous to the functional used to generate the θ-vacua in instanton theory (see, for example, Coleman 1977).

Now, under the action of a local Lorentz transformation Ω, chosen from the class k say, equation (3.11) becomes

$$\begin{aligned} Z^{\chi}[e\Omega] &= \sum_{h \in H} \chi(h) Z_h[e\Omega] \\ &= \sum_{h \in H} \chi(h) Z_{kh}[e] \\ &= \sum_{h \in H} \chi(h)\chi(k) Z_h[e] \\ &= \chi(k) Z^{\chi}[e] \end{aligned} \tag{3.12}$$

Thus $Z^{\chi}[e]$ is invariant, up to an irrelevant sign, under local Lorentz transformations.

More generally, take any product $Z_{h_1}[e] \times \ldots \times Z_{h_n}[e]$ – with $h_1 = 1$ and $h_i \neq h_j$ except when $i = j$ – and form the composite

$$Z^{\chi}[e](h_1,\ldots,h_n) = \sum_{h \in H} \chi(h) Z_{h_1 h}[e] \times \ldots \times Z_{h_n k}[e] \tag{3.13}$$

where χ is as before a $\mathbb{Z}_2$ character of H except that it is necessary to restrict one's attention to those characters for which (3.13) is not identically zero. Clearly (3.11) is a special case of (3.12).

As an example, consider the metrically flat space-time with topology $\mathbb{R}^3 \times S^1$. Then there are precisely two different spin-structures since $H = H^1(\mathbb{R}^3 \times S^1;\mathbb{Z}_2) = \mathbb{Z}_2 = \{1,-1\}$. Thus the above construction of locally Lorentz invariant generating functionals yields three different candidates, the product functional

$$Z^{\pi}[e] = Z_1[e] \cdot Z_{-1}[e] \tag{3.14}$$

and the two sum functionals

$$Z^{\pm}[e] = Z_1[e] \pm Z_{-1}[e] \tag{3.15}$$

Some of the implications of using the generating functionals given in (3.14) and (3.15) have been calculated by Ford (1979).

ACKNOWLEDGEMENT

Most of the work mentioned in this seminar was carried out in collaboration with C.J. Isham, and I should like to thank him for many enlightening discussions.

REFERENCES

Abramowitz, M. and Stegun, I.A. (1964). *Handbook of mathematical functions*. National Bureau of Standards.
Avis, S.J. (1980). Ph.D. thesis of London University.
—— and Isham, C.J. (1978). *Proc. R. Soc.* **363**, 581.
—— —— (1979*a*). in *Proc. 1978 Cargèse summer inst. recent devel. gen. relativity*. Plenum press, New York.
—— —— (1979*b*). *Nucl. Phys.* **B156**, 441.
—— —— (1980). *Commun. Math. Phys.* **72**, 103.
Back, A., Freund, P.G.O., and Forger, M. (1978). *Phys. Lett.* **77B**. 181.
Coleman, S. (1977). in *Proc. 1977 int. school subnucl. phys. Ettore Marjorana*.
Daniel, M. and Viallet, C.M. (1980). *Rev. Mod. Phys.* **52**, 175.
De Witt, B.S., Hart, C.F., and Isham, C.J. to be published in the schwinger Festschrift.
Ford, L.H. (1979). *Vacuum polarization in a non-simply connected space-time*. Kings College London preprint.
Forger, M. and Hess, H. (1979). *Commun. Math. Phys.* **64**, 269.
Geroch, R. (1968). *J. Math. Phys.* **9**, 1739.
——(1970). *J. Math. Phys.* **11**, 343.
Hawking, S.W. (1978). *Nucl. Phys.* **B144**, 349.
——(1979*a*). in *General relativity: an Einstein centenary*. Cambridge University Press.
——(1979*b*). in *Proc. 1978 Cargèse summer inst. recent devel. gen. relativity*. Plenum Press, New York.
—— and Pope, C.N. (1978). *Phys. Lett.* **73B**, 42.
Hitchin, N.J. (1974). *Adv. Math.* **14**, 1.
Husemoller, D. (1966). *Fibre bundles*. Springer-Verlag, New York.
Isham, C.J. (1978*a*). *Proc. R. Soc.* **A362**, 383.
——(1978*b*). *Proc. R. Soc.* **A364**, 591.
Milnor, J. (1963). *Enseign. Math.* **9**, 198.
Petry, H.R. Exotic spinors in superconductivity. to appear in *J. Math. Phys*.

Steenrod, N. (1951). *Topology of fibre bundles*. Princeton University Press.
Wheeler, J.A. (1964). *Relativity, groups and topology*. Gordon and Breach, London.
Whiston, G. (1975). *Gen. Relativ. Grav.* **6**, 463.

7

INTERACTING QUANTUM FIELD THEORY IN CURVED SPACE-TIME

N.D. Birrell

Department of Mathematics, King's College, London WC2

1. INTRODUCTION

Since the first Oxford Quantum Gravity Conference, an enormous amount of effort has been devoted to the study of quantum fields propagating in classical, curved background space-times, in the expectation that such investigations will shed light on the as yet unknown full quantum gravity theory (for a comprehensive review see Birrell and Davies 1981). The preponderence of these studies have been of *free* fields in curved space-time, which, despite their trivial nature in flat space-time, can have a considerable effect in curved space-time, due to the fact that the gravitational field couples to all other fields (see, for example, Fischetti, Hartle, and Hu 1979; Hartle and Hu 1979; Hartle in this volume). However in every-day laboratory physics it is interacting fields which give rise to interesting phenomena in experiments, and one must ask what happens in these experiments if one's laboratory is situated in a strong gravitational field. This leads to the study of interacting quantum field theory in curved space-time, which is the topic of this paper.

Before embarking upon such a study, one must ask whether interacting field theories in curved space-time can in any sense be viewed as consistent theories; in particular, can they be taken as consistent approximations to full quantum gravity theory? This is a particularly relevant question in the light of the paper of Duff (this volume). In the case of free fields, Duff is essentially pointing out that (within the context of a full quantum gravity treatment) the free matter fields are being treated to one-loop level, while, by treating the metric of the space-time classically, one is inconsistently working only to the zero-loop level in the gravitational field. It is not, however, difficult to restore consistency in this case, as one can simply include the graviton as one of the matter fields and calculate its contribution to the one-loop level (in DeWitt's 1976 background field method; see, for example, Hartle 1977; Critchley 1978).

In the case of interacting field theories in curved space-time, the problem is worse: one is now including arbitrary numbers of matter field loops, but only treating the gravitational field to, at best, the one-loop level. The inconsistency present in this situation manifests itself in the properties of such theories under field transformations as discussed by Duff. The only way to remove this inconsistency is to work to a consistent number of loops (i.e. a consistent order in $\hbar$ (Nambu 1966)). But this would mean working with arbitrarily many graviton loops, in which case one is faced with the problem of the non-renormalizable nature of quantum gravity.

So how should one view the study of interacting quantum fields in curved space-time? The answer is that one must take a pragmatic point of view, as is indeed adopted by the vast number of physicists who are not so perverse as to want to work in curved space-time. To these physicists, the background space-time of their laboratory is taken to be Minkowski space, and if they perform, for example, quantum electrodynamics calculations and compare the results with experiment they obtain excellent agreement. If one were to tell these flat space physicists that one is free to perform transformations which change their perfectly well behaved Minkowski space quantum electrodynamics into a non-renormalizable theory in a space-time with a complicated metric, one would doubtless be met with a great deal of scepticism.

The reason for the success of the pragmatic terrestrial physicist lies in the size of quantum gravity corrections. Consider first the magnitude of the effect of a single isolated graviton loop. This depends roughly on the magnitude of the curvature of the background space-time (in the background field method), as do the curved background corrections to matter field processes. On Earth, the curvature is so small compared with any other quantity of physical interest with the same dimensions that both the effects of isolated graviton loops and the fact that matter fields are propagating in curved space-time can be ignored.

In a region where the curvature is large compared with other quantities of physical interest with the same dimensions (e.g. the inverse square of the electron Compton wavelength), the fact that one is in a curved space-time cannot be ignored. On the other hand, neither can single isolated (free) graviton loops, and this is why they *must* be included for consistency. But what about higher order graviton loops, which should in principle be included in the interacting case?

For every graviton loop in a connected Feynman diagram, there is a factor of Newton's constant G. Assuming that in the renormalizable quantum theory of gravity, the finite contribution of each graviton loop is still comparable with the curvature, then a measure of magnitude of the effect of each graviton loop will be the dimensionless quantity ($\hbar = c = 1$) G × the size of the curvature. Provided this quantity is very much smaller than the fine structure constant, which is the coupling constant for loops in quantum electrodynamics, then higher graviton loop corrections can be neglected in comparison with interacting matter field loops. It is most likely that the assumption made regarding the unknown full quantum gravity theory in this argument is correct, or else we would have seen signs of its effects in the solar system.

Thus the pragmatic approach is that interacting quantum field theory in curved space-time is a useful approximation to the full theory of quantum gravity, provided that one takes account of free graviton loops, and that the curvature of space-time is much smaller than the product of the matter field interaction coupling constant with the inverse square of the Planck length (i.e. with G^{-1}).

In this regime, what are the possible causes of differences from flat space interacting quantum field theory? Just as in the case of free field theory these causes can be classified as

(A) lack of Poincaré invariance,

(B) possible non-uniqueness of the 'physical' vacuum and associated Fock space, and

(C) the fact that the topology need not necessarily be R^4 (as it is in the Minkowskian case).

The first of these three causes is self-explanatory, and one should only note that in certain circumstances there will be other isometries replacing Poincaré invariance, such as invariance under the de Sitter group in de Sitter space.

The second point requires some elucidation: in the conventional LSZ (Lehmann, Symanzik, and Zimmerman 1955) treatment of an interacting quantum field ϕ, one assumes the asymptotic conditions

$$\phi(x) \xrightarrow{x^0 \to -\infty} \phi_{in}(x) \tag{1.1}$$

$$\phi(x) \xrightarrow{x^0 \to -\infty} \phi_{out}(x) \tag{1.2}$$

(in the sense of weak operator convergence), where x^0 is the time coordinate, and ϕ_{in} and ϕ_{out} are free fields. In applying the LSZ formalism in curved space-time, one should be careful of the following:

1. One must be able to choose a time coordinate x^0. This requirement implies that the space-time is globally hyperbolic (see, for example, Kay 1980). Although free quantum field theory in a non-globally hyperbolic space-time has been investigated successfully by Avis, Isham, and Storey (1978), no similar treatment has been given for interacting fields and the construction of interacting theories in such space-times remains a major problem.

2. In Minkowski space, the motivation for (1.1) and (1.2) above is that in any experiment involving the scattering of interacting particles, the particles initially and finally are so far apart (in principle, infinitely far) that they are effectively outside the range of one another's interaction. However, in curved space-time, it may not be possible (even in principle) for the particles to be outside each other's range; for example, if the space-time has closed spatial sections, or if it is singular in the distant past. In the former case, provided the volume of the spatial sections in the distant past and future is sufficiently large, one can regard (1.1) and (1.2) as holding approximately (as is the case in a practical terrestrial experiment, where the particles are never infinitely separated). In the latter (singular)

case, which applies in realistic cosmological models, one can either adopt (1.1) as an approximate condition, letting $x^0 \to -T$, rather than $x^0 \to -\infty$ where time $-T$ is far removed from the 'time of interaction', or one can use a quite different approach to interacting theories in curved space-time, which makes no reference to an asymptotic condition in the past. Such an alternative approach has been given by Kay (1980), based on work of Hajicek (1978). As yet, no actual calculations have been performed using this method.

Assuming that the asymptotic conditions (1.1) and (1.2) do hold in some sense, one can construct asymptotic vacua and associated Fock spaces using the decomposition of the *free* fields ϕ_{in} and ϕ_{out} in terms of creation and annihilation operators and a complete set of positive and negative frequency mode solutions of the free field equation. As is well known from the study of free quantum fields in curved space-time, the definition of positive frequency modes cannot, in general, be unambiguously fixed, and even if there is good reason for defining them one way in the distant past, there may well be an equally good reason for defining them another way in the distant future. Thus, one denotes the vacuum defined using ϕ_{in} as $|\text{in}\rangle$, and the vacuum defined using ϕ_{out} as $|\text{out}\rangle$, and labels the corresponding Fock space elements in a similar way. One of the underlying principles of Minkowski space interacting quantum field theory (Wightman 1956; Garding and Wightman 1965) is that the vacuum is unique; in particular, $|\text{in}\rangle = |\text{out}\rangle$ (up to a phase factor). The essence of (B) above is that, in general, in curved space-time

$$|\text{in}\rangle \neq |\text{out}\rangle \quad . \tag{1.3}$$

As well as being due to possibly different definitions of positive frequency modes in the past and future, it will emerge later that (1.3) also occurs because $|\text{in}\rangle$ is unstable against decay into many particle 'out' states due to interaction processes allowed in curved space-time because of (A) above.

The phenomena arising from (A) and (B) in the case of free fields in curved space-time are now well known, and they, of course, persist in the interacting case. There are also new effects arising in what can be broadly classified as,

(a) the renormalization process, and

(b) the outcome of physical processes.

Changes in (a) and (b) from Minkowski space theory, because of (A) and (B), will be discussed in sections 2 and 3 respectively. In section 4 the effect of (C) on (a) and (b) will be considered separately, as it can give rise to interesting phenomena even in flat, but non-Minkowskian space-times.

2. RENORMALIZATION IN CURVED SPACE-TIME

Just as in Minkowski space, when one attempts to calculate S-matrix elements for interacting theories as a perturbation series in the interaction strength, one encounters infinities. These infinities must be removed from the calculation in a consistent way by renormalization of constants in the Lagrangian of the theory. For example, one might typically have to renormalize a bare mass m_{B} by absorbing into it an infinite 'constant' δm, to form a renormalized mass

$$m_{\mathrm{R}} = m_{\mathrm{B}} + \delta m \qquad (2.1)$$

which is to be measured by experiment.

To implement such a renormalization procedure in practical calculations, one usually adopts a regularization scheme. Regularization involves the replacement of infinite S-matrix elements by functions of a regularization parameter ε, such that for $\varepsilon = 0$ the functions are formally equal to the original S-matrix elements, and for ε away from zero, the functions are finite. These functions (regularized S-matrix elements) can then be divided into a part which is finite and a part which is infinite as $\varepsilon \to 0$. The purpose of regularization is to make sense of such a split, and to allow the removal of the latter part in renormalization. For example, with the regularization scheme in force, (2.1) would become

$$m_{\mathrm{R}} = m_{\mathrm{B}} + \delta m(\varepsilon) \qquad (2.2)$$

$\delta m(\varepsilon)$ is finite for ε away from zero, and $\delta m(0) = \delta m$ is infinite.

The only restriction that need be applied in the choice of a regularization scheme is that it preserves as many of the symmetries of the unregularized Lagrangian as possible. For example, general covariance and gauge invariance should be maintained. Such a regularization scheme is generally provided by dimensional regularization (t'Hooft and Veltman 1972), in which $\varepsilon = n - d$, where d is the physical dimension of space-time. This regularization scheme has been used in many of the calculations described below, and can be used in the renormalization process in precisely the way described by t'Hooft (1973) for Minkowski space.

Given a particular choice of a regularization scheme, a theory is said to be renormalizable if only a finite number of constants require normalization. It is crucial that this be the case if a theory is to be capable of making physical predictions. Since all interacting theories of interest (apart from gravity of course) are renormalizable in Minkowski space, one must ask whether they remain so in curved space-time. Without careful thought, one might think that the answer to this question is obviously 'yes', because the infinities arise from ultra-violet properties of Feynman integrals in momentum space which, in coordinate space, are short distance properties, and locally (over short distances) all curved space-times look like Minkowski space. There are two objections to this argument, which are borne out in actual calculations. Firstly, it is not clear that even short distance divergences will not probe sufficiently far as to pick up effects due to local curvature, and secondly, there are so-called overlapping divergences, which are manifestly non-local; resulting from a local (short distance) infinity from one part of a Feynman graph, multiplying a finite, but non-local, quantity from another part of a graph.

One must therefore prove that these effects do not spoil the renormalizability of interacting theories in curved space-time. Ideally, one would hope to do this without having to choose a particular regularization scheme. Otherwise one could not be sure that if a theory proved to be renormalizable in one regularization scheme, it might not be so in another. One should accordingly try to show renormalizability by imposing only those conditions demanded of any regularization scheme; namely, general covariance and gauge invariance. Such proofs have existed for many years in Minkowski space, where general covariance is replaced by Poincaré invariance (Ward 1951; Symanzik 1960; Taylor 1960, 1963; for a brief text-book treatment, see Roman 1969, sections 4.2 and 5.1). The method of Taylor (1960) has the additional advantage of showing that well defined solutions to the theories considered can exist *outside perturbation theory*. However, in attempting to extend such proofs to curved space-time (Birrell and Taylor 1980), one is met with the full force of (A) above; that is, one can no longer appeal to Poincaré invariance, but must use general covariance, which is very much less restrictive.

Consider the following rough example of this difficulty: in Minkowski space, δm is argued to be constant, and hence removable by mass renormalization, because a constant is the only form invariant scalar available (see, for example, Weinberg 1972, section 13.4). This argument also applies in de Sitter space, where invariance under the de Sitter group can be used to take through the Minkowski space proofs virtually unchanged. (That this is the case is confirmed in the explicit calculations of Drummond (1975).) In a general curved space-time, however, all one can say is that δm is a scalar field, and, for example, as well as being a constant, δm could be proportional to the Ricci scalar $R(x)$. Indeed, actual calculations of $\lambda\phi^4$ theory in curved space-time (Birrell and Ford 1979; Birrell 1980; Bunch, Panangaden, and Parker 1980; Bunch and Panangaden 1980; Bunch and Parker 1980) show that divergent terms proportional to $R(x)$ do indeed occur in first and second order perturbation theory. Such terms cannot be removed by mass renormalization, but necessitate the renormalization of a new bare constant ξ_B, introduced into the Lagrangian in a term $\xi_B R(x)\phi^2(x)$ (Callan, Coleman, and Jackiw 1970).

The Ricci scalar is not the only scalar field that could arise in the term δm. Indeed, $R(x)$ is especially simple in that it depends only on the metric at the point x, i.e. it is local. There might also be non-local divergent scalars. As an example, consider the coincident point limit of the dimensionally regularized Feynman propagator for a scalar field in four dimensional (d = 4) curved space-time; this has the form (see any of the 1980 references just cited):

$$G_{\mathrm{F}}(x,x) = -\,\mathrm{i}\,\langle 0|T(\phi(x)\phi(x))|0\rangle \tag{2.3}$$
$$= -\frac{\mathrm{i}}{8\pi^2(n-4)}\left[m^2 + \left(\xi - \frac{1}{6}\right)R(x)\right] + G_{\mathrm{F}}^{\mathrm{finite}}(x,x) + 0(n-4)$$

where $G_{\mathrm{F}}^{\mathrm{finite}}(x,x)$ is finite as $n \to 4$. Since the left hand side of (2.3) is a scalar, as is the divergent, first term on the right hand side, so must be $G_{\mathrm{F}}^{\mathrm{finite}}(x,x)$. However, unlike $R(x)$, $G_{\mathrm{F}}^{\mathrm{finite}}(x,x)$ is not only non-local, depending on the geometry at points other than x (as is seen explicitly in the expression of Birrell 1980), but also depends on the choice of state $|0\rangle$ appearing in (2.3). General covariance alone cannot rule out the appearance in S-matrix elements of divergent terms proportional to $G_{\mathrm{F}}^{\mathrm{finite}}$, or of other state dependent, non-local scalars. In fact, in the second order calculations mentioned above, individual Feynman diagrams involving overlapping divergences do give rise to divergent terms proportional to $G_{\mathrm{F}}^{\mathrm{finite}}$. However, in the sum of all second order diagrams all such non-local divergences cancel, and one is not compelled to take the undesirable step of adding a state dependent counter-term into the action. Can one be sure that the cancellation of non-local divergences will continue to all orders?

As yet, there is no regularization independent proof that this is the case, and it is clear that such a proof will require stronger arguments than can be provided using general covariance alone. One is thus forced to look for proofs using particular regularization schemes. Such a proof has recently been given by Bunch (1980) for $\lambda\phi^4$ theory using dimensional regularization, and leads one to expect that a regularization independent proof probably does exist. Of course, until such a proof is given, one cannot be sure that there does not exist some other covariant regularization scheme in which the cancellations mentioned above do not occur. Nor can one be sure that solutions exist outside perturbation theory.

It should also be mentioned that a number of authors (Utiyama 1962; Callan, Coleman, and Jackiw 1970; Freedman, Muzinich, and Weinberg 1974; Collins 1976) have studied the renormalizability of interacting quantum fields in curved space-time by expanding the *S*-matrix elements (or Green functions) as a functional power series (Volterra series) in the metric around Minkowski space. In so doing they have found that the theories considered do remain renormalizable (with perhaps the additional renormalization of ξ_B in $\lambda\phi$ theory, as mentioned above). This method is open to the criticism that it gives no information about the global (i.e. non-local) effects of the space-time on the field theory, and in particular, takes no account of (B) and (C) above. It is, for example, a particularly inappropriate method for treating interacting quantum field theory in Schwarzschild space-time.

One especially important consequence of (B), following from (1.3) is that one must consider the renormalization of the vacuum to vacuum amplitude $\langle \text{out}|\text{in}\rangle$. In Minkowski space, this amplitude is consistently set to unity by normal ordering the interaction Hamiltonian. In the gravitational domain (where energy differences are measurable), such a heavy handed approach is no longer acceptable. Not all of the divergences in $\langle \text{out}|\text{in}\rangle$ can be removed by the renormalization of constants in the matter field Lagrangian, and, for the theory to be renormalizable, the additional divergences must be removed by renormalization of constants in the gravitational Lagrangian (the Einstein Lagrangian augmented by higher derivative terms). This is the case even for free fields (see, for example, Bunch 1979; Christensen 1978; Birrell and Davies 1981) and the interaction simply causes the relation between the bare and renormalized gravitational constants to depend on the interaction strength (Bunch and Panangaden 1980; Shore 1980). Full information about such renormalizations *cannot* be obtained by expansion about Minkowski space (Drummond and Hathrell 1980*a*).

3. PHYSICAL PROCESSES INVOLVING INTERACTIONS

If one accepts that theories which are renormalizable in Minkowski space will remain so in curved space-time, then one can turn to consideration of the effect of curvature on the outcome of physical processes involving interactions.

S-matrix elements, which give the probability amplitude for a process to occur, are usually calculated from Green functions by the use of so-called reduction formulae (see, for example, Bjorken and Drell 1965, Ch. 16). In curved space-time, these reduction formulae (Birrell and Taylor 1980) are considerably complicated by (B) above. As well as involving the Green functions

$$\frac{\langle \mathrm{out}|T(\phi(x_1)\ \ldots\ \phi(x_n))|\mathrm{in}\rangle}{\langle \mathrm{out}|\mathrm{in}\rangle}\ , \tag{3.1}$$

the reduction formulae depend on the Bogolubov coefficients, which relate positive frequency modes in the past to those in the future.

Just as in the case of free fields (Parker 1968, 1969), a non-trivial Bogolubov transformation between past and future positive frequency modes implies that particles are created from the vacuum and this is one of the reasons for (1.3). However, in the case of interacting fields there is another cause of particle production from the vacuum, stemming directly from (A). That is, certain processes, which are forbidden in Minkowski space by energy—momentum conservation, can occur in curved space-time. Two such processes, representing particle production from the vacuum in quantum electrodynamics, are illustrated by the Feynman diagrams in Fig.1. Actual calculations of similar diagrams have been given by Scarf (1962); Lotze (1978); Birrell and Ford (1979); Birrell, Davies, and Ford (1980).

It is perhaps worth remarking that none of the recent studies of asymmetric, cosmological baryon production in grand unified field theories (see, for example, Nanopoulos and Weinberg 1979, or the review of Ross in this volume) take account of (A) and (B) in the way described above. The investigation of grand

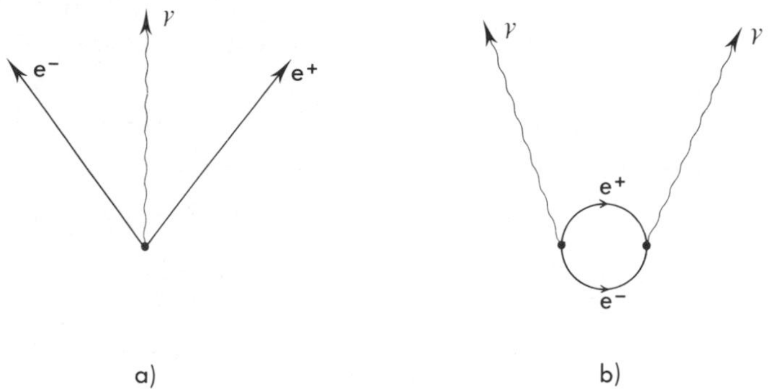

FIG.1a) An electron positron pair and a photon being created from the vacuum.

1b) Two photons being created from the vacuum.

unified field theories in curved space-time, though technically difficult, should prove most rewarding.

The calculation of S-matrix elements only provides one with information about the state of the system in the distant future and past. To obtain information about the intervening region, one must calculate the expectation value of the various currents of the theory. Of particular interest is the expectation value of the stress tensor $T_{\mu\nu}$, as it is this quantity which, in the semi-classical theory, plays the role of the source of the classical background gravitational field.

Although the author is unaware of any papers calculating $\langle T_{\mu\nu}\rangle$ for an interacting theory in curved space-time (Bunch, Panangaden, and Parker (1980) calculate it to first order in $\lambda\phi^4$ theory in the asymptotic region only), a number of studies have been made of the conformal trace anomaly for interacting theories in curved space-time (Birrell and Davies 1978; Drummond and Shore 1979; Drummond and Hathrell 1980a; Shore 1980). In theories whose action is invariant under conformal transformations ($g_{\mu\nu}(x) \to a(x)\, g_{\mu\nu}(x)$, accompanied by a re-scaling of the fields), one can show that $T_{\mu}^{\ \mu} = 0$. In such theories, the appearance of a non-zero trace of the expectation value of the stress tensor, i.e. $g^{\mu\nu}\langle T_{\mu\nu}\rangle \neq 0$, is known as a conformal anomaly. In free theories the conformal anomaly is a state independent, local, geometrical quantity (for reviews see Parker 1979; Birrell and Davies 1981), but for interacting theories it will in general be non-local, non-geometrical and state dependent.

From a knowledge of the trace of the stress tensor in a Robertson—Walker space-time, one can determine whether particle production occurs (Parker 1979, Eq. (6.8)). In the case of free fields, the trace anomaly implies that no production will occur, which is a statement of a result of Parker (1973) that in conformally flat space-time no particle production can take place for theories with conformally invariant actions. This latter result holds formally for interacting theories as well, if one ignores the effects of renormalization. However, use of the $\lambda\phi^4$ theory trace anomaly in Parker's (1979) Eq. (6.8) shows that in general there will be particle production in the case of interacting theories (Birrell and Davies 1980). This particle production arises directly from the need to renormalize constants in the Lagrangian of the interacting theory, and is, in particular, due to the non-local, non-geometrical nature of the anomaly. The renormalization has broken the conformal symmetry of the system in a non-trivial way.

Other examples of symmetry breaking (of different types) in interacting theories in curved space-time can be found in the papers of Dreitlein (1974); Domokos, Janson, and Kovesi-Domokos (1975); Bludman and Ruderman (1977); Canuto and Lee (1977); Frolov, Grib, and Mostepanenko (1977, 1978); Gibbons (1978); Kennedy (1980), within which references to related work may be found.

Consideration of the effect of interactions on particle production raises the question of their effect on the Hawking (1975*a*,*b*) radiation from black holes. In particular, it is of prime importance that the black hole continues to emit *thermal* radiation in the presence of interactions, since, otherwise, the generalized second law of thermodynamics (Bekenstein 1974; Hawking 1975*a*,*b*) could be transcended (see, for example, the argument in Birrell and Davies 1978). That interacting Hawking radiation is thermal when calculated in perturbation theory is strongly supported by the arguments of Gibbons and Perry (1976, 1978) and Hawking (this volume). That this remains the case outside perturbation has been verified in the particular exactly solvable case of the Thirring model (Birrell and Davies 1978) and is suggested by the arguments of Sewell (1980) based on considerations of axiomatic quantum field theory.

Finally, one of the most remarkable results to arise from the study of physical processes involving interacting fields in curved space-time must be mentioned: by calculating the one-loop vacuum polarization contribution to the photon effective action in quantum electrodynamics, Drummond and Hathrell (1980*b*) have shown that in most space-times the photon will propagate at speeds 'greater than that of light'. Although Drummond and Hathrell argue that there is no 'logical or experimental inconsistency' in this result, this is a phenomenon which must deserve further investigation.

4. TOPOLOGICAL EFFECTS

Although this final topic is being treated separately, topological effects are implicit in many of the results discussed in the previous sections, since wherever *global* properties of the space-time play a role, one must consider the topology of space-time. For example, it was remarked that the use of expansion about Minkowski space does not provide information about the effect of the global properties of space-time on the renormalization process. In particular, Drummond and Hathrell (1980*a*) have shown that in such a treatment of the renormalization of $\langle \text{out} | \text{in} \rangle$, no information is given concerning the renormalization of the term in the gravitational action proportional to a certain *topological invariant* (the Euler number of the Euclideanized manifold). Renormalization of such a term is found to be necessary if one does not expand about Minkowski space (Shore 1980).

Topological effects also manifest themselves in the non-local divergences arising from individual Feynman diagrams containing overlapping divergences. In general such terms will contain both metric and topological structure, as was found explicitly by Birrell and Ford (1980), who considered the second order renormalization of $\lambda\phi^4$ theory in spatially flat Robertson-Walker space-times with topology $R^3 \times S^1$. As in the topologically simple case, the renormalization (of the self-energy and vertex functions) was successfully effected with the addition of a ξ_B renormalization to the renormalizations already necessary in Minkowski space. In particular, no topology dependent renormalization was necessary, all divergent topological terms cancelling.

One of the interesting aspects of the effect of topology on interacting theories is that it can be studied in flat, non-Minkowskian, space-times. This removes the formidable technical difficulties associated with interacting quantum fields in curved space-time.

The renormalization of interacting theories in flat space-times with non-trivial topologies is an essentially trivial matter, because the regularization independent proofs continue to hold. The reason for this is that these proofs use only the infinitesimal isometries of the space-time (Killing vectors), and all flat space-times, regardless of topology are maximally symmetric (see, for example, Weinberg 1972, section 13.1), thus allowing only the same maximally form invariant renormalizations as in Minkowski space. For example, δm, mentioned in section 2, can still only be a constant, removable by mass renormalization. What these proofs cannot reveal is whether δm is independent of the space-time topology. This latter point can only be investigated using a particular regularization scheme. To see this, consider flat space-time with topology $R^3 \times S^1$, where the 'circle' S has circumference L. One might find that the regularized δm depends on the topology through the length L, even though $\delta m(\varepsilon;L)$ is still constant. This possibility has been investigated to second order in perturbation theory in $\lambda\phi^4$ theory by Birrell and Ford (1980) and Toms (1980*a*,*b*,*c*) who find that δm and other counter-terms remain independent of topology. An argument that this is a general result has been proposed by Banach (1980).

Space-time topology can also have a profound effect on the physical processes that occur due to interaction. To start with, Isham (1978*a*,*b*) has shown that a new type of field configuration (usually called a *twisted field* is generally possible in space-times with non-trivial topology (see, for example, the review of Avis in this volume). Not only is it *possible* to combine twisted and untwisted fields in interacting theories, but, in the case of spinor fields, Avis and Isham (1979) have argued that, to preserve SL(2,C) gauge invariance, one *must* combine them.

One of the most interesting phenomena to be found in the study of interacting twisted and untwisted fields in space-times with non-trivial topology is the generation of topology dependent particle masses, and the possibility of symmetry breaking without the need for a tachyonic mass in the original Lagrangian (Ford and Yoshimura 1979; Birrell and Ford 1980; Denardo and Spallucci 1980*a*,*b*; Ford 1980; Percacci 1980; Toms 1980*a*,*b*,*c*,*d*). In the case of space-times with topology $R^3 \times S^1$, this phenomenon is very closely related to the corrections due to interactions to the Casimir effect (Ford 1979; Kay 1979) and to symmetry breaking in finite temperature field theory (Kirzhnits and Linde 1972, 1976; Dolan and Jackiw 1974; Weinberg 1974). The study of quantum effects in symmetry breaking in the case of twisted fields is severely complicated (Toms 1980*b*,*c*) by the fact that if a twisted field develops a non-zero vacuum expectation value, it must be non-constant (Avis and Isham 1978). Thus it is not possible to consider symmetry breaking by expansion of the effective action about a constant field as in the usual treatment (see, for example, Abers and Lee 1973, section 16). This problem is still open to investigation.

5. CONCLUSION

The topics discussed in the previous sections show that, although much has been achieved in the subject under review, there are still many areas needing further study. In particular, much more can be done to make more rigorous the results and formulations already given. There are also sure to be other, as yet unknown, results of physical importance to arise from the study of interacting quantum fields in curved space-time.

ACKNOWLEDGEMENTS

I have enjoyed discussions on the subject of this article with many researchers, especially my past or present colleagues T.S. Bunch, P.C.W. Davies, L.H. Ford and J.G. Taylor. I am grateful to M.J. Duff for sending me a preliminary copy of his paper and for conversations concerning the same.

This work was supported by an S.R.C. research grant.

REFERENCES

Abers, E.S. and Lee, B.W. (1973). *Phys. Reps.* **9C**, 1.
Avis, S.J. and Isham, C.J. (1978). *Proc. R. Soc.* **A363**, 581.
—— —— (1979). *Nucl. Phys.* **B156**, 441.
——, Isham, C.J., and Storey, D. (1978). *Phys. Rev.* **D18**, 3565.
Banach, R. (1980). Topology and Renormalizability. University of Manchester preprint.
Bekenstein, J.D. (1974). *Phys. Rev.* **D9**, 3292.
Birrell, N.D. (1980). *J. Phys.* **A13**, 569.
—— and Davies, P.C.W. (1978). *Phys. Rev.* **D18**, 4408.
—— —— (1980). *Phys. Rev.* **D22**, 322.
—— —— (1981). *Quantum Fields in Curved Space-time*. Cambridge University Press (to be published).
—— and Ford, L.H. (1979). *Ann. Phys. (N.Y.)* **122**, 1.
—— —— (1980). *Phys. Rev.* **D22**, 330.
—— and Taylor, J.G. (1980). *J. Math. Phys.* **21**, 1740.
——, Davies, P.C.W., and Ford, L.H. (1980). *J. Phys.* **A13**, 961.
Bjorken, J.D. and Drell, S.D. (1965). *Relativistic Quantum Fields*. McGraw-Hill, New York.
Bludman, S.A. and Ruderman, M.A. (1977). *Phys. Rev. Lett.* **38**, 255.
Bunch, T.S. (1979). *J. Phys.* **A12**, 517.
—— (1980). BPHZ renormalization of $\lambda\phi^4$ field theory in curved space-time. University of Liverpool preprint.
—— and Panangaden, P. (1980). *J. Phys.* **A13**, 919.
—— and Parker, L. (1980). *Phys. Rev.* **D20**, 2499.
——, Panangaden, P., and Parker, L. (1980). *J. Phys.* **A13**, 901.
Callan, C.G., Coleman, S., and Jackiw, R. (1970). *Ann. Phys. (N.Y.)* **59**, 42.
Canuto, V. and Lee, J.F. (1977). *Phys. Lett.* **B72**, 281.
Christensen, S.M. (1978). *Phys. Rev.* **D17**, 946.
Collins, J.C. (1976). *Phys. Rev.* **D14**, 1965.
Critchley, R. (1978). *Phys. Rev.* **D18**, 1849.
Denardo, G. and Spallucci, E. (1980*a*). Dynamical Mass Generation in $S^1 \times R^3$. University of Trieste preprint.
—— —— (1980*b*). Gauge Fields in $S^1 \times R^3$. University of Trieste preprint.
DeWitt, B.S. (1967*a*). *Phys. Rev.* **160**, 1113.
—— (1967*b*). *Phys. Rev.* **162**, 1195, 1239.
Dolan, L. and Jackiw, R. (1974). *Phys. Rev.* **D9**, 3320.
Domokos, G., Janson, M.M., and Kovesi-Domokos, S. (1975). *Nature*. **257**, 203.
Dreitlein, J. (1974). *Phys. Rev. Lett.* **33**, 1243.
Drummond, I.T. (1975). *Nucl. Phys.* **B94**, 115.

Drummond, I.T. and Hathrell, S.J. (1980*a*). *Phys. Rev.* **D21**, 958.
—— —— (1980*b*). QED vacuum polarization in a background gravitational field and its effect on the velocity of photons. DAMTP Cambridge preprint.
—— and Shore, G.M. (1979). *Phys Rev.* **D19**, 1134.
Fischetti, M.V., Hartle, J.B., and Hu, B.L. (1979). *Phys. Rev.* **D20**, 1757.
Ford, L.H. (1979). *Proc. R. Soc.* **A368**, 305.
—— (1980). *Phys. Rev.* **D21**, 933.
—— and Yoshimura, T. (1979). *Phys. Lett.* **70A**, 89.
Freedman, D.Z. and Pi, S.-Y. (1975). *Ann. Phys. (N.Y.).* **91**, 442.
—— and Weinberg, E.J. (1974). *Ann. Phys. (N.Y.).* **87**, 354.
——, Muzinich, I.J., and Weinberg, E.J. (1974). *Ann. Phys. (N.Y.).* **87**, 95.
Frolov, V.M., Grib, A.A., and Mostepanenko, V.M. (1977). *Teor. Math. Fiz. (USSR).* **33**, 42. *(Theor. Math. Phys. (USA)* **33**, 869.)
—— —— —— (1978). *Phys. Lett.* **65A**, 282.
Garding, L. and Wightman, A.S. (1965). *Ark. Fys.* **28**, 129.
Gibbons, G.W. (1978). *J. Phys.* **A11**, 232.
—— and Perry, M.J. (1976). *Phys. Rev. Lett.* **36**, 985.
—— —— (1978). *Proc. R. Soc.* **A358**, 467.
Hajicek, P. (1978). A new generating functional for expectation values of field operator products. University of Berne preprint.
Hartle, J.B. (1977). *Phys. Rev. Lett.* **39**, 1373.
—— and Hu, B.L. (1979). *Phys. Rev.* **D20**, 1772.
Hawking, S.W. (1975*a*). *Commun. Math. Phys.* **43**, 199.
——(1975*b*). in *Quantum Gravity.* (ed C.J. Isham, R. Penrose, and D. Sciama). Clarendon, Oxford.
t'Hooft, G. (1973). *Nucl. Phys.* **B61**, 455.
—— and Veltman, M. (1972). *Nucl. Phys.* **B44**, 189.
Isham, C.J. (1978*a*). *Proc. R. Soc.* **A362**, 383.
—— (1978*b*). *Proc. R. Soc.* **A364**, 591.
Kay, B.S. (1979). *Phys. Rev.* **D20**, 3052.
—— (1980). *Commun. Math. Phys.* **71**, 29.
Kennedy, G. (1980). paper in preparation.
Kirzhnits, D.A. and Linde, A.D. (1972). *Phys. Lett.* **42B**, 471.
—— (1976). *Ann. Phys. (N.Y.).* **101**, 195.
Leymann, H., Symanzik, K., and Zimmermann, W. (1955). *Nuovo Cim.* **1**, 1425.
Lotze, K.-H. (1978). *Acta Phys. Pol.* **B9**, 665, 677.
Nambu, Y. (1966). *Phys. Lett.* **26B**, 626.
Nanopoulos, D.V. and Weinberg, S. (1979). *Phys. Rev.* **D20**, 2484.
Parker, L. (1968). *Phys. Rev. Lett.* **21**, 562.
—— (1969). *Phys. Rev.* **183**, 1057.
—— (1973). *Phys. Rev.* **D7**, 976.
—— (1979). in *Proc. NATO Adv. Study Inst. Gravitation: Recent Devel.* (ed M. Levy and S. Deser). Plenum Press, New York.
Percacci, R. (1980). Global definition of non-linear sigma model and some consequences. University of Trieste preprint.
Roman, P. (1969). *Introduction to Quantum Field Theory.* Wiley, New York.
Scarf, F.L. (1962). in *Les Theories Relativistes de la Gravitation.* Centre National de Recherche Scientifique, Paris.
Sewell, G.L. (1980). Queen Mary College seminar.
Shore, G.M. (1980). *Phys. Rev.* **D**, (in press).
Symanzik, K. (1960). *Commun. Math. Phys.* **18**, 48.
Taylor, J.G. (1960). *Nuovo Cim.* **17**, 695.
—— (1963). *Nuovo Cim. Suppl.* **1**, 857.
Toms, D.J. (1980) *Phys. Rev.* **D21**, 928.
—— (1980*b*). *Phys. Rev.* **D21**, 2805.
—— (1980*c*). *Ann. Phys. (N.Y.).* (in press).
—— (1980*d*). *Phys. Lett.* **A**, (in press).

Utiyama, R. (1962). *Phys. Rev.* **125**, 1727.
Ward, J.C. (1951). *Phys. Rev.* **84**, 897.
Weinberg, S. (1972). *Gravitation and Cosmology*. Wiley, New York.
—— (1974). *Phys. Rev.* **D9**, 3357.
Wightman, A.S. (1956). *Phys. Rev.* **101**, 860.

8

IS THERMODYNAMIC GRAVITY A ROUTE TO QUANTUM GRAVITY?

P.C.W. Davies

Department of Theoretical Physics, University of Newcastle upon Tyne

1. THE IMPORTANCE OF THE THERMODYNAMIC CONNECTION

It is frequently remarked that the effects of quantum gravity will manifest themselves only on Planck scales (10^{-33} cm, 10^{-43} s). While this reasoning is based on a linearized theory, and may even then not be strictly true (the graviton contribution to the energy loss by a microscopic black hole would be observable), nevertheless the subject of quantum gravity offers little prospect for experimental validation. It is therefore all the more important that our theories should (i) be internally consistent (ii) consistent with other branches of physics. Indeed, it may be that consistency will emerge as the only criterion by which a fully developed theory can be elevated over its rivals.

The motivation for constructing a complete theory of quantum gravity is, in my opinion, threefold. Firstly, it is clearly essential that two areas of physics as central as quantum theory and gravitation should at least peacefully coexist, and preferably marry. Secondly, the recent progress towards a unified description of all non-gravitational forces within the context of quantum gauge field theory encourages the speculation that gravity too is but a connected facet of a single quantum interaction. Thirdly, although the effects of quantum gravity as such may be beyond all forseeable technological access, it is to be hoped that the enormous extension in understanding that would accompany the achievement of a full theory would have an impact on other aspects of theoretical physics that are accessible to observation and experiment.

One is reminded of the early days of kinetic theory when the atomic hypothesis could not be directly tested by experiment. Nevertheless, the further development of atomic concepts and their application to gas dynamics enabled definite observational predictions to be made. In a similar vein, it might be hoped that although the Planck regime is still beyond us, the application of quantum gravity ideas to bigger systems may lead to observable effects.

Especially promising in the latter respect is the thermodynamic connection. It has long been noticed that self-gravitating systems display a tendency to increase their degree of aggregation and inhomogeneity – they grow ever more clumpy. We can observe the growth of clumpiness in the heirarchy of gravitational structure around us in the universe: the solar system, stars, star clusters, galaxies and galactic clusters. These structures have apparently evolved from the formless gases that erupted from the big bang several billion years ago.

How can the tendency for self-gravitating systems to irreversibly clump together be encompassed within the traditional framework of thermodynamics that describes time-asymmetric change in the language of entropy and the second law? The application of entropy to gravitational fields does not make any immediate sense; entropy is a concept rooted in ideas about arrangements of discrete entities and our imperfect observation of them. Yet, as we shall see, there is at least one type of self-gravitating system for which a gravitational entropy does seem to be meaningful.

In the early days of the development of quantum theory, thermodynamics played a crucial part. Indeed, the work of Planck and Einstein in groping towards the notion of photons emerged from studies of black body radiation and statistical mechanics (see, for example Kuhn 1978). Can we expect a similar tight relationship when moving on from the photon to the graviton? One elementary fact at least can guide us: gravitational entropy, like its electromagnetic counterpart, requires a quantum description to give it substance. It is via quantum gravity that we must attempt to bring gravitating systems within the framework of thermodynamics. So it may be that this wider 'gravitational thermodynamics' may have important consequences well beyond quantum gravity proper, even though it will have arisen from it.

Further speculations and discussion along these lines can

be found in the article by Penrose in this volume (see also Davies 1974; Penrose 1979).

2. BLACK HOLE THERMODYNAMICS

We already have one beautiful example of thermodynamics facilitating the development of quantum gravity – the black hole. As in more conventional systems, the study of the equilibrium situation is simpler than non-equilibrium. The black hole represents the equilibrium end state of gravitational collapse, so on general grounds we might expect it to be the state of maximum entropy for a self-gravitating system.

In the early 1970s it came to be appreciated that a close relationship existed between the dynamics of black holes and good, old fashioned, heat-engine style thermodynamics (Bardeen, Carter, and Hawking 1973; see also the review by Carter 1973). Processes such as that due to Penrose (1969) indicated that energy could be extracted from rotating black holes, but only with the strictly limited efficiency reminiscent of the entropic limitations on the efficiency of heat-engines (Christodoulou 1970; Christodoulou and Ruffini 1971). Further analysis suggested a complete parallel between the four laws of thermodynamics and certain counterparts in the organization of black hole states.

Briefly, the zeroth law of thermodynamics (the existence of a temperature parameter that is constant throughout an equilibrium system) has its counterpart in the existence of the constant surface gravity parameter, denoted κ, that is constant over the event horizon (i.e. the surface) of a black hole (Carter 1973). The first law of thermodynamics finds a ready counterpart in the conservation of mass-energy, which is, of course, an integral part of the energy interchange that can take place between black holes and their environment.

The second law of thermodynamics – the irreversible increase of entropy – finds the most dramatic parallel in black hole physics. Hawking's area theorem (Hawking 1972) states that in any process involving black holes the total event horizon area cannot decrease. This already strongly suggests the association of entropy with black hole area.

Finally, the third law of thermodynamics, which states, crudely speaking, that the absolute zero of temperature cannot be attained in a finite sequence of processes, seems to be paralleled by the assumption (there is no proof) that a Kerr—Newman black hole, with electric charge Q, angular momentum J and mass M cannot ever achieve the extreme values of Q and J such that

$$\frac{Q^2}{M^2} + \frac{J^2}{M^4} = 1 \tag{2.1}$$

at which the surface gravity κ vanishes. The reason why condition (2.1) is assumed to be unattainable is because it lies on the (J,Q) parameter boundary separating black hole solutions from those of naked singularities. The latter are regarded as abhorrent, and are usually excluded under the cosmic censorship hypothesis (Penrose 1969).

Although laws zero and three clearly point to the association of κ with temperature, because this early work was dealing with classical black holes, their temperature was thought to be absolute zero (i.e. they are completely black), and the similarity between the two sets of four laws was regarded as at most an analogy. Indeed, the assumption of zero temperature even confounds attempts to associate entropy with event horizon area, because the naive formula d (entropy) = d (heat) / temperature implies that the entropy of a black hole is infinite.

There is another way in which black holes can be seen to possess apparently infinite entropy: the use of information theory. When a star implodes to form a hole, the information content of its internal microstates is completely lost to an outside observer. The only vestige of information that is retained is in the gross parameters of M, J and Q. These parameters alone completely characterize the black hole state as viewed externally. If black hole entropy is identified with the total information loss, according to the ideas of Shannon and Weaver (1949), then it is easily seen to have no upper bound in classical physics. The mass of the imploding object can be divided up among an unlimited number of degrees of freedom, each of which carries, say, one bit of information, so the total information content of the star need have no upper bound.

It was pointed out by Bekenstein (1973) that this reasoning has to be amended when quantum effects are taken into account. In order for a particle to fall into a black hole its Compton wavelength should be smaller than the size of the hole, or it will tend simply to scatter off the gravitational field. This sets a lower bound to the mass of the particles of about $\hbar c/GM$, and hence limits the information content to about $GM^2/\hbar c \sim Ac^3/\hbar G$ where A is the event horizon area. In the classical limit ($\hbar \to 0$) this quantity diverges.

Not only does Bekenstein's analysis suggest that quantum theory damps out the divergent black hole entropy, but it indicates the connection

$$S \propto A$$

which is already implied by the association of Hawking's area theorem with the second law of thermodynamics.

These vague ideas were placed on a sound footing by the work of Hawking (1975) on quantum black holes. The application of quantum field theory in a given, classical background gravitational field (sometimes called the one-loop approximation to quantum gravity) led him to discover that black holes are not black after all, but emit radiation with a spectrum characteristic of a body in thermal equilibrium (due to the presence of space-time curvature the spectrum is not Planckian). Thus, a black hole is a black body, and the temperature turns out to be, appropriately enough, proportional to the surface gravity κ. Specifically

$$T = \frac{\kappa}{2\pi k} = \frac{1}{8\pi kM}\left[1 - \frac{(J^2 + \frac{1}{4}Q^4)}{A^2/64\pi^2}\right] \tag{2.2}$$

$$S = \tfrac{1}{4}kA$$

$$= 2\pi k\left[M^2 - \tfrac{1}{2}Q^2 + M^2\left(1 - \frac{Q^2}{M^2} - \frac{J^2}{M^4}\right)^{\frac{1}{2}}\right] \tag{2.3}$$

in natural units $\hbar = c = G = 1$. The factor k is Boltzmann's constant. This is not the place to give a detailed discussion of the Hawking effect as such. The reader is referred to Birrell and Davies (1981) for a review.

Having established a reliable thermodynamic basis for black holes, Hawking's work revealed that the four laws of black hole dynamics are really only specific examples of the four laws of thermodynamics. One may therefore proceed to investigate the thermodynamic consequences of black holes in the familiar way by application of these laws.

It is possible to develop a detailed theory of black hole thermodynamics (Davies 1977, 1978; Hut 1977; Landsberg and Tranah 1980*a*,*b*; Tranah and Landsberg 1980). For example, the stability of black holes confined to the interior of a perfectly reflecting box emerges from a search for entropy minima in the combined system of black holes plus surrounding heat bath. A single Schwarzschild hole in a box will remain in stable equilibrium if the volume V of the box satisfies the inequality

$$V < \frac{2^{20}\pi^4}{5^5}\,\frac{E^5}{a} \tag{2.4}$$

where a is the Stefan–Boltzmann radiation constant (Hawking 1976*a*). For boxes larger than this the hole will start to evaporate. Its fate may be deduced by computing the specific heat at constant J,Q:

$$C_{J,Q} = T\left(\frac{\partial S}{\partial T}\right)_{J,Q} = \frac{MS^3T}{8\pi^2k^2(J^2 + \tfrac{1}{4}Q^4) - 8\pi kT^2S^3}\,. \tag{2.5}$$

For a Schwarzschild hole, $J = Q = 0$, $C = -M/8\pi kT = -M^2$. The fact that C is negative indicates that the black hole is inherently unstable. As it radiates it gets hotter, and thereby radiates more strongly, the process escalating until the evaporation of the hole turns into an explosion. We shall return to the question of the remnant of this explosion later.

The appearance of a negative specific heat is not unfamiliar in self-gravitating systems. A star, for example, if deprived of its internal energy, will rapidly shrink and become hotter. However, closer inspection of (2.5) reveals that $C_{J,Q}$ is not always negative. In the limit $T \to 0$ (achieved if J and Q

approach the values consistent with condition (2.1), it is clear that $C_{J,Q} \to 0$ from *positive* values. In Fig. 1, the

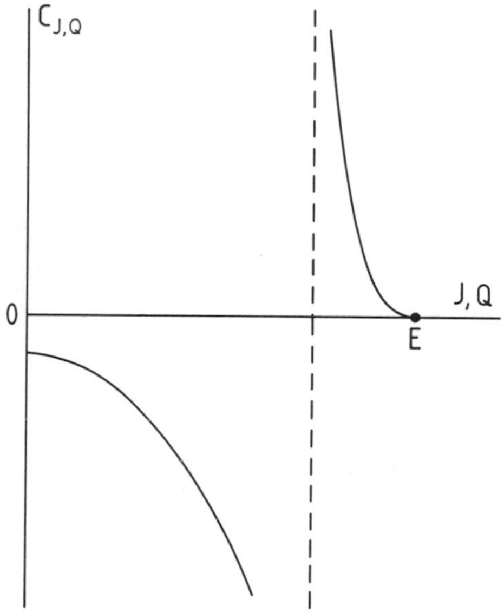

FIG. 1.

specific heat is plotted schematically against J and Q. It passes through an infinite discontinuity at the paremeter values J, Q that lie along the line

$$\alpha^2 + 6\alpha + 4\beta = 3 \tag{2.6}$$

where $J^2 = \alpha M^4$, $Q^2 = \beta M^2$. To the left (low J,Q) side of this discontinuity, $C_{J,Q}$ is negative, but to the right it is positive, indicating that a high J,Q black hole could exist in stable equilibrium with a surrounding heat bath of unlimited volume. The change from positive to negative specific heat is characteristic of a second order phase transition (Davies 1977), though quite what that implies for the external appearance or behaviour of the hole is not clear.

One may also investigate idealized processes such as Carnot

cycles and the reversible transfer of energy between black holes using heat pumps, heat reservoirs and all the other paraphernalia of classical thermodynamics (Davies 1977). The presence of both J and Q as adjustable parameters leads to a very large number of possible processes of this type.

The status of the third law in black hole thermodynamics is curious. In a conventional system one expects $S \to 0$ as $T \to 0$. This expectation has a readily visualizable origin. At the absolute zero of temperature all energy is removed from the system, and it settles into its ground state. Assuming that this is unique it implies that the absolute zero state can be achieved by only *one* microstate. As S is defined in statistical mechanics in terms of the logarithm of the number of microstates consistent with a given macrostate, S clearly vanishes at absolute zero.

In contrast to this, the entropy of the black hole, according to (2.3), tends to $(2M^2 - Q^2)\pi k \neq 0$ as $T \to 0$, a result which already suggests that there is something a bit odd about $T = 0$ for a black hole. A careful analysis (Davies 1977) of the way in which various thermodynamic quantities behave as $T \to 0$ reveals that the usual obstacles that prevent a finite number of operations achieving absolute zero do not apply to black holes. Might it therefore be possible to cool down a black hole to absolute zero, or to induce gravitational collapse in such a way that the hole forms with zero temperature? Does this threaten cosmic censorship?

One strategy to cool a black hole is the following. Take a non-rotating hole of mass $> 10^5 M_\odot$ and drop electric charges into it until it approaches the limiting value of $Q^2 = M^2$. Because of its high mass, and hence low temperature, superradiant (Klein paradox) creation of electron–positron pairs is not energetically possible (Gibbons 1975) during this process. Hence the hole is not able to discharge itself spontaneously. It will continue to radiate (feebly) massless, uncharged quanta (photons, gravitons and neutrinos). In this condition, therefore, it can lose mass, but not charge. If this continues until $M^2 = Q^2$ then the hole will have reached zero temperature. Of course it is not clear (i) how to do this thermodynamically (though, as remarked, there is no *thermodynamic* objection to reaching $T = 0$) (ii) whether such a path is otherwise *dynamically* permitted. Nevertheless, this example does call into question the validity of the third law as applied to black holes.

Another example has been discussed by Farrugia and Hajicek (1979). In this case a black hole is created with the limiting value of charge by projecting a spherical charged shell onto an existing charged black hole. Outside the shell a new event horizon forms, representing a hole with $Q^2 = M^2$.

A naive attempt to reach absolute zero by spinning a black hole ever faster until $J^2 = M^4$ is to drop co-rotating particles along the spin axis of the hole. The intention is to deliver more angular momentum than (mass)^2 to the hole, thus raising the ratio J^2/M^4. As already remarked, there is a minimum mass delivery if the particle is to enter the hole, because the Compton wavelength must be $\lesssim M$, i.e. mass of particle $\gtrsim M^{-1}$. For a particle of spin n units, the ratio J^2/M^4 is altered to

$$\frac{(J + n)^2}{(M + M^{-1})^4} \simeq \frac{M^2 + 2n}{M^2 + 4}$$

if $J^2 \simeq M^4$. This ratio is always $\leqslant 1$ if $n \leqslant 2$. Quantum field theory, which seems to exclude particles with spin > 2 in nature, comes to the rescue of cosmic censorship.

A further line of enquiry concerns non-equilibrium thermodynamics. In the case of small perturbations about thermodynamic equilibrium much progress can be made. For example, black holes are subject to the fluctuation-dissipation theorem. Further details on this topic are given in the article by Sciama in this volume.

3. FUNDAMENTAL SIGNIFICANCE OF BLACK HOLE ENTROPY

The fact that quantum gravity (albeit in a questionable approximation) confirms the classical analogy that exists between black holes and thermodynamics is compelling evidence in favour of the theory behind quantum black holes, even if we never directly confirm the black hole radiance predicted by Hawking. However, before rushing to claim that Hawking's discovery opens the door to an intimate connection between gravity and thermodynamics it is vital to check that the nature of the Hawking radiation really is thermal under all circumstances, and that his result was not just a mathematical quirk arising from an idealized model.

Strong evidence that the Hawking effect is fundamental comes from the fact that it can be derived in several different ways. Perhaps the most elegant route is to forsake the collapsing star altogether and work with the maximally extended manifold, such as Kruskal space-time which is everywhere a solution of the vacuum Einstein equations. Indeed the role of the star seems to be only to establish certain boundary conditions in the past on the quantum state; the details of the collapse do not enter into the final result.

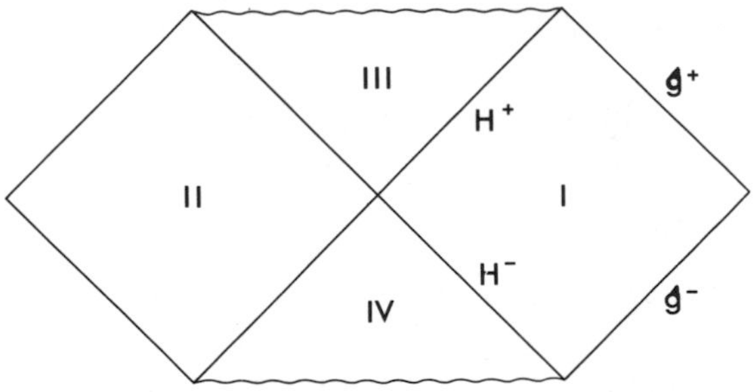

FIG. 2.

A Penrose conformal diagram of the Kruskal manifold is shown in Fig. 2. Region I corresponds to 'our universe' outside the hole, region II is a sort of 'mirror' universe connected to ours through the hole, region III is the black part of the hole and region IV the time-reversed white hole. The wiggly lines are the singularity at $r = 0$ in the past and future. Note that the two universes I and II are completely causally disconnected by the event horizons marked H.

The quantum field, which for simplicity we take to be a massless scalar field, satisfies the wave equation

$$\Box\phi = 0 \qquad (3.1)$$

on this manifold. To solve the wave equation a choice of coordinates must be made. In terms of Schwarzschild coordinates r,t the metric of this space-time is

$$ds^2 = \left(1-\frac{2M}{r}\right)dt^2 - \left(1-\frac{2M}{r}\right)^{-1}dr^2 - r^2(d\theta^2+\sin^2\theta d\phi^2) \qquad (3.2)$$

which is clearly singular at the horizons $r = 2M$. If these coordinates are used to separate the variables in (3.1) then the solutions will also blow up at the horizons $H^{\pm}$.

On the other hand, if we use Kruskal coordinates defined by (Kruskal 1960)

$$\bar{u} = - e^{-u/4M} \qquad (3.3)$$

$$\bar{v} = e^{v/4M} \qquad (3.4)$$

$u = t - r^*$, $v = t + r^*$, $r^* = r + 2M \ln\left|\frac{r}{2M} - 1\right|$, then (3.2) becomes

$$ds^2 = \frac{32M^3}{r} \exp(-r/2M)d\bar{u}d\bar{v} - r^2(d\theta^2 + \sin^2\theta d\phi^2) \qquad (3.5)$$

which can be analytically continued across the entire manifold shown in Fig. 2. The coordinate singularity at $r = 2M$ has been removed.

Rather than discuss the full four dimensional Kruskal manifold it is convenient to drop the angular part and treat the two-dimensional analogue. The wave equation (3.1) can then be solved exactly (Davies, Fulling, and Unruh 1976) in terms of simple exponential modes of the form $e^{-i\omega\bar{u}}$ and $e^{-i\omega\bar{v}}$. If these modes are denoted $\bar{u}_k$ $(\bar{u},\bar{v})$, where $-\infty < k < \infty$ and $\omega = |k|$, then a general solution ϕ may be expanded

$$\phi = \sum_k (a_k\bar{u}_k + a_k^{\dagger}\bar{u}_k^*) \qquad (3.6)$$

and a vacuum state $|\bar{0}\rangle$ constructed by the requirement

$$a_k |\bar{0}\rangle = 0, \qquad \forall k. \qquad (3.7)$$

Had we instead solved the wave equation in terms of Schwarzschild null coordinates u, v, and expanded ϕ

$$\phi = \sum_k (b_k u_k + b_k^\dagger u_k^*) \tag{3.8}$$

where u_k are Schwarzschild modes of the form $e^{-i\omega u}$ and $e^{-i\omega v}$, then the vacuum state defined by

$$b_k|0\rangle = 0, \quad \forall k \tag{3.9}$$

would differ from $|\overline{0}\rangle$ because the u_k are a superposition of positive and negative frequencies $\overline{u}_k$ modes.

What is the physical significance of the difference between $|0\rangle$ and $|\overline{0}\rangle$? A computation of $\langle 0|T_{\mu\nu}|0\rangle$ reveals that the stress tensor diverges at the horizons $H^{\pm}$, as might be expected, because the modes u_k are singular there (Christensen and Fulling 1977; Candelas 1980). On the other hand $\langle\overline{0}|T_{\mu\nu}|\overline{0}\rangle$ is finite $\forall r \neq 0$. However at large r, the metric (3.2) goes over to the Minkowski metric, and the associated modes u_k reduce to the standard exponential modes of ordinary Minkowski space quantum field theory. In contrast, the Kruskal metric behaves very differently at large r, and the modes $\overline{u}_k$ bear no simple relation to those of the conventional theory. One may conclude that, far from the hole (in either region I or II), the state $|0\rangle$ approximates to the usual vacuum state of ordinary quantum field theory.

What state, then, does $|\overline{0}\rangle$ represent far from the hole? One way of finding out is to compute the positive frequency Wightman functions for the two vacua:

$$G^+(x,x') = \langle 0|\phi(x)\phi(x')|0\rangle \tag{3.10}$$

and

$$\overline{G}^+(x,x') = \langle\overline{0}|\phi(x)\phi(x')|\overline{0}\rangle. \tag{3.11}$$

These are easily evaluated (Birrell and Davies 1981)

$$G^+ = \frac{-1}{2\pi} \ln \Delta u \Delta v \tag{3.12}$$

$$\bar{G}^{+} = \frac{-1}{2\pi} \ln \Delta\bar{u}\Delta\bar{v} \tag{3.13}$$

from which an infra-red divergence has been discarded.

A substitution for $\bar{u}$ and $\bar{v}$ from (3.3) and (3.4) into (3.13) yields

$$\bar{G}^{+}(x,x') = \frac{-1}{2\pi} \ln\{\cosh[(t-t')/4M] - \cosh[(r^{*}-r^{*\prime})/4M]\} + \text{function of } (r,r'), \tag{3.14}$$

which manifestly satisfies the relation

$$\bar{G}^{+}(t,r;\ t',r') = \bar{G}^{-}(t + 8\pi Mi, r;\ t',r') \tag{3.15}$$

where $\bar{G}^{-}(x,x') = \bar{G}^{+}(x',x)$. This is the condition that characterizes a *thermal* Green function (Gibbons and Perry 1978) with temperature $T = (8\pi Mk)^{-1}$. One concludes that an observer in the asymptotic ($r \to \infty$) part of region I would observe the state $|0\rangle$ as a thermal distribution of quanta. This interpretation can be confirmed by studying the experiences of a particle detector (Unruh 1976; De Witt 1979) confined to region I.

The thermal character of the state $|\bar{0}\rangle$ is really a consequence of the causal structure of the Kruskal manifold. The crucial feature is that $|\bar{0}\rangle$ has been constructed using modes $\bar{u}_k$ that are analytic across the horizons $H^{\pm}$. In this respect, the exact form of the modes is not important. The associated entropy can be envisaged as due to the fact that an observer permanently confined to region I must forsake all information about the quantum state in the region II (the mirror universe) which is rendered inaccessible by the horizons. Thus, a pure state on the whole manifold must be regarded as a mixed state if all measurements are confined to the submanifold I. An observer in I is therefore obliged to resort to a density matrix description (Hawking 1976*a*,*b*; Israel 1976). One may therefore see that the thermal properties of black holes are rooted in the causal structure, and not the precise space-time geometry. (Indeed, very similar results are obtained for accelerated observers in Minkowski space (Davies 1975; Unruh 1976).) This implies that black hole entropy is a rather fundamental feature.

The use of time-symmetric coordinates $\bar{u},\bar{v}$ leads to a quantum state $|\bar{0}\rangle$ that in region I reproduces the effects of thermally distributed quanta. Because of the time symmetry, this must be regarded as a *bath* of thermal radiation, i.e. both outflowing and incoming, representing a steady state in which the black hole is in thermal equilibrium with its environment. Had we used coordinates $\bar{u},v$ the associated quantum state would have represented a thermal *flux* from the black hole ($u,\bar{v}$ would give a flux into the hole). Although such a state (Unruh 1976) results in a divergent $\langle T_{\mu\nu}\rangle$ on the past horizon H^- in I, it accurately mimics near the future horizon H^+ the physical state that would have resulted from the implosion of a 'star' (in which case the past horizon does not exist).

The thermal nature of black hole radiation goes beyond the mere spectrum. It can be shown that all correlations between different emitted quanta are absent (Hawking 1976*a*,*b*; Parker 1975; Wald 1975). That is, the black hole behaves like a black body in every detail. Furthermore, the expectation value of any observable A restricted to region I, in the state $|\bar{0}\rangle$, can be shown (Israel 1976) to have the form

$$\langle\bar{0}|A|\bar{0}\rangle = \frac{\sum_{n=0}^{\infty}\langle n|A|n\rangle \exp(-E_n/kT)}{\sum_{n=0}^{\infty}\exp(-E_n/kT)} \tag{3.16}$$

where $T = (8\pi Mk)^{-1}$ and $\langle n|A|n\rangle$ is the expectation value for A in the *Schwarzschild* state $|n\rangle$ containing n quanta, E_n being the energy of such a state. This form is the same as that obtained by averaging A over a grand canonical ensemble representing thermally distributed quanta at temperature T.

The appearance of a periodicity in *imaginary* time in (3.15) suggests that one examines quantum field theory in Euclidean (i.e. Riemannian as opposed to pseudo-Riemannian) space. Euclidean quantum field theory enjoys certain mathematical advantages, and it is usually possible to 'rotate' back to space-time at the end of a calculation. This approach has been developed in great detail by Hawking and co-workers (for a review see Hawking 1979). It leads naturally to the construction of the state $|\overline{0}\rangle$ (Hartle and Hawking 1976) and to the identification of black holes with gravitational instantons (Hawking 1977).

If the thermal nature of black holes is as fundamental as these results suggest, then one would expect it to survive the presence of non-gravitational interactions. It is vital to know if, for example, a self interacting field such as $\lambda\phi^4$ is thermally radiated by a black hole, or that QED on a black hole background does not destroy the thermal character of the emissions when $e \neq 0$. If for any reason the emission deviated from the strictly thermal then it would be possible to violate the second law of thermodynamics in the following way. A Schwarzschild black hole is immersed in a bath of thermal radiation, the temperature of which is adjusted so that the rate of black hole energy emission is exactly balanced by absorption. However, the emitted radiation would be non-thermal and the absorbed radiation would be thermal. The black hole would thus gain more radiation entropy than it loses and yet, by remaining constant in mass, its own entropy (see (2.3)) does not change. The hole has therefore succeeded in reducing the entropy of its environment without itself experiencing any change.

Gibbons and Perry (1978) have pointed out that, within the context of perturbation theory, the periodicity properties of the free field propagators (e.g. relation (3.15)) ought to be sufficient to ensure the thermal character of interacting field quanta from a radiating black hole. However, Gass and Dresden (1980) have investigated an explicit model field theory based on the Lagrangian

$$L = L_0 - \frac{\lambda_1}{2}\,\phi_1\phi_2{}^2 - \frac{\lambda_2}{2}\,\phi_2\phi_1{}^2 \qquad (3.17)$$

where L_0 is the free field Lagrangian for the two coupled scalar fields ϕ_1 and ϕ_2. They analyze many perturbative terms, illustrated by Feynman diagrams of the form shown in Fig. 3. The circle represents the event horizon of the black hole. If the fields ϕ_1 and ϕ_2 are uncoupled then the emission of their respective quanta (denoted 1 and 2) are quite

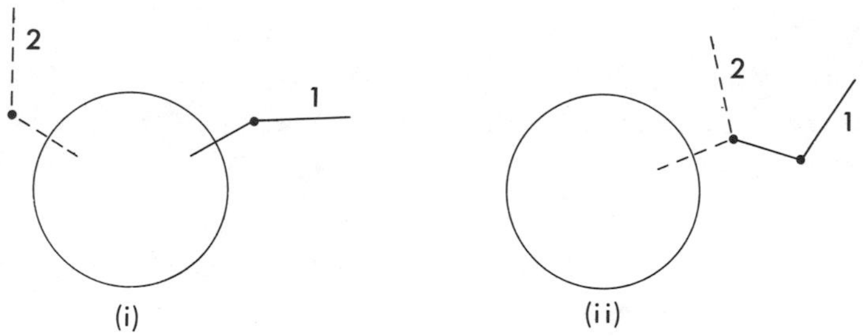

FIG. 3.

uncorrelated. These emissions can be represented as pair creation events at a point (denoted by a dot) just outside the horizon, in which one particle travels to infinity to contribute to the Hawking flux while the other 'tunnels' into the black hole and is lost. When interactions are permitted, diagrams of the sort shown in Fig. 3 (ii) can result, in which the combined emission of quanta 1 and 2 are evidently correlated. Of course, there will be many such diagrams and one hopes (and perhaps expects) that the sum total of all such processes will cancel the apparent correlations introduced by the interaction. Gass and Dresden were not, however, able to conclusively demonstrate this cancellation in their analysis.

Birrell and Davies (1978) have investigated an exactly soluble model of a self-coupled spinor field in a two-dimensional Schwarzschild background. The so-called massless Thirring model (Thirring 1958) is based on the flat space Lagrangian

$$= \frac{i}{2}[\bar{\phi}\gamma^{\mu}\phi,_{\mu} - \bar{\phi},_{\mu}\gamma^{\mu}\phi] + \lambda \mathcal{J}^{\mu}\mathcal{J}_{\mu} \qquad (3.18)$$

where the fermion current $\mathcal{J}^{\mu}$ is

$$\mathcal{J}^{\mu} = \bar{\phi}\gamma^{\mu}\phi \qquad (3.19)$$

as usual. Because of the conformal triviality in two dimensions, one readily computes the Wightman functions in the state analogous to $|\overline{0}\rangle$, even when the self-coupling is present ($\lambda \neq 0$). The result, when compared to a previous analysis by Dubin and Tarski (1967) and Dubin (1976) of the Thirring model in a thermal state, confirms that the black hole emission of Thirring fermions remains thermal, though the spectrum deviates somewhat from the usual form due to the effect of the interaction.

In all the treatments mentioned so far, the black hole itself is not dynamical. For example, no mention has been made of the recoil of the hole to the fluctuations that must occur in the thermal quanta emitted. Page (1980) has pointed out that a mini-hole of 10^{15} gm (10^{-13} cm radius) will recoil significantly (i.e. through 10^{-13} cm) when only 10^{-13} of its total mass has been emitted. This recoil will inevitably introduce correlations in the emitted quanta. However, this would be the case for any hot body, and is merely the result of a lack of thermal equilibrium between the body and its environment.

An important question about the entropy associated with the Hawking process is whether it more properly belongs to the quantum field or to the black hole. Gibbons and Hawking (1977) have investigated this point in connection with tree level quantum gravity calculations. They point out that in the Euclideanized path integral formulation of quantum gravity, the generating functional for the path integrals is also the partition function for the thermodynamics of the system. They construct the gravitational action associated with a Schwarzschild black hole and use it to compute the partition function and hence entropy of the hole. The calculation is purely gravitational, no additional 'matter' fields are present. Their result, the same as (2.3), confirms that the entropy is an intrinsic property of black holes, and does not depend on the type or number of quantum 'matter' fields that may be present in nature.

In spite of the Gass and Dresden analysis, it seems extremely likely that the second law of thermodynamics will survive black holes. Just how quantum field theory 'knows' about the second law and comes to the rescue (classical black holes transcend the second law) is a mystery, but Ford (1978) has shown that this can remain true even when one makes a deliberate attempt to manipulate the quantum fields. Noting that negative energy fluxes can occur in quantum field theory (e.g. the radiation produced by an accelerating mirror can carry negative energy, (Fulling and Davies 1976)) Ford observes that such a flux will reduce the mass, hence entropy of a black hole, should one intervene. He finds that the duration $\hat{\tau}$ of a pulse of negative energy is related, for reasons of quantum field theory, to the (negative) energy change ΔE, by a sort of Heisenberg relation

$$|\Delta E| \simeq \tau^{-1}.$$

For the notion of black hole entropy to have meaning it is necessary for its new energy-robbed state to settle down to equilibrium, which must take at least the time M, i.e. the light travel time across the hole. Hence, to regard the entropy of the black hole as having diminished, one requires $\tau \gtrsim M$, so $|\Delta E| = |\Delta M| \lesssim M^{-1}$. Consequently the reduction in entropy $\Delta S = 8\pi k M \Delta M$ is

$$|\Delta S| \lesssim 8\pi k$$

representing a few bits of information only, i.e. consistent with ordinary statistical fluctuations about thermal equilibrium. This remarkable result indicates an as yet dimly perceived consistency between quantum field theory, black holes and thermodynamics.

4. BEYOND THE BLACK HOLE

As remarked, black holes are the equilibrium end states of self-gravitating systems. The extension of the entropy concept to them appears to be brilliantly successful. Yet in ordinary thermodynamics one can extend the theory beyond the equilibrium case and treat a more general system. Can one hope to similarly extend gravitational entropy to other types of gravitational field?

A number of questions remain unanswered concerning black hole entropy. The original Bekenstein treatment, mentioned in Section 2, began with the information content of the 'star' that imploded to form the hole. This approach is close to the spirit of statistical mechanics, where entropy is defined in terms of the arrangements of microstates. It is now generally believed that the 'star' itself has little to do with either the Hawking emission or the black hole entropy (the Gibbons–Perry derivation of black hole thermal properties did not even include a collapsing star, but worked directly with the vacuum Kruskal manifold). So one wonders what precisely, if anything, corresponds to the 'internal microstates' of the black hole, the arrangements of which provide a measure of its entropy. (See also Bekenstein and Meisels 1977.)

Another curious puzzle about black hole entropy concerns its apparently 'objective' nature. In ordinary statistical mechanics, entropy arises as a result of some sort of coarse-graining which reflects the fact that the observer, being macroscopic, cannot distinguish between closely similar arrangements of microstates. The equilibrium state is then that macrostate which can be achieved in a maximum number of ways by indistinguishable microstates. There is thus an element of subjectivity in this concept, involving as it does the limitations on the observational powers of human beings. In the black hole case, the only limitation seems to be the crude one that the observer be restricted to a submanifold of space-time. Thus, black hole entropy seems, if anything, to have a more concrete physical status than its laboratory counterpart!

One can go further than this. The obligation for an observer outside the black hole to use a density matrix rather than a pure state to describe the quantum properties of a field implies that the presence of a black hole effectively converts pure quantum states into mixtures. This is highly relevant to the old problem (von Neumann 1955) of the so-called collapse of the wave function during a quantum measurement process. Specifically, according to conventional epistemology (see, for example, Davies (1980) for a heuristic introduction or D'Espagnat (1971) for a technical discussion) the observed microsystem only 'collapses' into reality when the off-diagonal terms of the density matrix describing the quantum state disappear. Physically this corresponds to destroying all the interference terms between different vectors, the superposition of which goes to make up the quantum state being probed. When a piece of macroscopic apparatus couples to the microsystem, the interference terms in the quantum state describing the microsystem do indeed disappear, but the *total* system, including the apparatus, is still in a pure state, i.e. interference terms still exist between macroscopic apparatus states. There seems to be no way of ridding the universe of these interference terms, a fact which has led some to propose radically alternative epistemological frameworks (see the articles by De Witt and Bell in this volume).

When the quantum system is allowed to interact with the gravitational field however, we do seem to have at hand a mechanism for collapsing the wave function from a pure state involving interference terms to a mixed state described by a diagonal density matrix. This collapse is, moreover, irreversible, and seems to endow the universe with a fundamental (if observationally insignificant) time asymmetry (see Davies (1974) for a full discussion of this aspect of time asymmetry). In this way it seems that black holes could play a crucial role in removing from the observer the responsibility of creating 'objective reality'.

At first sight it appears that the ability of a black hole to convert a pure quantum state into a mixed one is a consequence of the semi-classical nature of the Hawking effect, in which the background gravitational field itself is not quantized. By allowing the quantum system to couple to a classical one it is no surprise if the wave function loses its status as a quantum superposition. Indeed, Page (1980) has argued that in a full theory of quantum gravity, the interference terms, and the time symmetry, would reappear, i.e. the total system of quantum gravitational field plus quantum 'matter' fields would be described by a pure state.

However, Wald (see the article in this volume) has argued that even a full theory of quantum gravity could not restore the apparently mixed state of the Hawking radiation to a pure state. The reason is, essentially, that by the time higher order quantum gravity effects begin to matter (i.e. when the black hole has evaporated to about one Planck mass) its information content is far too low to be able to restore all the correlations apparently absent between all the enormous number of quanta emitted up to that time. Whether this reasoning is valid when the presence of the singularity (which swallows up vast numbers of degrees of freedom when the imploding body reaches it) is taken into account is questionable.

In the previous section it was remarked that black hole entropy is ultimately associated with the causal structure of the space-time. In particular, the existence of the future event horizon seems to play a crucial role. Gibbons and Hawking (1977) have argued that in the absence of an event horizon, the entropy of a self-gravitating system is simply, and unremarkably, the entropy of the matter.

Does this mean that there is no 'gravitational entropy' for gravitational fields without horizons? Certainly one would not expect the entropy of non-black hole gravitational fields to be defined in terms of the simple concepts used in the equilibrium black hole case. This is true in the case of ordinary entropy. In equilibrium systems S may be defined in terms of macroscopic parameters such as temperature and pressure, but in general, away from equilibrium, these concepts are meaningless, and a more elaborate definition is necessary, in terms of the statistical properties of the microstates. The issue of the consistency of these two definitions is a highly non-trivial matter.

During the collapse of a star to form a black hole, the ordinary entropy of the star is obliterated as the red-shift escalates. In its place appears the (much larger) black hole entropy. Yet this transition, rapid though it may be in practice, is nevertheless not instantaneous. There is an intermediate phase when the matter entropy must be dwindling, and the gravitational entropy growing, towards their final, equilibrium, values. The event horizon, so crucial to the black hole (equilibrium) entropy, is not created at some moment — it is an asymptotic concept. It ought to be possible to study the irreversible approach of a collapsing star to black hole equilibrium using some sort of gravitational analogue to the Boltzmann equation.

Perhaps the most useful suggestion for a formulation of the

gravitational entropy idea is due to Penrose (1979) who directs attention to the way in which the universe as a whole evolves in a time-asymmetric way from a reasonably smooth initial state to a final state involving a great deal of clumpiness (see also the article by Penrose in this volume). This suggests that the Weyl curvature is some sort of measure of the gravitational entropy; the universe begins with most of its curvature in the Ricci part of the Riemann tensor, and the Weyl part accumulates with time, as distortions from homogeneity and isotropy set in. Unfortunately nobody has succeeded in actually writing down a believable formula for gravitational entropy as a functional of Weyl curvature. Moreover, as pointed out by Barrow (1980)*, there exist cosmological models that evolve to end states with vanishing Weyl curvature (though these could be of an insignificant measure in the space of all initial conditions).

If one can indeed regard an anisotropic space-time as possessing more gravitational entropy than an isotropic one, then there is at least one consistency check that can in principle be made. Allowing gravity to couple to matter, and taking into account back reaction effects, an anisotropically expanding universe will induce quantum particle creation and vacuum polarization effects which will in turn exert their own gravitational influence and operate to damp down the anisotropy. If the second law of thermodynamics is to remain intact, then the loss in entropy by the gravitational field during the partial isotropization cannot exceed the entropy of the quanta that are created as a consequence, and the latter is calculable. The quantum dissipation of anisotropy has been studied by several authors, e.g. Zeldovich (1970); Zeldovich and Starobinski (1972); Hu and Parker (1978); Hartle and Hu (1979).

A perturbation technique that enables the number and energy density of created particles to be computed for a wide class of space-times with (small) anisotropy has been given by Birrell and Davies (1980). As an illustration of the sort of consistency calculation I have in mind, consider the case of the metric

---

* Barrow, J. (1980). Private communications.

$$ds^2 = a^2(\eta)\left[d\eta^2 - \sum_{i=1}^{3}(1+h_i(\eta))dx_i^{\,2}\right] \tag{4.1}$$

with

$$h_i(\eta) = e^{-t/t_I}\cos(\nu t+\delta_i)$$

$$a \propto t^{\frac{1}{2}}$$

$$(t_I,\ \nu,\ \delta_i \text{ constants})$$

and where the δ_i are chosen to differ by $2\pi/3$, so that $\sum_i h_i = 0$. This is reminiscent of the mixmaster model universe (Misner 1969).

The energy per unit volume of created quanta at epoch t is computed (Birrell and Davies 1980) to be

$$\left(C^{\alpha\beta\sigma\delta}C_{\alpha\beta\sigma\delta}\right)^{5/4}\frac{t_I^{\,3}}{t^2}\bigg/\left[\left(\frac{3}{2}\right)^{5/4}\times 46080\pi\right] \tag{4.2}$$

where the answer has been expressed in terms of the Weyl tensor at $t = 0$ using

$$C^{\alpha\beta\sigma\delta}C_{\alpha\beta\sigma\delta} = \frac{a^{-4}}{2}\sum_i (h_i'')^2 \ .$$

Although the spectrum of created particles is not thermal, the assumption of a thermal spectrum supplies an upper bound. In the units $\hbar = c = k = 1$, the entropy per unit volume is, to within small numerical factors, the energy density to the power 3/4:

$$\sim \left(C^{\alpha\beta\sigma\delta}C_{\alpha\beta\sigma\delta}\right)^{15/16} t_I^{\,9/4}/t^{3/2} \ .$$

This result should provide an upper bound on the loss of gravitational entropy. The fact that it does not involve G is no surprise for we have not yet taken any account of the back reaction that causes isotropization. This will come from the gravitational dynamics. No attempt is made here to solve the full dynamical problem self-consistently; the metric (4.1) has been simply prescribed as a background space-time in which isotropization gradually occurs over a time scale t_I. The next step is to compute t_I from the gravitational dynamics. Unfortunately this cannot be done from knowledge of the particle density alone, for during the important phase of back reaction, vacuum polarization effects will be just as important as particle creation. The theoretical machinery exists to perform this calculation (Birrell and Davies 1981) but perhaps not the stamina!

5. A ROUTE TO QUANTUM GRAVITY?

Enough has been said about the thermodynamic connection of gravity in general, and black holes in particular, to suggest an intimate relationship between them. At the very least one expects consistency, if not unification. But what can these, sometimes vague, comments on thermodynamic gravity tell us about the more central goal of quantum gravity? Do they enable us to glimpse something of its form?

If one is convinced that the thermodynamic structure of black holes remains paramount, then it is possible to glean at least limited information about quantum gravity. For example, one can follow the arguments of Page and Wald, already discussed, to try and determine the properties of quantum gravity under time reversal. If it turns out that CPT is violated, it may indicate that almost all of the cherished conservation laws of particle physics have to be abandoned when quantum gravity is included. This accords well with the general rule that the weaker the interaction, the more conservation laws it violates.

Another glimpse of quantum gravity is obtained from a study of the quantum black hole. In the original Hawking derivation, incoming field modes are traced through the interior of a collapsing 'star', from which they emerge exponentially red-shifte[d.] The Bogolubov transformation between these red-shifted modes and standard 'out' region modes predicts a thermal spectrum. In evaluating the Bogolubov transformation one is obliged to integrate over all mode frequencies (i.e. to infinity).

If a cut-off is introduced at the Planck energy, then the

resulting spectrum is not thermal, and on physical grounds one would not expect to obtain a steady flux of radiation from the hole at all, but rather a short-lived pulse. This suggests that the compelling thermodynamic properties of black holes would be disrupted if the structure of space-time were drastically altered, e.g. became 'foamy' (Hawking 1979), on the scale of the Planck length.

Nevertheless it is hard to imagine that the Planck length is not of some significance to quantum gravity. Most workers assume that space-time would be highly irregular over Planck dimensions, so perhaps we should abandon altogether the concept of a differentiable space-time manifold and replace it with a non-differentiable structure such as a fractal (Mandelbrot 1978; Berry and Lewis 1980). A good example of a fractal is Weierstrass' function

$$W(t) = \sum_{n=-\infty}^{\infty} \frac{(1-\exp(i\gamma^n t)\exp(i\phi_n)}{\gamma^{(D-2)n}}$$

where γ and D are parameters ($\gamma > 1$) and ϕ_n is an arbitrary phase. Functions such as this are continuous, but nowhere differentiable. Because of the extreme 'jitter' of such curves, their Hausdorff dimensionality, D, is greater than one ($1 < D < 2$). They seem to approximate a wide class of irregular structures found in nature (Mandelbrot 1978).

The parameter D is a measure of the degree of irregularity. Thus $D = 1.5$ produces a Brownian curve, while $D \to 2$ gives the characteristics of $1/f$ noise, which is much more 'spiky' and chaotic. If gravitational entropy is to be identified with escalating irregularity, then perhaps S could be defined in terms of some sensitive function of D near $D = 2$ (e.g. $(D-2)^{-1}$) in this one dimensional case. Indeed, one may even compute the particle production rate and spectrum induced by a fractalized space-time 'shaking' quanta out of the vacuum. For example, by letting the Ricci structure become a fractal, it will generate a non-trivial Bogolubov transformation between smooth in and out space-time regions. (It will then be necessary to abandon covariant conservation of $T_{\mu\nu}$ because $G^{\mu\nu}$ will no longer be defined.) Alternatively, one could fractalize the gravitational action. The fractal paths, because of their 'peculiar' measure, might then be expected to dominate the Feynman path integral. It is too soon to say whether the thermal properties of black holes will survive fractalization.

The thermodynamic connection can never be a substitute for a proper theory of quantum gravity. Fascinating as it may be, it is at best a consistency framework within which such a theory might develop. Nevertheless it does serve to direct attention to some peculiar properties of gravity, and to indicate that space-time structure is intimately linked with the organization of matter in a way that Einstein did not guess.

REFERENCES

Bardeen, J., Carter, B., and Hawking, S.W. (1973). *Comm. Math. Phys.* **31**, 161.

Bekenstein, J.D. (1973). *Phys. Rev.* **D7**, 2333.

—— and Meisels, A. (1977). *Phys. Rev.* **D15**, 2775.

Berry, M. and Lewis, Z.V. (1980). *Proc. Roy. Soc.* **A370**, 459.

Birrell, N.D. and Davies, P.C.W. (1978). *Phys. Rev.* **D18**, 4408.

—— —— (1980). *J. Phys.* **A13**, 2109.

—— —— (1981). *Quantum fields in curved space.* Cambridge University Press.

Candelas, P. (1980). *Phys. Rev.* **D21**, 2185.

Carter, B. (1973). in *Black holes.* (ed B.S. De Witt and C. De Witt) Gordon and Breach, London.

Christensen, S.M. and Fulling, S.A. (1977). *Phys. Rev.* **D15**, 2088.

Christodoulou, D. (1970). *Phys. Rev. Lett.* **25**, 1596.

—— and Ruffini, R. (1971). *Phys. Rev.* **D4**, 3552.

Davies, P.C.W. (1974). *The physics of time asymmetry.* Surrey University Press/California University Press.

—— (1975). *J. Phys.* **A8**, 609.

—— (1977). *Proc. R. Soc.* **A353**, 499.

—— (1978). *Rep. Prog. Phys.* **41**, 1313.

—— (1980). *Other Worlds.* Dent, London.

——, Fulling, S.A., and Unruh, W.G. (1976). *Phys. Rev.* **D13**, 2720.

D'Espagnat, B. (1971). *Conceptual foundations of quantum mechanics.* Menlo Park.

De Witt, B.S. (1979). in *General Relativity.* (ed S.W. Hawking and W. Israel) Cambridge University Press.

Dubin, D.A. (1976). *Ann. Phys. (N.Y.)* **102**, 71.

—— and Tarski, J. (1967). *Ann. Phys. (N.Y.)* **43**, 263.

Farrugia, C.H. and Hajicek, P. (1979). *Comm. Math. Phys.* **68**, 291.

Ford, L.H. (1978). *Proc. R. Soc.* **A364**, 227.

Fulling, S.A. and Davies, P.C.W. (1976). *Proc. R. Soc.* **A348**, 393.

Gass, R. and Dresden, M. Personal communications.

Gibbons, G.W. (1975). *Comm. Math. Phys.* **44**, 245.

—— and Hawking, S.W. (1977). *Phys. Rev.* **D15**, 2752.

—— —— (1978). *Proc. R. Soc.* **A358**, 467.

Hartle, J.B. and Hawking, S.W. (1976). *Phys. Rev.* **D13**, 2188.
—— and Hu, B.L. (1979). *Phys. Rev.* **D20**, 1772.
Hawking, S.W. (1972). *Comm. Math. Phys.* **25**, 152.
—— (1975). *Comm. Math. Phys.* **43**, 199.
—— (1976*a*). *Phys. Rev.* **D13**, 191.
—— (1976*b*). *Phys. Rev.* **D14**, 2460.
—— (1977). *Phys. Lett.* **A60**, 81.
—— (1978). *Nucl. Phys.* **B144**, 349.
—— (1979). in *General Relativity*. (ed S.W. Hawking and W. Israel) Cambridge University Press.
Hu, B.L. and Parker, L. (1978). *Phys. Rev.* **D17**, 933.
Hut, P. (1977). *Mon. not. R. astr. Soc.* **180**, 379.
Israel, W. (1976). *Phys. Lett.* **A57**, 107.
Kruskal, M.D. (1960). *Phys. Rev.* **119**, 1743.
Kuhn, T.S. (1978). *Black body theory and the quantum discontinuity.* 1894-1912. Clarendon Press, Oxford.
Landsberg, P.T. and Tranah, D. (1980*a*). *Collective Phenom.* **3**, 73.
—— —— (1980*b*). *Phys. Lett.* **78A**, 219.
Mandelbrot, B.B. (1977). *Fractals*. Freeman, London.
Misner, C.W. (1969). *Phys. Rev. Lett.* **22**, 1071.
von Neumann, J. (1955). *The mathematical foundations of quantum mechanics*. Princeton University Press.
Page, D.N. (1980). *Phys. Rev. Lett.* **44**, 301.
Parker, L. (1975). *Phys. Rev.* **D12**, 1519.
Penrose, R. (1969). *Nuovo Cim.* **1**, Special, 252.
—— (1979). in *General Relativity* (ed S.W. Hawking and W. Israel). Cambridge University Press.
Shannon, C.E. and Weaver, W. (1949). *A mathematical theory of communication*. University of Illinois Press.
Thirring, W.E. (1958). *Ann. Phys. (N.Y.)* **3**, 91.
Tranah, D. and Landsberg, P.T. (1980). *Collective Phenom.* **3**, 81.
Unruh, W.G. (1976). *Phys. Rev.* **D14**, 870.
Wald, R.M. (1975). *Comm. Math. Phys.* **45**, 9.
Zeldovich, Ya. B. (1970). *JETP Lett.* **12**, 307.
—— and Starobinski, A.A. (1972). *JETP.* **34**, 1159.

9

THE IRREVERSIBLE THERMODYNAMICS OF BLACK HOLES

D.W. Sciama

Department of Astrophysics, Oxford University
and

Center for Relativity, University of Texas, Austin, Texas 78712

1. INTRODUCTION

The thermodynamic properties of black holes are now well established, and have been reviewed in a number of places (Bekenstein 1980; Davies 1978; Frolov 1976; Sciama 1976; Sciama, Candelas, and Deutsch 1981). These properties relate mainly to equilibrium states of static or stationary black holes. The associated thermodynamics would then be reversible or equilibrium thermodynamics. There exists also a well-developed irreversible or non-equilibrium thermodynamics, and the question arises whether this theory also applies to black holes. It was argued briefly in Candelas and Sciama (1977) that this is indeed the case, and we now revert to this question and present the argument in more detail. One reason for doing this here is that there may be a lesson to be learnt from such a discussion which would help in the search for a quantum theory of gravity.

The essential difference between the two kinds of thermodynamics is that one is concerned entirely with the relations between different *equilibrium* states of a system, while the other is concerned with the actual *irreversible transitions* of a system from one state to another. This difference is reflected in the fact that equilibrium thermodynamics is not concerned with the *rates* of processes; in particular the time taken for a system to reach thermal equilibrium is completely irrelevant. It is only the properties of the system after it has reached equilibrium which are of concern.

By contrast, irreversible thermodynamics is dominated by a consideration of the rates of processes. In particular the manner in which a system approaches the equilibrium state is of central concern. This theory is naturally much simpler if one is dealing with states that, in some suitable sense, are already close to equilibrium, and we shall confine ourselves to such states in this article. This restriction leads to the appearance of rate or transport co-efficients (or their reciprocals impedances) which also appear in a *linear* relation between small disturbing forces acting on an equilibrium system and the *response* of the system to the disturbing force.

Irreversible thermodynamics in this restricted sense is dominated by a discovery of Einstein's (1905*b*). *Impedances are determined by the equilibrium properties of the system.* One of our main concerns in this article will be to explain this result which is now enshrined in the *fluctuation-dissipation theorem*. Our other main concern will be to show how these ideas provide a simple physical explanation of the thermodynamic properties of black holes, both reversible and irreversible. For simplicity we shall conduct the discussion in terms of the related problem of an observer moving with constant acceleration through Minkowski space-time. The manifold accessible to such an observer is bounded by an event horizon, and he finds that this manifold also possesses thermal properties. Since these properties are quantum mechanical in origin, what we are seeking is a relation between quantum mechanics and thermodynamics in the presence of a time-independent event horizon. We shall find that this relation has its ultimate origin in a property of the quantum vacuum in an inertial frame: *the zero point fluctuations in the standard vacuum state of a quantum field have many of the properties of a state in thermal equilibrium.* This has been known in essence for a long time, the key papers being Bloch (1932); Callen and Welton (1951); Senitzky (1959, 1960, 1961), but its significance for black holes, and for the related system of observers accelerating in Minkowski space-time, was not appreciated until the now classic papers of Hawking (1974, 1975) and Unruh (1976) (a discussion of the related problem of an accelerating mirror was first given by Davies (1975)).

2. THERMAL FLUCTUATIONS IN BLACK BODY RADIATION AND THE FLUCTUATION-DISSIPATION THEOREM

To understand the role of thermal fluctuations in irreversible thermodynamics we consider with Einstein (1909) a perfectly reflecting plane mirror immersed in a field of black body radiation at temperature T. This radiation field defines an *inertial rest frame* at each point, namely, that inertial frame in which the radiation field appears to be isotropic, and to possess the temperature T. If the mirror were moving uniformly with velocity v relative to such a frame, then because of the Doppler effect and aberration, the radiation field relative to the mirror would have different temperatures in the direction of, and opposed to, the mirror's motion. There would then result a different radiation pressure on each face of the mirror, leading to a net force tending to retard the motion of the mirror. This drag force F was evaluated by Einstein for a mirror of unit area reflecting radiation in the frequency range ν, $\nu + d\nu$ and moving with a small velocity v, with the result

$$F = \eta_\nu v,$$

where the drag co-efficient η_ν is given by

$$\eta = \frac{3}{2c}\left(\rho - \frac{1}{3}\,\nu\frac{\partial\rho}{\partial\nu}\right) \qquad (2.1)$$

and $\rho d\nu$ is the radiant energy density in the frequency range ν, $\nu + d\nu$.

If this drag force were the only force acting the mirror would be brought permanently to rest relative to the radiation field. However, this outcome would not be consistent with the state of thermal equilibrium which we are supposing to prevail. Since the mirror is coupled to a reservoir at temperature T, we would expect its motion to reflect this fact. In other words, for a mirror large enough to be regarded classically, we would expect it to move in an irregular fashion with a mean kinetic energy $\frac{1}{2}kT$ and a Maxwellian distribution of velocities. To drive this motion against the drag force Einstein invoked the fluctuations of radiant energy density and so of radiation pressure which he had considered earlier (1905*a*) in connexion with his photon picture of the radiation field.

These thermal fluctuations can be regarded as arising from interference effects between different modes of the radiation field. In black body radiation these modes would have a random distribution of phases, associated with the independent emission processes of atoms lining the walls of a large cavity at temperature T, which could be regarded as the origin of the black body radiation field. By the law of large numbers these interference effects would give rise to Gaussian fluctuations at an interior point of the cavity. Einstein used thermodynamic arguments to show that the mean square deviation of this Gaussian distribution for the energy density is given by

$$\overline{\Delta E_\nu{}^2} = -k\left(\frac{\partial\rho}{\partial(1/T)}\right)_V$$

In the classical regime $kT \gg h\nu$, substitution of the Rayleigh-Jeans approximation for ρ leads to

$$\overline{\Delta E_\nu{}^2} \propto \rho^2 ,$$

a result which also follows directly from an interference calculation (Lorentz 1911, 1916) for classical waves with a random distribution of phases.

When Einstein used the full Planck distribution for ρ he obtained instead

$$\overline{\Delta E_\nu{}^2} = A\rho + B\rho^2 ,$$

and he interpreted the linear term in ρ as arising from the Poisson statistics of independent particles (photons) of the radiation field. Today we would interpret the two terms together on *either* the particle picture or the wave picture. On the particle picture the quadratic term would arise from the photons being indistinguishable (Einstein 1925) and on the wave picture the linear term would again arise from interference, this time between *zero point* and thermal fluctuations (Born, Heisenberg, and Jordan 1926).

As a result of these combined fluctuations the mirror would be subject to a random driving force as well as to a systematic drag force. Einstein had no difficulty in working out the steady state motion of the mirror under this combination of forces since this problem was familiar to him from his earlier work on Brownian motion in a fluid. He confined himself to working out the mean kinetic energy $\overline{E}$ of the mirror, and found as required that

$$\overline{E} = \tfrac{1}{2}kT \ .$$

Of course, one wants more than this, namely the full Maxwellian (Gaussian) velocity distribution for the mirror's translational motion. This can also be derived from stochastic arguments, as was shown by Milne (1926).

Since both the systematic drag force $\eta_\nu v$ and the irregular driving force $F(t)$ arise from the same black body radiation field, it is not surprising that there is a relation between them which can be obtained by eliminating the Planck distribution ρ. One then finds that

$$\eta_\nu = \overline{\Delta E_\nu{}^2}/kT$$

$$= \frac{1}{kT}\int_0^\infty \overline{F(t_0)F(t_0+t)}\,e^{-i\nu t}dt \qquad (2.2)$$

by the Wiener–Khinchin theorem (Wang and Uhlenbeck 1945). We thus obtain Einstein's basic result referred to at the beginning of this article. The drag co-efficient η_ν, which determines the *rate* at which a mirror moving with arbitrary small velocity achieves equilibrium, is itself determined by the *equilibrium fluctuations* of the radiation field. Generalising this, we may say that the equilibrium properties of a thermal system determine the rate at which small departures from equilibrium die away, that is, the rate at which irreversible processes lead to the final stages of the achievement of thermal equilibrium. Nowadays this relation is called a fluctuation-dissipation theorem. The formal statement of it for the mirror problem is given by (2.2).

Summarising this discussion we may describe thermal equilibrium as the result of a *conspiracy*. Each mode of the system in equilibrium is *driven* and *damped* by the *reservoir* constituted of all the other modes. Under these combined forces each mode has its appropriate mean energy and (for a sufficiently weakly coupled system) a Gaussian probability distribution. The same damping also operates when the system is slightly perturbed from the equilibrium state.

3. A FLUCTUATION-DISSIPATION THEOREM FOR THE QUANTUM VACUUM

These results are clearly very general, but it is also important to recognise that in classical physics they have a serious limitation. Even a 'black body radiation field' at zero temperature, that is the vacuum, is a dissipative system, but there is now no question of fluctuations, at least classically. We know that the vacuum is dissipative from the behaviour of an arbitrarily moving classical charged particle in Minkowski space-time. Such a charge is acted on by a radiation damping force F_R which in the non-relativistic limit is given by

$$F_R = \frac{2}{3}\frac{e^2}{c^3}\ddot{v} \; . \tag{3.1}$$

There is, however, no associated classical fluctuating force.

In quantum theory things are different. In particular, the quantum vacuum of the electromagnetic field contains zero point fluctuations, and *these fluctuations are related to the radiation damping force by a fluctuation-dissipation theorem* (Callen and Welton 1951). At finite temperatures one form of the quantum version of this theorem is

$$\langle V^2 \rangle = \frac{2}{\pi}\int_0^\infty R(\nu)E(\nu,T)\mathrm{d}\nu \; ,$$

where $\langle V^2 \rangle$ is the mean square force associated with the fluctuations, $R(\nu)$ is the frequency dependent impedance and

$$E(\nu,T) = \frac{1}{2} h\nu + h\nu/\exp(h\nu/kT) - 1$$

is the mean energy of a harmonic oscillator at temperature T, *including the zero point energy*. For the accelerating charge, the radiation damping force takes the frequency dependent form $R(\nu) \propto \nu^2$. Physically this arises because the density of states in the vacuum, which together with the coupling constant e determines the dissipation rate, has the form $\sim \nu^2 d\nu$.

To explore these ideas further and to make contact with the thermodynamics of black holes and of accelerating observers, we now consider not a single charge, but rather an accelerating mirror. This has the advantage that we do not have to take into account the Coulomb field of the charge. Of course a mirror contains many charges, but their Coulomb fields cancel out outside the mirror, while their motion guarantees that the parallel component of the total electric field at the surface of the mirror is zero.

Alternatively we can regard the mirror as reflecting the zero point modes that fall on it. If the mirror were at rest in an inertial frame the radiation pressure resulting from this reflection would be the same on each side, and no net force would act. In this case there would also be no Einstein-type fluctuations in the zero point energy and pressure, since the system is in an eigenstate of the energy. The mirror would thus remain permanently at rest. Associated with this vanishing of the fluctuation force is the vanishing of the Einstein drag coefficient for uniform motion of the mirror. This follows immediately from Einstein's expression (2.1) for η_ν since the energy density ρ for the zero point fluctuations has the well-known form

$$\rho \propto \nu^3.$$

The fluctuation-dissipation theorem is then trivially satisfied in this case. The result itself is far from trivial, however, since it expresses the *Lorentz invariance of the vacuum state*. This is the main respect in which the vacuum state differs from a thermal state at finite temperature, for which an inertial rest frame is defined at each point.

If the mirror is accelerating relative to an inertial frame the Lorentz invariance is no longer relevant, and the situation is more interesting. We now have to impose the boundary condition of a vanishing field at an accelerating boundary. This boundary condition can be used to solve for the field outside the mirror without further consideration of the effect of the mirror (De Witt 1975; Fulling and Davies 1976; Davies and Fulling 1977; Candelas and Deutsch 1977). One finds in general that radiation is present, which one can regard as resulting from the reflexion of the zero point modes by the mirror. Because the reflexion occurs at an accelerating boundary there will in general be positive frequency components in the reflected waves, superposed on the usual negative frequency components of the incident zero point waves. These positive frequency components correspond to a flux of particles and (except in the case of constant acceleration) of radiant energy (the relation between the fluxes of particles and of energy, can, however, be non-intuitive in these cases, as stressed by Davies and Fulling (1977)). Associated with this emission of radiation is a radiation damping force F acting on the mirror, which for non-relativistic velocities Fulling and Davies (1976) found to be given by

$$F_R = \frac{4}{3}\frac{h}{c^2}\ddot{v} , \tag{3.2}$$

in close analogy to the classical radiation damping force (3.1).

There is also an analogy with the Einstein damping force $\eta_\nu v$ which arises in the presence of black body radiation at a finite temperature. In the vacuum case the damping force would tend to bring the mirror to a state of constant acceleration ($\ddot{v} = 0$) instead of to rest as in the finite temperature case ($v = 0$). However the analogy is actually much closer in view of the Unruh result that, relative to an observer with

$\ddot{v} = 0$ (or more exactly the relativistic version of motion with constant acceleration) the quantum vacuum has a temperature T given by

$$kT = \frac{\hbar \dot{v}}{2\pi c} .$$

This is a truly thermal state in that *it also possesses the Einstein fluctuations appropriate to the temperature T*. (This result is usually stated (Unruh 1976) in terms of the incoherence of the excited modes, but this amounts to the same thing.) If the mirror is being dragged by a constant external force, then the combination of the Einstein-type fluctuating force and the drag force (3.2) would ensure that in an accelerating frame of reference the mirror has the correct irregular translational motion to be in thermal equilibrium at temperature T. Thus the standard concepts of thermal equilibrium and of the fluctuation-dissipation theorem apply to the Unruh heat bath observed in the accelerating frame. The thermal radiation emitted by a black hole with an absolute event horizon also possesses the appropriate Einstein fluctuations (Parker 1975; Wald 1975; Hawking 1976).

4. THE PHYSICAL ORIGIN OF BLACK HOLE THERMODYNAMICS

We now apply the point of view which we have been developing to an understanding of the physical origin of the thermal radiation observed in a uniformly accelerating frame, and by implication in the general case of a real time-independent black hole. There are three questions to be answered:

(a) Why does a detector accelerating through a quantum vacuum become excited?

(b) Why is this excitation thermal when the acceleration is constant?

(c) Why in case (b) are there fluctuations at the appropriate thermal level?

We deal with these questions in turn.

(a) We first consider the behaviour of an inertial detector. We know, of course, that if it is in its ground state initially it remains in its ground state even if it is coupled to the zero point fluctuations of a quantum vacuum. There are several ways of deriving this result. The one which is most useful for us starts from the recognition that the detector in its ground state is also undergoing zero point fluctuations. The detector can be regarded as emitting radiation due to these fluctuations, and also as absorbing from the fluctuations of the ambient vacuum field, with these two processes being in detailed balance so that there is no net energy transfer in either direction (Fain and Khanin 1969; Milonni 1976). Thus the detector and the vacuum both remain permanently in the ground state. Since there are no Einstein-type fluctuations in the vacuum this statement is true at all times and not just statistically.

We can express this situation in a language borrowed from Planck's (1900) discussion of the equilibrium between a harmonically oscillating dipole and a stochastic radiation field. The emission rate of the dipole is proportional to the noise power in its fluctuation, while the absorption rate is proportional to the noise power of the ambient field. If the oscillator is in its ground state and is at rest in an inertial frame then each mode of the system has the same noise power, namely, $\frac{1}{2}h\nu$. To understand what happens when the oscillator accelerates it is helpful to re-phrase the inertial situation using the Wiener–Khinchin theorem. This tells us, in particular, that the noise power per mode $P(\nu)$ in the ambient field is the one-dimensional Fourier transform of the auto-correlation function (or Wightman function in the language of quantum field theory)

$$P(\nu) = \int_0^\infty \langle \phi(0)\phi(t) \rangle \exp(-i\nu t)\,dt \ ,$$

for the simpler case of a scalar field ϕ. The origin of time is irrelevant since the stochastic process involved is stationary in time. The important point is that $\phi(0)$ and $\phi(t)$ refer to values of the field *at points on the world-line of the detector*. In the present case this world-line is a geodesic, and if ϕ is in the vacuum state the resulting noise power per mode $P(\nu)$ is the same as the noise power of the detector, namely $\frac{1}{2}\, h\nu$.

When the detector accelerates a number of complications arise, which are discussed in detail by Sciama, Candelas, and Deutsch (1981). (See also Sciama 1977; Meyer 1978; Boyer 1980.) The essential point is that the noise power in the field must now be evaluated via an auto-correlation function for which $\phi(0)$ and $\phi(t)$ are evaluated at points on the *curved* world-line of the accelerated detector. *The resulting noise power per mode in the field now exceeds that of the detector*, and so the detector becomes excited. The energy for this excitation comes from the work done by the force accelerating the detector, so that the zero point fluctuations of the vacuum are acting as a passive dissipative system. If the acceleration were not constant the detector would also radiate, as would a similarly moving mirror, but this is not our concern here.

(b) We now consider why the excitation is thermal when the acceleration is constant in time. The essential property of the constant acceleration is that the associated world-line is the trajectory of a time-like Killing vector (a boost) so that the world still looks *static* relative to a constantly accelerating observer. This static property is not, however, enough to guarantee the thermal nature of the excited state. We need in addition a *stability* property of the vacuum. (Recall the physicist's speck of dust in a radiation cavity which would convert a general time-independent radiation field into a thermal one.)

To study this question of stability we switch on a weak coupling between the vacuum modes of the radiation field. The resulting situation is very similar to several that we have already discussed. Each mode of the field is now driven and damped by the reservoir constituted by all the other modes. As first emphasized by Senitzky (1959, 1960), the resulting mean energy of each mode is now $\frac{1}{2}\, h\nu$ rather than $\frac{1}{2}\, kT$. (The frequency ν itself suffers a small Lamb shift, of course). Thus the quantum noise of the reservoir guarantees that each driven mode has its correct zero point energy, as required by the Heisenberg uncertainty principle. In addition, the law of large numbers guarantees that the probability distribution for the amplitude and momentum of each harmonic oscillator mode is Gaussian, just as in the thermal case (Senitzky 1960). But the uncoupled modes also have a Gaussian probability distribution, according to the elementary quantum mechanics of a single uncoupled harmonic oscillator. Thus the uncoupled vacuum is stable to the switching on of a weak coupling between the modes. We can express this by saying that the uncoupled vacuum is *already equilibrated*. This pre-established harmony is no accident. That an uncoupled mode should have the same Gaussian distribution as is imposed by the law of large numbers is a compatibility relation demanded by the *uniqueness* of the vacuum for simple field theories. Of course this uniqueness is not given *a priori* but is itself a consequence of quantum theory.

(c) The presence of thermal fluctuations in the constantly accelerating frame is another consequence of the Gaussian character of the vacuum state. We thus have the interesting situation that quantum fluctuations become converted into thermal fluctuations as a result of the motion of the observer. We can express this by saying that the distinction between quantum and thermal fluctuations is not an invariant one but depends on the motion of the observer. Similar remarks apply to the thermal state associated with real black holes.

5. CONCLUSIONS AND FUTURE PROSPECTS

This completes our brief attempt to give a physical explanation for the thermal properties of a quantum vacuum in the presence of a time-independent event horizon. Apart from the physical insight obtained in this way, our simple picture of the processes involved may be helpful in the much more complex problem of understanding quantum gravity. In so far as one can think at all in terms of modes for the quantum gravity vacuum, there is no doubt that these modes will be coupled, perhaps weakly perhaps strongly. Our discussion suggests that a detailed understanding of the consequences of this coupling should be a helpful ingredient in the total picture. My own belief is that a *physical* picture of these processes should be built up. This work has still to be done. I hope that it can be reported on at the next Oxford Conference on Quantum Gravity.

REFERENCES

Bekenstein, J.D. (1980). *Physics today*. **32**, 24.
Bloch, F. (1932). *Zeit. f. Phys.* **74**, 295.
Born, M., Heisenberg, W., and Jordan, P. (1926). *Zeit. f. Phys.* **35**, 557.
Boyer, T.H. (1980). *Phys. Rev.* **D21**, 2137.
Callen, H.B. and Welton, T.A. (1951). *Phys. Rev.* **83**, 34.
Candelas, P. and Sciama, D.W. (1977). *Phys. Rev. Lett.* **38**, 1372.
—— and Deutsch, D. (1977). *Proc. R. Soc.* **A354**, 79.
Davies, P.C.W. and Fulling, S.A. (1977). *Proc. R. Soc.* **A356**, 237.
—— (1978). *Rep. Prog. Phys.* **H**, 1313.
—— (1975). *J. Phys.* **A8**, 609.
De Witt, B. (1975). *Phys. Rep.* **19C**, 295.
Einstein, A. (1905*a*). *Ann. d. Phys.* **17**, 132.
—— (1905*b*). *Ann. d. Phys.* **17**, 549.
—— (1909). *Phys. Zeit.* 10, 185, 323.
—— (1925). *Berl. Ber.* 3.
Fain, V.M. and Khanin, Y.M. (1969). *Quantum Electronics*. (transl. H.S.H. Massey) Pergamon, Oxford.
Frolov, V.P. (1976). *Sov. Phys. Usp.* **192**, 244.
Fulling, S. and Davies, P.C.W. (1976). *Proc. R. Soc.* **A348**, 393.
Hawking, S.W. (1974). *Nature* **248**, 30.
—— (1975). *Comm. Math. Phys.* **43**, 199.
—— (1975). in *Quantum Gravity: an Oxford Symposium* (ed C.J. Isham, R. Penrose, and D.W. Sciama) Oxford University Press.
—— (1976). *Phys. Rev.* **D14**, 2460.
Lorentz, H.A. (1911). Solvay Conference.
—— (1916). *Theories Statistiques en Thermodynamiques*, Appendix IX, Leipzig.
Meyer, P. (1978). *Phys. Rev.* **D18**, 609.
Milne, E.A. (1926). *Proc. Camb. Phil. Soc.* **23**, 465.
Milonni, C.W. (1976). *Phys. Rep.* **25C**, 1.
Parker, L. (1975). *Phys. Rev.* **D12**, 1519.
Planck, M. (1900). *Ann. d. Phys.* **1**, 69.
Sciama, D.W. (1976). *Vistas in Astron.* **19**, 385.
—— (1977). *Ann. N.Y. Acad. Sci.* **302**, 161.
——, Candelas, P., and Deutsch, D. (1981). *Adv. Phys.*
Senitzky, I.R. (1959). *Phys. Rev.* **115**, 227.
—— (1960). *Phys. Rev.* **119**, 670.
—— (1961). *Phys. Rev.* **124**, 642.
Unruh, W.G. (1976). *Phys. Rev.* **D14**, 870.
Wald, R.M. (1975). *Comm. Math. Phys.* **45**, 9.
Wang, M.C. and Uhlenbeck, G.E. (1945). *Rev. Mod. Phys.* **17**, 323.

10

BLACK HOLES, THERMODYNAMICS, AND TIME REVERSIBILITY

R.M. Wald

Enrico Fermi Institute, University of Chicago, Chicago, Illinois 60637

Although many interesting developments have occurred in the past decade and some notable progress has been made, the fundamental outstanding issues of the quantum theory of gravitation still concern the basic nature of the theory. It is generally agreed that a purely classical theory of gravity is inadequate, and that a more fundamental theory which incorporates quantum principles must replace it. But it is less clear that this 'quantization' should proceed (as many approaches assume) by making the space-time metric a quantum field on a classical space-time manifold. It is even less clear what types of questions we should expect quantum gravity to answer. For example, the quantity of prime interest in most quantum field theories is the S-matrix, which gives the amplitudes for scattering from free particle 'in' states to free particle 'out' states. Quantum gravity should also provide us with an S-matrix describing processes like graviton—graviton scattering, but the description of the universe as a whole clearly falls outside the usual context of scattering theory; gravity was not weak in the past and a description of the 'big bang' condition in terms of an incoming graviton state does not seem appropriate. In what framework should these cosmological issues be analyzed? Is quantum gravity supposed to 'predict' the 'big bang' initial conditions? If so, it will be unlike any theory of physics we presently have, where initial conditions are always part of the input; if not, must we appeal to an unexplainable mystery of creation? More generally, how will quantum gravity describe situations corresponding to the singularities of classical general relativity? It is possible (and is the hope of many) that quantum gravity will manage to avoid singularities entirely, but it is also possible that they will play a fundamental role in the theory.

The answer to these and other similar questions will undoubtedly require a major breakthrough in the formal developments of one or more of the approaches to the quantization of gravity. This article will not present such a breakthrough nor even describe recent progress in these approaches. Rather, we will describe some developments arising out of the investigation — initiated by Hawking at the time of the first Oxford Conference — of particle creation in the strong gravitational field of a black hole. While providing no definitive answers to any of the above questions, in a rather remarkable way these investigations have given us some hints and suggestions of the kind of features that may be present in quantum gravity. It is my hope that they will help guide us to a complete, quantum theory of gravitation.

1. PARTICLE CREATION NEAR BLACK HOLES

All the results and speculations described in this article will be based on the study of particle creation near black holes. We begin by summarizing the nature and conclusions of this work.

Although we have no satisfactory quantum theory of gravitation itself, we may construct a semi-classical theory of quantum fields interacting with gravitation by treating gravity classically. This is analogous to semi-classical radiation theory in atomic physics (where the electrons in atoms are treated quantum mechanically but the electromagnetic field is treated classically) and is precisely analogous to external field problems in flat space-time quantum field theory (where the quantum field interacts with a classical external potential). In the gravitational case, we prescribe a space-time metric and may ask questions such as what is the S-matrix of the quantum field or what is the expected stress-energy of the quantum field in a given state? We will restrict our attention to fields which are free except for their interaction with gravity. For space-time geometries which become flat in the past and future, there is no problem defining 'in' and 'out' particle states, so the S-matrix is certainly physically meaningful in this case. In particular, S applied to the 'in' vacuum state tells us the particle creation. For space-time curvature of compact support, it can be proven that a unique, consistent, physically sensible quantum scattering theory exists (Wald 1979*a*). Thus, predictions of particle creation in this case are completely well-defined and unambiguous. On the other hand, the definition of the stress-energy operator $T_{\mu\nu}$ for the quantum field in the semi-classical theory suffers from the same 'non-renormalizability' problems as quantum gravity itself. Although it can be proven (Fulling, Sweeny, and Wald 1978) that the 'point-splitting' prescription gives a mathematically well-defined formula for $T_{\mu\nu}$ (at least in simple cases such as curvature of compact support), there appears to be no natural way of fixing the ambiguity of adding multiples of the two conserved local curvature terms of the right dimension (times the identity operator) to $T_{\mu\nu}$. (Axiom 5 of Wald (1977) cannot be satisfied (Wald 1978).) This is unfortunate, since the next logical step in carrying the semi-classical theory beyond calculations of particle creation would be the calculation of 'back-reaction' effects via the semi-classical Einstein equation $G_{\mu\nu} = 8\pi\langle T_{\mu\nu}\rangle$. The ambiguity in $T_{\mu\nu}$ means that the dynamics given by this equation is not well-defined. There are also strong indications of serious difficulties with this dynamics (specifically, run-away solutions) for any choice of the parameters governing the ambiguity in $T_{\mu\nu}$ (Horowitz 1980).

The study of particle creation by black holes is basically a straightforward application of this semi-classical theory. Here one prescribes the classical space-time geometry of a black hole. One can take the gravitational field to become weak in the past by considering a black hole formed by the collapse of originally dispersed matter. By doing this, one can avoid ambiguities in the definition of 'in' states. However, the presence of the black hole means the gravity is not weak in the future. On both physical and mathematical grounds it seems clear that the correct 'out' Hilbert space must include both the ordinary space of particles propagating out to infinity as well as particles which enter the black hole. But it is not clear either physically or mathematically precisely how these 'black hole particle states' are to be defined. Furthermore, even after one has chosen a mathematically consistent definition of such states, the physical interpretation of these states in terms of particles is far from clear. Indeed, the work of Unruh (1976) shows that the response of a particle detector near a black hole depends critically on whether it is stationary or freely falling. Nevertheless, one can proceed by temporarily ignoring these ambiguities, choosing a convenient definition of black hole particle states, and calculating the S-matrix (and, thus, in particular, the particle creation from the vacuum).

The mathematical expression for the 'out' state associated with a given 'in' state, such as $\langle 0 \text{ in} \rangle$, will depend on the definition of black hole particle states. However, from this state vector, one can construct a density matrix describing the particles which emerge to infinity. This density matrix is independent of the definition of black hole particle states (Wald 1975). Thus, a completely unambiguous prediction is made for the particles a distant observer would see emerging from the vicinity of the black hole.

Since this density matrix plays a key role in much of the subsequent discussion, it is perhaps worthwhile to review the definition and interpretation of density matrices. One is often interested in cases where the Hilbert space of states $\mathcal{H}$ is the tensor product of two Hilbert spaces $\mathcal{H}_1$ and $\mathcal{H}_2$, $\mathcal{H} = \mathcal{H}_1 \otimes \mathcal{H}_2$. In such cases we may only be interested in measuring properties of the first space $\mathcal{H}_1$, i.e., all the observables O of interest may be of the form $O = O_1 \otimes I_2$ where $O_1 : \mathcal{H}_1 \to \mathcal{H}_1$ is an operator on $\mathcal{H}_1$ and I_2 is the identity operator on $\mathcal{H}_2$. For example, we may have a two particle system and may only be interested in measuring properties of the first particle. In our case, the 'out' Hilbert space is the tensor product of the Fock space of 'black hole particles' with the Fock space of particles emerging to infinity and we wish to consider only observations made on the emerging particles. A vector $\Psi \in \mathcal{H} = \mathcal{H}_1 \otimes \mathcal{H}_2$ describes the complete quantum state of the system. By tensor producting Ψ with its dual vector Ψ, we can form the simple linear map $\Psi \otimes \overline{\Psi}$ taking $\mathcal{H}$ into itself.

In index notation, we could write $\Psi \in \mathcal{H}_1 \otimes \mathcal{H}_2$ as Ψ^{Ab}.

Here capital letters correspond to $\mathcal{H}_1$, lower case letters correspond to $\mathcal{H}_2$, raised indices correspond to vectors, and lowered indices correspond to dual vectors. We would then write our map $\Psi \otimes \overline{\Psi}$ as $\Psi^{Ab}\overline{\Psi}_{Cd}$. The density matrix ρ describing observations on $\mathcal{H}_1$ is formed by taking the trace of this map with respect to a basis of $\mathcal{H}_2$. In index notation, we have

$$\rho^A{}_C = \Psi^{Ab}\overline{\Psi}_{Cb} \tag{1.1}$$

We may view ρ as a map from $\mathcal{H}_1$ to $\mathcal{H}_1$. It follows in a straightforward way that ρ is a positive, self-adjoint, trace class operator (with $\mathrm{tr}\rho = 1$ if Ψ is normalised). It is easy to verify that for any observable O of the form $O_1 \otimes I_2$ we have

$$\langle \Psi | O | \Psi \rangle = \mathrm{tr}(\rho O_1) \tag{1.2}$$

Since the probabilities of outcome of any measurements can be expressed in terms of expectation values of projection operators, this shows that ρ contains all the information about measurements of the first system.

By the Hilbert–Schmidt theorem (Reed and Simon 1972) ρ can be spectrally resolved in terms of orthonormal eigenvectors α_i,

$$\rho = \sum_i p_i \alpha_i \otimes \overline{\alpha}_i \; . \tag{1.3}$$

Since ρ is positive, each p_i is positive and we may interpret p_i as the probability that the system is in state α_i. Thus, the probabilities of outcomes of measurements of the system described by density matrix ρ is identical to that of a statistical ensemble of pure states α_i weighted with probability p_i. The point of view is sometimes expressed that the system described by ρ 'really is' in one of the states α_i but we have a lack of information as to which one. I disagree with this viewpoint. One reason is that if there is degeneracy in the eigenvalues p_i of ρ, the spectral decomposition, equation (1.3) becomes non-unique, making it difficult to maintain that the system 'really is' in one of these states. But more fundamentally, the viewpoint that the system described by ρ 'really is' in state α_i is reminiscent of the hidden variables viewpoint that a system in quantum state ψ 'really is' in a state where all observables have definite values. I do not believe that either of these viewpoints is correct. Thus, I favour the view that density matrices should be regarded as descriptions of states of a quantum system on an equal footing with pure states. Of course, given only the above remarks, one could also take the point of view that pure states are really more fundamental; that the entire universe is described by a pure state, and that density matrices are just a mathematically convenient way of describing the part of the universe we are actually interested in observing. However, we will very shortly present evidence against this view.

We have now described in some depth the theoretical framework for the calculation of particle creation by black holes. When the calculation is carried out for a Schwarzschild black hole, one gets the by now famous result of Hawking (1975): the density matrix describing 'out' particles which emerge to infinity is *exactly* (Wald 1975) that of black body emission at temperature

$$kT = \kappa/2\pi \tag{1.4}$$

where $\kappa = 1/4M$ is the surface gravity of the black hole. (Planck units, $G = c = \hbar = 1$, are used here and throughout the article.) In particular, the expected number of particles in a mode of frequency ω is given by the Planck factor times the absorption factor of the black hole for that mode. In this and all other respects the black hole 'emits' exactly like a perfect black body of finite size. The black hole also continues to behave like a black body in the presence of incident radiation (Panangaden and Wald 1977).

At this level of approximation, the semi-classical theory predicts that the black hole radiates forever at the steady rate given by equation (1.4). Obviously this cannot be the case if energy is conserved, and it is clear that we get this answer because we have not yet taken into account the back-reaction of the quantum field on the gravitational field. Unfortunately, although the renormalized stress-tensor on the horizon of a Schwarzschild black hole has been calculated

(Candelas 1980), as mentioned above there are some difficulties of principle involved in properly carrying out a semi-classical back-reaction calculation (and even if there were none, the practical difficulties would be enormous). However it seems clear on physical grounds that the main back-reaction effect should be the mass decrease of the black hole; it must lose mass at the same rate at which energy is radiated to infinity. Assuming that this is the case, since T is inversely proportional to M one gets an ever increasing rate of energy emission (and hence rate of black hole mass loss) resulting in the total loss of mass by the black hole within a finite time (Hawking 1975). Thus, back-reaction should cause the black hole to completely 'evaporate' in a finite time.

By making this small extrapolation from solidly based calculations, we now encounter a phenomenon which is new to quantum theory. As discussed above, in the particle creation calculations the emission to infinity is described by a density matrix. The use of a density matrix in this and all previous quantum mechanical contexts is entirely for convenience; the joint black hole-infinity final state is pure (assuming, of course, that the initial state was pure). However, we have now concluded that as a result of back-reaction the black hole will disappear. When it does so, the *complete* state of the quantum system will be described by a density matrix. The correlations which previously existed between the particles at infinity and those entering the black hole will have been swallowed by the space-time singularity within the black hole and be lost forever. Thus, in the process of black hole formation and evaporation, an initial pure state should evolve to a final (mixed) density matrix.

This conclusion could be criticized on the grounds that is is based on a semi-classical particle creation calculation together with an extrapolation to account for back-reaction. But, aside from truly radical proposals, I know of only two basic ideas for attempting to avoid the conclusion that pure states evolve to density matrices, and there are strong arguments against each of them. The first idea is that perhaps at the Planck scale – when the semi-classical approximation should no longer be applicable – the particle creation ceases, thus leaving behind a Planck mass remnant. Thus, the black hole may not evaporate completely and one could argue that its remnant could correlate with the emitted radiation to yield a pure state for the joint system. However, while it is certainly possible that a Planck scale remnant could be left behind – indeed, it is beyond our present powers to rule out any possibilities on the Planck scale – the number of 'internal states' which the remnant would need to totally correlate with the emitted radiation is absolutely enormous. It seems highly implausible that a Planck size remnant could have this many states. Indeed, black hole thermodynamics suggests that the number of internal states should be roughly $\exp(\frac{1}{4}A)$ (where A is the horizon area measured in Planck units) which is approximately 1. Thus, even if the black hole does not totally evaporate, this should not alter the conclusion that an initial pure state evolves to a final density matrix.

The second idea is that – although the black hole may totally evaporate – the radiation emitted at late times is correlated with the early radiation so that the final state is pure. Page (1980) has expressed some sympathy with this view and indeed, this is what does happen when an ordinary hot body, initially in a pure state, cools to absolute zero by radiating photons. The photons emitted at early times are correlated in a very detailed way with the states of the atoms of the hot body. As the body continues to cool, these correlations are transferred to the photons emitted at later times. However, if the notion of a black hole retains any validity in quantum theory, this type of behaviour should not occur in black hole emission. In the black hole case, the particles emitted at early times should be correlated in a very detailed way with the internal states of the black hole. But if the black hole truly is a black hole, these detailed correlations should not be transferable to particles emitted at later times. All the particles reaching infinity are produced outside the black hole and the internal states of the black hole should not be able to exert any causal influence on them. There will, of course, be some correlation between the early and late radiation. For example, the mass and angular momentum of the black hole is correlated with the external black hole geometry, so the total energy and angular momentum of the early emitted radiation can be correlated in this aspect with the radiation emitted later (as must be the case if these quantities are to be conserved). But the 'no hair' theorems of classical general relativity suggest that these quantities (along with the location and momentum of the black hole) may be the only properties of the internal states which affect the external geometry. Thus it does not seem plausible that the very highly detailed correlation required to produce a pure state can occur.

Indeed, from an abstract point of view, it seems that the evolution of a pure state to a density matrix expresses the quantum mechanical analogue of the classical notion of a black hole. Classically, a black hole is a space-time region which cannot causally communicate with future null infinity, $\mathscr{I}^+$. This means that if a black hole is present, knowledge of a classical field at $\mathscr{I}^+$ does not give complete information about the field throughout the space-time, i.e. one loses information in passing from a description of the field on an initial Cauchy surface to its final state on $\mathscr{I}^+$. In quantum theory, the evolution of an initial pure state at $\mathscr{I}^-$ to a final density matrix at $\mathscr{I}^+$ expresses an analogous kind of information loss, as it suggests that what can be observed in the final state at infinity cannot provide complete knowledge of processes occurring throughout the space-time. Thus, if one did not have evolution from pure states to density matrices, it would be doubtful that quantum theory would tolerate the existence of objects which correspond to our classical notion of black holes. For the remainder of this article, I will assume that black holes do exist and that the picture of particle emission and black hole evaporation outlined above is correct.

2. BLACK HOLES AND THERMODYNAMICS

The most striking and immediate consequence of Hawking's prediction of particle creation by black holes was its implications for black hole thermodynamics. The mathematical analogy between the laws of black hole mechanics and the ordinary laws of thermodynamics had been explored earlier (Bardeen, Carter, and Hawking 1973) but was viewed by most researchers (with the notable exception of Bekenstein) as merely a curiosity. Hawking's prediction changed that situation dramatically, as it showed that black hole surface gravity κ is not merely *analogous* to a thermodynamic temperature, it *is* a temperature – namely the temperature of the emitted radiation. Furthermore, it made viable the 'generalised second law of thermodynamics', previously proposed by Bekenstein (1973), which states that the sum of ordinary entropy of matter outside black holes plus 1/4 times the black hole event horizon area (expressed in Planck units) never decreases. Without quantum particle creation, violations of this law could be produced by immersing a black hole in a sufficiently cold radiation bath.

The subject of black hole thermodynamics has been discussed extensively by many authors and I have presented my views in detail elsewhere (Wald 1979*b*), so I will confine myself here to some brief comments. It seems generally agreed that black hole thermodynamics simply is ordinary thermodynamics applied to a self-gravitating quantum system. 'Ordinary thermodynamics' arises because the observables of a complicated system which we actually measure fail by a wide margin to determine the detailed inner workings of the system. Thus, such measurements generally only distinguish between large regions of phase space in the classical case and between subspaces of large dimension in the quantum case. Because we do not see the complicated internal dynamics, such systems may appear to be in equilibrium, i.e. the observables we do measure do not change with time (over relatively long time scales). For a given observable O, the entropy of a state of the system can be defined in the classical case as the logarithm of the volume of the region of phase space having that value of O and in the quantum case (for an eigenstate of O), as the logarithm of the dimension of the subspace having that value of O (Wald 1979*b*). Since classical systems tend to spend 'equal times in equal volumes' of phase space and quantum systems tend to spend 'equal times in subspaces of equal dimension' of the Hilbert space, the entropy of a state with a given value of O is a measure of the (logarithm of the) fraction of time that O has that value. Thus, systems in states of low entropy are likely to change their macroscopic appearance (since that value of $\mathcal{O}$ is maintained for a relatively short time), while systems in a state of maximum entropy generally appear to be in equilibrium (since the time scale on which $\mathcal{O}$ changes is typically much larger than observation time scales). This yields an argument for the second law of thermodynamics, and similar arguments for the other laws can also be given (Wald 1979*b*).

Black hole thermodynamics can, in essence, be explained by postulating that the proper quantum description of a black hole will view it as a complicated state of the gravitational field, with many different 'internal states' of the black hole producing the same external macroscopic state. If one further postulates that the number, N, of such quantum internal states is $\exp(\frac{1}{4}A)$, where A is the horizon area, then $\frac{1}{4}A$ would simply be the 'ordinary entropy' of the black hole. The 'generalized second law' would then follow as the ordinary second law applied to a system consisting of a black hole and ordinary matter, and the other laws of black hole thermodynamics would follow similarly (Wald 1979*b*).

I believe that this explanation of black hole thermodynamics is basically correct. However, there are some puzzling aspects of this picture which remain to be fully explored. In general relativity there is no natural, general prescription for choosing a time function on a space-time; the proper formulation of a statement concerning the state of a system 'at an instant of time' generally requires the specification of the space-like hypersurface which represents that instant of time. Thus, one would expect that in order to formulate the question 'What is the number, N, of black hole internal states at a given instant of time?' it would be necessary to specify a time slice within the black hole. Therefore, it is puzzling that, above, the formula $N = \exp(\frac{1}{4}A)$ was proposed without such a time slice being specified. Furthermore, the general explanation of the validity of thermodynamics of an ordinary quantum system is based on an ordinary Schrödinger evolution applying to that system (Wald 1979*b*). However, as discussed above, ordinary Schrödinger dynamics apparently does not apply to a quantum black hole system, as the information contained in the black hole internal states propagates into the space-time singularity and is lost, resulting in evolution from a pure state to a density matrix. It is not clear exactly how thermodynamics would arise from an evolution of this sort. These issues appear to be worthy of further investigation.

Another intriguing aspect of black hole thermodynamics is its relation to the thermal properties of the ordinary vacuum state as seen by accelerating observers (Sciama in this volume). Unruh (1976) has shown that an accelerating observer in flat space-time sees a thermal bath of particles at temperature

$$kT = a/2\pi \tag{2.1}$$

where a is the acceleration. But, by definition, the surface gravity of a black hole is the limit as one approaches the horizon of the acceleration an observer would need to remain stationary (or, more precisely, co-rotating with the horizon) times the redshift factor for that position. Thus, comparing equations (1.4) and (2.1), we see that the particle emission from a black hole could be viewed as effectively being the escape to infinity of the particles an observer just outside the black hole would see on account of his acceleration. This clearly shows that black hole thermal emission is closely related to the thermal properties of the vacuum, but a complete understanding of the deep reasons which must underlie this relationship remains for future investigations.

In summary, black hole thermodynamics has already provided us with some intriguing insights into the nature of quantum gravitational dynamics and it seems certain to provide us with more in the future.

3. BLACK HOLES AND TIME REVERSIBILITY

The ideas of the previous two sections yield an attractive picture of the dynamics of a self-gravitating gas, say, confined by a box. The gas may clump together and collapse to form a black hole. Indeed, for fixed total energy if the box is small enough, it will be unstable with respect to forming a black hole by this process. On the other hand, as we have discussed above, this black hole may evaporate. Indeed, for fixed total energy if the box is large enough it will be unstable with respect to evaporation. The dynamics of these processes is apparently sufficiently well-behaved that ordinary thermodynamics is applicable.

Indeed, perhaps the most disturbing aspect of this picture is that it appears to be a satisfactory, complete description as it stands. The reason this might be disturbing is that the picture appears to grossly violate time reversal invariance: black holes may occur, but no mention is made of their time reverse, namely white holes. This raises the issue of whether quantum gravity satisfies time reversal or CPT invariance.*

The time reversibility of quantum gravity recently has been questioned on entirely different grounds by Penrose (1979). It is well known that we commonly see apparently time irreversible behaviour; specifically, we observe entropy increase but do not observe entropy decrease. As already indicated in the previous section, one can account for this behaviour using time reversal invariant laws of physics from the fact that systems spend most of their dynamical history in states of high entropy. If the entropy of a system is low, we should expect it to increase rapidly, but we should have to wait a very long time to see it decrease again. Thus, the apparent arrow of time we observe stems from the fact that the entropy of the present universe is remarkably low.†

The present low entropy of the universe can be explained in turn by the even lower entropy of the very early universe. Now, in the standard 'big bang' model of the universe, it is assumed that, early on, the matter was in thermal equilibrium and was homogeneously distributed. Thus, it may seem that the

---

* One should not expect quantum gravity to satisfy the Wightman axioms (Streater and Wightman 1964) since it should not be a Poincaré invariant theory (in any meaningful sense) so the CPT theorem cannot be invoked.

† More precisely, it should be said that the entropy of various subsystems of the universe — viewed as isolated systems — is remarkably low. A satisfactory definition of the gravitational contribution to entropy in general relativity has not yet been given, so we cannot really speak of the total entropy of the universe and indeed it is not even clear that a useful notion of total entropy will exist in non-asymptotically flat situations.

entropy of the early universe was very high. However, this is not so. For a self-gravitating Newtonian system, one can always increase the entropy by clumping the matter. Although one loses some entropy by making the distribution inhomogeneous, one gains more entropy by using the binding energy to heat the matter to a higher temperature. Thus, for a Newtonian gas in a box, there is no global entropy maximum (Antonov 1962) (i.e., one can always further increase entropy by more clumping) and, for fixed gas density there is not even a local entropy maximum if the box is large enough (Lynden-Bell and Wood 1968). In general relativity if one accepts black hole thermodynamics, a self-gravitating gas in a box will have a global entropy maximum, but for fixed energy per unit volume this state will contain a black hole if the box is sufficiently large (Gibbons and Perry 1977). Thus, for a large self-gravitating system, it is indeed true that a homogeneous distribution of matter is a state of relatively low entropy.

Why was the initial entropy of the universe so low? Penrose (1979) has speculated that perhaps this is because there *is* a fundamental time asymmetry in the laws of quantum gravity. Specifically, Penrose conjectured that perhaps quantum gravity puts stringent restrictions on initial singularities – possibly requiring the vanishing of the Weyl tensor – but it places no such restrictions on final singularities. In this way, the initial state of the universe would have low entropy and would lead to the arrow of time we observe at present.

A self-gravitating system confined to a box provides an ideal testing ground for the issue of time reversibility in quantum gravitational physics. As already mentioned above, one seems to have a successful picture which incorporates black holes but not white holes. Can white holes be added to the picture in a consistent way to restore time reversibility? Is there any other evidence for or against the time reversal invariance of this system?

Recently, Ramaswamy and Wald (1980) have presented evidence that the answer to the first question is 'no'. The problem with white holes is that they are subject to an instability first analyzed in the classical context by Eardley (1974).

In the quantum case, an infinite burst of particles reaches infinity at the disappearance time of the white hole (presumably resulting from the 'blue sheet' instability of the white hole) for any 'in' state which is a product of a 'white hole state' with an 'infinity state' (Ramaswamy and Wald 1980). Thus, for white holes to behave in a non-singular fashion, one would have to postulate that they are always 'born' in states with tremendously high correlation with the states of the incoming particles from infinity. But even if this were so, by stationing an observer at past null infinity to measure the incoming particles, it should be possible to 'knock' the 'in' state into a product state and thereby induce the white hole instability. Thus, white holes do not seem to fit in naturally into a consistent picture of the dynamics of a self-gravitating quantum system.

Further evidence of an even more clear-cut nature against time reversibility follows directly from the above conclusion that in the process of black hole formation and evaporation, an initial pure state evolves to a final (mixed) density matrix. If time or CPT reversal invariance holds, then it must be possible for an initial (mixed) density matrix to evolve to a final pure state. However, this can be shown to be impossible by the following argument (Wald 1980). (A similar argument has been given by Page (1980).)

The scattering of initial density matrices to final density matrices can be described by a superscattering matrix (Hawking 1976) \$. General arguments show that \$ must be linear (Wald 1980). If a density matrix σ evolves to a pure state ψ, we have

$$S(\sigma) = \psi \otimes \overline{\psi}. \tag{3.1}$$

Expanding σ in terms of its eigenvectors ϕ_i,

$$\sigma = \sum_i p_i \phi_i \otimes \overline{\phi}_i \tag{3.2}$$

and using the linearity of \$, we have

$$\sum_i p_i \ \$(\phi_i \otimes \overline{\phi}_i) = \psi \otimes \overline{\psi} \ . \tag{3.3}$$

Let χ be orthogonal to ψ. Taking the expectation value of the operators of equation (3.3) in state χ, we find,

$$\sum_i p_i \langle \chi | \$(\phi_i \otimes \overline{\phi}_i) | \chi \rangle = 0. \tag{3.4}$$

But each term in the sum is non-negative: each p_i is positive and $\$(\phi_i \otimes \overline{\phi}_i)$ is a density matrix and thus is a positive operator. Thus, we must have

$$\langle \chi | \$(\phi_i \otimes \overline{\phi}_i) | \chi \rangle = 0 \tag{3.5}$$

for all χ orthogonal to ψ, which, in turn, implies

$$\$(\phi_i \otimes \overline{\phi}_i) = \psi \otimes \overline{\psi} \tag{3.6}$$

Thus, each ϕ_i by itself evolves to ψ. Time reversal (or CPT) invariance then implies that the time (or CPT) reverse of ψ evolves to the time (or CPT) reverse of *each* ϕ_i, which is a manifest contradiction. Thus, the evolution of a pure state to a density matrix is incompatible with time or CPT reversal invariance, and we have definite evidence for an arrow of time in quantum gravitational dynamics.

However, it does not appear that this is the end of the story. The question arises as to how much of this time reversibility failure arises from *true* time asymmetry (i.e., an arrow of time apparent in sequences of physical measurements) and how much merely arises from a time asymmetric *description* of the processes. As will be discussed below, it is quite possible that a large portion (and conceivably all) of the time asymmetry results from the way we are describing nature. What I mean by this is perhaps best illustrated by the following example suggested to me by R. Sorkin:

Consider an object (say a ball) which can be in one of two possible positions. Consider, now, a probabilistic dynamical evolution whereby after each interval of time Δt the ball either remains where it is with probability ½, or changes its position, also with probability ½. This dynamics clearly has no arrow of time built into it. A motion picture made from a sequence of measurements of the ball's position would look equally plausible if run backwards in time. However, suppose we describe the dynamical evolution of the ball by its probability distribution function $(p_1(t),\ p_2(t))$ for being in each of the two positions. Then we would find that any initial condition (say the 'pure states' (1,0) or (0,1)) evolves at late times to the distribution $(\frac{1}{2},\frac{1}{2})$. On the other hand, the distribution $(\frac{1}{2},\frac{1}{2})$ remains unchanged and never evolves to (1,0) or (0,1). Thus, in this example there is time asymmetry in our probabilistic *description* of the dynamics, though not in the physical dynamics itself.

The above example is classical and it could be objected that the time asymmetry in our description results from the fact that it is incomplete; the ball 'really is' in a definite position at each time but we are throwing away this information. The time asymmetry results from the fact that we are losing more and more information in our description as time proceeds. On the other hand, in quantum mechanics, the probability distributions determined by the state vector (or density matrix) *are* supposed to be the complete description of the system. Unless one believes in hidden variable theories, one is not throwing away any information. Nevertheless, even though our description is complete, it is possible for the description to manifest attributes which are not reflected in the physical measurements. A good example of this is the Einstein–Podolsky–Rosen experiment, where our description manifests causality violation – a measurement of the state vector in one region causes a change in the state vector in a space-like separated region – yet no physical information can be communicated from one region to the other. Thus, we pose the question as to whether the time asymmetry we have found is any more physical than the causality violation of the Einstein–Podolsky–Rosen experiment.

It turns out that there is a simple condition which is necessary and sufficient for physical time (or CPT) reversal symmetry in gravitational scattering theory. Since CPT appears to be a more fundamental symmetry of nature, I will state the condition for it. We assume that the 'in' and 'out' Hilbert spaces $\mathcal{H}_{in}$ and $\mathcal{H}_{out}$ can be identified as the CPT reverses of each other; let $\theta : \mathcal{H}_{in} \to \mathcal{H}_{out}$ be that map. Above, we proved that the dynamics fails to satisfy CPT symmetry in the sense that if pure state ψ

evolves to density matrix ρ, then density matrix $\theta^{-1}\rho\ \overline{\theta}^{-1}$ cannot evolve to $\theta\psi$. However, we could ask for the probability that, starting from state ψ, we would find the final system to be in the pure state ϕ if we made a measurement. This probability is given by

$$P(\psi \to \phi) = \langle \phi | \rho | \phi \rangle \ . \tag{3.7}$$

The condition we want is,

$$P(\psi \to \phi) = P(\theta^{-1}\phi \to \theta\psi) \tag{3.8}$$

for all $\psi \in \mathcal{H}_{in}$, $\phi \in \mathcal{H}_{out}$.

If equation (3.8) holds then no 'arrow of time' will be apparent in physical measurements. If an observer performing scatterin experiments CPT reverses the results of his measurements of the precise state of the system, this new sequence will be equally ir accord with the laws of dynamics. A second physicist given the actual data and the fabricated data would have no way of telling which was which.

Does equation (3.8) hold in scattering dynamics in quantum gravity? Equation (3.8) is compatible with the type of scattering dynamics considered here, so it cannot be ruled out b any general arguments like those given above. One reason for believing that equation (3.8) is valid comes from the fact that it implies that the microcanonical ensemble is preserved under scat tering dynamics (Wald 1980), a condition which is important for the existence of thermodynamics. However, this is not a very strong argument, since this thermodynamic condition is much weaker than equation (3.8). I think the strongest supporting evidence for equation (3.8) comes from the 'randomicity' of blac hole emission displayed in semi-classical calculations. It is po sible to argue that if equation (3.8) holds, then the space of a black hole evaporation states must be the CPT reverse of the col lapse states (Wald 1980). Therefore, since a huge number of initial states can collapse to a black hole – most of the internal details of the collapsing matter are irrelevant – it is nece sary that the black hole be able to emit a huge number of possib final states. But this is exactly what one finds in semi-classic calculations, as the spectral decomposition of a thermal density matrix involves a huge number of pure states. Furthermore, the very tiny probability of occurrence of any final pure state make plausible – or, at least, difficult to argue against – the validity of equation (3.8), as it is quite difficult to disprove the equality of the probabilities of two occurrences, each of which is extremely unlikely. Thus, I feel that there is a very good chance that equation (3.8) does hold, although it seems clear that belief in equation (3.8) will remain largely a matter of personal prejudice until we know more about quantum gravity.

Suppose equation (3.8) is valid. Does this mean that the arrow of time of quantum gravity appears only in our description of it? It certainly does mean this for the 'armchair observer' who performs only scattering experiments. But can an observer measure the time asymmetry of the space-time geometry – corresponding to the fact that black holes can occur but white holes do not – by directly performing strong field measurements? It first should be pointed out that even the armchair observer can deduce that black holes occur but white holes do not: if he drops a probe in a hole, it will (almost) always return for a white hole and (almost) never return for a black hole, so he can tell the difference. But this is no more significant than the fact that the armchair observer can also deduce that pure states evolve to density matrices but not vice versa, in violation of CPT symmetry. The problem in both cases is that, if equation (3.8) holds, the CPT violation occurs only in the *deductions*, not in the measurements.

However, the observer who makes measurements in the strong field region should be able to measure the arrow of time which appears in the theoretical framework of quantum gravity.* Suppose we have observers measure, say, the tidal forces they feel in traversing through the space-time (staying outside the black hole and returning to infinity so as not to bias initial conditions over final conditions). From this data, one could then fabricate new time reversed data by recording the same pattern of tidal forces but in the opposite order. A second physicist handed (many sets of) the real data and the fabricated data *should* be able to tell which is which, corresponding to the difference in the inner workings of the collapse and evaporation processes. For example, an observer

* CPT invariance in scattering measurements does not imply CPT invariance of the internal measurements; see footnote 17 of Wald (1980).

in the strong field region (near the black hole) should (almost always) measure tidal forces corresponding to positive energy density early (during the collapse phase) and negative energy density late (during the evaporation phase). The time reverse of these measurements should almost never occur. However, I say 'should' because our knowledge of the measurement process in the strong field region is sufficiently uncertain that it remains conceivable that when all effects (such as those analyzed by Unruh (1976)) are properly taken into account, one would find that the measurements do not show this arrow of time. Thus, it remains a possibility that even with regard to the strong field region, the time and CPT asymmetry of quantum gravity appears only in our theoretical framework.

In summary, if one accepts the basic picture of black hole particle creation and evaporation outlined in section 1, one is forced to accept some form of time reversal and CPT violation in quantum gravity. However, the nature of this violation could take one of the following three forms:

1. CPT violation occurs in an unmitigated fashion. Scattering data displays an arrow of time.
2. CPT violation occurs but is hidden from observers who make measurements only at times when the gravitational field is weak. Scattering data fails to show an arrow of time, but it can be seen from data taken by observers who enter the strong field region.
3. The CPT violation occurs only in our description of quantum gravity in much the same way as causality violation occurs in our description (and only in our description) of the Einstein–Podolsky–Rosen experiment. No physical measurements display an arrow of time.

In possibilities 1 and 2 we have definite evidence for the arrow of time which Penrose (1979) has sought for quite different reasons. In possibility 3 (and, to a lesser extent, possibility 2) we are brought face to face with issues involving measurement and the interpretation of quantum states as describing the underlying physical reality. As already indicated by the above discussion, I favour possibility 2. But whichever possibility holds, it is certain to tell us a great deal about the nature of quantum gravity.

ACKNOWLEDGEMENTS

I wish to thank D. Page, R. Penrose, and R. Sorkin for helpful discussions. This research was supported in part by NSF grant PHY 78-24275 to the University of Chicago and by the Alfred P. Sloan Foundation.

REFERENCES

Antonov, V.A. (1962). *Vest. leningr. gos. Univ.* **7**, 135.
Bardeen, J., Carter, B., and Hawking, S.W. (1973). *Commun. Math. Phys.* **31**, 161.
Bekenstein, J. (1973). *Phys. Rev.* **D7**, 2333.
Candelas, P. (1980). *Phys. Rev.* **D21**, 2185.
Eardley, D. (1974). *Phys. Rev. Lett.* **33**, 442.
Fulling, S.A., Sweeny, M. and Wald, R.M. (1978). *Commun. Math. Phys.* **63**, 257.
Gibbons, G. and Perry, M. (1977). *Proc. R. Soc.* **A358**, 467.
Hawking, S.W. (1975). *Commun. Math. Phys.* **43**, 199.
—— (1976). *Phys. Rev.* **D14**, 2460.
Horowitz, G. (1980). *Phys. Rev.* **D21**, 1445.
Lynden-Bell, D. and Wood, R. (1968). *Mon. Not. R. Astr. Soc.* **138**, 495.
Page, D. (1980). *Phys. Rev. Lett.* **44**, 301.
Panangaden, P. and Wald, R.M. (1977). *Phys. Rev.* **D16**, 929.
Penrose, R. (1979). in *General Relativity, an Einstein Centenary Volume*. (ed S.W. Hawking and W. Israel). Cambridge University Press.
Ramaswamy, S. and Wald, R.M. (1980). *Phys. Rev.* **D21**, 2736.
Reed, M. and Simon, B. (1972). *Functional Analysis*. Academic Press. New York.
Sciama, D. Contribution to this volume.
Streater, R.F. and Wightman, A.S. (1964). *PCT, Spin, Statistics, and All That*. W.A. Benjamin Inc., New York.
Unruh, W.G. (1976). *Phys. Rev.* **D14**, 870.
Wald, R.M. (1975). *Commun. Math. Phys.* **45**, 9.
—— (1977). *Commun. Math. Phys.* **54**, 1.
—— (1978). *Phys. Rev.* **D17**, 1477.
—— (1979*a*). *Ann. Phys.* **118**, 490.
—— (1979*b*). *Phys. Rev.* **D20**, 1271.
—— (1980). *Phys. Rev.* D21, 2742.

11

TIME-ASYMMETRY AND QUANTUM GRAVITY

R. Penrose

Mathematical Institute, University of Oxford

At our previous Oxford Quantum Gravity Conference, held in 1974, Stephen Hawking put forward his remarkable theory of black-hole evaporation (Hawking 1975*a*,*b*). It seemed clear at the time that these ideas were destined to play a key role in any future development in which quantum field theory and general relativity were to be successfully united. Two years later, in 1976, we held a conference which was specifically devoted to the furtherance of these ideas. But by then, there had begun to emerge a certain divergence of opinion between Hawking and myself – something which we still have not found ourselves able to resolve. A point on which we are certainly agreed, however, is that this conflict has proved immensely stimulating to us both. My own line of argument has been driven, in part under the sting of Hawking's penetrating criticisms, into some dangerous but fascinating territory. I hope my audience will bear with me if I may seem to be drawing some insufficiently supported, and perhaps somewhat radical conclusions concerning the nature of quantum-mechanical observations. But to me, though they are yet very rudimentary, these conclusions seem to be pointing in a necessary direction for future physics. I hope, also, that I may be forgiven if I do not adequately describe Hawking's own viewpoint – and undoubtedly I shall do it considerably less than justice. But, however inadequately, I shall need to give some indications of it in order to present the reasons behind my own very different point of view more clearly.

1. HOW SPECIAL WAS THE BIG BANG?

Let us first recall the Bekenstein–Hawking formula (Bekenstein 1973, 1974; Hawking 1975*a*) for the entropy of a stationary black hole with a horizon of surface area A:

$$S_{bh} = \frac{kAc^3}{4\hbar G}$$

$$= \tfrac{1}{4}A \text{ (with } k = c = \hbar = G = 1\text{)}.$$

(k = Boltzmann's constant; c = velocity of light; $2\pi\hbar$ = Planck's constant; G = Newton's gravitational constant.) One way in which this formula can be used is to get an estimate of how 'special' the big bang was (Penrose 1979). In order to fix ideas, let us consider a closed universe containing $\sim 10^{80}$ baryons. The standard figure, assuming our universe to be closed, for the observed entropy per baryon (essentially, the number of photons per baryon and taking $k = 1$) is $\sim 10^8$. This does not take into account the possibility of an appreciable contribution to the total entropy arising from the presence of black holes. We shall see how black holes could affect the discussion in a moment. Another problem with this figure is the strong possibility that in the very early stages there may have been an appreciable non-conservation of baryon number. Such ideas have become very fashionable recently, particularly in connection with the 'grand unified theories'. For this reason it is sometimes useful to talk in terms of the *total* entropy of the universe, the 'observed' figure therefore being 10^{88}.

Now consider what the Bekenstein–Hawking formula tells us should be the entropy per baryon for a single solar mass black hole. The figure is $\sim 10^{20}$. This is some twelve orders of magnitude larger than the 'observed' figure. Thus if our (assumed closed) universe had consisted, at our epoch, entirely of solar-mass black holes, the *total* entropy would have been about twelve orders of magnitude larger than that which is apparently observed. Of course, our universe is not so constituted! But these figures indicate something about the 'observed' entropy per baryon, namely that in a gravitational context it must be regarded as an exceedingly small value.

Before the discovery of the Bekenstein–Hawking formula it was perhaps possible to maintain that this 'observed' figure represented a surprisingly high value for the entropy per baryon in the universe. Indeed, the programme of chaotic cosmology was proposed (Misner 1968, 1969; Zel'dovich 1972) according to which an initially chaotic geometry was supposed to be responsible for the 'high' present entropy figure, dissipation effects serving to iron out most of the irregularities in the initial geometry and transferring whatever entropy was contained in these irregularities into the black-body radiation. We shall see in a moment how absurdly huge these constraints on the initial geometry would have had to have been in order not to have grossly exceeded the present observed figure for the entropy (cf. Penrose 1974, 1979 and also Barrow and Matzner 1977 for a more specific detailed argument). But even without performing any numerical calculations, one may have strong reservations about the chaotic cosmology idea if, in some sense, the initial chaos was supposed to be at a maximum. For then the universe would have had to have started in a 'maximum entropy state', dissipation would not occur, and there would be no second law of thermodynamics.* The entropy of the universe would have already reached its maximum and there would be no further way in which the universe could 'relax'.

Though the phenomenon of the expansion of the universe serves to obscure this discussion somewhat, it is *not* true that the mere fact of expansion alone could lead to an entropy increase and itself provide us with our second law of thermodynamics. (I have gone into this question at some length elsewhere (Pęnrose 1979) and have no wish to repeat the argument here, though some further comments are given in the next section.) In is necessary, in fact, that the structure of the big bang singularity was severely constrained in order that thermodynamics as we know it should have arisen.

In order to form an impression of just *how* constrained it

---

* It should be made clear here that the second law *is* something in need of explanation, in our universe. For while it is not surprising that a system in a given state of low entropy should normally proceed, in the future direction, to a state of higher entropy, it is a remarkable fact of our world that this argument does *not* apply in the *past* direction. As we work backwards to earlier and earlier epochs we find that the entropy goes down and down. For the second law to operate as we know it, and to have operated as we know it at all times in the past, it is necessary that the earliest stages of the universe should have been constrained in some very precise way. The puzzle is: where do these constraints come from?

must have been, let us proceed further in our application of the Bekenstein–Hawking formula. Rather more realistic than galaxies populated entirely with solar mass black holes would be, say, $10^{11}\ M_{\odot}$ galaxies with $10^{6}\ M_{\odot}$ black-hole cores. Here the entropy per baryon would be 10^{21} – an order of magnitude higher again than the solar mass black hole case – and the total entropy for a 10^{80} baryon universe entirely constituted in this way would be 10^{101}. (Having large black holes enables us to gain enormously in the entropy and to compensate easily for the fact that now most of the mass is in the form of ordinary stars.) Of course we can gain much more in the total entropy by allowing each entire galaxy to be swallowed by its black hole core. For the resulting $10^{11}M_{\odot}$ black hole the entropy per baryon is now 10^{31}, and for a 10^{80} baryon universe entirely composed of such objects the total entropy would be 10^{111} – some 23 orders of magnitude higher than the supposedly high 'observed' figure of 10^{88}.

In the final collapsing phase of the universe, irrespective of the particular one of these ways in which it might happen to have been constituted, we expect smaller black holes to unite into larger and larger ones until ultimately a figure of the order of 10^{43} for the entropy per baryon is achieved, giving 10^{123} for the total entropy of the universe. To obtain this figure, we cannot quite apply the Bekenstein–Hawking formula directly since the final state of the universe does not at that stage, at all geometrically resemble the isolated black-hole model to which the Bekenstein–Hawking formula applies. However, the figure seems a reasonable estimate, since only a little before that stage the universe would have consisted essentially of a few black holes more or less separated from one another, the figure of 10^{123} for the total entropy being approached from below as the holes all finally unite together. Thus it would seem that the figure of 10^{123} is a plausible estimate for the maximum entropy state of a universe of this type. (See Table 1.1)

If we accept this figure, and compare it with the original 'observed' figure of 10^{88} for the entropy which, according to the chaotic cosmology view, should have been produced out of initial irregularities of the big bang, then we realize

TABLE 1.1

Entropy values for a closed universe with B = 10^{80} *baryons (Boltzmann's constant = 1)*

	S/B	corresponding total S
'observed'	10^{8}	10^{88}
1 $M_{\odot}$ b.h.	10^{20}	10^{100}
10^{6} $M_{\odot}$ b.h. in 10^{11} $M_{\odot}$ galaxy	10^{21}	10^{101}
10^{11} $M_{\odot}$ b.h.	10^{31}	10^{111}
collapsed universe	10^{43}	10^{123}

how absurdly tiny this 'observed' figure is in comparison with what it 'might have been'. This provides us with a measure of the degree to which the initial state was special. Indeed, even the figure 10^{88} totally ignores the entropy increase that would have had to have been involved in the dissipation processes responsible for converting the irregular initial geometry into background radiation. According to the chaotic cosmology view, the figure of 10^{88} for the 'observed' total entropy must be considerably higher than the 'actual' value for the entropy in the initial chaotic geometry.

To get a quantitative picture of the degree which the big bang was special imagine a manifold* $\mathscr{W}$ whose points represent the various possible initial configurations of the universe. Imagine the Creator, armed with a pin which is to be placed at one spot in $\mathscr{W}$ thereby determining the state of our actual universe. How accurate must the Creator's aim have been in order that a universe of the 'observed' type could have arisen? Since entropy is, after all, related to a phase-space volume w in $\mathscr{W}$ by the logarithmic formula $S = k \log w$ ($k = 1$), we are led to estimate that the accuracy of the Creator's aim must

* I am using the word 'manifold' loosely here. No requirement is implied that it be a manifold in the technical sense.

have been at least of the order* of $10^{10^{88}}$ parts in $10^{10^{123}}$ (this being the ratio of the volume to be aimed at to the total volume of $\mathscr{W}$) i.e. one part in†

$$10^{(10^{123}-10^{88})} \doteq 10^{10^{123}} \quad .$$

Even if the universe had consisted almost entirely of $10^{11}\, M_{\odot}$ black holes, the Creator's aim would have had to have been accurate to effectively the same degree as before since replacing '88' by '111' in this expression will make no difference to it. Without wishing to denigrate the Creator's abilities in this respect, I would insist that it is one of the duties of science to search for physical laws which explain, or at least describe in some coherent way, the nature of the phenomenal accuracy that we so often observe in the workings of the natural world. Moreover, I cannot even recall seeing anything else in physics whose accuracy is known to approach, even remotely, a figure like one part in $10^{10^{123}}$. So we need a *new law of physics* to explain the specialness of the initial state!

At this point I should make clear something which might otherwise be misleading. It may be felt that, in a chaotic cosmology viewpoint the reason for the time-asymmetry of the world has merely to do with the Creator's choice being made

---

* Some peculiarities of operating with such large numbers should be pointed out (see Littlewood 1953). It makes no essential difference, for example, whether one writes

$10^{10^{88}}$, or $e^{10^{88}}$, or even $1000^{10^{88}}$, for that matter, since $1000^{10^{88}} = 10^{3\times 10^{88}} \doteq 10^{10^{88.5}}$, while $e^{10^{88}} \doteq 10^{0.43\times 10^{88}} \doteq 10^{10^{87.6}}$. In such expressions, the symbol '$\doteq$' must be interpreted suitably. It does not now mean that the ratio of the two sides of the equation is close to unity, but that it is not large enough to imply a substantial change in the uppermost exponent. (Note, that, in this sense, even cubing the number $10^{10^{88}}$ makes no essential difference to it!) The uncertainties in the figure '88' will quite swamp any of these other considerations.

† In an earlier work (Penrose 1979) there is, implicitly, an oversight which amounts to writing the figure erroneously as $10^{(10^{120}/10^{88})} = 10^{10^{32}}$. Thus the Creator's aim would need to be enormously more accurate even than is suggested in that reference.

in the remote past rather than in the remote future. However, if we accept the standard picture of classical mechanics, this is not the case. I am doubtful, also, that it would make any real difference if a conventional quantum mechanical description is adopted instead of a classical one. But let us, for definiteness, suppose that the universe evolves classically according to standard Hamiltonian-type equations. Then the phase-space measure on the manifold $\mathscr{W}$ remains constant as the universe evolves (Liouville's theorem), where we now envisage that each point of $\mathscr{W}$ represents not just the initial conditions, but the entire 4-dimensional universe-history (i.e. space-time) that evolves from this initial state. It makes no difference to the remarkableness of the Creator's selection, whether this is in choosing initial or final conditions, or simply selecting one entire space-time history out of all the vast collection of alternative possibilities. The figure of one part in $10^{10^{123}}$, which represents the improbability of the Creator's selection, still applies whether the choice is made at the beginning or at the end, or as a selection of an entire universe history, since it simply represents a ratio of phase-space volumes.

In this, I am taking that figure as a measure of the improbability of our actual universe (assumed closed, for the purposes of this argument), if we *assume* that the Creator's choice was made entirely at random – for this is the essential hypothesis of the chaotic cosmology viewpoint. I think that it should be clear from the above discussion that this viewpoint cannot actually be valid, in our world, if we assume an unconstrained initial state. The 'pure' chaotic cosmology viewpoint has, indeed, fallen out of favour in recent years. But even if we adopt a viewpoint of 'partially constrained chaos' according to which *some* geometrical constraint is placed on the initial singularity, we still have the problem of finding how such a constraint can produce something so absurdly close to uniformity while not yielding *exact* uniformity in the initial geometry. In my view, this is a clear indication that chaos in the initial geometry was *not* what was responsible for the present entropy of the universe. The 'observed' figure of 10^8 for the entropy per baryon of the present universe must have its origin in some part of physics other than gravitation theory. The most likely candidate would seem to be particle physics, but I shall make no attempt to pursue the matter further here.

2. IS A TIME-SYMMETRIC UNIVERSE TENABLE?

In Fig. 2.1 is a crude qualitative picture which contrasts the 'universe as we know it' (assumed closed, for our expressed purpose of convenience only) with a 'more probable universe' having a chaotic big bang. The big crunch is, in each case, a high entropy ($\sim 10^{123}$) complicated unconstrained final singularity, while, for the left-hand picture the big bang is a low entropy ($\leqslant 10^{88}$) Friedmann-like, highly constrained initial singularity, the right-hand picture having an unconstrained 'much more probable' big bang. The diagrams are suggestive only, and not drawn strictly according to the rules of standard conformal diagrams whereby null cones would be drawn as sloping at 45°. The 'stalactites' represent the singularities of black holes, while the 'stalagmites' of the second picture represent the singularities of white holes.

Universe 'as we know it' *A 'more probable universe'*

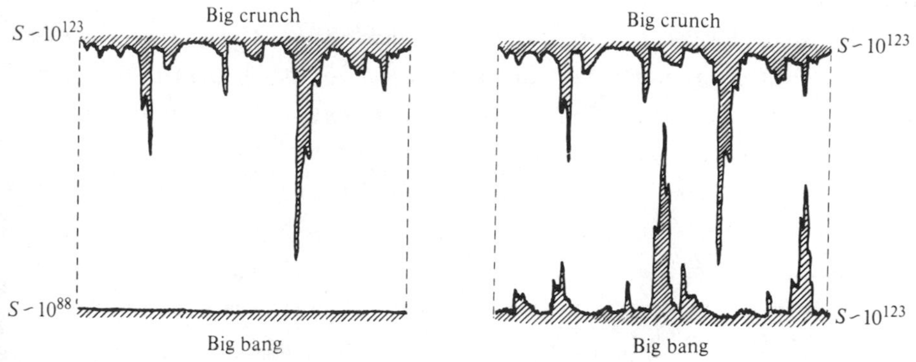

FIG. 2.1. If there were no severe constraint placed on the big bang, as there is in the diagram on the left, we would be driven to the picture of a white-hole-riddled universe on the right. (Not drawn strictly according to conformal diagram conventions.)

Note that it is primordial white holes, rather than primordial black holes that seem to occur most naturally in a chaotic big bang. For the 'chaos' is simply the time-reverse of that which we expect to arise naturally in the generic collapse of a universe such as our own (assumed closed). Primordial black holes could occur too, but they represent a more complicated geometrical situation (see Fig. 2.2).

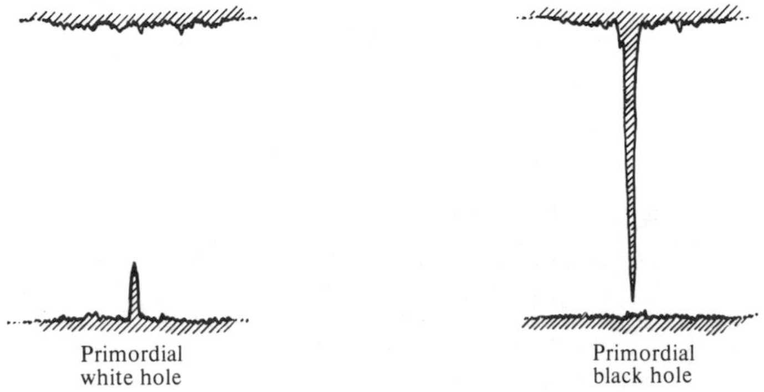

FIG. 2.2. The geometry of the singularity structure for primordial white and black holes.

It should be pointed out, also, that the various arguments in the literature which indicate that white holes ought, in some sense to be violently *unstable* (Eardley 1974; Zel'dovich 1974; Ramaswamy and Wald 1980) depend upon assuming that there is an otherwise 'normal' thermodynamic time-arrow and that the white hole can be treated as isolated. Although not directly applicable in the present context, these arguments seem to imply that a white-hole singularity which can be well separated from the remainder of the big-bang singularity is a highly 'improbable' type of initial singularity, requiring very precisely organised correlations in the incoming fields (though even so, it is not as 'improbable' as the Friedmann-type singularity itself, among all unconstrained big bangs). The reason that the time-reverse of this – namely an isolated *black* hole – can occur as a common feature of a 'normal' universe is that correlations in the outgoing fields are to be *expected* owing to the fact that such a 'normal' universe has a low-entropy Friedmann-type big bang (and hence a second law of thermodynamics), this low entropy being entirely reflected in such correlations when the final state is reached. In time-reversed form, the correlations needed to keep an isolated white hole in existence would require a low-entropy *final* state. I certainly do *not* want to assume a low-entropy final state, but this should not restrict the freedom in the choice of structure for the initial singularity (quite the reverse, in fact!), so the white hole type of initial singularity should be a common feature of unconstrained big bangs even if it does not result in an isolated white hole persisting with time.

The view is sometimes expressed, as an attempt to preserve time-symmetry, that the universe should finally return to a (fairly) low entropy state in the big crunch, the normal arrow of time being reversed in the collapsing phase of the universe (Gold 1962; Wheeler 1979). It may be pointed out that, in addition to all the other extreme difficulties that this viewpoint encounters (cf. Davies 1974; Penrose 1979) is the fact that if black holes are to be expected at all in such a universe model (and it is difficult indeed, to see how to avoid them) then a considerable population of white holes, and some very large ones too, would be expected to persist from the big bang until our present epoch and well beyond it – accompanied by highly correlated non-thermodynamic behaviour. I do not believe that the implications of such a model resemble, in the slightest degree, the universe as we know it, and I would challenge present supporters of this view to give a serious discussion of such matters!

Another issue which is often injected into the entropy–time-asymmetry discussion is the *anthropic principle*. The argument appears to date back to the time of Boltzmann (1895) and it asserts, in effect, that the low entropy that we observe about us is a consequence merely of our existence. The second law is, indeed, an essential ingredient of the workings of the metabolic processes upon which our lives depend, a high degree of order (i.e. low entropy) in our surroundings and in ourselves being necessary for our existence, in order that we may think, feel and perceive. Might it not be, so the argument runs, that only in those universes in which there is an operative second law and a high degree of order at some stage (though perhaps only as part of a huge fluctuation), will the conditions suitable for life and consciousness arise?

The main difficulty with such a view is that only a very very tiny proportion of the 'specialness' of the initial state is actually of relevance to life. Indeed, since entropy is continually increasing, it would have been vastly 'cheaper', in terms of lowness of entropy, for the Creator merely to have selected a universe whose entropy was adequately low, say one million years ago, and to have allowed it to go up into the past before that time. In terms of the sorts of improbability that we have been discussing, the long tedious(?!) years of evolution which preceded this period were quite unnecessary! The fluctuation could simply have produced all the earthly living creatures intact! As a rough and extremely liberal calculation, for example, I estimate that all the particles in the solar system could simply 'happen' to come into their present positions with an improbability of one part in much less than $10^{10^{62}}$. If we require a healthy peppering of such inhabited solar systems throughout the universe — and let us be overwhelmingly generous by allowing every single star to have such a solar system, $\sim 10^{23}$ in all — then we obtain one in $10^{10^{85}}$ as an over-estimate of the improbability of it all happening simply by pure chance!

If, for example, we now allow the remaining inter-galactic gas to be collected into $10^{11}M_\odot$ black holes, then the Creator need only aim for a region of $\mathcal{W}$ whose volume (in absolute units) is $10^{10^{111}} \div 10^{10^{85}} \doteq 10^{10^{111}}$. Clearly this is a vastly 'cheaper' method than the one which appears actually to have been used (volume $\sim 10^{10^{88}}$)! And it is obvious, also, that one could improve considerably upon this figure. While I am happy to believe that most of $\mathcal{W}$ may well consist of uninhabited universes, because of a totally inhospitable chaotic geometry, there must nevertheless be a huge portion of it (with volume $10^{10^{115}}$ being surely a gross underestimate) in which conscious life abounds (assuming that physical laws not implying the hypothesis of Section 3 would allow life!) and in which the appearance of the universe is not at all close to the one in which we find ourselves, with its very closely Friedmannian beginnings (cf. Rees 1981).

I hope that it is clear from this that the anthropic principle cannot be what is responsible for the time-asymmetry of the second law. Indeed, it would appear from this that the Creator was not particularly 'concerned' about our existence, but was constrained in some very precise time-asymmetric way for some quite other reason. From this point of view, our present existence would arise merely as a by-product!

3. THE WEYL CURVATURE HYPOTHESIS

I have stated elsewhere (Penrose 1977; 1979) that a plausible hypothesis for the constraint on the big bang is that the Weyl conformal curvature vanishes as this singularity is approached. This would have the effect that 'gravitational clumping' is effectively absent in the initial state. The entropy in the matter can be assumed to be at a maximum, but the entropy in the gravitational field starts at zero. As the universe evolves, gravitational clumping increases, and this enables the entropy to rise above its original value, the greatest increases in entropy occurring when matter collapses into black holes.

The singularities of black holes, on the other hand, are not at all constrained in this way. In the spherically symmetrical case, the Weyl curvature diverges to infinity (according to the inverse cube of the radial distance) as the singularity is approached; and in the generic case, also, we expect the Weyl curvature to diverge to infinity at the singularity. In this respect, the 'final' singularities of black holes do not at all resemble the time-reverse of the structure that I am postulating for the big bang.

It would seem unnatural, however, merely to postulate one law for the big bang singularity and to have all other singularities unconstrained. It would seem that *all* 'initial' singularities (e.g. those occurring in white holes) ought to be constrained in the same way — and so I am claiming that they should have vanishing Weyl curvature — but this constraint should not apply to 'final' singularities. Indeed, if the *ideal point* definition of a singularity is used (Geroch, Kronheimer, and Penrose 1972; Penrose 1974) then singularities do in fact appear as being of two types, namely the singular TIPs ('final' singularities) and singular TIFs ('initial' singularities). (See Fig. 3.1.) If strong cosmic censorship (Penrose 1974) is assumed then the TIPs and TIFs remain separated from one another, while without this hypothesis some singularities may have existence both as TIPs and TIFs.

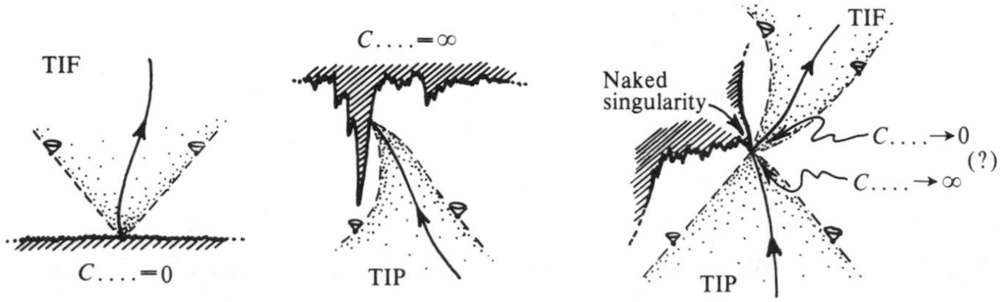

FIG. 3.1. A TIF describes an 'initial' singularity; and a TIP, a 'final' singularity. If strong cosmic censorship fails, a singularity can be represented both by a TIF and by a TIP.

Even when this occurs (such as in the presumed final ephemeral naked singularity of a Hawking-evaporating black hole) (Fig. 3.2) it seems not unreasonable that the curvature structure for the TIPs may differ from that for the TIFs (Penrose 1979). Accordingly, I am proposing:

Hypothesis: The Weyl curvature in any singular TIF tends to zero as the ideal point representing the TIF is directly approached from within the TIF.

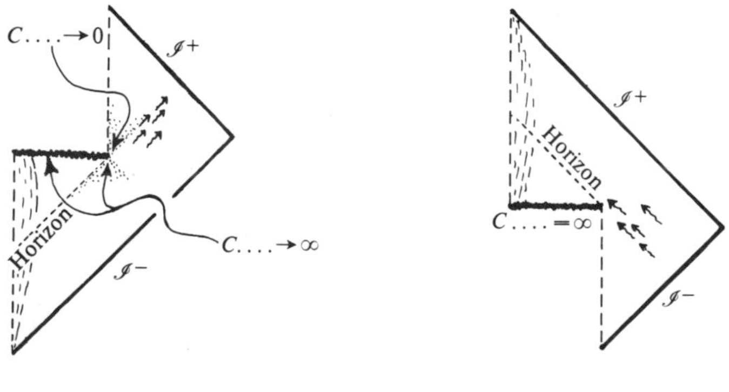

FIG. 3.2. Hawking-evaporating black hole (left) and its time-reverse (right).

Other slightly different statements could also be given (for example, the omission of the word 'singular' might be reasonable), and the statement is being kept deliberately a little vague. (The statement about approaching from within the TIP is being made in order that Hawking's ephemeral naked singularity should not necessarily violate the hypothesis.) See Figs. 3.1 and 3.2.

The manifold $\mathscr{W}$, that we considered earlier, must now be restricted to a tiny subset $\mathscr{U} \subset \mathscr{W}$. Thus, while in the definition of $\mathscr{W}$ we allowed all possible geometries for the initial singularity, the subset $\mathscr{U}$, to which the Creator's choices are now restricted, represents only those universes in which all initial singularities are constrained by the Weyl curvature hypothesis. Though the world-view that I am proposing (Penrose 1979) differs very greatly in its implications from that of chaotic cosmology, there is one essential feature of the chaotic cosmology viewpoint that I wish to retain (at least provisionally), namely that the Creator's pin is to be considered to be placed *at random* in the manifold of all allowed universe models — but in this case, in the manifold $\mathscr{U}$. This has the implication that the matter, in the early stages, is in thermal equilibrium, so one would anticipate that the Ricci tensor is initially very closely uniform, but that some (thermal) fluctuations are allowed — which of course, would be necessary in order that irregularities can arise. The exact nature and magnitude of these fluctuations would presumably ultimately depend upon some detailed particle physics.

It is an essential feature, also, that the final singularities should *not* be constrained in this way. The geometrical constraints on the initial geometry have the *effect* of producing a state of very low entropy — though I should emphasize that the Weyl curvature hypothesis is not in any way a statistical one, but is intended to be a precise physical law governing the structure of singularities (so it is not in any way necessary, for example, that a definition of gravitational entropy should be at hand). Likewise, any such constraints on final singularities would have the *effect* of producing a state of very low entropy in the future. That would have innumerable totally unacceptable physical implications, one of the clearest being that all sorts of very precise correlations would exist in particle motions which would conspire to prevent any collapse to a black hole. These correlations could produce effects that would appear to us like pure magic, and physics would not at all resemble the physics that we know. The absence of such correlations in the initial state (e.g. the law of conditional independence of O. Penrose and Percival 1962) is, according to the world-view that I have been presenting, a consequence of the *absence* of constraints on final singularities, together with the randomness of the Creator's pin.

The other main implications of this world-view strongly involve the Weyl curvature hypothesis. These are:

(a) the second law of thermodynamics,*

(b) a closely Friedmannian early universe,†

(c) the absence of white holes,

and it would seem, also, that the absence of chaos in the early stages should preclude any appearance of black mini-holes.

It is my impression that one reason people may have felt uncomfortable with a big bang as uniform as the one that I am proposing, is that it *seems* to run counter to the appealing idea that all matter may have originated out of the vacuum, by means of particle-creation processes incurred by the initial highly curved state. If the Weyl curvature was initially zero (and if the big bang was indeed a curvature singularity), then all the initial curvature must have already resided in the Ricci tensor, i.e. by Einstein's equations, matter must always have been present right from the start! However, there is no reason (provided that we can accept non-conservation of baryon number) that the initial 'matter' should not have been something with the quantum numbers of the vacuum. The initial Ricci tensor *could*, perhaps, have been entirely an 'effective energy tensor', produced because vacuum fluctuations give rise to an effective modification of Einstein's equations in regions of very high curvature.

---

* The retardation of radiation is one aspect of this.

† Exactly *how* closely Friedmannian it is, would depend upon the statistical uniformity of the Ricci tensor.

Another reason that people may find the Weyl curvature hypothesis hard to accept is that it is grossly time-asymmetric. Apart from K°-decay, there seems to be nothing else in the local laws of physics which violates time-symmetry. However I would insist that to explain the second law of thermodynamics we do need a grossly time-asymmetric law of physics (cf. Section 2). The local laws of physics that we know and understand are, it is true, all time-asymmetric. But the very presence of space-time singularities in realistic models, together with the fact that these singularities are 'regions where the presently known laws of physics break down', suggests that we are in need of laws *other* than those that we know and understand. These new laws are required in order that the physics of what we now refer to (in the absence of a proper theory) as 'space-time singularities' may eventually be understood.

4. HAWKING'S BLACK = WHITE HYPOTHESIS

Let me now turn to another quite different point of view which has been put forward by Hawking (1976*a*,*b*). Again, the starting-point is the Bekenstein—Hawking formula, but employed in a quite different context. Hawking envisages a box of volume V, with perfectly reflecting walls, in which there is a certain amount, E, of energy in some form. The box is left undisturbed for an indefinite period and its contents allowed to find equilibrium. Now provided that E is large enough (but not so large as to collapse the entire box), given V, we find that the maximum entropy is achieved when the box contains a black hole in equilibrium with a small amount of thermal radiation. According to Hawking's argument, the physics which has entered into the discussion of this problem is entirely time-symmetric, namely general relativity, the standard formalism of quantum field theory, and Maxwell theory. Thus, it is claimed, the configuration of maximum entropy ought also to be time-symmetric. (Hawking also presents another argument, based on the second law. A modified version of this will have its role to play in the next section.) Since the maximum entropy configuration is a black hole, with radiation, and since the time-reverse of a black hole is a white hole, then the conclusion is that black holes and white holes are physically indistinguishable!

Alarming as such a suggestion may seem, at first, there is a certain elegance and economy in the idea. The Hawking radiation emitted from a black hole can now be re-interpreted as matter-creation at the white hole's singularity. Likewise, the matter or radiation which the black hole swallows may be re-interpreted as the time-reversed Hawking radiation from the white hole. The apparent difference between the random Hawking radiation and the coherent objects (e.g. television sets) that can be swallowed by black holes is explained (reasonably convincingly) in terms of the time-asymmetric boundary conditions that are normally imposed on external objects (that is, it is normally considered reasonable to impose a low-entropy input, e.g. a television set, but not to select for a low-entropy output).

However, I have never found myself able to accept this view that black holes and white holes should be physically the same. A glance at Fig. 3.2 should make clear what are my difficulties in this respect. The left-hand diagram represents the formation of a black hole by classical gravitational collapse followed, ultimately, by its disappearance as Hawking radiation; the right-hand diagram represents the formation of a white hole by time-reversed Hawking radiation, followed by its disappearance by the classical disgorging of some ordinary matter. Even if the entire portions inside the horizons are deleted, we obtain two space-time geometries which are quite different. Hawking's answer to this criticism is that it is not appropriate to insist upon the existence of a uniquely defined objectively determined space-time geometry when considering processes of this nature. Since the back-reaction on the space-time, of the quantum mechanical Hawking radiation, is an essential ingredient of these pictures, the space-time structure has itself become subject to quantum-mechanical rules. Thus, Hawking envisages that the space-time geometry does not have objective existence but has become an essentially observer-dependent construct.

For myself, I find this line of argument very hard to accept. It seems to me that if we are talking about reasonably large black holes of a solar mass or more, then the space-time geometry ought to be sufficiently well-defined that an objective general-relativistic picture is valid and that the black hole and white hole ought to be distinguishable by physical experiment. As a further objection, it had seemed to me clear that the most probable mode of formation of a black hole in the box (by classical collapse) was not at all the time-reverse of its most probable mode of disappearance — even if all that one was allowed to 'observe' was the radiation reaching (or leaving) the walls of the box. However, these arguments have become involved in much dispute, the main difficulty being that no-one seems yet to have been able to perform the relevant calculations of the probabilities involved in a sufficiently trustworthy manner.

A problem is that one is dealing with relaxation times that are ridiculously long when the mass is, say $M_{\odot}$ or more, if we wish to wait for the hole to disappear and to reform merely by chance fluctuations. Even if $V = \infty$, it would take 10^{56} Hubble times, or so, for a $M_{\odot}$ black hole to disappear, and in this case it will almost certainly never reform. For a smaller box containing $M_{\odot}$, with V say the size of a galaxy, the black hole would represent maximum entropy, so its occasional disappearance by a fluctuation would take far far longer. Also, for it to form again out of radiation would involve time-scales and a precise mode for formation that are not easy to estimate in any detail. I do not have reason to doubt the qualitative arguments against time-symmetry that I have presented earlier (Penrose 1979), but I accept that there is considerable room for argument and that the discussion is perhaps not so conclusive as I had previously thought.

In Fig. 4.1, I have indicated the three different situations: (i) large box with pure radiation as maximum entropy and unstable hole, (ii) medium-sized box with a locally stable hole but pure radiation still as maximum entropy and (iii) smallish box with the hole as maximum entropy (cf. Hawking 1976*a*,*b*; Gibbons and Perry 1978). In Table 4.1 are listed various familiar objects, providing the mass-energy E and the corresponding volumes V for which the maximum entropy is about equally balanced between the pure radiation state and the black hole, so that some feeling may be obtained for the scales involved.

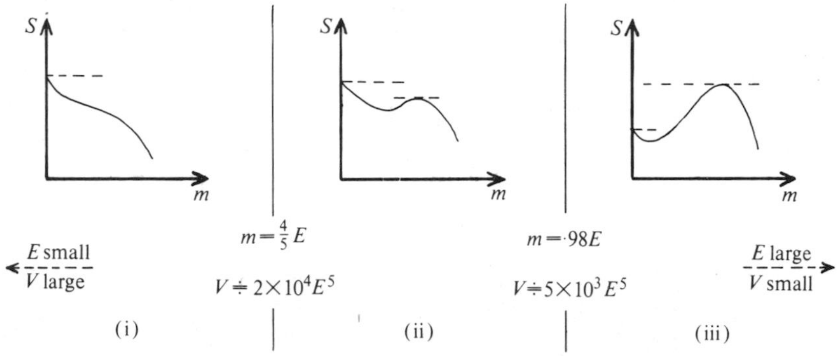

FIG. 4.1 i-iii. The entropy S plotted against the mass m which is contained in a black hole, for total energy E in a box of volume V ($c = \hbar = G = k = 1$).

TABLE 4.1

Volumes V and mass-energies E for which pure radiation or a black hole have roughly equal entropies

(Note: a black hole of 10^{15}g has roughly the radius of a proton; but the mass of a small mountain. $M_\oplus$ is the mass of the Earth.)

E	V
pyramid	10^{-7}cc
10^{15}g	10^{3}cc (ordinary box)
comet	Vol. of Moon
$M_\oplus \doteqdot 6 \times 10^{27}$g	10^{67}cc (galactic dim.)
$M_\odot \doteqdot 2 \times 10^{33}$g	10^{95}cc (bigger than obs. universe)

5. SELF-COLLAPSING WAVE FUNCTIONS?

From my own point of view, Hawking's suggestion that black and white holes might be physically the same would not be acceptable for another reason. For the Weyl curvature hypothesis of Section 3 rules out white holes as having unacceptable singularity structure, whereas it allows black holes. In fact, if this hypothesis is admitted, it invalidates the argument that the physical laws involved in the discussion of Section 4 are all time-symmetric. Part of the physics involved inside the box is the physics of space-time singularities, since this is a necessary part of the discussion of black or white holes. Thus, I would maintain that when maximum entropy is achieved by the black hole state, this state need not (and, indeed *cannot*) be time-symmetric, since some time-asymmetric physics (namely the Weyl curvature hypothesis) is crucially involved in the discussion.

However, this does not absolve me from considering Hawking's other objection to time-asymmetry (Hawking 1976*a*,*b*) which was based on considerations of the second law of thermodynamics. I prefer, however, not to consider Hawking's argument in its original form (since the second law's validity seems to me to be somewhat questionable in any case, when time-scales as long as the ones needed here are involved). Instead I shall give a related argument which seems to me to contain the essence of Hawking's objections.

Consider the phase space $\mathscr{V}$ of the contents of the box. I shall be somewhat vague as to whether this is to be a classical phase space or a quantum one. If classical, it is a space of finite volume. If quantum-mechanical, it is a finite-dimensional Hilbert space. This is because the volume V and energy E are both bounded finite quantities, the standard rules of classical or quantum mechanics being assumed to apply. Let us now consider the flow F on the space $\mathscr{V}$, which gives the time-evolution of the contents of the box. It will be assumed that it is valid to consider an essentially non-relativistic description, with the contents of the box given as a function of time – even when the box contains one or more black or white holes. As the system evolves, the corresponding phase-space point P moves about $\mathcal{V}$, as determined by the flow F. At certain times the box will contain at least one hole, either black or white (or both), and at those times the point P lies in a special subset of $\mathcal{V}$ labelled $\mathcal{H}$. (I am ignoring any difficulty about defining the precise moments at which the box is considered to contain such a hole, since this does not seem to have much effect on the discussion.)

Now there are various viewpoints that one may consider. One of these is Hawking's, according to which black and white holes are physically the same and there is an essential time-symmetry in the whole set-up. Another, in which there is also time-symmetry, is the 'common-sense' viewpoint, according to which black holes can exist and white holes can exist but they are simply different objects from each other, and furthermore many 'zebroid' combined holes might also exist. In each of these viewpoints, because of the time-symmetry, there are 'as many' flow lines entering the region $\mathcal{H}$ as there are leaving it. The situation is depicted in the left-hand diagram of Fig. 5.1.

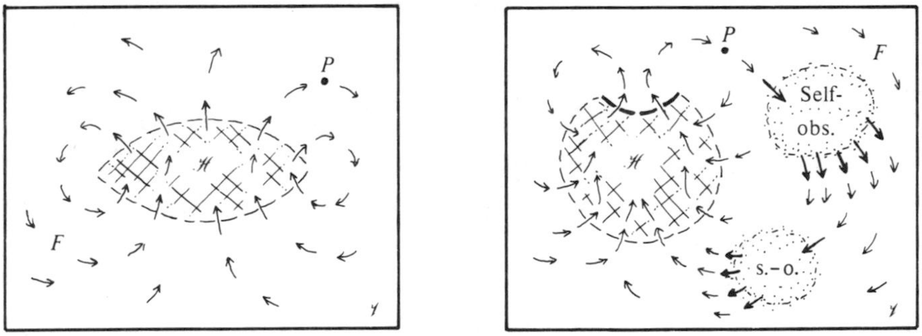

Hawking viewpoint, or 'common-sense' viewpoint

A viewpoint involving the Weyl curvature hypothesis

FIG. 5.1. The phase space $\mathcal{V}$ of the contents of the box, $\mathcal{H}$ being the region of $\mathcal{V}$ for which the box contains at least one hole (black or white or both).

My own viewpoint is depicted in the right-hand diagram, and here there are 'more' flow lines entering $\mathcal{H}$ than there are leaving it. The reason for this is that there are, in my viewpoint, many more ways of creating a black hole than there are ways of destroying it. For a black hole can only disappear quantum-mechanically by means of the Hawking process, while it can be produced either quantum mechanically or by means of classical collapse. Since a black hole is not a white hole, it cannot simply disappear by means of those evolutions which represent the time-reverse of collapse. Those motions are not physically allowed since they involve a space-time singularity which does not satisfy the Weyl curvature hypothesis.

Another way of roughly seeing this is to look again at Fig. 3.2. Only the left-hand diagram is allowed, while to specify data for geometries of that type, information may be given* on $\mathscr{I}^-$, or equivalently *both* on $\mathscr{I}^+$ *and* on the singularity. Thus, while the data on $\mathscr{I}^-$ may be freely specified, only a restricted class of data on $\mathscr{I}^+$ is allowed because this must be supplemented by extra data on the singularity in order to give as many states as those allowed on $\mathscr{I}^-$. (Indeed, it was considerations of this kind which led Hawking to discover his radiation in the first place.) The disallowed data on $\mathscr{I}^+$ would, when evolved backwards, give geometries resembling that of the right-hand diagram of Fig. 3.2, or also with final singularity data, geometries resembling combinations of both diagrams in Fig. 3.2. All of these are excluded in my viewpoint.

It is here that my viewpoint runs into some trouble. According to Liouville's theorem the phase-space volume should be preserved as we follow the flow. However, phase-space volume is lost within $\mathcal{H}$ because 'fewer' flow lines leave $\mathcal{H}$ than enter it (more accurately, there is a net loss of volume per unit time leaving the region $\mathcal{H}$). There is no difficulty about contradicting Liouville's theorem for the region $\mathcal{H}$ since for that region a space-time singularity exists within the box so we expect that the ordinary laws of physics must be partially suspended. The essential difficulty comes with the region *outside* $\mathcal{H}$. In *that* region normal physics is supposed to hold good, so it would come as more of a shock if we were to be

* Strictly speaking, there is also some information in the final singular TIF of the Hawking-evaporated black hole, but (at least in my own viewpoint) this information counts for very little and will be ignored in this discussion.

confronted with a contradiction to Liouville's theorem there. But if we *assume* that Liouville's theorem holds good in $\mathcal{V} - \mathcal{H}$ we seem driven to conclude that certain regions within $\mathcal{V} - \mathcal{H}$ must become inaccessible once $\mathcal{H}$ has been cycled through. These regions, in fact, correspond to physical motions that, when extrapolated backwards, would lead to time-reversed gravitational collapse in which there is a singularity violating the Weyl curvature hypothesis (e.g. to white holes). These regions need not be excluded when the box and its contents are *first* formed, but seem to be disallowed as part of whatever 'equilibrium' configurations the contents of the box finally settle into. This gives us a rather strange picture of the ultimate thermodynamics of the contents of the box.

One can also look at this difficulty from the quantum-mechanical point of view. Here Liouville's theorem translates into conservation of probability. If no 'observations' take place, the Schrödinger evolution holds good and single pure states are supposed to evolve into single pure states. The generalization of this view given by Hawking (1976*b*) to cover his own viewpoint allows single pure states to evolve into density matrices. This, in effect, allows flow lines to 'spread' into wider regions, but a probability weighting is given, so that the total spread-out region counts just the same as a single line. In my own viewpoint, then, there is classically a volume-reduction taking place whenever $\mathcal{H}$ is cycled through, and this is interpreted quantum-mechanically as a reduction in the (finite) dimension of the Hilbert space of allowable states, as compared with the dimension of the space of states initially available. In neither Hawking's viewpoint nor the 'common sense' one does the difficulty that I am encountering with my own seem to occur. My objections to those other viewpoints do not spring from the present considerations but from all the other several difficulties that allowing white holes to exist seem to involve us in.

However, some of those other difficulties do, in my opinion, provide us with a clue. In Hawking's viewpoint, the geometry begins to get inextricably involved with quantum mechanics as the system evolves. (Recall the 'observer dependence' of the geometry that we seemed to be forced into.) After the box has been left in isolation for a very long time, its contents would not resemble a classical geometry at all. It seems that an 'observer' would be needed to peer into the box and so collapse the state into that of a well-defined geometry. I have always had difficulties with that aspect of quantum mechanics. It seems to me that something objective must actually be taking place within the box when such gross features such as the existence or location of a huge black hole are being considered. Perhaps quantum linear superposition fails when space-time geometry is involved? Indeed, that is the kind of line that I am proposing to take. My point of view is not so unconventional as it may seem at first. The normal Copenhagen interpretation requires, after all, that 'Rule 1' (Schrödinger evolution of the state vector) is sometimes replaced by 'Rule 2' (discarding the old state vector and instituting a new one which is an eigenstate of an appropriate operator, these being the times when the quantum mechanical probabilities arise). It is nowhere stated exactly *when* Rule 2 is to be used, but it involves the presence of some large 'classical' structure. Might it not be (cf. Karolyhazy 1966; Komar 1969) that this rule comes into play when systems become sufficiently 'macroscopic' that they significantly affect the space-time geometry?

There is a relation also to entropy. An 'irreversible process' is normally considered necessary in order that Rule 2 be brought into play. This would be stated as: 'the entropy of the system has gone up'. But entropy is normally considered to be a rather subjective concept, dependent upon someone's notion of coarse-graining. It is hard to see how an objective change in the state of the world can take place according to Rule 2 unless 'entropy increase' can sometimes be objective. The best example in physics, so far, in which an entropy seems to have acquired an objective meaning, is in the Bekenstein–Hawking formula. Here the entropy is equated to a precise space-time-geometric quantity, namely an absolute event horizon's surface area. So if space-time geometry is to be regarded as in some sense objective, then entropy also can apparently acquire objectivity in these circumstances.

I regard these considerations as representing encouragement for the following picture: for much of the time the evolution of a system can be very accurately treated as proceeding according to Rule 1. But when macroscopic space-time geometry becomes inextricably involved, Rule 2 comes objectively into play as an accurate representation of reality. The precise boundary between the application of these two rules appears to be effectively unimportant (von Neumann 1955) — though ultimately we shall require some new theory which tells us what 'really' happens. No 'conscious observer' is required to intervene. Rule 2 just takes place *spontaneously* when the circumstances of the coupling of the system to space-time geometry are appropriate. I regard this picture as an approximation to some undiscovered theory of 'quantum gravity' in which both the Rules 1 and 2 arise as approximations to this more correct, but yet undiscovered theory. Thus, my own view of what the term 'quantum gravity' means appears to differ from that of most physicists. In particular, I believe that the correct quantum gravity must be a *time-asymmetric theory!*

Let us return to the discussion of the right-hand diagram of Fig. 5.1. What is required, in order that we can obtain a reasonable phase-space picture, is that exactly compensating for the loss of phase-space volume in $\mathcal{H}$ must be other regions of $\mathcal{V}$ in which phase-space volume is regained. These regions I envisage as representing the transitions from the application of Rule 1 to Rule 2 and then back to Rule 1 again. So long as Rule 1 applies, we follow along a unique trajectory of the flow F. But then, somewhere along the line, *self-observation* of the system in the box takes place, so that Rule 2 operates. A number of different trajectories emerge from this rather vague region in $\mathcal{V}$, these representing the different possible outcomes of the result of this self-observation. The system could, indeed, 'select' any one of these outcomes. This is where the non-determinism of quantum mechanics comes in. The system has to 'make its choice' and thereafter it continues following some definite trajectory of the flow F (Rule 1 back in operation again). Now, after wandering round and round the space $\mathcal{V}$ for an enormous length of time, our phase-space point P will finally come back in along the same trajectory that we considered before. The self-observing system is confronted with exactly the same 'choice' as before, but now it may well be that a different selection is made, and P goes wandering off along some quite different path. Each time, the point comes in along some definite path and it goes out along some definite path. The (quantum mechanical) Liouville theorem has the *appearance* of not being violated. Yet because many more alternatives leave the 'self-observation' regions than enter them we do have an *effective* net volume increase, in violation of Liouville's theorem.

I shall try to be a little more explicit about what is involved, in this picture, when Rule 2 comes into play. I have asserted that whenever the region $\mathcal{H}$ is cycled through, the dimension of the Hilbert space of allowable states gets reduced, the non-allowable states being those which when evolved backwards would lead to geometries involving a violation of the Weyl curvature hypothesis. Now, as time progresses, it will become harder and harder to ascertain precisely which these non-allowed states are: an initial specialness of geometry (an apparently low entropy) will develop into correlations between the detailed particle motions (an apparently higher state of entropy). The Hilbert space $\mathcal{V}$ of allowable states may be thought of as a subspace of a larger space $\mathcal{Y}$ for which the Weyl curvature constraint is not imposed. (Compare the corresponding spaces $\mathcal{U} \subset \mathcal{W}$ of Section 3.) We may regard $\mathcal{Y}$ as a Hilbert space of constant dimensionality, while $\mathcal{V}$ now has a dimension that changes with time. After a Hawking evaporation, the subspace $\mathcal{V}$ is partly discernible in terms of macroscopic features of the geometry*, but this becomes progressively less so with time, as some gross macroscopic features of geometry get 'lost' in particle correlations. At some stage Rule 2 comes into operation as a consequence (?) of this and the

* The point of final explosion is clearly a state of less than maximum entropy. This is not unreasonable since a large fluctuation was needed at an earlier stage in order for this situation to arise.

state vector, which initially lies in the subspace $\mathscr{V}$ (because of the Weyl curvature hypothesis) gets projected into some state vector in $\mathscr{Y}$ with components lying outside $\mathscr{V}$. (The 'natural' or 'macroscopically discernible' Hilbert space bases in $\mathscr{Y}$ are no longer specially related to $\mathscr{V}$.) The space of allowable states has now grown in dimensionality – and we have to *relabel* this larger subspace of $\mathscr{Y}$ as our new $\mathscr{V}$. This dimensionality increase is just the net volume increase that we require, in an effective violation of Liouville's theorem.

The idea, of course, is that this volume increase should exactly compensate for the volume loss in the black hole region $\mathscr{H}$. In fact we now have the intriguing possibility of actually performing a calculation which estimates the amount of self-observing that a system of energy E in a volume V ought to be indulging in, according to this point of view. Since the formation of black holes is an entirely gravitational phenomenon, we now have, in principle, a quantitative link between gravity, as represented by black holes, via the Bekenstein–Hawking formula, and the quantum-mechanical observation process. It seems that all one need do is to calculate the volume loss in the region $\mathscr{H}$, for given E and V, and use this to obtain a measure of averaged effective volume-spreading due to self-observation. An appealing aspect of this is that while it is obvious that black holes need not be present in any particular observation process, their theoretical existence *is* necessary, in this picture of things, to produce the correct overall balance.

All this will repay closer examination. But as things stand, I have found considerable difficulty in estimating what the suggested self-observation rate ought to be. The calculation of the volume-loss within the region $\mathscr{H}$ is complicated by the fact that there is also some volume-gain within $\mathscr{H}$ owing to applications of Rule 2 whilst black holes are present. There is also something a little disturbing, in this respect, about Table 4.1. The mass-energy contained in a pyramid is certainly ample for a laboratory and a team of physicists performing quantum interference experiments. If they were suddenly encased in a box with perfectly reflecting walls, they would certainly be able to continue to perform such experiments for a long while to come, presumably employing many applications of Rule 2 before finally expiring. Is it reasonable to suppose that this needs to be compensated for by the occasional collapse of the mass-energy in their (sometime) laboratory into a black hole considerably smaller than a proton? This is very very far from the maximum entropy configuration since here $V \gg 10^{-7}$cc.

This all sounds rather absurd, yet I feel that it does not invalidate the earlier discussion. The observations which are taking place when the box is first formed may have more to do with the big-bang specialness $\mathcal{U} \subset \mathcal{W}$, than $\mathcal{V} \subset \mathcal{Y}$. But perhaps one must also be much more careful about the idealizations involved in the idea of a box with perfectly reflecting walls. In practice — and also in principle — there are severe difficulties with that concept. Our physicists in the above discussion could *not* in fact be isolated, by such a box, to the required degree.

These arguments remain inconclusive as they stand. I do, however, strongly believe that any successful quantum gravity theory will have something to say about such matters. Space-time singularities, thermodynamical time-asymmetry and quantum mechanical observations all seem to be interlinked with quantum gravity, the Bekenstein–Hawking formula forming one cornerstone of present discussion of such matters.

ACKNOWLEDGEMENTS

I am grateful to Stephen Hawking, Bob Wald and a number of other colleagues for many inspiring conversations. Most particularly, I am indebted to Don Page for some penetrating criticisms and also for supplying many of the numerical values that I have used.

REFERENCES

Barrow, J.D. and Matzner, R.A. (1977). *Mon. Not. R. astr. Soc.* **181**, 719.
Bekenstein, J.D. (1973). *Phys. Rev.* **D7**, 2333.
—— (1974). *Phys. Rev.* **D9**, 3292.
Boltzmann, L. (1895). *Nature*, **51**, 413.
Davies, P.C.W. (1974). *The Physics of Time Asymmetry*. Surrey University Press.

Eardley, D.M. (1974). *Phys. Rev. Lett.* **33**, 442.
Geroch, R., Kronheimer, E.H., and Penrose, R. (1972). *Proc. R. Soc.* **A327**, 545.
Gibbons, G.W. and Perry, M.J. (1978). *Proc. R. Soc.* **A358**, 467.
Gold, T. (1962). *Am. J. Phys.* **30**, 403.
Hawking, S.W. (1975*a*). *Comm. Math. Phys.* **43**, 199.
—— (1975*b*). In *Quantum Gravity, an Oxford Symposium.* (ed C.J. Isham, R. Penrose, and D.W. Sciama). Oxford University Press.
—— (1976*a*). *Phys. Rev.* **D13**, 191.
—— (1976*b*). *Phys. Rev.* **D14**, 2460.
Karolyhazy, F. (1966). *Nuovo Cim.* **A42**, 390.
Komar, A. (1969). *Int. J. Theor. Phys.* **2**, 157.
Littlewood, J.E. (1953). *A Mathematician's Miscellany.* Methuen, London.
Misner, C.W. (1968). *Astrophys. J.* **151**, 431.
—— (1969). *Phys. Rev. Lett.* **22**, 1071.
Neumann, J. von (1955). *Mathematical Foundations of Quantum Mechanics.* Cambridge University Press.
Penrose, O. and Percival, I.C. (1962). *Proc. R. Soc.* **79**, 509.
Penrose, R. (1974). In *Confrontation of Cosmological Theories with Observational Data (IAU Symp. 63).* (ed M.S. Longair). Reidel, Boston.
—— (1977). In *Proc. First Marcel Grossmann Meet. Gen. Relativ. (ICTP Trieste).* (ed R. Ruffini). North-Holland, Amsterdam.
—— (1979). In *General Relativity, An Einstein Centenary Survey.* (ed S.W. Hawking and W. Israel). Cambridge University Press.
Ramaswamy, S. and Wald, R.M. (1980). *Phys. Rev.* **D21**, 2736.
Rees, M.J. (1981). (This volume).
Wheeler, J.A. (1979). *Frontiers of Time* (North Holland, for Società Italiana di Fisica, Amsterdam).
Zel'dovish, Ya.B. (1972). In *Magic Without Magic: John Archibald Wheeler.* (ed J. Klauder). Freeman, San Francisco.
—— (1974). In *Gravitational Radiation and Gravitational Collapse (IAU Symp. 64).* (ed C.M. De Witt). Reidel, Boston.

12

INHOMOGENEITIES FROM THE PLANCK LENGTH TO THE HUBBLE RADIUS

M.J. Rees

Institute of Astronomy, Cambridge

1. INTRODUCTION

To the specialist in quantum gravity the most remarkable feature of our universe is its scale – the fact that it has lasted for $\sim 10^{60}$ Planck times – and its overall symmetry. Superposed on this large scale isotropy are the inhomogeneities on the scale of stars, galaxies and clusters, which may be a necessary requirement for our own existence. My aim in this talk will be to discuss the various quantitative limits on the universe's homogeneity, and to consider what inferences can be drawn about initial conditions. There is as yet no explanation for the degree of small scale structure – allied with large scale uniformity – which our universe apparently displays. The structure may indeed have been laid down at the initial 'quantum phase' of the big bang. Maybe this possibility justifies the inclusion of the topic on the programme for this conference; indeed, one could claim, more assertively, that the explanation of the structural features of the cosmos may offer the only empirical challenge to theories of quantum gravity.

I shall briefly review the state of play in classical cosmology (i.e. the programme to determine the overall geometry and mean density of the universe); then discuss the evidence on homogeneity on various scales; and finally discuss various hypotheses regarding the form of fluctuations. Many of the observations, and their implications, are still controversial among astronomers, but since the issues are peripheral to the interests of quantum gravity specialists, I shall gloss over many uncertainties in this report.

2. 'CLASSICAL' COSMOLOGY: q_0, Ω_0 AND THE 'UNSEEN MASS'

The microwave background provides the most impressively accurate evidence that the observable universe is highly isotropic. Searches for anisotropies have revealed no effect exceeding ~ 0.1 per cent on any angular scale down to ~ 1 minute. (Being relative measurements these are much more precise than the intensity measurements.)

According to the standard 'hot big bang' model the material in the early universe would be strongly coupled to a thermal radiation field. When adiabatic expansion had cooled the material to approximately 4000 K, the plasma could recombine and the radiation would no longer undergo substantial scattering or absorption. The observed microwave photons would not then be scattered since an epoch when the matter was ~ 10^9 times denser than it is now – long before any galaxies came into existence. The effective source of the background is thus a 'cosmic photosphere' at a redshift $z \simeq 1000$.

More general (and perhaps more realistic) cosmologies make this picture of the thermal history more complicated. The gas may have been reionized by an early generation of stars, resulting in a more blurred 'cosmic photosphere' at a more recent epoch. Heavy chemical elements (rather than just electron scattering) may contribute to the opacity. The spectrum of the microwave background may be distorted from an exact black body, and the input from stars, etc, may augment its intensity. However, although these complications affect the quantitative details, it is still true that the microwave background is a 'relic' from an epoch earlier than any discrete objects yet observed.

The measured isotropy implies that the observable part of the universe can be described by a Robertson–Walker metric with a precision of $\lesssim$ 1 part in 10^3. The clearest anisotropy so far detected, at the ~ 0.1 per cent level (Smoot, Gorenstein, and Muller, 1977; Cheng, Saulson, Wilkinson, and Corey 1979; Smoot 1980), can be attributed *either* to a ~ 500 km s^{-1} velocity of our galaxy (and its neighbours) relative to the mean velocity of distant matter on the cosmic photosphere, *or* to a ~ 0.1 per cent deviation from Robertson–Walker.

(I shall mention later how we might distinguish between these two options.) For the moment, the noteworthy inference is that the gross kinematics of the universe are exceedingly symmetrical; the microwave background establishes this with greater precision than the data on which the Hubble law is based. Unless we adopt an anti-Copernican viewpoint, this forces us to adopt a Robertson—Walker metric: the expansion is described by a single scale factor $R(t)$; the only other parameter is a curvature constant k, such that the spacelike 3-surfaces of homogeneity at time t have a curvature $\propto k(R(t))^{-2}$. Of course, it is only because the universe possesses this gross simplicity that observational cosmology is feasible at all. Our empirical evidence is essentially limited to observations along our past light cone, and to inferences about the history of the region near our world-line. Only because of the homogeneity can we infer any resemblance between distant events along our past light cone and events in our own galaxy's past.

Field equations are required to derive the form of $R(t)$ and to relate this function and the Robertson—Walker curvature to the content of the universe. General relativity tells us that, for $\Lambda = 0$, a homogeneous universe is ever-expanding (negative curvature of 3-surfaces), or destined to recollapse (positive curvature of 3-surfaces), according as $\rho \gtrless \rho_{crit}$, where $\rho_{crit} = (3H_0^2/8\pi G)$. It has become conventional to define a density parameter $\Omega = (\rho/\rho_{crit})$; the suffix zero denotes the present epoch.

Determining q_0 *from the magnitude—redshift relation*

The best known 'classical' cosmological test involves measuring the deviations from linearity at large redshifts in the Hubble law for the brightest galaxies in clusters, and thereby determining the deceleration parameter $q_0 = -(R_0\ddot{R}_0/\dot{R}_0^2)$. In Friedmann models where the pressure is $\ll \rho c$ (and with $\Lambda = 0$) q_0 is related to the present density parameter Ω_0 by $\Omega_0 = 2q_0$. If this relationship failed to hold, a non-zero Λ would be indicated.

The most obvious difficulty with the classical programme for determining q_0 is that the galaxies become almost invisibly faint – even with the best telescopes and detectors – before one can probe deep enough into space for the effects of deceleration to really show up. This makes it hard to measure magnitudes accurately for relevant galaxies. But there are several further problems, among which are these:

(i) No class of galaxies constitutes a 'standard candle' with a precision better than ~ 30 per cent; this introduces an inevitable scatter into the magnitude–redshift relation.

(ii) Galaxies are not point sources, nor do they have sharp edges: their surface brightness falls with increasing radius, and eventually merges into the background. Thus the measured magnitude must be corrected to allow for the fraction of the galaxy's total light contained within the aperture. This correction is dependent on the redshift z, but what is worse is that it depends also on q_0 (because the angular size corresponding to a given linear dimension and redshift is a function of q_0).

(iii) A systematic error may stem from the inclusion in the chosen sample of progressively more exceptional and intrinsically luminous galaxies at larger z (the 'catchment volume' rises roughly as z^3).

(iv) Galaxies with large redshifts are being seen at a younger stage in their evolution; this may make the luminosities depend on z. There are two competing evolutionary effects. (a) The evolution of the stellar population probably causes elliptical galaxies to fade with age, as progressively lower-mass stars evolve off the main sequence and onto the giant branch (the importance of this effect depends on the mass function of the stars). *But* (b) massive galaxies tend to gain mass, and therefore luminosity, by capturing and swallowing smaller galaxies in the same cluster. The importance of these 'mergers' and 'galactic cannibalism' has only recently been appreciated.

There are additional uncertainties which will need to be assessed before we can determine q_0 precisely. But at the moment the corrections due to evolution ((iv) above) are so uncertain that they would change q_0 by an amount $\pm$ 1.

The situation may improve when better data are available on large samples of galaxies with $z \simeq 0.5$. Pushing the Hubble diagram to still larger redshifts will not in itself help, because the evolutionary corrections then become bigger and still more uncertain. The space telescope (to be launched in 1985) should be able to resolve the profiles of individual galaxies out to $z \simeq 0.5$, and thereby elucidate the evolutionary effects. The astrophysical and geometrical problems are so intertwined that there is little hope of determining q until we understand galaxies better.

The density parameter Ω_0

The prospects of determining Ω_0, the other key parameter of the Friedmann models, are somewhat brighter.

The directly observed content of galaxies – stars and gas within ~ 10 kpc of the centre – would contribute a smoothed-out density corresponding to $\Omega \simeq 0.01$. There is X-ray evidence for plasma in clusters of galaxies, and this may amount to a similar contribution. Thus, a plausible 'floor' to the density parameter is $\Omega \simeq 0.02$, contributed mainly by baryons. There are, however, dynamical indications for additional 'unseen' matter. The evidence comes from the rotation curves for neutral hydrogen in the outlying parts of galaxies, from the dynamics of binary galaxies, and from the virial theorem as applied to clusters of galaxies. There is still a lively debate among astronomers on this question, but it is widely accepted that there may be 5–10 times as much 'unseen' material as 'luminous' material. The 'unseen' matter is in diffuse halos around galaxies, or in intergalactic space; it is less concentrated in the central potential wells of galaxies than is the 'luminous' matter. The data are however consistent with the idea that any region of space whose dimensions exceed a few million light years contains the two kinds of material in the consistent ratio of (5–10):1. This suggests that $\Omega_0 \gtrsim 0.1$. A value substantially higher cannot be ruled out, particularly if there is some form of matter that does not participate in clustering.

The recognition that 90 per cent of the gravitating content of the universe is in some unknown form (not even necessarily composed of baryons) should quench any optimism that we shall quickly understand galaxies and clusters. The unseen mass could be low mass stars, or the dead remnants of an early generation of massive or supermassive stars: various 'scenarios' along these lines are now being investigated in the hope that they can be confronted with some observational tests. But more exotic options cannot be excluded: indeed the individual masses of the unseen objects could lie anywhere between 10 ev/c^2 and 10^6 $M_\odot$ (70 order of magnitude uncertainty!)

I will discuss two such options not because they are necessarily the most likely, but because they are of special interest to participants in this conference — namely the possibility that the unseen mass is in black holes or in neutrinos.

Primordial black holes

Constraints on possible populations of black holes with individual masses in the 'astrophysical' range (> 1 $M_\odot$) have been summarized by Carr (1978). There are no significant restrictions on the mass range 1 - 10^6 $M_\odot$: black holes of this type could collectively contribute the inferred dark mass in galactic halos and clusters without being readily detectable in any other way. Larger individual masses would entail some consequences — e.g. dynamical friction, tidal effects, and luminosity energized by accretion of diffuse gas — incompatible with what is observed.

If a black hole lies sufficiently close to our line of sight to a quasar (or other compact intense source of radiation) gravitational lens amplification can in principle be observed. The characteristic angular scale of the image is ~ $10^{-6}(M_h/M_\odot)^{\frac{1}{2}}$ arc seconds. For large M_h, the characteristic image structure might be resolvable; for small M_h when the angles involved are too small to resolve, the transverse motion of 'lens' or source may cause detectable variability within a few years. The probability of observing the effect along a typical line of sight out to a redshift $\gtrsim$ 1 is of order Ω_{holes}, irrespective of the individual mass (the cross section $\propto M_h$ and the number required to yield a given Ω_{holes} is M_h^{-1}). Radio or optical techniques may eventually detect 'unseen mass' in this way if it is constituted by compact objects, and even narrow down the predominant mass within the broad range that now seems possible.

Black holes of 1–$10^6 M_\odot$ could be the remnants of an early

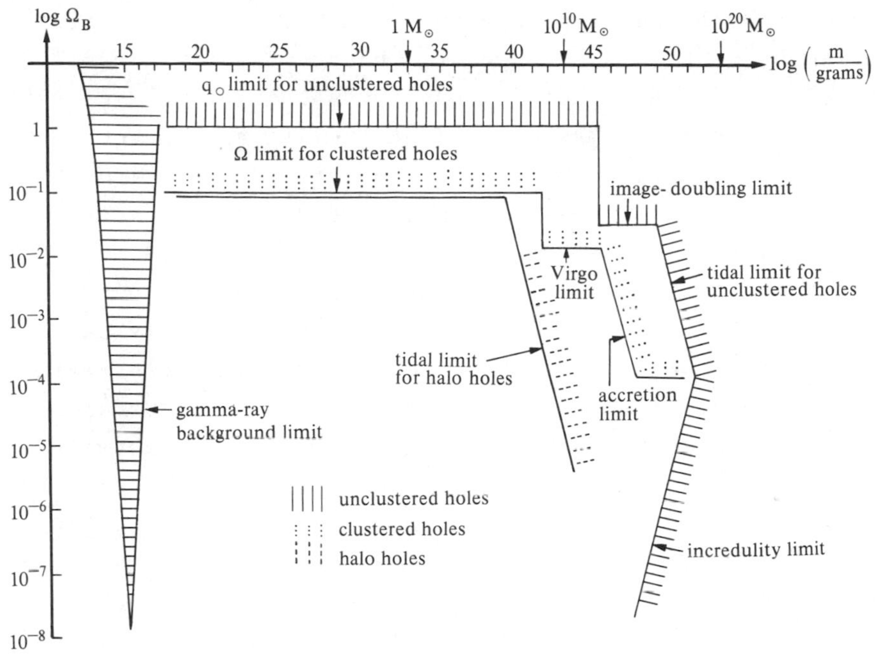

FIG. 1. This diagram, reproduced from Carr (1978), shows the maximum contributions to Ω that could come from black holes in various mass ranges. If the holes are clustered in galactic halos (including our own), tidal and dynamical friction effects constrain the range $M_h \gtrsim 10^6\ M_\odot$. If the holes are *not* clustered in halos, the only significant limits pertain to much larger holes (of galactic masses and above) which would manifest themselves via the gravitational lens effect, and by conspicuous radiative output fuelled by accretion even when immersed in a diffuse intergalactic gas.

Over the entire mass-range 10^{17}–10^{40} g, there are no constraints other than those set by the dynamics of galaxies and the overall deceleration of the universe. The various limits on evaporating 'miniholes' are discussed more fully in the text.

generation of stars or supermassive objects; alternatively they could be a consequence of irregularities in the very early universe. Only the latter option is open for formation of holes with masses $\ll 1\ M_\odot$: these cannot be imploded by any mechanism now occurring; they must be fossils of an era when the universe was at supernuclear density. It is appropriate at this conference to discuss the constraints on miniholes in rather more detail. The quantum evaporation process, first expounded by Hawking at the Oxford Quantum Gravity Conference in 1974, makes such objects of unsurpassed importance as missing links between the macro-world of gravity and the quantum micro-world of particle physics; moreover, the conspicuous consequences of black hole explosions renders these very small holes much less elusive than the passive bigger ones.

The evaporation is important only for 'miniholes' of mass $\lesssim 10^{15}$ g and dimensions comparable with an elementary particle; it is only for these holes that the observational constraints are interestingly tight. Holes with evaporation timescales $\lesssim 10^{10}$ years have temperatures so high that their thermal radiation is at $\gtrsim 10$ MeV. The observed γ-ray background contributes a mass—energy density amounting to only 10^{-8} of the critical 'closure density' for the universe. This sets a corresponding limit to the density of evaporating miniholes (Carr 1976; Chapline 1975); the resulting constraints on the formation mechanism then make their existence very unlikely (Carr 1975, 1976).

The prospects of detecting a burst produced by the evaporation of a single nearby minihole depend sensitively on the particle physics: in the Hagedorn picture, when the number of radiating species diverges as the temperature approaches a limiting value of ~ 200 MeV, the final 10^{14}g ($\approx 10^{35}$ erg) emerges almost instantaneously; in other schemes, according better with present ideas (where the number of radiating species remains finite, or diverges only at a much higher energy than in Hagedorn's model), the final disappearance is more gradual (Carter, Gibbons, Lin, and Perry 1976; Page and Hawking 1976).

The possibility of detecting γ-ray bursts from evaporating miniholes has been discussed by several authors, but the prospects seem poor, even if the spatial density has the highest value compatible with the γ-ray background limit. *Radio* and *optical* techniques are, however, capable of sensing a much lower energy flux than any foreseeable γ-ray detectors; it is therefore interesting that under some conditions the explosion could cause a radio or optical pulse. Suppose that, when its mass falls to $M_{\rm crit}$, the hole 'instantaneously' evaporates into a fireball where 50 per cent of the energy is in e^+-e^- pairs. Then Hawking's (1974) relation between temperature and mass implies that the fireball expands with a Lorentz factor $\gamma_f \approx (M_{\rm crit}/10^{17}g)^{-1}$. If the fireball behaves like a conducting sphere and 'sweeps up' the ordered interstellar magnetic field, its energy can be converted into a coherent polarized electromagnetic pulse (Rees 1977; Blandford 1977). The value of $M_{\rm crit}$ (and hence γ_f) depends on the uncertain particle physics. If $\gamma_f \approx 10^5$, 10^{32} erg emerges at radio wavelengths; if $\gamma_f \approx 10^7$ an optical flash results.

The best present upper limits to the rate of γ-ray bursts attributable to exploding miniholes are $\leqslant 0.04$ explosion pc^{-3} yr^{-1} in the Hagedorn picture, and $\leqslant 7 \times 10^5$ explosions pc^{-3} yr^{-1} on the 'elementary' particle picture where only the final 10^{30} erg is supposed to emerge in the final second (Weekes and Porter 1978). The predicted coherent radio (or optical) bursts are very model-dependent, and thus the failure to detect such events cannot, in our present ignorance of the relevant particle physics, be used to place any general constraints on the density of exploding miniholes. However, if the radio bursts occurred in an optimally detectable way, the present limits (obtained as a by-product of observations intended for other purposes) would be $\leqslant 5 \times 10^{-7}$ explosion pc^{-3} yr^{-1} (Meikle 1977). A simple array with large-area coverage could improve such a limit to 10^{-12} explosion pc^{-3} yr^{-1}. If $\gamma_f \approx 10^5$, an Arecibo-type telescope could detect a single pulse from a minihole in the Andromeda galaxy.

Barring really spectacular refinements of searches for γ-ray bursts, our only hope of detecting exploding miniholes (assuming they exist, but with densities well below Carr's limit) would prevail if the particle physics was such that each explosion generated an e^+-e^- fireball with $\gamma_f \gtrsim 10^5$. For a Schwarzschild hole this requires a sharp increase in the number of independent radiating species where the temperature exceeds a critical value; if rotating holes are subject to instabilities, sudden outbursts may occur under more general conditions (Zouros and Eardley 1979).

Novikov, Starobinsky, and Zeldovich (1979) extend the limits downward to cover primordial holes in the mass range 10^9–10^{13} g. These holes would have evaporated at early epochs, but could have affected the processes of nucleosynthesis. Holes with masses 10^9–10^{10}g are 'hot' enough to emit nucleon-antinucleon pairs; the antinucleons so produced would annihilate with nucleons, thereby increasing the n/p ratio, yielding more primordial ^{4}He. On initial mass scale 10^{10}–10^{13}g, the limit comes from spallation of primordial ^{4}He by relativistic nucleons from the final stages of evaporation. This can yield excessive primordial D. (Contrariwise, this could be invoked as a possible origin of D in a high-density universe.) To set analogous limits below $\sim 10^9$ g one would need to adopt a specific model for behaviour of matter at superhigh energies.

On mass scales exceeding 10^{15} g the constraints are less stringent. The evaporation time then becomes $\gg 10^{10}$ yrs and kT falls below a few MeV. Some significant constraints can be set to the 10^{15}–10^{17} g mass range from the maximum permissible production rate for positrons in our galaxy (Okeke and Rees 1980); but there are no direct limits to miniholes of above 10^{17} g.

I shall return to these limits in the next section in connection with their implications for the initial fluctuation.

Massive neutrinos

Theorists are now (April 1980) taking increasingly seriously the possibility that neutrino rest masses are non-zero; there are even some experiments favouring masses of the general order ~ 10 ev. This has great ramifications for cosmology, because the thermal neutrinos from the big bang would be almost as abundant as the photons, and would therefore outnumber the baryons by a factor $\sim 10^8\ \Omega^{-1}_{\text{baryon}}$. They are therefore dynamically important if the mass exceeds a few ev; moreover, they would have cooled adiabatically in the expanding universe to such low thermal speeds that they would have participated in gravitational clustering.

If we define ρ_ν as the mean mass–energy density contributed by the neutrinos at the present epoch, and $\rho_{\text{crit}} = 3H_0{}^2/8\pi G$ is the critical density, then

$$\Omega_\nu = \left(\frac{\rho\nu}{\rho_{\text{crit}}}\right) = 0.04 \sum_{\text{species}} (m_\nu c^2/1\text{ev}) \times (H_0/50 \text{ km sec}^{-1}\text{Mpc}^{-1})^2$$

This means that the neutrinos are certain to be dynamically important if their masses exceed a few electron volts. Moreover, the 'hot big bang' model would be in serious difficulties if the masses of any neutrino species exceeded 50–100 ev: one would then be forced to give greater attention to other cosmological models in which the photons (and their associated entropy) were generated at later epochs when $kT \lesssim 10$ Mev, so that there would be no reason for a comparable density in thermal neutrinos.

If $m_\nu \neq 0$, primordial thermal neutrinos will stop being relativistic when $3kT$ falls below $\sim m_\nu c^2$; thereafter $v_\nu \propto T_\gamma$. If the universe had remained homogeneous, the present neutrino velocities would be very low if their masses were high enough for Ω_ν to be dynamically significant:

$$v_\nu \simeq 100\left(\frac{m_\nu}{1\text{ev}}\right)^{-1} \text{ km sec}^{-1} .$$

These velocities are so low that the neutrinos would be influenced by the gravitational fields of galaxies and clusters. This means that neutrinos are candidates for the 'dark mass' in halos and in clusters of galaxies; equally it means the discussion of galaxy formation and clustering must allow for the possibility that the universe may be dynamically dominated by collisionless neutrinos at the time the clustering occurs.

Other constraints on Ω_0

An often-cited review by Gott, Gunn, Schramm, and Tinsley (1974) deployed a range of arguments favouring a low Ω_0. The most stringent of these stems from the requirement that deuterium should emerge from the big bang with at least the observed galactic abundance of $(1–5) \times 10^{-5}$ relative to H. D is an intermediate product on the route to helium formation, and the amount which survives is adequate only if the baryon density is low (corresponding to $\Omega_0 < 0.04$ for $H_0 = 50$ km sec^{-1} Mpc^{-1}). Note however that the present mass–energy in neutrinos or small black holes is not constrained by this argument: such entities do not augment the baryon density, nor do they affect the expansion rate at the nucleosynthesis epoch (where their rest mass is swamped by other mass–energy with $p = 1/3\ \rho c^2$).

3. IMPLICATIONS FOR THE DEGREE OF INITIAL INHOMOGENEITY

On the assumption that the overall shape of the universe can be approximately described by a Robertson–Walker metric, we can ask what evidence we have on perturbations with various scales.

Galaxies are aggregated in groups and clusters; the statistics of the clustering have been studied by many astronomers over the years, culminating in the elaborate analyses by Peebles and his associates (see Peebles (1980*a*) and bibliography therein). The claimed message of these investigations is that there is no preferred scale for a cluster of galaxies; the galaxies are grouped in such a fashion that $\xi(r) \underset{\sim}{\propto} r^{-1.8}$, the probability of a galaxy lying in a volume element δV a distance r from a given galaxy being proportional to $(1 + \xi(r))\ \delta V$. The characteristic radius r_0 for which $\xi(r_0) = 1$ is 8.4 Mpc (for a Hubble constant of 50 km sec^{-1} Mpc^{-1}). On scales $r \gtrsim r_0$ the inhomogeneities in the density of galaxies generally have amplitudes less than unity. There is no firm evidence that $\xi(r)$ follows the same straight power law when $r > r_0$ (and reasons for suspecting that it does not); there is good evidence for some clumping on scale of 50 Mpc. But the universe does seem to be more uniform on scales > 100 Mpc than on any scales < 10 Mpc. The evidence for this uniformity comes from counts of faint galaxies in different directions, and to different depths in the same direction. Analyses of the sources in radio surveys (many of which are quasar-type objects with $z \simeq 1$) fail to reveal any anisotropy. The X-ray background, which is largely due to discrete sources with $z \simeq 1$, or to diffuse gas, is isotropic to a precision of 1 to 2 per cent. There is radio, optical, and X-ray evidence that the distribution on the sky of all kinds of luminous objects becomes smoother as we look deeper. This suggests, given a 'Copernican' assumption, that the spatial distribution becomes smoother as we average over cells running from ~ 10 Mpc to ~ 3000 Mpc in size (and is inconsistent with any Charlier-style hierarchy).

One must of course be cautious about inferring the distribution of *all* mass from observations of luminous objects which may comprise only 10 per cent of the total amount of gravitating material — the more so because within individual galaxies the dark material is less centrally concentrated than the visible star and gas. However, while it is easy to envisage how the segregation could have occurred in the course of an individual galaxy's formation and evolution, it is less easy to envisage how the two kinds of matter could be segregated on scales exceeding a few megaparsecs. Thus on large scales it is probably justifiable to regard galaxies as valid tracers for the mass distribution. This inference would *not* hold, however, if a significant contribution to Ω came from a relativistic fluid (or a fluid whose pressure was too high for the gravitational fields of galaxies and clusters to effect its distribution).

The conventional view is that the galaxies and clusters of galaxies have developed primarily via a process of gravitational clustering. As the expansion proceeds, 'overdense' regions lag behind more and more, and eventually condense out and virialize. The growth is straightforward to analyse in the case of perturbations small compared to the current particle horizon: the problem is then a Newtonian one. Recent N-body simulations, particularly by Aarseth and his associates, have shown that if point masses are sprinkled down at random at some early epoch, then they eventually cluster in a fashion that strongly resembles the actual clustering of galaxies.

Gravitational clustering probably played a dominant role in the 'late' stages of the universe ($R/R_0 > 0.1$, $z \lesssim 10$, $t \gtrsim 10^9$ yrs) when large scale structures such as clusters were separating out. It is less evident that individual galaxies formed in this way — complex gas dynamical processes and dissipative effects may have been crucial. However, it is at least a tenable view that galaxies formed from small-amplitude density enhancements that were already present in the early universe.

When pressure gradients are negligible (i.e. for perturbations with masses well above the Jeans mass), growth of linear perturbations proceeds according to the law $\left(\frac{\delta\rho}{\rho}\right) \underset{\sim}{\propto} R$ (in a low density universe the growth saturates, however, when $(R/R_0) \simeq \Omega_0$). The growth is of power-law rather than exponential form because the growth rate and the expansion rate for a Friedmann model both scale as $(G\rho)^{\frac{1}{2}}$. The actual situation in the hot big bang universe is complicated by the effects of pressure, radiative viscosity, etc.

Two classes of perturbation can be envisaged: isothermal (or entropy) perturbations, in which the radiation pressure is unaltered; or adiabatic (isentropic) perturbations in which the photon/baryon ratio is unperturbed. Isothermal perturbations are essentially 'frozen in' before recombination; a general perturbation whose oscillatory component is damped by viscosity can leave behind an isothermal component. The oscillatory behaviour and damping of these various modes prior to recombination has been extensively discussed in the literature (see, for instance, the various contributions to the 'Universe at Large Redshifts' Conference in *Phys.Scr.* **21**, No. 5 1980). After recombination, when the gravitational instability of the matter is opposed only by gradients in the gas pressure (less than radiation pressure by a factor $\sim 10^8\ \Omega_{\text{baryon}}^{-1}$), all scales of $\gtrsim 10^6\ M_\odot$ grow, at least until bound systems form and generate enough energy to heat and reionize the gas. To obtain a bound system by the present time, the necessary amplitude at recombination must still be at least 10^{-3}. The growth of perturbations on the mass scales of galaxies and clusters is inhibited before recombination by radiation drag and viscosity; this means that amplitudes of $\gtrsim 10^{-3}$ are necessary even at the (earlier) epoch when such scales are first encompassed within the particle horizon.

If the universe is now dynamically dominated by massive neutrinos, the growth of perturbations is further impeded. The neutrinos introduce a new characteristic mass M_{crit} of order $10^{17}(m_\nu/1\text{ ev})^{-2}M_\odot$. This is the horizon mass at the epoch when the neutrinos stop behaving as relativistic particles (i.e. when $3kT$ falls below $m_\nu c^2$). Since the neutrinos are collisionless at this stage, any initial inhomogeneities in their distribution on mass scales below M_{crit} will be phase-mixed away. Furthermore, when $\rho_\nu > \rho_{\text{baryon}}$, the presence of the homogenized 'neutrino sea' inhibits the growth of inhomogeneities in the baryon component: the growth timescale is ~ $(G\rho_b)^{-\frac{1}{2}}$, whereas the expansion timescale is only a $(G(\rho_b + \rho_\nu))^{-\frac{1}{2}}$. This means that the growth proceeds at a negligible rate until the neutrinos have cooled adiabatically to such low thermal speeds that they become trapped in the regions of enhanced baryon density.

This means that a larger initial amplitude is needed in order to form galaxies by the appropriate time. To give one specific example, suppose that $\Sigma(m_\nu c^2) = 10$ ev (i.e. $\Omega_\nu = 0.4$ for $H_0 = 50$ km sec^{-1} Mpc^{-1}), and that $\Omega_{baryon} = 0.04$ so that $\rho_\nu/\rho_b = 10$. Then a $10^{11} M_\odot$ perturbation in the baryons does not start growing according to the $(\delta\rho/\rho) \propto R \propto (1 + z)^{-1}$ law until z has dropped below 100. In order for a galaxy to have formed by $z \simeq 5$ one requires a perturbation in the *baryon* density that was initially of order unity (yielding an overall density enhancement at $z = 100$ of ~ 0.1).

Granted the assumption of a big bang model subject to gravitational instability, the observed properties of galaxies and clusters tell us – at least in order of magnitude – the amplitude of the necessary perturbations on these particular scales. The most convenient measure of the 'lumpiness' of a Friedmann universe on a particular scale is the 'curvature fluctuation' ε (M), which is essentially equivalent to the amplitude of the fluctuation on mass-scale M when that scale is first encompassed within a particle horizon. In a pressure-free universe, a 'constant curvature fluctuation' spectrum translates into a spectrum $(\delta\rho/\rho) \propto M^{-2/3}$ at a given epoch. (In a hot big bang model, the situation is more complicated because pressure and radiative damping cause a scale-dependent growth for perturbations smaller than the horizon.)

It is then interesting to ask what evidence we have on other scales. I shall mention first the very small scales (where, not surprisingly, the constraints are weak), and then consider the large scales up to and beyond the Hubble radius.

Constraints at small scales

The only constraints that we can set on the form of $\varepsilon(M)$ on small mass-scales are predicated on a particular model for the very early universe. On scales $\ll 1\ M_\odot$ some limits follow from the consideration that the content of the universe was not all incorporated into small black holes at $t \ll 1$ sec. If the equation of state at supernuclear densities is radiation-like or stiffer ($p \gtrsim 1/3\ \rho c^2$), this requires $\varepsilon < 1$; for a soft equation of state, which present ideas on particle physics render unlikely, the constraint on ε would be more stringent, because miniholes could form even from small fluctuations if these grew unimpeded by pressure after entering the horizon. If helium is produced primordially as in a standard hot big bang model, the local expansion timescale must not differ from the Friedmannian value; this implies $\varepsilon < 1$ on the scale of the horizon at relevant epochs ($t \simeq 1$ sec) and also sets somewhat model-dependent constraints at larger scales. (Hogan (1978) has discussed some additional constraints on large-amplitude isothermal fluctuations where $\rho_{\mathrm{rad}} \gtrsim \Delta\rho_{\mathrm{matter}} \gg \rho_{\mathrm{matter}}$. For these, $\varepsilon < 1$, but the local photon/baryon ratio, which affects nucleosynthesis, can differ greatly from the standard model. Another restriction on large-amplitude isothermal fluctuations comes from the possibility that they may deflate, and collapse into black holes, at epochs prior to recombination.)

The fraction of mass–energy collapsing into primordial holes at any epoch must be

$$\lesssim \left(\frac{\rho_{\mathrm{rad}}}{\rho_{\mathrm{matter}}}\right)^{-1} \simeq \left(\frac{t}{10^{12}\mathrm{s}}\right)^{\frac{1}{2}} \quad (\text{for } t \lesssim 10^{12}\mathrm{s}) \quad .$$

Otherwise they would contribute too much mass ($\Omega > 1$) at the present time.

Adiabatic fluctuations on scales $\ll 10^6 M_\odot$ which do not have sufficient amplitude to collapse but instead enter an oscillating phase would be damped by photon viscosity (or shock waves) at such an early time that the excess energy could be fully thermalized; on scales $10^6 M_\odot \lesssim M \lesssim 10^{13} M_\odot$ the damping occurs sufficiently late that the energy input cannot (in the standard hot big bang) be properly thermalized, and shows up as a distortion in the microwave background.

Figure 2 summarizes the rather complicated (though somewhat

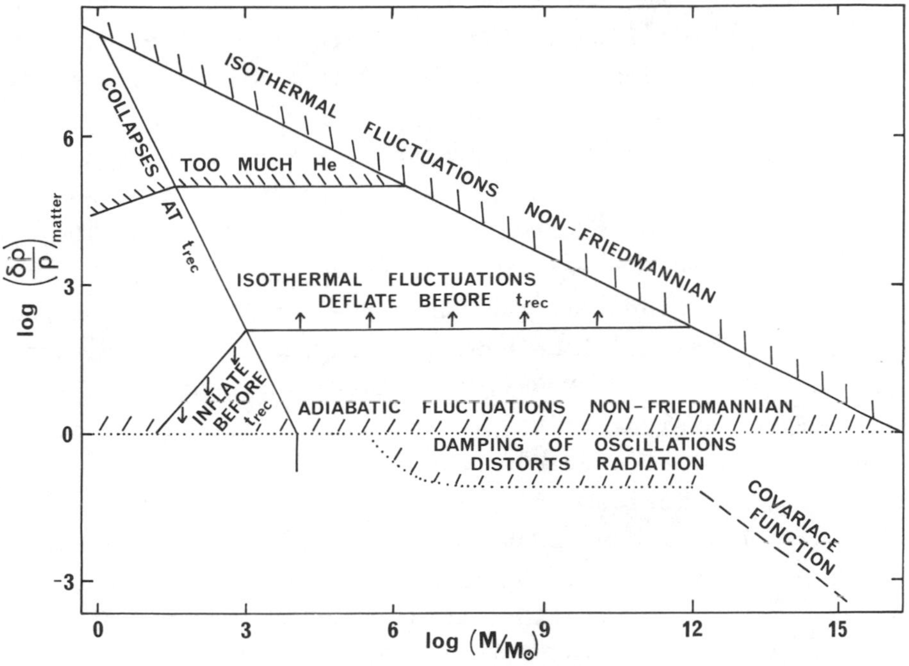

FIG. 2. This diagram illustrates various constraints on the amplitude of inhomogeneities on various 'small' mass-scales: isothermal (continuous lines) and adiabatic (dotted lines); based on work by Hogan (1978) and Sunyaev and Zeldovich (1970) respectively. The covariance function provides evidence on $\delta\rho/\rho$ on mass scales of galaxies and larger; on smaller scales, however, the constraints are much less stringent, particularly for isothermal fluctuations.

For adiabatic fluctuations, only $\delta\rho/\rho \lesssim 1$ is consistent with a Friedmannian assumption; on scales $\gtrsim 10^6\, M_\odot$ a stricter limit comes — on the hot big bang hypothesis — from observations that the microwave background is not measurably distorted. Isothermal fluctuations in a Friedmann model, on the other hand, may have $(\delta\rho/\rho)_{\text{matter}} \gg 1$ provided that $\rho_{\text{rad}} \gg \rho_{\text{matter}}$ when they enter the horizon. The primordial He abundance would be $\gtrsim 40$ per cent if $(\delta\rho/\rho)_{\text{matter}} \gtrsim 2 \times 10^5$ at the epoch of nucleosynthesis. Fluctuations of amplitude $\gtrsim 100$ are not rigidly 'frozen in' before recombination; by t_{rec} they would have deflated or inflated according to whether their mass $\gtrless$ the value of the Jeans mass calculated taking gas pressure alone into account. If the very early universe was Friedmannian and small black holes did not form prolifically then adiabatic fluctuations on even the smallest scales were less than unity. Note however that when ε becomes ~ 1 the Friedmannian assumption breaks down. One cannot exclude the possibility that the universe is 'chaotic' on small scales, locally resembling an anisotropic model, since the rapid expansion would inhibit gravitational instability and black hole formation.

model-dependent) constraints that can be placed on ε on sub-galactic scales. Note that the constraints are less stringent for 'isothermal' than for 'adiabatic' perturbation — in other words, the degree of deviation from strict Friedmannian dynamics entailed by the existence of galaxies is *less* if the fluctuations are isothermal than if they are purely adiabatic.

In the early universe when the mass—energy of radiation dominates that of the baryons, a large-amplitude isothermal perturbation causes only a small curvature fluctuation. There may however be good theoretical reasons for focussing on adiabatic perturbations. If the photon/baryon ratio is established by, for instance, non-equilibrium processes at the grand unification era ($kT \simeq 10^{15}$ Gev) then the value depends only on basic micro-physical constants and on G (which determines the expansion rate of a Friedmann model). Small amplitude curvature fluctuations on galactic scales would have a negligible effect on the expansion rate at the grand unification era, so the photon/baryon ratio would be undisturbed, yielding purely adiabatic fluctuations. Only if there were to be a change in expansion rate, correlated over regions much larger than the scale (at that time) of the particle horizon, would entropy perturbations arise.

Scales 100 Mpc $\lesssim \lambda \lesssim$ 1000 Mpc

Fluctuations on scales $\gtrsim$ 100 Mpc are probably only of small amplitude. Their influence may nonetheless be significant. This is because they induce velocity perturbations of order

$$V_{\text{pec}} \simeq c\Omega\left(\frac{\delta\rho}{\rho}\right)\left(\frac{\lambda}{\lambda_H}\right)$$

and fractional metric perturbations of order

$$\Delta g \simeq \Omega\left(\frac{\delta\rho}{\rho}\right)\left(\frac{\lambda}{\lambda_H}\right)^2 .$$

Peculiar velocities relative to the mean Hubble flow are consequently dominated by the largest scales unless $(\delta\rho/\rho)$ falls off more steeply than λ^{-1} $(\propto M^{-1/3})$; the large-scale metric fluctuations dominate unless the fall-off is steeper than λ^{-2} $(\propto M^{-2/3})$.

Constraints on large-scale inhomogeneities come from the isotropy manifested by surveys of any class of extragalactic discrete source; but these are of limited value. Inferences drawn from optical counts of galaxies are bedevilled by patchy galactic obscuration; for radio sources, absorption is negligible, but the problem here is the broad luminosity distribution (such that intrinsically faint nearby sources and powerful remote ones appear in surveys in comparable numbers at the same flux density). The best limits amount to $\lesssim$ 5 per cent on scales of ~ 1/3 the Hubble radius (Webster 1976).

The X-ray background offers greater promise: it is predominantly from cosmological distances ($z \gtrsim 1$); the galactic contribution, away from the plane, is only a small contamination and absorption is negligible above ~ 2 kev. X-ray data can therefore provide useful evidence on the large-scale distribution of matter, if we assume that the X-ray emission per unit volume (at a given epoch) scales with the overall matter density. The present isotropy limits have a precision of ~ 1 per cent on angular scales $\gtrsim 20°$ (Warwick, Pye, and Fabian 1980).

Clumping of matter, on any scale, would cause a corresponding anisotropy or 'graininess' in the observed X-ray background if the X-ray source distribution followed the overall density. Suppose, for illustration, that the clumping has a characteristic (comoving) scale λ (smaller than the present Hubble radius) and that the amplitude of the variations is $\delta\rho/\rho$. In general $(\delta\rho/\rho)$ will be a function of z. The inhomogeneities nearest to us will cause anisotropies in the X-ray background. The precise amplitude depends on the configuration of the irregularities around us, but the characteristic expected amplitude is obviously

$$\left(\frac{\Delta I}{I}\right)_{\mathrm{x}} \simeq \left(\frac{\delta\rho}{\rho}\right)\left(\frac{\lambda}{\lambda_{\mathrm{H}}}\right) .$$

Additionally, there will be fluctuations on small angular scales $\sim (\lambda/\lambda_H)$ due to similar irregularities at redshifts $z \gtrsim 1$.

The peculiar velocity indicated at the boundary of a lump of scale λ and amplitude $(\delta\rho/\rho)$ depends on the value of Ω. The problem is discussed by Silk (1974), who presents graphs for V_{pec} as a function of δ, for different choices of Ω. When $\lambda \ll \lambda_H$, our peculiar velocity will give a 24 hour X-ray anisotropy of amplitude

$$\left(\frac{\Delta I}{I}\right)_x = (3 + \alpha)\left(\frac{V_{pec}}{c}\right)$$

the X-ray spectrum being assumed a power law $I(\nu) \propto \nu^{-\alpha}$. (When $\lambda \simeq \lambda_H$ the situation is more complicated because the sources of a substantial fraction of the X-ray background are themselves given a peculiar velocity.) The microwave background, if observations are made on the Rayleigh–Jeans portion of the spectrum ($\alpha = -2$), would yield

$$\left(\frac{\Delta T}{T}\right) = V_{pec}/c\ .$$

Comparison of microwave and X-ray measurements yields an estimate of Ω since V_{pec} depends on Ω as well as $(\delta\rho/\rho)$: if Ω is *very* small the galactic peculiar velocity of ~ 600 km s^{-1} reported by Smoot *et al.* (1977) and Cheng *et al.* (1979) could not be induced by a large-scale inhomogeneity on any scale between ~ 100 Mpc and the Hubble distance, without the corresponding anisotropy in the X-ray source distribution exceeding that is observed.

Figure 3 shows the relative sensitivity of the microwave, X-ray and other limits in restricting the inhomogeneity of the universe on large scales. The amplitude of the inhomogeneities on a scale λ is expressed in terms of the present value of $(\delta\rho/\rho)$. The large scales ($\lambda >$ 100 Mpc) would have grown since entering the horizon unimpeded by radiation pressure effects, so for them $(\delta\rho/\rho)$ is related to the curvature fluctuation by

$$\varepsilon \simeq \left(\frac{\delta\rho}{\rho}\right)\left(\frac{\lambda}{\lambda_H}\right)^2\ .$$

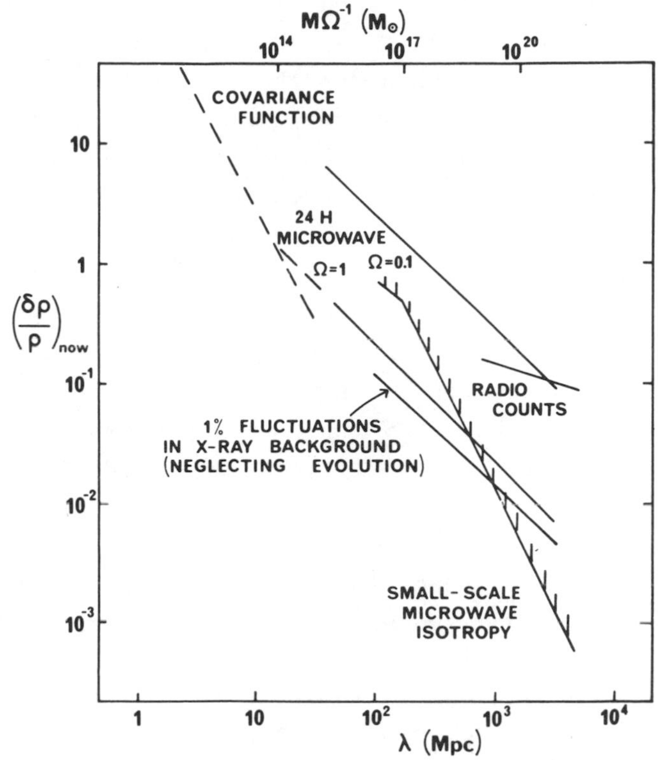

FIG. 3. Constraints on density perturbations ($\delta\rho/\rho$) on various length scales λ exceeding $\sim$ 10 Mpc (adapted from Warwick, Pye, and Fabian, 1980). The lines labelled 'covariance function' and 'radio counts' assume that the galaxy and radio source distributions mimic the underlying mass distribution. The 'microwave velocity' lines are chosen to give a peculiar velocity 600 km s^{-1} and assume that we lie at the edge of such a perturbation. The radio count limits can be improved to 3 per cent on scales of 10^3 Mpc, but depend upon the radio luminosity function, etc. The microwave background observations imply upper limits to the Doppler and gravitational perturbations at the epoch of last scattering, and these yield approximately the limits indicated. Limits of $\sim$ 1 per cent on the isotropy of the X-ray background would yield (apart from an evolutionary correction) the line shown: note that this is potentially the most sensitive constraint on scales 100–1000 Mpc; on larger scales the microwave limits are likely to remain the best.

(Note that Ω does not enter into this expression.) The upper limit on any 24 hour component in the X-rays sets a limit shown as a line in Fig. 3. Data on the X-ray background are likely to improve when larger-area detectors are put into orbit; the interpretation of X-ray anisotropies will clarify as we learn more about the sources that contribute to the background. X-ray observations are likely to be our most sensitive probe for inhomogeneities on scales 10^2–10^3 Mpc.

In principle, complementary information comes from the small-scale isotropy of the microwave background, which probes inhomogeneities on the cosmic photosphere. The data are summarized by Partridge (1980); on scales $\lesssim 1°$ the best limits correspond to $\Delta T/T \lesssim$ a few times 10^{-4}. The implied limits to $\delta\rho/\rho$ on various mass-scales are however sensitive to 'blurring' due to the finite thickness of the cosmic photosphere, reheating, compton distortions, etc (cf. Hogan 1980*a*).

Scales 1000 Mpc

On scales exceeding 1000 Mpc, the microwave background anisotropy limits are less ambiguous; these scales are in general larger than the horizon size at the epoch of the last scattering, so blurring of the cosmic photosphere is not a serious uncertainty. The dominant effect is due to gravitational potential fluctuations on the cosmic photosphere. Figure 3 shows the limits to $(\delta\rho/\rho)$ corresponding to $|\Delta T/T| \lesssim 3 \times 10^{-3}$ (Smoot 1980).

In Fig. 3, inhomogeneities of characteristic amplitude $(\delta\rho/\rho)$ are assumed to pervade the whole universe - $(\delta\rho/\rho)$ measures the amplitude of the Fourier components over a particular range of length scales. If, on the other hand, the universe contains isolated 'lumps' on large scales, embedded in a much smoother general background, then, as Fabian and Warwick (1979) have pointed out, there may be detectable consequences for the X-ray background, for observations with presently attainable sensitivity, even though the influence of the 'lump' may be undetectable as far as the microwave background is concerned. The effect on the X-ray background can be analysed into three components;
(i) A 24 hr effect due to the enhanced density of sources in the 'lump'. (This effect would not exist if the lump lay beyond the redshift at which the X-ray background originates.)

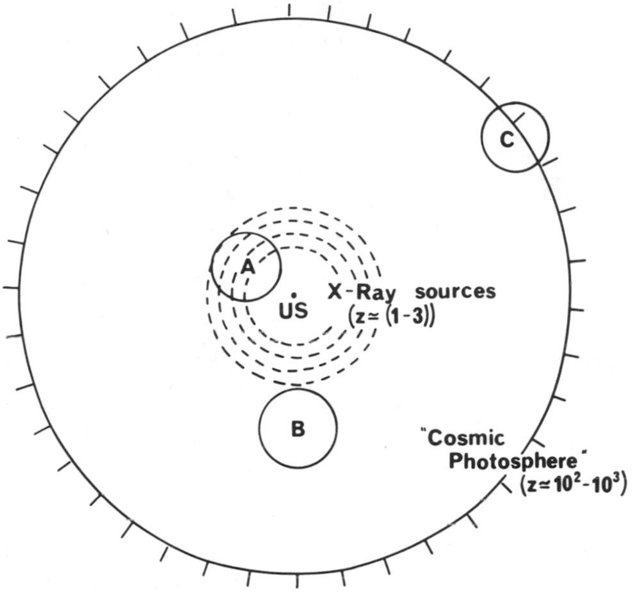

FIG. 4. This diagram is intended to illustrate schematically the effect of a single isolated 'lump' of dimensions comparable with the Hubble radius, placed in various positions around us. The radial co-ordinate is the 'co-moving r' of the Robertson–Walker metric; the 'cosmic photosphere' is assumed to be at $z \simeq 1000$, and the sources of the X-ray background at $z \lesssim 3$. (Note that, in terms of the co-ordinate r, $z \simeq 3$ is about half-way to $z \simeq 1000$ if $\Omega = 1$. If $\Omega < 1$ it is less than half way; moreover, volumes then increase faster than r^3.) The effects on the X-ray and microwave background are as follows (cf. Fabian and Warwick 1979).

A. (i) 24 hr effects in microwave and X-ray backgrounds due to our peculiar motion. (ii) 12 hr effect of similar magnitude in X-ray background owing to shear induced by lump, (iii) 24 hr effect, Ω^{-1} times larger than others, due to excess sources in region of lump.

B. As compared with case A, effect (ii) is more important relative to (i); but (iii) is now absent because the lump lies beyond the X-ray sources.

C. There will in this case be a small 12 hr effect in the X-rays; our peculiar velocity shows up in the microwave background but not significantly in X-rays, since all the 'sources' are falling at almost the same rate as us. The dominant effect, however, would be in the microwave background, due to the gravitational perturbation on the 'cosmic photosphere'.

(ii) A 24 hr effect due to our motion towards the 'lump'.
(iii) A 12 hr quadrupole effect, with a minimum along the axis of symmetry towards the 'lump', due to the shear induced within the volume whence the bulk of the X-ray background originates.

These various effects are illustrated in Fig. 4. When $\lambda \simeq \Lambda_H$ there are contributions to the microwave anisotropy even from transparent lumps lying between us and the cosmic photosphere (Rees and Sciama 1968; Dyer 1974; Kaiser 1980).

It is the scales $\gtrsim$ 100 Mpc, probed by microwave measurements on angular scales $\gtrsim 1°$ and by the X-ray background, that offer the most direct evidence on the homogeneity of the early universe. One believes this even though the present data – with only one exception – offer merely an assortment of upper limits. On these large scales, the initial acausally-determined fluctuations have not yet grown to large amplitude, and are unlikely (in contrast to those on galactic scales) to have been modified by secondary astrophysical processes, dissipation, non-linearities, etc.

The one convincing positively-measured anisotropy is the 24 hr effect in the microwave background (Smoot *et al.* 1977; Cheng *et al.* 1979). This could be caused by a relatively local 'lump' on a scale ~ 100 Mpc (cf. White 1980). At the other extreme, it could be a manifestation of an irregularity with scale $\gg \lambda_H$, which produces a gradient across the 'cosmic photosphere'. There are two ways to decide which of these interpretations is nearest the truth.
(i) An associated 24 hr effect in the X-rays would be expected if a local lump were responsible; whereas the X-ray background would not necessarily be affected to a comparable extent if the relevant perturbation had a scale $\gg \lambda_H$ (because all the X-ray sources at $z \lesssim 1$ would be 'falling' in the same way as us).
(ii) If the spectrum of fluctuations extends continuously upward to scales greater than λ_H, as envisaged by Grischuk and Zeldovich (1978), then high precision observations of the microwave background should reveal quadrupole, octupole and all higher harmonics. On the otherhand, if the 24 hr microwave anisotropy is induced 'locally', and the amplitudes are

undetectably small on scales $\gtrsim \lambda_H$, there is no reason for expecting any associated quadrupole effect.

Scales exceeding the present Hubble length λ_H

Curvature fluctuations can in principle, in an open universe, have a spectrum which extends to wavelengths much larger than our present particle horizon. Such fluctuations can however be detected because they yield gradients across the presently-observable part of the universe, and temperature anisotropies on the last scattering surface. Grishchuk and Zeldovich (1978) show that, subject to a random phase assumption, growing mode fluctuations would have $\delta\rho/\rho \lesssim 10^{-4}$ on all scales. The corresponding limit on the curvature fluctuations is $\varepsilon \lesssim 10^{-4} (\lambda/\lambda_H)^2$. (Limits can also be placed on large scale gravitational waves, on overall vorticity, etc.)

These calculations assume that our location is typical within an infinite expanse. But perhaps, even if our universe expands for ever, we can legitimately be sceptical about whether its material content is unbounded: could we be in an 'island' embedded in Minkowski space?

Global homogeneity?

The homogeneity implicit in Friedmann models is broadly vindicated by all observations in the part of the universe accessible to us. But our direct evidence is restricted – even in principle – to the domain within our present particle horizon, which does not (except in some models with $\Lambda \neq 0$) encompass the entire content of the universe; one may therefore query the plausibility of extrapolating into the unknown region.

In a finite closed Friedmann universe the fraction of the matter content lying within our particle horizon at time t_0 is $\gtrsim (t_0/t_{cyc})$, where t_{cyc} is the time for the entire cycle. Unless Ω exceeds unity by only a very tiny amount, we have thus already seen a significant fraction of the whole hypersphere, so only a limited extrapolation is involved in inferring genuine global homogeneity. If we have already seen half way round the universe, it is not too implausible to conjecture that the other half is similar, and meshes smoothly onto the part we see.

On the other hand, an ever-expanding Friedmann model with $\Omega < 1$ contains an infinite amount of matter. The amount within our present horizon is of course finite, but grows with t (the growth being $\propto t^2$ as $t \rightarrow \infty$ in a $k = -1$ $(\Omega < 1)$ model and $\propto t$ if $\Omega = 1$ $(k = 0)$.)

The large-scale isotropy of the microwave background sets constraints on the amplitude of fluctuations on scales larger than the present Hubble radius, which would manifest themselves as a shear (12 hr effect) or a gradient (24 hr effect) within our horizon. But the extrapolation to global homogeneity is a far greater one for an open field and infinite universe, and less evidently warranted. Even though we know the deviations from homogeneity are small on the scale of our present horizon (and can infer *something* about even larger scales) there is no prohibition on adopting a finite model whose edge may be seen in some future era — this would however require that the spectrum of $(\delta\rho/\rho)$ should eventually *increase* on some scales $\gg \lambda_H$.

In the foregoing speculations I have implicitly assumed that the topology of the universe is straightforward. But one cannot — at least on purely empirical grounds — exclude multiply connected topologies in which we see the same finite volume over and over again in a lattice structure (Ellis 1971; Sokolov and Schvartsman 1974; Gott 1980). In a $k = 0$ universe, the cell-size of the lattice is arbitrary; the lack of kaleidoscopic repetitiousness in the pattern of galaxies in the sky requires that the cells be at least 100 Mpc. (The best constraint comes from the fact that the Coma Cluster has no obvious 'twin' elsewhere in the sky. Rarer objects such as bright quasars cannot be used to strengthen this limit since their lifetimes may be shorter than the light travel time across one cell.) If $k \neq 0$, geometrical constraints set a maximum to the number of cells that can be packed within a hypersphere. This implies a lower limit to the cell size in a cosmological model with given H_0 and Ω_0 (Gott 1980).

4. THE FORM OF THE FLUCTUATION SPECTRUM: WHEN WAS IT 'IMPRINTED'?

If curvature fluctuations with $\varepsilon \simeq 10^{-3}$–$10^{-4}$ are 'given' on the appropriate scales, one can explore scenarios whereby gravitational clustering and dissipative processes gradually transform them into the galaxies and clusters we now observe (see McCrea and Rees (1980) for fuller discussions of the physics of galaxy formation). It is then natural to inquire whether the curvature fluctuations required for galaxy formation can be part of a spectrum extending over a much broader range of scales. Inspection of Figs. 2 and 3 shows that there are some (albeit rather insensitive) constraints on subgalactic scales, and strong limits on the amplitudes on very big scales.

The simplest assumption would be that the amplitude of curvature fluctuations depends on a power law: $\varepsilon \propto \lambda^{x}$. If $x \neq 0$, the amplitude 'blows up' leading to a non-Friedmannian universe, on scales which are small or large according to $x \gtrless 0$. The absence of effects due to miniholes suggests that the universe is Friedmannian on scales down to 10^9g, implying that $x > -0.1$. We cannot use the data to set such a strict limit on positive values of x because there is a shorter 'lever arm' between cluster scales and λ_H than there is on the small-scale side.

The 'constant curvature fluctuation' spectrum ($x = 0$) is obviously especially attractive, and has been advocated particularly by Sunyaev and Zeldovich (1970). It is scale-independent: there is no characteristic mass where ε becomes of order unity. This spectrum is entirely consistent with the small-scale constraints (cf. Fig. 2). However it is a delicate matter to decide whether ε can be big enough to yield galaxies and clusters without violating the limits on large scales set by the microwave background. Detailed discussions by Press and Vishniac (1980) and Silk and Wilson (1980) suggest that the constant curvature spectrum may be *in*consistent with those scenarios for galaxy formation where the initial fluctuations are purely adiabatic.

The present microwave and X-ray background measurements are tantalisingly close to the sensitivity level at which they can usefully delineate the fluctuations on scales from 100 Mpc to λ_H. In the next few years we can expect improved techniques (keeping pace, one hopes, with theoretical progress aimed at predicting what the spectrum should be).

Could the fluctuations have been generated in the post-quantum era?

If the amplitudes of protogalactic fluctuations are already 10^{-3}–10^{-4} when they are first encompassed within the particle horizon, then they must have been imprinted acausally into the initial data. If we consider very tiny scales containing only (say) 10^6–10^8 particles, which enter the horizon soon after the grand unification era, then some fluctuations may arise from discreteness effects. However, such effects yield a spectrum $\delta\rho/\rho \propto M^{-7/6}$ (Zeldovich 1965; Carr 1976) and fall off too rapidly with increasing scale to give significant effects on astronomical mass-scales. So again their origin is pushed back further (to the era of quantum cosmology?).

The equation of state at supernuclear densities is sufficiently uncertain that phase transitions cannot be ruled out; at the grand unification epoch, at the era of symmetry-breaking in the Weinberg–Salam theory, or related to the quark-nucleon transition density (see Linde (1979) and Ruderman (1980) for clear reviews). This offers the prospect that macroscopic fluctuations can suddenly appear, attaining amplitude unity in a single expansion timescale. The most optimistic assumption on the scale of these irregularities is that it corresponds to the horizon mass (or maybe the sound speed) when the transition occurs. The latest stage at which a phase transition could occur is when the density has dropped to a few times nuclear density; the horizon mass is then $\sim 1\ M_\odot$. However, even this 'optimistic' case is not good enough to yield galaxies: Hogan (1980*b*) shows that – even if the matter is cold so that gravitational clustering proceeds unimpeded by pressure gradients – the non-linear mass scale still cannot reach cluster order by the present era. This is because the $M^{-7/6}$ law applies, with the result that the non-linear scale grows only as $t^{4/7}$ and would still be $\lesssim 10^{10}\ M_\odot$ at $t \simeq 10^{10}$ yrs.

Preferred masses

If the fluctuations follow the Sunyaev–Zeldovich prescription, *and if* $\Omega = 1$, there is in effect only a single parameter involved: the amplitude ε characterizing the roughness of the initial conditions on all scales. But if $\Omega \neq 1$, the specification of an unperturbed Friedmann model inevitably involves a (large) characteristic mass: the mass contained within a region whose scale is set by the overall radius of curvature. This mass is part of the initial prescription of a Friedmann model – fixed no later than the end of the quantum era, when the classical particle horizon is smaller by 10^{60}. The realization that there is one preferred mass reduces inhibitions about invoking a second one (maybe related to the first by some dimensionless ratio of microphysical constants). This mass could in principle be of galactic (or cluster) order. The data are not incompatible with the idea that all structures are initiated in this way. (Even though the covariance function $\xi(r)$ has a power law form, it is only empirically well-determined in the non-linear domain when it could have been modified by N-body dynamical effects.) Nor is the preferred mass concept even 'unaesthetic': two masses related by a (potentially calculable) factor would seem no more *ad hoc* than a spectrum where the power law and amplitude must each be specified.

The characteristic mass derived from the curvature scale of a $\Omega \neq 1$ Friedmann model is of interest in the context of power-law spectra (Peebles 1980*b*). There is no understanding of how these spectra could be set up, but ideas aired in the literature relate the characteristic $\delta\rho/\rho$ within a comoving sphere to some negative power of: (number of particles in sphere); (surface of sphere); or (radius of sphere in Robertson–Walker comoving co-ordinate r). On small scales, these prescriptions are indistinguishable, since the three quantities in brackets are related by simple power laws. But for scales exceeding the overall curvature radius, this is not so – for instance a power law in r does not give a power law in M, and vice versa – so any evidence on how the spectrum extends to these large scales could elucidate the underlying mechanism.

When $\Omega \ll 1$ these scales can be studied by the microwave background. The critical mass subtends $\sim (\log \Omega^{-1})^{-1}$, and $\Delta T/T$ can be probed on a range of angles larger than this.

Friedmannian background versus 'chaotic' initial conditions

The approach adopted here has been to postulate a basically Friedmannian model and ask what initial 'imperfections' must have been present in order that the galaxies and clusters should arise by $t \simeq 10^{10}$ yrs, and what amplitude *may* exist on other scales. This is the simplest hypothesis – it also accords with the views of Penrose (and I have perhaps been influenced by his lecture on the explanation of time asymmetry). But one should recall the opposite viewpoint, advocated by Misner (1968) especially, that it would have been more 'natural' for the universe to have availed itself of all the macroscopic degrees of freedom open to it.

The chaotic cosmology programme – the idea that the horizon structure and dynamics of an early non-Friedmannian stage could have facilitated homogenization and led to the present overall homogeneity – may not have succeeded (cf. MacCallum 1979): there seems no way of understanding the overall homogeneity, on the basis of classical relativity, without postulating special initial conditions. On the other hand, one cannot rule out 'chaos' on all scales below galactic or cluster masses (i.e. on all but three of the many orders of magnitude between the Planck length and the Hubble radius). The chaotic cosmology programme provided an impetus for the extensive body of work on anisotropic (but strictly homogeneous) models. A crucial question is the extent to which these models mimic the (more realistic) *in*homogeneous models when the scale of the inhomogeneities far exceeds the particle horizon.

REFERENCES

Blandford, R.D. (1977). *Mon. Not. R. astr. Soc.* **181**, 489.
Carr, B.J. (1975). *Astrophys. J.* **201**, 1.
—— (1976). *Astrophys. J.* **206**, 8.
—— (1978). *Comments Astrophys. Space Phys.* **7**, 161.
Carter, B., Gibbons, G.W., Lin, D.N.C., and Perry, M.J. (1976). *Astron. Astrophys.* **52**, 427.
Chapline, G.F. (1975). *Nature* **253**, 251.

Cheng, E.S., Saulson, P.R., Wilkinson, D.T., and Corey, B.E. (1979). *Astrophys. J. Lett.* **232**, L139.
Dyer, C.C. (1976). *Mon. Not. R. astr. Soc.* **175**, 429.
Ellis, G.F.R. (1971). *Gen. Relativ. Gravitat.* 2, 7.
Fabian, A.C. and Warwick, R.S. (1979). *Nature* **280**, 39.
Gott, J.R. (1980). *Mon. Not. R. astr. Soc.* **193**, 153.
—— , Gunn, J.E., Schramm, D.N., and Tinsley, B.M. (1974). *Astrophys. J.* **194**, 543.
Grishchuk, L.P. and Zeldovich, Y.B. (1978). *Sov. Astr.* **22**, 125.
Hawking, S.W. (1974). *Nature* **248**, 30.
Hogan, C.J. (1978). *Mon. Not. R. astr. Soc.* **185**, 889.
—— (1980*a*). *Mon. Not. R. astr. Soc.* **192**, 891.
—— (1980*b*). *Nature* **286**, 360.
Kaiser, N. (1980). in preparation.
Linde, A. (1979). *Rep. Prog. Phys.* **42**, 390.
MacCallum, M.A.H. (1979). In *General Relativity: an Einstein Centenary Survey* (ed. S.W. Hawking and W. Israel). Cambridge University Press.
McCrea, W.H. and Rees, M.J. (ed.) (1980). *Phil. Trans. Roy. Soc.* **A296**.
Meikle, W.P.S. (1977). *Nature* **269**, 41.
Misner, C.W. (1968). *Astrophys. J.* **151**, 431.
Novikov, I.D., Polnarev, A.G., Starobinsky, A.A., and Zeldovich, Y.B. (1979). *Astron. Astrophys.* **80**, 104.
Okeke, P. and Rees, M.J. (1980). *Astron. Astrophys.* **81**, 263.
Page, D.N. and Hawking, S.W. (1976). *Astrophys. J.* **206**, 1.
Partridge, R.B. (1980). *Physica scripta* **21**, 624.
Peebles, P.J.E. (1980*a*). *The Large-Scale Structure of the Universe.* Princeton University Press (in press).
—— (1980*b*). *Physica scripta* **21**, 720.
Press, W.H. and Vishniac, E.L. (1980). *Astrophys. J.* **236**, 323.
Rees, M.J. and Sciama, D.W. (1968). *Nature* **217**, 511.
—— (1977). *Nature* **266**, 333.
Ruderman, M.A. (1980). *Proc. Acad. Lincei* (in press).
Silk, J.I. (1974). *Astrophys. J.* **193**, 525.
—— and Wilson, M.L. (1980). *Physica scripta* **21**, 708.
Smoot, G.F. (1980). *Physica scripta* **21**, 619.
—— , Gorenstein, M.V., and Muller, R.A. (1977). *Phys. Rev. Lett.* **39**, 898.
Sokolov, D.D. and Shvartsman, V.F. (1974). *J. expl theor. Phys.* **39**, 196.
Sunyaev, R.A. and Zeldovich, Y.B. (1970). *Astrophys. Space Sci.* **7**, 3.
Warwick, R.S., Pye, J.P., and Fabian, A.C. (1980). *Mon. Not. R. astr. Soc.* **190**, 243.
Webster, A.S. (1976). *Mon. Not. R. astr. Soc.* **175**, 71.
Weekes, T.C. and Porter, N.A. (1978). *Mon. Not. R. astr. Soc.* **183**, 205.
White, S.D.M. (1980). *Physica scripta* **21**, 640.
Zeldovich, Y.B. (1965). *Adv. Astron. Astrophys.* **3**, 241.
Zouros, T.J.M. and Eardley, D. (1979). *Ann. Phys.* **118**, 139.

13

BARYON NUMBER ASYMMETRY IN GRAND UNIFIED THEORIES

G.G. Ross

Theoretical Physics Department, University of Oxford

1. INTRODUCTION

In this talk I want to review recent progress that has been made in understanding the observed baryon number asymmetry via processes predicted by grand unified theories.

Cosmology deals with macroscopic phenomena and particle physics deals with microscopic phenomena. One of the most exciting developments in recent years has been the interplay between these two disciplines and it now appears that theories developed in elementary particle physics may shed light on some of the fundamental puzzles in cosmology. Amongst these a major challenge is to explain the enormous differences in the observed particle abundances. In particular it appears there is much more matter than antimatter (locally more than 10^4 to 1) (Steigman 1976) and the baryon to photon ratio n_B/n_γ is $10^{-9\pm1}$. In the big bang model of the universe, which assumes that at early times the universe was in thermal equilibrium, baryons and antibaryons will annihilate as the temperature drops. If we start with equal numbers almost all will have annihilated by a temperature of 50 MeV giving (Weinberg 1972; Wagoner 1979; Barrow 1980) $n_{\bar{B}}/n_\gamma = n_B/n_\gamma \simeq 10^{-18}$, grossly in conflict with the observed ratios. How can this puzzle be resolved? In the next section we discuss the conditions necessary to give the observed particle ratios. In Section 3 we review grand unified theories and show how they satisfy these conditions. Finally we give estimates for the particle ratios in Section 4 and comment on the possible role of mini black holes in producing the asymmetry in Section 5.

2. CONDITIONS FOR B ASYMMETRY

There are three possible ways to generate a baryon asymmetry.

1. If $n_B \neq n_{\overline{B}}$ initially, the asymmetry may persist. This is aesthetically unsatisfactory for it requires the initial conditions should be such that

$$\frac{n_B - n_{\overline{B}}}{n_B + n_{\overline{B}}} = 10^{-9}$$

i.e. initially there are $10^9 + 1$ baryon to every 10^9 antibaryons! Moreover if B violating processes exist (see Omnes 1970) any initial B asymmetry will be destroyed in a system in thermal equilibrium.

2. There may be a separation of baryons and antibaryons (Omnes 1970) leading to matter and antimatter galaxies (the constituents of other galaxies have not yet been firmly established (Steigman 1976)). However there is a severe theoretical problem with this explanation. Separation must have happened at times $t \leqslant 0.1$ sec before baryon—antibaryon annihilation is complete.

However in this time light signals could only traverse a region containing 10^{56} particles (equivalent to 0.1 solar masses) and it is hard to maintain that $B,\overline{B}$ separation could occur for a larger size. This would not explain the absence of antimatter in our galaxy.

3. There exist baryon number violating processes and the baryon number excess has been generated since the big bang. (Sakharov 1967; Ignatiev, Krosnikov, Kuemin, and Tavkhelidze 1978; Dimpoulos and Susskind 1978; Toussaint, Treiman, Wilczek, and Zee 1979; Ellis, Gaillard, and Nanopoulos 1979; Sakharov 1979; Weinberg 1979). In fact the existence of B number violation is not enough since, if there exists a baryon number violating process changing the baryon number of a state A, there will, by CP invariance, be a corresponding baryon number violating process operating on the state $\overline{A}$ which cancels the net baryon number change. Thus CP must also be violated by the B violating interaction. Even this is not enough for if the

system is in thermal equilibrium all states of the system with a given energy, not distinguished by a conserved quantum number, will be equally populated. CPT guarantees that the states A and $\bar{A}$ have the same energy and so in thermal equilibrium the system will have zero net baryon number. Even if the initial conditions are asymmetric, in the approach to thermal equilibrium B number violating processes will reduce the asymmetry ultimately to zero. Thus the conditions for a production of baryon number asymmetry after the big bang are (Sakharov 1967)

(a) B number violating processes
(b) CP violating processes
(c) Departure from thermal equilibrium while (i) and (ii) are in action (for a theory with CPT invariance).

3. GRAND UNIFIED THEORIES

Recently there has been considerable interest in a class of gauge theories, known as grand unified theories (GUTs) which seek to unify the strong, the weak and the electromagnetic interactions within a single semi-simple non-abelian gauge group. These have the remarkable property that they satisfy all of the above conditions and can generate B number asymmetry at the observed level. In the remainder of this talk I would like to briefly review GUTs and discuss the predictions they make for B number asymmetry.

The 'standard' gauge theory for strong, weak and electromagnetic interactions is based on the group $SU(3)_C \times SU(2) \times U(1)$. The gauge vector bosons associated with these groups generate the strong interactions (8 vector boson 'gluons') and the weak and electromagnetic interactions (4 vector bosons, the $W^{\pm}$, Z and γ). The gauge bosons interact with leptons and quarks according to the transformation properties of the fermions under the gauge group. These form families of left handed fields transforming as

$$\ell_{a_L} = \begin{pmatrix} \nu_a \\ e_a \end{pmatrix}_L \qquad (1,2,\tfrac{1}{2})$$

$$\bar{e}_{a_L} \qquad (1,1,-1)$$

$$q_{a_L} = \begin{pmatrix} u_a \\ d_a \end{pmatrix}_L \qquad (3,2,-\tfrac{1}{\gamma})$$

$$\bar{u}_{a_L} \qquad (\bar{3},1,\tfrac{2}{3})$$

$$\bar{d}_{a_L} \qquad (\bar{3},1,-\tfrac{1}{3})$$

where $e_a = e$, μ and τ leptons and $u_a = u$, c and t and $d_a = d$, s, and b quarks respectively.

The numbers in parentheses denote the SU(3), SU(2) multiplicities and the U(1) quantum number respectively.

Grand unified theories seek to unify the strong, weak and electromagnetic forces by embedding the three separate gauge groups in a single semi-simple gauge group G with a single gauge coupling g. In these models the quarks and leptons transform in one (or more) representation of the group G. As a result there are processes which connect quarks to leptons and in general these processes violate baryon number conservation.

The simplest GUT is given by the smallest group which can accommodate $SU(3)_C \times SU(2) \times U(1)$. This is SU(5) (Georgi and Glashow 1974; Buras, Ellis, Gaillard, and Nanopoulos 1978) and each of the families fit neatly into two representations of SU(5) the 5 and the $\overline{10}$. These are

$$\begin{pmatrix} \bar{d}_R \\ \bar{d}_W \\ \bar{d}_B \\ e \\ \nu \end{pmatrix}_{a_L} \qquad \frac{1}{\sqrt{2}} \begin{pmatrix} 0 & \bar{u}_B & -\bar{u}_W & -u_R & -d_R \\ \bar{u}_B & 0 & u_R & -u_W & -d_W \\ \bar{u}_W & -\bar{u}_R & 0 & -u_B & -d_B \\ u_R & u_W & u_B & 0 & -e^+ \\ d_R & d_W & d_R & e^+ & 0 \end{pmatrix}_{a_L}$$

where R, W, B denote the $SU(3)_C$ colour quantum number. In addition to these fermions in SU(5) there are 24 gauge bosons; the 8 gluons, the $W^{\pm}$, Z and γ and 12 new gauge bosons the X_V bosons. These new bosons couple quarks to leptons and violate B number. In addition to the gauge bosons the theory requires at least a 5 and a 24 of new scalar bosons, the Higgs bosons, X_S, which also violate B in their couplings to fermions. Thus condition (a) is satisfied.

In these models CP violation may occur in several ways. In SU(5) the fermion mass matrix may acquire relative phases giving rise to CP violation in gauge boson exchange. CP may also be violated via Higgs boson exchange. Thus condition (b) is also, in general, satisfied by GUTs.

Finally what about condition (c) and the need for departure from thermal equilibrium. In discussing B number asymmetry we will find it generated via the B number violating decays of X_V and X_S. As the universe evolves after the big bang the temperature of the universe will fall below $kT = m_{X_V}, m_{X_S}$ and these bosons will decay. If their decay rate is less than the expansion rate H of the universe they will not have enough time to achieve the equilibrium Boltzmann distribution and, as a result, they will depart from thermal equilibrium. The expansion rate $H = 1.66\ (kT)^{2/1} N^2 m_p^{-1}$ where $m_p = 1.22 \times 10^{19}$ GeV is the Planck mass and N is the number of particle states. For X_V bosons the coupling is g and their decay rate is of order $\alpha m_X N$ where $\alpha = g^2/4\pi$. Thus the decay rate is less than the expansion rate and condition (c) is satisfied when (Yoshimura 1978, etc; Kolb and Wolfram 1979)

$$kT > (N^{1/2} \alpha m_X m_p)^{1/2} \ . \tag{3.1}$$

Since the X bosons must decay during this non-equilibrium situation we require $kT < m_X$ which holds for

$$m_X > N^{1/2} \alpha m_p \ . \tag{3.2}$$

For vector bosons $\alpha \sim \frac{1}{50}$ and the bound gives

$$m_{X_V} \geqslant N^{1/2} \ .\ 10^{17} \text{ GeV} \ . \tag{3.3a}$$

For scalar bosons the coupling is predicted to be much smaller and this bound gives

$$m_{X_S} \geqslant 10^{13}\ N^{1/2} \text{ GeV} \ . \tag{3.3b}$$

Thus a condition for thermal equilibrium departure during

X boson decay is that the bosons should be very heavy. Remarkably this is what is expected in many GUTs. The baryon number violating processes discussed above would also mediate proton decay. In order to be consistent with the non-observation of this process ($\tau_p > 10^{30}$ years!) the mediating X bosons must be very heavy. In SU(5) this requires $m_{X_V} \geqslant 10^{15}$ GeV and $m_{X_S} \geqslant 10^{13}$ GeV. (Georgi, Quinn, and Weinberg 1974; Goldman and Ross 1979; Marciano 1979, 1980; Ellis, Gaillard, Nanopoulos, and Rudaz 1980; Binetray and Schücker 1980; Llewellyn Smith, Ross, and Wheater 1981).

That the boson mass should be this large is actually a prediction in SU(5) and many other GUTs. It follows from demanding that the strong, weak and electromagnetic couplings of $SU(3)_C \times SU(2) \times U(1)$ should all be related to g of SU(5). Experimentally these couplings are very different and the only way this difference can be explained is through large radiative corrections to g, which are different for the different couplings of $SU(3)_C \times SU(2) \times U(1)$, and give a logarithmic variation with the energy scale at which the couplings are measured. Experiment measures the couplings at a scale $\leqslant 1$ GeV. The GUT predictions for these couplings is at the scale $\sim M_X$. Since logarithmic variation with energy is very slow M_X must be very large to explain why the different observed strong, weak and electromagnetic couplings should be approximately equal at M_X. For SU(5) this gives $M_X \simeq 6 \times 10^{14}$ GeV. By comparison with eqns (3.2) and (3.3a,b) we see that condition (c), departure from thermal equilibrium, is satisfied, at least for Higgs scalar decays.

4. BARYON NUMBER ASYMMETRY IN GUTs

We may now apply the ideas discussed above to a more detailed prediction of the baryon number asymmetry in SU(5). As we discussed above, in the decay of X_V and X_S to fermions only the X_S bosons depart significantly from thermal equilibrium during their decay. So we concentrate on them. The decay modes are $X_S \to QL$ and $\bar{Q}\bar{Q}$ with ratios r and $(1-r)$ and $\bar{X}_S \to \bar{Q}\bar{L}$ and QQ with ratios $\bar{r}$ and $(1-\bar{r})$. The mean net baryon number produced is

$$\Delta B = \frac{1}{2}\left[\frac{1}{3}r - \frac{2}{3}(1 - r) - \frac{1}{3}\bar{r} + \frac{2}{3}(1 - \bar{r})\right]$$

$$= \frac{1}{2}(r - \bar{r}) \tag{4.1}$$

If CP is conserved $r = \bar{r}$ and so no asymmetry may develop. In fact non-zero r-$\bar{r}$ can first arise in order α^2 via virtual scalar or vector exchange (Nanopoulos and Weinberg 1979). The actual value depends on the phases giving rise to CP violation. Unfortunately there is no prediction for the magnitude of these phases and the observed CP violation at low energies in K decay cannot be simply related to these phases. Thus we can only produce estimates based on reasonable ranges for those of these phases. Typical estimates for $|\Delta B|$ for graphs with vector exchange give

$$|\Delta B| \simeq 10^{-10} \text{ to } 10^{-6} \tag{4.2}$$

and for graphs with scalar exchange give

$$|\Delta B| \simeq 10^{-6} \text{ to } 10^{-2}\ . \tag{4.3}$$

For SU(5) only scalar exchange can give rise to CP violation (Nanopoulos and Weinberg 1979) so eqn (4.3) should then be used.

Using this the asymmetry is, roughly,

$$\frac{n_B}{n_\gamma} \approx \frac{n_x(t_x)}{n_\gamma(t_x)}\Delta B \approx \frac{Nx}{N}\ \Delta B \tag{4.4}$$

$$\simeq 10^{-2}\ \Delta B.$$

This number can be improved by including the effects of 2-2 scattering processes and 2-X production above $kT = M_X$. These have the effect of decreasing the B asymmetry both above and below M_X. Thus any initial baryon number asymmetry will be diminished as will any B asymmetry produced if B violating processes exist far below M_X (Kolb and Wolfram 1980).

5. BLACK HOLE PRODUCTION

An alternative scenario, still involving X boson decay, but avoiding the conditions (3.2) and (3.3) on the grand unified masses, has been suggested (Barrow 1980). In this, mini black holes decay producing X bosons. The X bosons subsequently decay again producing a B excess as discussed above. However in this case the decays occur at a much later time for the black holes with an effective Hawking–Beckenstein temperature $\sim M_X$ have a mass $\sim m_p^2 m_x^{-1}$ and decay at a time $t \sim N^{-1} m_p^2 m_X^{-3}$. At this time the background temperature of the universe is $< M_X$ and thus the X bosons produced by black hole decay are not in thermal equilibrium and automatically satisfy condition (c). This means that GUTs of the Pati–Salam type (Pati and Salam 1973) with $M_X \simeq 10^5$ GeV may be able to generate the observed B excess even though the free decay production discussed in section 4 is negligible and any pre-existing B excess is wiped out. The actual asymmetry and the baryon to photon ratio depends on the abundance of mini black holes which is at present unknown, although bounds may be obtained for reasonable distributions of mini black holes (Barrow and Ross 1981).

6. SUMMARY

Grand unified theories, constructed to describe elementary particle physics, remarkably possess the ingredients necessary to generate baryon number asymmetry. Detailed calculations show that they can explain the observed asymmetry and the baryon to photon ratio. At present neither an absolute prediction of the magnitude of the asymmetry nor even a prediction of the sign is possible due to uncertainties in CP violating phases. Much work remains to be done but it seems there is good reason to hope that one of cosmology's puzzles may at last have an answer.

REFERENCES

Barrow, J.D. (1980). *Surv. high energy phys.* **1**, 183.
—— (1980). *Mon. Not. R. astr. Soc.* 192, 198; 192, 427.

Barrow, J.D. and Ross, G.G.(1981). *Nucl. Phys.* **B181**, 461.
Binétruy, P. and Schüker, T. (1980). CERN preprints TH-2802 and TH-2857.
Buras, A.J., Ellis, J., Gaillard, M.K., and Nanopoulos, D.V. (1978). *Nucl. Phys.* **B135**, 66.
Dimopoulos, S. and Susskind, L. (1978). *Phys. Rev.* **D18**, 4500.
Ellis, J., Gaillard, M.K., and Nanopoulos, D.V. (1979). *Phys. Lett.* **80B**, 360.
—— —— ——, and Rudaz, S. (1980). Annecy preprint LAPP-TH-14.
Georgi, H. and Glashow, S.L. (1974). *Phys. Rev. Lett.* **32**, 438.
——, Quinn, H.R., and Weinberg, S. *Phys. Rev. Lett.* **33**, 451.
Goldman, T.J. and Ross, D.A. (1979). *Phys. Lett.* **84B**, 208; Caltech preprint CALT-68-759 (1980).
Ignatiev, A. Yu., Krosnikov, N.V., Kuzmin, V.A., and Tavkhelidze, A.N. (1978). *Phys. Lett.* **76B**, 436.
Kolb, E.W. and Wolfram, S. (1980). *Phys. Lett.* **91B**, 217; *Nucl. Phys.* **B172**, 224.
Llewellyn Smith, C.H., Ross, G.G., and Wheater, J. (1981). *Nucl. Phys.* **B177**, 263.
Marciano, W.J. (1979). *Phys. Rev.* **D20**, 274.
—— (1980). Rockefeller University preprint COO-2232B-195.
Nanopoulos, D.V. and Weinberg, S. (1979). *Phys. Rev.* **D20**, 2484.
Omnes, R. (1970). *Phys. Rep.* **C3**, 1.
Pati, J.C. and Salam, A. (1973). *Phys. Rev. Lett.* **31**, 661; (1973). *Phys. Rev.* **D8**, 1240; (1974) *Phys. Rev.* **D10**, 275.
Sakharov, A.D. (1967). *Pisma Zh. eksp. teor. Fiz.* **5**, 32.
—— (1979). *Zh. eksp. teor. Fiz.* **76**, 1172.
Steigman, G. (1976). *Ann. Rev. Astrom. Astrophys.* **14**, 339.
Toussaint, B., Treiman, S.B., Wilczek, F., and Zee, A. (1979). *Phys. Rev.* **D19**, 1036.
Wagoner, R.V. (1979). *The Early Universe*. Les Houches Lectures.
Weinberg, S. (1972). *Gravitation and Cosmology*. Wiley, New York.
—— (1979). *Phys. Rev. Lett.* **42**, 850.
Yoshimura, M. (1978). *Phys. Rev. Lett.* **41**, 28.

14

PARTICLE PRODUCTION AND DYNAMICS IN THE EARLY UNIVERSE

J.B. Hartle

Department of Physics, University of California, Santa Barbara, California 93106

1. INTRODUCTION

The central problem in cosmology is to find general principles for initial conditions and a physically complete dynamics which will explain the large scale structure of the universe we see today. Broadly speaking the features of the universe we would like to explain are the observed isotropy and homogeneity, the approximate spatial flatness, the spectrum of density fluctuations, the entropy per baryon, the existence of a thermal background radiation, and, if one regards it as a dynamical variable, the present small value of the cosmological constant. The lesson of all recent work in this area is, I think, that one must go to very early times to seek explanations for these features and that in these explanations quantum processes and quantum gravity will ultimately play a fundamental role. Indeed, there is no avoiding it. Between us and the initial conditions there stands an epoch about 10^{-43} sec long in which curvatures vary significantly on the Planck scale and in which quantum phenomena are of dominant importance dynamically. To connect any initial conditions with present day observations we have to learn how to do dynamics in this era. If one asks for the observational problem for which quantum gravity is important, I believe that this is it.

2. THE PAIR CREATION PROCESS

We are very far from having a complete calculable dynamics of the quantum epoch not least because we do not yet have a complete calculable quantum theory of gravity. Still preliminary investigations have been made and it seems useful to do so. A process of particular interest is the creation of elementary particle pairs in the strongly time dependent geometry of the big bang. As first suggested by Zel'dovich (1970) pair creation will play an important role dynamically in the dissipation of initial anisotropy. The nature of the initial singularity and the early causal structure could also be significantly affected. In addition the pair creation process will certainly influence the observed distributions of matter and radiation at later epochs. It is this process and its associated quantum effects that I would like to focus on today.

Because of its importance, the pair creation process has been widely studied in model cosmologies. Much of this work has concentrated on the dissipation of anisotropy in homogeneous cosmologies because of the conceptual and mathematical simplicity of these models. One might mention in particular the pioneering work of Parker (1969) and Zel'dovich (1970) and the calculations of Zel'dovich and Starobinsky (1971), Lukash and Starobinsky (1974) and Hu and Parker (1978). [For more extensive references to the large literature on this subject see the above references and the reviews by Parker (1977), Lukash, Novikov, Starobinsky and Zel'dovich (1976), and Hu (1980).] In one respect these calculations are conclusive. Significant anisotropies existing at the Planck time are rapidly dissipated by pair creation on a time scale comparable to the Planck time.

In another important way, however, these calculations are limited. Calculations of the pair creation process encounter not only the usual ultraviolet divergences of quantum field theory but also infinities arising from the cosmological singularity itself. Consider for example the total probability P to produce a pair of conformally invariant scalar particles in a world tube T of constant comoving cross section extending over the whole history of a homogeneous universe with small anisotropy. In the test field approximation where the gravitational effects of the produced particles are neglected and to quadratic order in the small anisotropy this probability is (Zel'dovich and Starobinsky 1977; Hartle and Hu 1980)

$$P = \frac{1}{1920} \int_T d^4x\,(-g)^{\frac{1}{2}} C_{\alpha\beta\gamma\delta} C^{\alpha\beta\gamma\delta} \quad . \tag{2.1}$$

Near the singularity in a radiation dominated universe

$C_{\alpha\beta\gamma\delta}C^{\alpha\beta\gamma\delta} \sim t^{-10}$ and $(-g)^{1/2} \sim t^{3/2}$ so this integral diverges strongly at the singularity. Physically this infinity simply indicates that the test field approximation is inconsistent. The gravitational effect of the produced particles has been assumed small, but in fact it is large. Because they have been carried out in the test field or related approximations all of the calculations of anisotropy dissipation mentioned above have had to be limited in an *ad hoc* fashion to avoid infinities of this type. The typical procedure is to limit the calculation (and hence the integral in eqn. (2.1)) a regime later than a time t_0 of order of the Planck time. This does not seem very satisfactory. What these calculations really argue for, it seems to me, is that the dynamically significant regime for the pair production process is *before* the Planck time and that the calculations should not be restricted to late times but should be pushed all the way back to the singularity.

3. THE EFFECTIVE ACTION METHOD

To go beyond the test field approximation quantum effects must be included in the dynamics. The effective action technique of relativistic quantum field theory provides a general framework in which to do this. The framework is a natural one for the cosmological problem because it leads to dynamical equations for matrix elements of the metric whose form is analogous to the field equations of classical gravity with which we are all familiar. It allows approximation schemes to the putative exact quantum theory to be systematically discussed and it possesses a number of technical advantages yielding, for example, a compact treatment of the regulation of ultraviolet infinities.

The central quantity in the method is the effective action $\Gamma[\tilde{g}]$ which is functional of a general argument geometry $\tilde{g}$. The solution g to the variational problem

$$(\delta\Gamma[\tilde{g}]/\delta\tilde{g}_{\alpha\beta})_{\tilde{g}=g} = -\mathscr{T}^{\alpha\beta} \tag{3.1}$$

we will call the effective geometry* in the presence of an

* We use the term 'effective geometry' as a name for the matrix element in (3.2) which seems less confusing than the term 'classical geometry' used elsewhere (Fischetti, Hartle, and Hu 1979; Hartle and Hu 1979, 1980).

external source $\mathcal{T}$. Typically, this source can be taken to vanish except where necessary to enforce boundary conditions on the model. The effective geometry is the (suitably gauge averaged) matrix element of the metric field operator $\hat{g}$ between the initial vacuum state $|0_-\rangle$ and the final vacuum state $|0_+\rangle$ in the presence of this source:

$$g_{\alpha\beta}(x) = \langle 0_+|\hat{g}_{\alpha\beta}(x)|0_-\rangle/\langle 0_+|0_-\rangle \ . \tag{3.2}$$

The effective action evaluated at the effective geometry gives the vacuum persistence amplitude in the presence of the source $\mathcal{T}$ through the relation

$$\langle 0_+|0_-\rangle = \exp(iW) \ , \tag{3.3}$$

where

$$W = \Gamma[g] + \int d^4x g_{\alpha\beta}\mathcal{T}^{\alpha\beta} \ . \tag{3.4}$$

The total particle production probability P is $1 - |\langle 0_+|0_-\rangle|^2$ so that for small P the pair production probability is just

$$P = 2 \ \mathrm{Im} W \ . \tag{3.5}$$

In general, when non-classical processes such as pair production are taking place, the effective action will be complex, the effective geometry will also be complex, the vacuum persistence probability will be less than unity, and the particle production probability will be non-vanishing. There is thus a complete prescription for calculating pair production probabilities including all quantum mechanical effects by solving the quantum dynamical equations (3.1).

The effective action functional can be defined formally in terms of the functional integral for quantum gravity coupled to external sources. A systematic approximation scheme for it, called the loop expansion, is to expand it in powers of $\hbar$. The zero and one loop term in this expansion are

$$\Gamma[\tilde{g}] = S_E[\tilde{g}] - \frac{1}{2} i\hbar \ \Sigma_f[\mathrm{Tr} \ \log G^f(x,x')]_{\mathrm{Reg}} + \ldots . \tag{3.6}$$

Here, S_E is the classical action for gravity and G^f is the Green's function for a field f propagating in the argument geometry $\tilde{g}$. The sum is over all fields existing in nature including gravity and the subscript Reg indicates that the infinite trace which includes space-time labels must be suitably regulated. The second term in this expression for the effective action contains the first order in $\hbar$ quantum corrections to the classical dynamics. Its variational derivative contains the contributions to this order of the trace anomalies and the back reaction of produced particles to the dynamical equations governing the effective geometry. In the next section we shall see how to calculate these effects in a simple cosmological model.

4. A MODEL OF ANISOTROPY DISSIPATION

The model I will now describe was investigated by M.V. Fischetti, B.L. Hu and myself at Santa Barbara (Fischetti, Hartle, and Hu 1979; Hartle and Hu 1979, 1980; Hartle, 1980). We consider the production of conformally invariant scalar particles and the consequent dissipation of anisotropy in homogeneous spatially flat cosmologies containing some classical radiation. Quantum effects are calculated to first order in $\hbar$ (the one loop order) and the anisotropy will be assumed small. More specifically we do the following:

We shall search for effective geometries whose line element is of the form

$$ds^2 = a^2(\eta)[-d\eta^2 + (e^{2\beta(\eta)})_{ij}dx^i dx^j] , \qquad (4.1)$$

where β is a 3×3 traceless matrix which is small in a sense to be made precise later. The problem is, of course, considerably simplified by the high symmetry which is thereby imposed but one is also investigating geometries which are not very different from that of our own universe at least in later stages and on larger scales.

Restricting attention to the production of scalar conformally invariant particles means retaining in the one loop contribution to the effective action only the Green's function for a single scalar field obeying the wave equation

$$\Box^2\phi(x) - \frac{R}{6}\phi(x) = 0 . \tag{4.2}$$

This is a reasonable kinematical model for matter fields which on the basis of current particle theories are believed to be exactly or approximately conformally invariant at the high temperatures of the early universe. It is a poor model for the gravitational field itself which is not conformally invariant.

In addition to the effects of the quantized scalar field we include in the model some classical radiation whose energy density obeys the law

$$\rho_r(\eta) = \tilde{\rho}_r/a^4(\eta) , \tag{4.3}$$

where $\tilde{\rho}_r$ is a constant. This classical radiation may be thought of as high frequency scalar quanta present in the universe from its earliest stages and is included to drive the expansion so that the anisotropy can be maintained small at all times.

The effective action is now a functional of the metric parameters a and β. The dynamical equations which determine the effective geometry are

$$\delta\Gamma/\delta a = 0, \quad \delta\Gamma/\delta\beta_{ij} = 0 , \tag{4.4}$$

with all classical sources vanishing other than those for the classical radiation described above and those on the boundaries of the space-time necessary to fix an initial anisotropy. Since the anisotropy is assumed small the one loop effective action may be expanded in powers of β. We write

$$\Gamma[a,\beta] = \Gamma_0[a] + \Gamma_2[a,\beta] + \ldots \tag{4.5}$$

where Γ_n is nth order in β and treat the dynamical equations order by order.

We first examine the isotropic limit. The classical action up to a divergence is

$$S_E = -V\int d\eta[6(a')^2/\ell^2] . \tag{4.6}$$

Here V is the spatial volume under consideration, a prime denotes a derivative with respect to η, $\ell = (16\pi G)^{\frac{1}{2}}$ is the Planck length and we, henceforth, are using units where $\hbar = c = 1$. Since the geometry is conformally flat in the isotropic limit, the Green's function, G, for the conformally invariant scalar field needed to compute the one loop correction to the effective action may naturally be taken to be conformally related to the flat space Feynman Green's function G_F,

$$G(x,x') = [a(\eta)]^{-1} G_F(x,x') [a(\eta')]^{-1} . \tag{4.7}$$

The infinite trace involved in computing the one loop effective action may be regulated by a variety of methods, for example, dimensional continuation coupled with the addition of counter-actions which are integrals of polynomials in the curvature. A variety of schemes give the following regulated effective action in the isotropic limit

$$\Gamma_0[a] = -V \int d\eta \left[\frac{6(a')^2}{\ell^2} + 3\lambda\left(\frac{a''}{a}\right)^2 + 3\lambda\left(\frac{a'}{a}\right)^4\right] . \tag{4.8}$$

Here $\lambda = (2880\pi^2)^{-1}$. The isotropic effective action is real, the extremizing effective geometry is also real so that no particles are produced and the vacuum persists. This is a general result for conformally invariant fields in conformally flat space-times (Parker 1969).

The effective geometry is nevertheless significantly altered by quantum effects at early times. The dynamical equation $\delta\Gamma_0/\delta a = 0$ is Einstein's equation for the scalar curvature with the scalar field trace anomaly as a source. This is a fourth order, non-linear, ordinary differential equation for $a(\eta)$. At large times the solution should become the classical solution of Einstein's equations for a radiation dominated universe

$$a(\eta) \sim (\tilde{\rho}_r/6)^{\frac{1}{2}} \ell\eta, \qquad \eta \to \infty . \tag{4.9}$$

Towards the past we demand only that the universe not become large without also becoming classical. Runaway expansions on the Planck time scale are thereby excluded. These are not only unphysical but probably would not exist if the assumptions of exact isotropy and conformal invariance were relaxed.

There is a one parameter family of effective geometries satisfying these two boundary conditions which are displayed in Figure 4.1. All of these geometries deviate significantly from the classical solution $b = \chi$ at early times. All are singular but the singularity is weaker than in Einstein's theory in the following sense: in the classical theory the curvature diverges with vanishing scale factor as $R \sim a^{-4}$. For these geometries $R \sim a^{-2}$ – a weaker behaviour. This can be stated differently in terms of the one loop matrix elements of the stress energy tensor between the initial and final vacua. These are readily calculated as variational derivatives of effective action. In the present case the only non-vanishing components have the following dominant behaviour near the singularity

$$T^0_0 \sim -\rho_r, \quad T^1_1 = T^2_2 = T^3_3 \sim -p_r , \tag{4.10}$$

where p_r is the pressure of the classical radiation and is related to the energy density by $p_r = (1/3)\rho_r$. The strong and dominant energy conditions are thus both violated by the stress-energy. The effect is to weaken the singularity as described above but not remove it.

A member of the class of isotropic effective geometries of particular interest is the one labelled A_m in Figure 4.1. The singularity is at $\chi = -\infty$ although only a finite proper time away for any comoving observer. The geometry is conformally related to a complete flat space-time and contains no horizons. Quantum effects can thus not only affect the strengths of singularities in cosmological models but significantly alter the causal structure as well. This would be a necessary element of any explanation of the homogeneity of the universe.

Let us now endow these models with some small anisotropy and study the resulting pair creation and evolution. Because of their attractive features we shall focus on those models which in their isotropic limit become the conformally complete effective geometry described above. To carry out the calculation we need the effective action to higher orders in the metric parameter β. The rules for constructing this perturbation series

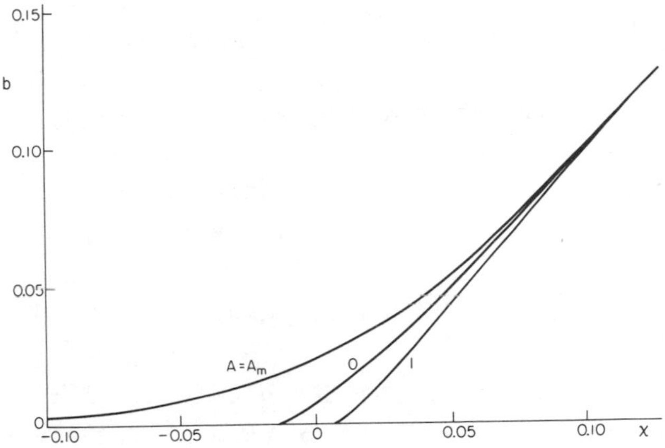

FIG. 4.1. The possible scale factors for the effective geometries of homogeneous, isotropic model cosmologies. Shown here are some members of the one parameter family of scale factors which extremize the isotropic effective action [e.g. (4.8)] and which satisfy the boundary conditions that at large times they approach the classical Friedmann solution for a radiation filled universe while towards the past they do not expand on the Planck time scale in a runaway fashion. The dimensionaless scale invariant variables b and χ are defined by $b = a/(\ell\tilde{\rho}_r^{\frac{1}{4}})$ and $\chi = 6^{-\frac{1}{2}}\tilde{\rho}_r^{\frac{1}{4}}\eta$ respectively. The parameter A which labels members of the family ranges over $A_m \leqslant A < \infty$ where A_m is a known minimum value.

The origin of the co-ordinate χ has been chosen so that at large χ all solutions approach the Friedmann solution for a radiation filled universe $b = \chi$. At early times the trace anomaly produces significant deviations from this classical behaviour. All of these geometries possess initial singularities. Those for $A > A_m$ are at finite values of χ. The singularity for $A = A_m$ is at $\chi = -\infty$ although still only a finite proper time in the past for any comoving observer. The effective geometry for $A = A_m$ is thus conformally related to a complete flat space-time and has no cosmological horizons.

can be read off the action for the conformally invariant scalar field

$$S[\tilde{g},\phi] = -\frac{1}{2}\int d^4x(-\tilde{g})^{\frac{1}{2}}[\tilde{g}^{\alpha\beta}\partial_\alpha\phi\partial_\beta\phi + \frac{1}{6}R\phi^2] \ . \tag{4.11}$$

For the geometries under consideration this action can be expanded in powers of β.

$$S = S_0[a] + S_1[a,\beta] + S_2[a,\beta] + \ldots \ , \tag{4.12}$$

where S_0 is the action in the isotropic limit and S_n represents a coupling of the scalar field to n powers of β. For example,

$$S_1 = \int d^4x a^2\beta^{ij}\partial_i\phi\partial_j\phi \ . \tag{4.13}$$

The resulting expansion of the Green's function for the conformally invariant scalar field in powers of β leads to a corresponding expansion of the effective action which is compactly expressed in terms of the Feynman diagrams in Figure 4.2. In these diagrams a solid line represents a propagator in the isotropic limit already discussed and ⓝ represents the coupling to the n powers of β generated by S_n. Because the isotropic propagator is conformally related to the flat space Feynman propagator these integrals are easily calculated and regulated by familiar methods. The result to quadratic order in β is

$$\Gamma_2[a,\beta] = V\int_{-\infty}^{+\infty} d\eta\Bigg\{\left[\left(\frac{a}{\ell}\right)^2 - \lambda\left(\frac{a''}{a}\right) - \lambda\left(\frac{a'}{a}\right)^2\right]\beta'_{ij}\beta'^{ij}$$
$$+ 3\lambda\left[\frac{i\pi}{2} + \log(\mu a)\right]\beta''_{ij}\beta''^{ij} - 3\lambda\beta''_{ij}K\beta''^{ij}\Bigg\} \ , \tag{4.14}$$

where indices are raised with the flat spatial metric and

$$Kf = Cf - \frac{1}{2}\int_{-\infty}^{+\infty} d\eta'\varepsilon(\eta-\eta')\log|\eta-\eta'|\frac{df}{d\eta'} \ , \tag{4.15}$$

C being Euler's constant, and $\varepsilon(x) = x/|x|$.

There are three points to notice about this expression for Γ_2. First, it is non-local and there was no reason to expect otherwise. Second, it is complex and this will mean non-vanishing particle production probabilities. Third, it involves an unfixed regularization scale μ reflecting the non-renormalizability of quantum gravity coupled to matter.

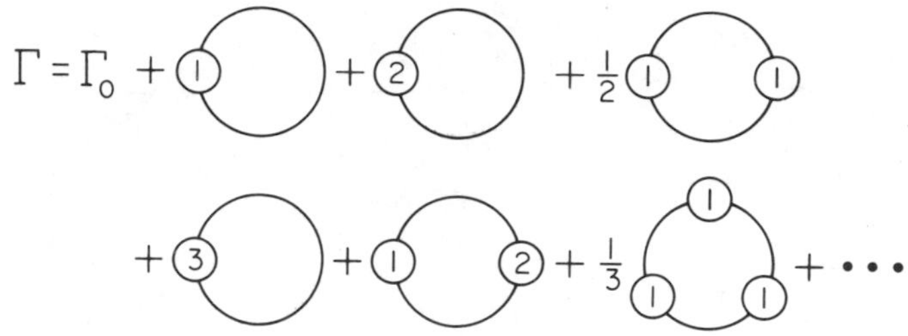

FIG. 4.2. The expansion of the effective action of homogeneous geometries in powers of the metric parameter β_{ij} which controls the anisotropy. Γ_0 is the effective action in the isotropic limit given in eqn. (4.8). A line in the Feynman diagrms represents the propagator in the isotropic limit which is conformally related to the flat space Feynman propagator through eqn. (4.7). ⓝ represents the interaction with n powers of β_{ij} generated by the term S_n in the expansion of the scalar field action in powers of β_{ij} [eqn. (4.12)]. All diagrams are understood to be regulated by the addition of some suitable counteraction.

To evaluate the particle production probabilities we find the extremum value of the effective action with respect to variations in β holding fixed some overall measure of the aniso tropy. For a calculation of the pair production probabilities to quadratic order in this anisotropy it is sufficient to consider this problem for Γ_2 evaluated at the already computed isotropic scale factor a. The resulting variational problem reduces to a linear integro-differential equation for β, whose solution is unique. Some typical results are illustrated in Figures 4.3 and 4.4 for various values of the regularization scale. The quantity h plotted there measures how β'_{ij} scales with time.

Further local properties of the effective geometry are then easily calculated. For example, Figure 4.5 shows a graph of a quantity W^2 proportional to the absolute square of the Weyl tensor versus proper time from the singularity for a variety of regularization scales. The anisotropy dissipates rapidly on a fraction of the Planck time scale. The rate for these small anisotropies is not strikingly different from that which would be predicted by the classical evolution of the model but still

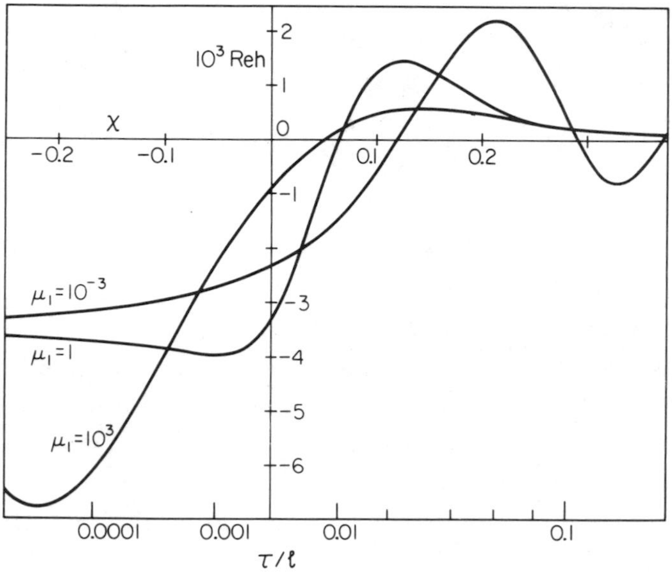

FIG. 4.3. The real part of a measure of the anisotropy in the metric of the effective geometry. The value of β_{ij} at a given instant of time contains little useful geometrical information because it can always be made to vanish by an appropriate redefinition of the spatial co-ordinates. Local geometrical invariants will involve the derivatives of β_{ij} and, indeed, the dynamical equation for the effective geometry can be completely expressed in terms of β'_{ij} as eqn. (4.14) shows. Since the dynamical equation is linear the solution for β'_{ij} may be written as $\beta'_{ij} = C_{ij}h(\chi)$ where C_{ij} is a matrix of constants which controls the overall magnitude and orientation of the anisotropy while $h(\chi)$ is a single scalar function which controls how it evolves in time. The real part of h in what for the present purpose may be taken as arbitrary units is shown in this figure plotted against the variable χ defined in Figure 4.1 for three values of the unfixed regularization scale μ given here by the dimensionaless quantity $\mu_1 = \mu\ell\tilde{\rho}_r^{\frac{1}{4}}$. In all cases the function h approaches a real constant at the singularity $(\chi \to -\infty)$ and evolves classically as (real const.)$/\chi^2$ at late times $(\chi \to +\infty)$. The bottom scale is the proper time from the singularity in units of the Planck time as measured by a comoving observer.

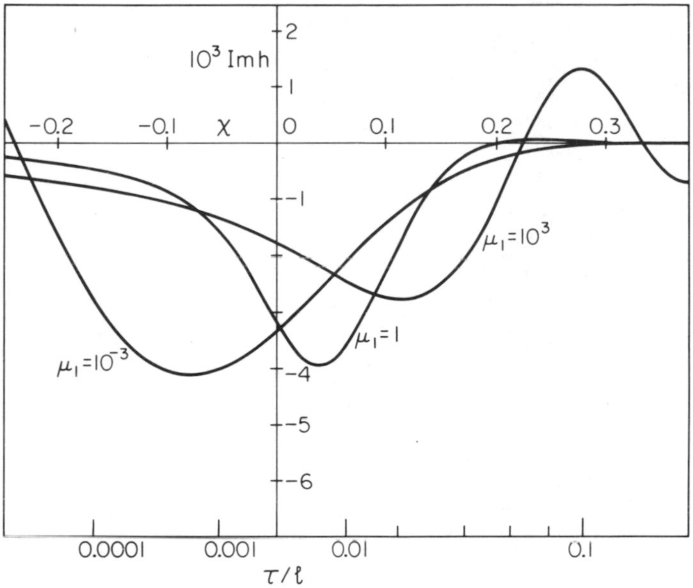

FIG. 4.4. The imaginary part of the measure of the anisotropic part of the effective geometry described in Figure 4.3. The imaginary part vanishes as $\chi \to \pm\infty$. For the values of the regularization scale shown here it is significant for τ less than a few tenths of the Planck time.

consistent with the idea of strong dissipation at very early times. Evaluated at one of these solutions the imaginary part of the effective action takes a simple form so that the total probability to produce a pair of particles in a co-ordinate volume V over the whole history of the universe has the following compact expression:

$$P = \frac{V}{960\pi} \int_{-\infty}^{+\infty} d\eta \bar{\beta}''_{ij} \beta''^{ij} \quad . \tag{4.16}$$

Equivalently to the quadratic order under consideration

$$P = \frac{1}{1920\pi} \int d^4x (-g)^{\frac{1}{2}} \bar{C}_{\alpha\beta\gamma\delta} C^{\alpha\beta\gamma\delta} \quad , \tag{4.17}$$

where the integration is over the world tube of cross section V.

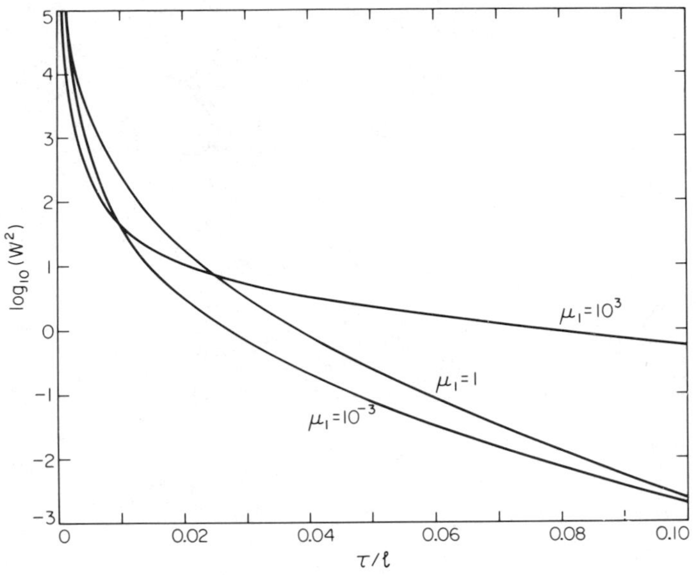

FIG. 4.5. A function W^2 which measures the absolute magnitude of the squared Weyl tensor of the effective geometry plotted against proper time from the singularity. The function $W^2 = 2|dh/d\chi|^2 b^{-4}$ is proportional to the absolute square of the Weyl tensor and is thus a measure of the anisotropy in the effective geometry. W^2 decays rapidly on the scale of a few tenths of a Planck time.

The probability is positive and vanishes in the isotropic limit. More importantly, it is finite. For example, for the probability P_r to produce a pair in the comoving volume occupied by one classical radiation quantum is for a regularization scale of 1 in natural Planck units ($\mu_1 = 1$)

$$P_r = 5.8 \times 10^2 \Delta^2 . \tag{4.18}$$

Here Δ^2 is a dimensionless, scale invariant measure of the anisotropy in the model which is constant in classical epochs and is given in terms of the average H and rms difference ΔH of the three principal Hubble constants by

$$\Delta^2 = \frac{1}{6}\left(\frac{\Delta H}{H}\right)^2 \frac{1}{\ell^2 \rho_r^{\frac{1}{2}}} \tag{4.19}$$

The inclusion of quantum effects to order $\hbar$ in the dynamics of the early universe has thus removed the divergence found in the test field approximation and enabled the calculation of pair creation to be pushed all the way back to the singularity with finite results and significant effects on the dynamics.

5. INITIAL CONDITIONS

The model I have just described is only a limited step in the direction of providing a quantum dynamics for the early universe but one consistent with the idea that quantum effects are important at a very early stage. The theory needs to be extended in many ways – to incorporate the effects of gravitons, to treat significant anisotropic inhomogeneities and spatial curvature, to incorporate more realistic theories of matter and its interactions and not least to incorporate a calculable theory of quantum gravity to all orders in $\hbar$. Even if all these elements could be included they would still yield only a theory of dynamics. The central issue of initial conditions remains to be addressed. In our model for example we took the no particle vacuum as an exemplary initial state from which to calculate further evolution but other states might be equally interesting initial conditions from some points of view and, I might add, equally calculable.

We seek a set of general principles which imply initial conditions which lead naturally via quantum and classical dynamics and the correct theory of matter interactions to the universe we see today. The most promising ideas on this question (ably expounded elsewhere in this conference) are statistical. The view is that the initial conditions which lead to today's universe would have some special property under the measure on all possible initial conditions. To name one, there is Penrose's idea that the universe starts in a state of zero gravitational entropy. The precise definition of this measure on initial states remains to be supplied but is essentially a quantum mechanical question as the Bekenstein–Hawking development of black hole thermodynamics already shows. This measure, a formulation of initial conditions, and a complete theory of cosmological dynamics seem to me not least among the questions to which a quantum theory of gravity must ultimately supply answers.

ACKNOWLEDGEMENT

This work was supported in part by the National Science Foundation.

REFERENCES

Fischetti, M.V., Hartle, J.B., and Hu, B.L. (1979). *Phys. Rev.* D20, 1757.
Hartle, J.B. (1980). *Phys. Rev.* D**22**, 2091.
—— and Hu, B.L. (1979). *Phys. Rev.* D**20**, 1772.
—— —— (1980). *Phys. Rev.* D**20**, 2756.
Hu, B.L. (1980). In *Recent Developments in Relativity*. Proc. Second Marcell Grossmann Meet., July 1979, ed. R. Ruffini, North-Holland Publishing Co., Amsterdam.
—— and Parker, L. (1978). *Phys. Rev.* D**17**, 933.
Lukash, V.N. and Starobinsky, A.A. (1974). *Zh. eksp. teor. Fiz.* **66**, 1515. [*Sov. Phys. - JETP* **39**, 742.]
——, Novikov, I.D., Starobinsky, A.A., and Zel'dovich, Ya.B. (1976). *Nuovo Cim.* **35B**, 293.
Parker, L. (1969). *Phys. Rev.* **183**, 1057.
—— (1977). In *Asymptotic structure of spacetime*, ed. F.P. Esposito and L. Witten. Plenum Press, New York.
Zel'dovich, Ya.B. (1970). *Pis'ma Zh. eksp. teor. Fiz.* **12**, 443. [*JETP Lett.* **12**, 307.]
—— and Starobinsky, A.A. (1971). *Zh. eksp. teor. Fiz.* **61**, 2161. [*Sov. Phys. - JETP* **34**, 1159.]
—— —— (1977). *Pis'ma Zh. eksp. teor. Fiz.* **26**, 373. [*JETP Lett.* **26**, 252.]

15

CANONICAL METHODS OF QUANTIZATION

K. Kuchař
Department of Physics, University of Utah

1. INTRODUCTION

When I was asked to review the canonical methods of quantization for this meeting, I started by rereading the rules of the canonization process. I must admit that I was not much encouraged in my endeavour. Two of the rules sounded especially ominous. First, one cannot canonize a person before he is deceased. Now, many people in my audience may feel that the canonical theory of quantization readily meets this condition. To those, I should read the second rule. At least two authentic miracles obtained through invocation after beatification must occur before the cause for canonization may be introduced. I fear that no quantum theory of gravitation can claim even a single miracle. Under these circumstances, I am not quite sure if my role here should be that of a postulator of the cause or rather that of a promoter of the faith, commonly known as the devil's advocate.

Still, what the canonical methods attempt to do, even if that task remains unaccomplished, is worth recalling. At least, these methods are straightforward, preserving the original simplicity of the primitive church of quantum mechanics. If they do not quite succeed in handling gravitation, it is good to know the reasons.

The outlines of the canonical quantization seem straightforward. First, the classical theory is cast into Hamiltonian form and the pairs of canonically conjugate variables are identified. Second, these variables are turned into operators satisfying the Dirac commutation relations. The substitution of the conjugate operators into the Hamiltonian should yield the Schrödinger equation which conserves an inner product during the dynamical evolution of the state. This product turns the space of solutions into a Hilbert space on which the probabilistic interpretation of the theory is based.

The special feature of the gravitational field is that it does not naturally lead to a Hamiltonian, but only to 'Hamiltonian constraints'. This circumstance incredibly complicates the implementation of the canonical quantization programme, in particular the construction of a satisfactory Hilbert space. Without such a space, however, the probabilistic interpretation of the theory remains largely obscure.

To clarify the nature of the difficulties, I shall compare the canonical theory of gravitation with simpler theories, namely, with non-relativistic and relativistic particle dynamics and with field theory on a given background. I shall follow the hints which these theories give me about how to quantize the gravitational field itself, but I conclude that most of those hints are misleading or at least elusive. In the end, I shall try to trace the difficulties to the basically different attitudes which the general theory of relativity and quantum mechanics take toward the concept of time.

2. CANONICAL GEOMETRODYNAMICS

Let me start by casting the Einstein law of gravitation into the canonical form. I cut the space-time by an arbitrary space-like hypersurface $X^\alpha = X^\alpha(x^a)$. The Latin indices shall always run through the values 1, 2, 3 and the Greek indices through the values 0, 1, 2, 3. At each point of the hypersurface, I have the basis consisting of three tangent vectors $X^\alpha_a \equiv X^\alpha_{\ ,a}$ to the hypersurface and of the unit normal vector n^α,

$$g_{\alpha\beta}X^\alpha_a n^\beta = 0, \quad g_{\alpha\beta}n^\alpha n^\beta = -1 . \tag{2.1}$$

I continuously deform the hypersurface through space-time, incorporating it into a one-parameter family of hypersurfaces $X^\alpha = X^\alpha(x^a, t)$. The deformation vector $N^\alpha \equiv \dot{X}^\alpha = \partial X^\alpha(x^a, t)/\partial t$ connecting the points with the same label x^a on two neighbouring hypersurfaces can be decomposed with respect to the basis vectors $\{n^\alpha, X^\alpha_a\}$,

$$N^\alpha \equiv Nn^\alpha + N^a X^\alpha_a \ ; \tag{2.2}$$

its components N and N^a are called the lapse and the shift functions.

To follow the dynamics of an arbitrary field, I project that field perpendicular and parallel to the hypersurface and observe how these projections change when the hypersurface is deformed through space-time. The metric field is no exception, but because the metric is needed to characterize what is meant by the normal direction (2.1), two of its projections, $g_{\perp\perp} = -1$ and $g_{\perp a} = 0$, are trivial. The only interesting projection of $g_{\alpha\beta}$ is that along the hypersurface

$$g_{ab}(x) = g_{\alpha\beta}(X(x))X^{\alpha}_{a}X^{\beta}_{b} \quad , \tag{2.3}$$

which is just the three-dimensional metric induced on the hypersurface. As this hypersurface is deformed through space-time, its 3-metric changes. Geometrodynamics formulates the laws which govern this change.

The rate of change of g_{ab} with respect to the label time t can be decomposed into the normal and tangential contributions,

$$\dot{g}_{ab} = -\,2NK_{ab} + 2N_{(a|b)} \; . \tag{2.4}$$

The normal change of g_{ab} is proportional to the extrinsic curvature K_{ab} of the hypersurface. The tangential change is given by the Lie derivative of g_{ab} along the shift vector, $L_{Nc}g_{ab} = 2N(a|b)$. The vertical stroke denotes the spatial covariant differentiation.

The formula (2.4) is valid in an arbitrary space-time. The dynamics of geometry, however, occurs only in the second order in time. The second derivative $\ddot{g}_{ab}$ of the spatial metric is to be determined from the Einstein law, or better, by varying the gravitational action. Dirac and Arnowitt, Deser and Misner (ADM) have rearranged the Hilbert action into a new form which is best adopted to this task:

$$S[{}^4g_{\alpha\beta}] = \int d^4X\,|{}^4g|^{\frac{1}{2}}\;{}^4R \longmapsto$$

$$S[g_{ab},N,N^a] = \int dt \int d^3x\; Ng^{\frac{1}{2}}[K_{ab}K^{ab} - K^2 + R] + \text{boundary terms.} \tag{2.5}$$

The ADM action is to be varied with respect to the spacial geometry g_{ab} and the lapse and the shift functions. The extrinsic curvature in eqn. (2.5) must be expressed as a function of these variables from eqn. (2.4).

It is remarkable that the time derivatives of N and N^a do not enter into the ADM action. This means that the lapse and the shift are not to be treated as dynamical variables, but merely as the Lagrange multipliers which specify how the hypersurfaces are to be drawn in space-time. It is only the spatial metric which must be treated as a dynamical variable. Consequently, I shall replace the geometrodynamical velocities $\dot{g}_{ab}$ by the geometrodynamical momenta p^{ab} when performing the Legendre dual transformation, but leave the Lagrange multipliers untouched:

$$S[g_{ab},p^{ab},N,N^a] = \int dt \int d^3x(p^{ab}\dot{g}_{ab} - N\mathcal{H} - N^a\mathcal{H}_a) \quad . \qquad (2.6)$$

The expressions $\mathcal{H}$ and $\mathcal{H}_a$ are constructed solely from the canonical variables g_{ab}, p^{ab},

$$\mathcal{H} = G_{ab\,cd}\, p^{ab}p^{cd} - g^{\frac{1}{2}}R \ , \qquad (2.7)$$

$$G_{ab\,cd} \equiv \frac{1}{2}g^{-\frac{1}{2}}(g_{ac}g_{bd} + g_{ad}g_{bc} - g_{ab}g_{cd}) \ ; \qquad (2.8)$$

$$\mathcal{H}_a = -\,2p_{a|b}^{b} \ , \qquad (2.9)$$

and do not contain the multipliers N, N^a. They are called the super-Hamiltonian and the supermomentum of the metric field.

The ADM canonical action principle (2.6)–(2.9) governs the dynamics of geometry. By varying the metric g_{ab} and the momentum p^{ab}, I get the evolution equations for the geometry, which are equivalent to the tangential projections ${}^4G_{ab} = 0$ of the Einstein law. The special feature of geometrodynamics is that I must vary the Lagrange multipliers N, N^a as well. Doing so, I learn that the conjugate variables g_{ab}, p^{ab} cannot be chosen arbitrarily, but must satisfy the constraints

$$\mathcal{H} = 0 = \mathcal{H}_a \ . \qquad (2.10)$$

These constraints are equivalent to the remaining projections ${}^4G_{\perp\perp} = 0 = {}^4G_{\perp a}$ of the Einstein law. Because the geometrodynamical Hamiltonian is a linear combination of the constraints (2.10) (with coefficients N, N^a) it is also constrained to vanish. This is a rather puzzling feature of the formalism. To understand its origin, it is useful to compare Hamiltonian geometrodynamics with simpler theories, namely, with parametrized particle dynamics and with field theory on a fixed metric background.

3. PARAMETRIZED PARTICLE DYNAMICS

The canonical action describing the motion of a single non-relativistic particle in the potential field $V(x^a,t)$ has the form

$$S[x^a,p_a] = \int dt[p_a\frac{dx^a}{dt} - H(x^a,p_a,t)] \ , \tag{3.1}$$

with the Hamiltonian

$$H = \frac{1}{2m}g^{ab}p_a p_b + V(x^a,t) \ . \tag{3.2}$$

The metric g^{ab} on the leaves of constant absolute time t is assumed to be flat in the Newtonian theory, but for the analogy I want to draw, it is useful to admit also curved spatial metrics.

I shall now parametrize the path which the particle takes in the phase space by an arbitrary label time τ and, moreover, adjoin the absolute time to the configuration variables x^a:

$$x^\alpha = \{t,x^a\}, \quad x^\alpha = x^\alpha(\tau), \quad p_a = p_a(\tau) \ .$$

The action (3.1) can be expressed in the new parametrization as

$$S[x^\alpha,p_a] = \int d\tau(p_a\dot{x}^a - H\dot{t}) \ , \tag{3.3}$$

the dot denoting the derivative with respect to the label time τ. Numerically, the expression (3.3) is equal to the expression (3.1), so that the variation with respect to the canonical variables x^a, p_a yields the same equations of motion. In addition, the variation of the parametrized action (3.3) with respect to t also yields a correct equation, namely, the energy balance equation $\dot{H} = H_{,t}\dot{t}$. The integrand of (3.3) is linear in the velocities $\dot{x}^\alpha = \{\dot{t},\dot{x}^a\}$. Introducing $p_0 \equiv -H$ as the momentum canonically conjugate to t, I can write it in the suggestive form $p_\alpha\dot{x}^\alpha$. However, the variables p_α cannot be varied freely, because p_0 is a mere abbreviation for the function $-H(x^\alpha,p_a)$. In other words, the variables x^α, p_α must satisfy the constraint

$$\mathcal{H} \equiv p_0 + H(x^\alpha,p_a) = 0 \; . \tag{3.4}$$

In order to be able to vary all the variables freely, adjoin the constraint (3.4) to the original action by means of a Lagrange multiplier N,

$$S[x^\alpha,p_\alpha;N] = \int d\tau\,(p_\alpha\dot{x}^\alpha - N\mathcal{H}) \; . \tag{3.5}$$

One of the equations of motion, namely, that obtained by varying p_0, endows N with a physical meaning:

$$N = \dot{t} = \frac{d[\text{absolute time}]}{d[\text{label time}]} \; . \tag{3.6}$$

The remaining equations, obtained by varying x^a, p_a and t, are the Hamilton equations in the arbitrary parametrization τ and the energy balance equation. After performing the variations, I can put $N = 1$ and return thus to the privileged parametrization by the absolute time t.

The procedure can be easily applied to a single relativistic particle moving in a curved spacetime $g_{\alpha\beta}(x^\gamma)$. The parametrized action has the same general form (3.5), but the constraint is somewhat different,

$$\mathcal{H} \equiv \frac{1}{2m}(g^{\alpha\beta}p_\alpha p_\beta + m^2) = 0 \; . \tag{3.7}$$

As one consequence, when I vary (3.5) with respect to p_0, I learn that

$$N = \frac{\mathrm{d(proper\ time)}}{\mathrm{d(label\ time)}} \ . \tag{3.8}$$

The analogue to the geometrodynamical action is evident. By writing the action of the particle in the parametrized form, I obtained several characteristic features of the geometrodynamical action – the presence of a constraint adjoint to the action by a Lagrange multiplier, and the vanishing Hamiltonian. The constraint arose because the (absolute or Minkowskian) time t and the energy of the particle $-p_0$ have supplemented the true dynamical variables x^a, p_a; it is nothing else but the definition of the energy as a function of x^a, p_a and t. It is tempting to conjecture that the geometrodynamical constraints have a similar meaning. Before drawing such a conclusion, I should explain why there is only one constraint in particle dynamics, but $(1 + 3)\infty^3$ constraints, four for each point of the hypersurface, in geometrodynamics. With this task in mind, I shall describe how to parametrize a field theory on a given background.

4. FIELD THEORY ON A BACKGROUND

There is no need to be general at this point, because I am solely interested in a single typical example to compare with canonical geometrodynamics. I shall thus describe the simplest possible model: a real scalar field which satisfies the Klein–Gordon equation on a curved background. I start from the space-time action

$$S[\phi] = \int \mathrm{d}^4X \ -|{}^4g|^{\frac{1}{2}}[\tfrac{1}{2}g^{\alpha\beta}\phi_{,\alpha}\phi_{,\beta} + \tfrac{1}{2}m^2\phi^2] \ . \tag{4.1}$$

I split the metric $g^{\alpha\beta}$ into N, N^a and g_{ab} and perform the Legendre transformation from the variables $\phi(x^a,t)$, $\dot{\phi}(x^a,t)$ to the canonical variables $\phi(x^a,t)$, $\pi(x^a,t)$. I cast thereby the action (4.1) into the canonical form

$$S[\phi,\pi] = \int \mathrm{d}t \int \mathrm{d}^3x[\pi\dot{\phi} - NH^{\phi} - N^aH^{\phi}_a] \ . \tag{4.2}$$

Here,

$$H^{\phi} = \frac{1}{2} g^{-\frac{1}{2}} \pi^2 + \frac{1}{2} g^{\frac{1}{2}} g^{ab} \phi_{,a} \phi_{,b} + \frac{1}{2} g^{\frac{1}{2}} m^2 \phi^2 \tag{4.3}$$

is the energy density of the field measured by an observer moving with the four-velocity n^{α} normal to the hypersurface, and

$$H_a^{\phi} = \pi \phi_{,a} \tag{4.4}$$

is the momentum density of the field measured by the same observer. The energy density (4.3) is positive definite. The action (4.2) has the same general form as the gravitational action (2.6). However, the space-time metric $g_{\alpha\beta}$ is not varied in the action (4.1), so that the lapse and the shift functions are not to be varied in the action (4.2). The expressions (4.3) and (4.4) are thus not constrained to vanish. If they were, the scalar field itself should vanish, $\phi = 0 = \pi$, due to the positivity of the energy density (4.3). If I arbitrarily prescribe the canonical variables ϕ, π on the initial hypersurface and specify the lapse and the shift functions, the Hamilton equations determine the canonical variables on the deformed hypersurface. When I continue specifying N and N^a, I evolve ϕ and π for a finite interval into the future.

The absence of constraints indicates that the scalar field action is not parametrized. Related to this is the fact that, by varying suitable momenta, I did not get equations fixing the meaning of the Lagrange multipliers N, N^a. Rather than parametrizing the space-time action directly, I shall follow an alternative procedure which leads to the same result.

5. KINEMATICS OF DEFORMATIONS

I ask how to reproduce the definition

$$\dot{X}^{\alpha} = N^{\alpha}$$

of the deformation vector from a canonical action. The answer is seemingly trivial. I prescribe the action

$$S[X^{\alpha}, P_{\alpha}, N^{\alpha}] = \int dt \int d^3x [P_{\alpha} \dot{X}^{\alpha} - N^{\alpha} P_{\alpha}] \tag{5.1}$$

and vary it in the variables X^α, P_α, N^α. I get thereby two sets of Hamilton equations and a constraint:

$$\begin{aligned} \delta P_\alpha &\Rightarrow \dot{X}^\alpha = N^\alpha , \\ \delta X^\alpha &\Rightarrow \dot{P}_\alpha = 0 , \\ \delta N^\alpha &\Rightarrow P_\alpha = 0 . \end{aligned} \tag{5.2}$$

The first Hamilton equation defines N^α in terms of $\dot{X}^\alpha$. The constraint $P_\alpha = 0$ tells me that the momentum P_α is trivial. The second Hamilton equation ensures that the momentum P_α remains trivial in the dynamical evolution.

Instead of varying N^α, I can replace it by the lapse and the shift functions from eqn. (2.2). The action (5.1) then assumes the form

$$S[X^\alpha, P_\alpha, N, N^a] = \int dt \int d^3x [P_\alpha \dot{X}^\alpha - N\mathcal{H}^K - N^a\mathcal{H}^K_a] . \tag{5.3}$$

The kinematical super-Hamiltonian $\mathcal{H}^K$ and supermomentum $\mathcal{H}^K_a$ are given by the expressions

$$\mathcal{H}^K = n^\alpha[X]P_\alpha, \quad \mathcal{H}^K_a = X^\alpha_a P_\alpha . \tag{5.4}$$

They are to be considered as functionals of X^α and P_β. The variation of P_α in the action (5.3) yields directly the lapse-shift decomposition (2.2).

6. PARAMETRIZED FIELD THEORY

I now add the kinematical action to the field action and write

$$S[X^\alpha, P_\alpha; \phi, \pi; N, N^a] = \int dt \int d^3x [P_\alpha \dot{X}^\alpha + \pi\dot{\phi} - N\mathcal{H} - N^a\mathcal{H}_a] , \tag{6.1}$$

with

$$\mathcal{H} = \mathcal{H}^K + H^\phi, \quad \mathcal{H}_a = \mathcal{H}^K_a + H^\phi_a . \tag{6.2}$$

The metric g_{ab} in H^ϕ is to be considered as a functional of X^α

according to eqn. (2.3). The variation of P_α still gives the lapse-shift decomposition. The variation of ϕ and π still leads to the Klein–Gordon equation for $\phi(X^\alpha)$. The momentum P_α, however, ceases to be trivial. By varying N, N^a, I get the constraints $\mathcal{H} = 0 = \mathcal{H}_a$ which inform me that $-n^\alpha P_\alpha$ is to be interpreted as the energy density and $-X^\alpha_a P^\alpha$ as the momentum density of the field. The remaining Hamilton equations, obtained by varying X^α, tell me how the energy and momentum changes from one hypersurface to another.

The resulting scheme is closely analogous to the parametrized particle mechanics which I discussed before. The 4 ∞^3 variables $X^\alpha(x^a)$ specifying the hypersurface play the role of the time t, and the 4 ∞^3 conjugate variables $P_\alpha(x^a)$ play the role of the energy $E = -p_t$. I have 4 ∞^3 variables replacing a single variable because of the 4 ∞^3 ways $\dot{X}^\alpha(x^a)$ how to deform the hypersurface. As a result, the single constraint $\mathcal{H} = 0$ defining the energy E in terms of the dynamical variables x^a, p_a and of t, is now replaced by the 4 ∞^3 constraints $\mathcal{H}(x) = 0 = \mathcal{H}_a(x)$ defining the energy-momentum desitites $P_\alpha(x)$ in terms of the dynamical variables $\phi(x)$, $\pi(x)$ and of $X^\alpha(x)$. The model which I want to compare with canonical geometrodynamics is now complete.

7. THE HAMILTON–JACOBI EQUATION

From the constraints, I can obtain the Hamilton–Jacobi equation by the substitution $p_\alpha = \partial S(x)/\partial x^\alpha$. The Hamilton–Jacobi equation is an important link between the classical theory and the quantum theory, because I can consider it as the first step in the WKB approximation to the Schrödinger equation.

For the particle models, the Hamilton–Jacobi equation is the familiar one. Thus, the non-relativistic constraint (3.4) yield

$$\mathcal{H}\left[x^\alpha, p_\alpha = \frac{\partial S(x)}{\partial x^\alpha}\right] \equiv \frac{\partial S}{\partial t} + H\left(x^\alpha, \frac{\partial S(x)}{\partial x^a}\right) = 0 , \tag{7.1}$$

while the relativistic constraint (3.7) leads to the equation

$$2m\mathcal{H}\left[x^\alpha, p_\alpha = \frac{\partial S(x)}{\partial x^\alpha}\right] = g^{\alpha\beta}\frac{\partial S}{\partial x^\alpha}\frac{\partial S}{\partial x^\beta} + m^2 = 0 . \tag{7.2}$$

For the parametrized field theory with the constraints (6.2) I get

$$\mathcal{H}(x)\left[X^{\alpha}(x),\phi(x);P_{\alpha}(x)=\frac{\delta S}{\delta X^{\alpha}(x)},\ \pi(x)=\frac{\delta S}{\delta\phi(x)}\right]$$
$$\equiv n^{\alpha}(x)[X^{\beta}]\frac{\delta S}{\delta X^{\alpha}(x)}+\frac{1}{2}g^{-\frac{1}{2}}(x)[X^{\beta}]\left(\frac{\delta S}{\delta\phi(x)}\right)^{2}$$
$$+\frac{1}{2}g^{\frac{1}{2}}g^{ab}(x)[X^{\beta}]\phi_{,a}(x)\phi_{,b}(x)$$
$$+\frac{1}{2}m^{2}g^{\frac{1}{2}}(x)[X^{\beta}]\phi^{2}(x)=0 \qquad (7.3)$$

and similarly

$$\mathcal{H}_{a}(x)\left[X^{\alpha}(x),\phi(x);P_{\alpha}(x)=\frac{\delta S}{\delta X^{\alpha}(x)},\ \pi(x)=\frac{\delta S}{\delta\phi(x)}\right]$$
$$\equiv X^{\alpha}_{a}\frac{\delta S}{\delta X^{\alpha}(x)}+\phi_{,a}\frac{\delta S}{\delta\phi(x)}=0. \qquad (7.4)$$

The Jacobi principal function $S[X^{\alpha},\phi]$ is considered as a functional of the hypersurface variables $X^{\alpha}(x^{a})$ and the field co-ordinates $\phi(x^{a})$. If we prescribe $S[X^{\alpha},\phi]$ on an initial hypersurface $X^{\alpha}=\underset{\circ}{X}^{\alpha}(x^{a})$, we can evolve it by means of eqns. (7.3) and (7.4) to an arbitrary hypersurface.

Finally, the Hamilton–Jacobi equation for geometrodynamics takes the form

$$\mathcal{H}(x)\left[g_{ab}(x),p^{ab}(x)=\frac{\delta S}{\delta g_{ab}(x)}\right]$$
$$\equiv G_{ab\,cd}(x)[g_{mn}]\frac{\delta S}{\delta g_{ab}(x)}\frac{\delta S}{\delta g_{cd}(x)}-g^{\frac{1}{2}}R(x)=0 \qquad (7.5)$$

and

$$\mathcal{H}_{a}(x)\left[g_{bc}(x),p^{bc}(x)=\frac{\delta S}{\delta g_{bc}(x)}\right]\equiv-2\left[g_{ab}(x)\frac{\delta S}{\delta g_{bc}(x)}\right]_{|c}=0\,. \qquad (7.6)$$

The Jacobi principal function is considered here as a functional of the metric $g_{ab}(x)$. However, a different choice of variables may be more appropriate.

8. THE ALGEBRA OF CONSTRAINTS

The Poisson brackets among the constraint functions $\mathcal{H}(x)$ and $\mathcal{H}_a(x)$ are themselves linear combinations of these functions,

$$[\mathcal{H}_a(x),\mathcal{H}_b(x')] = \mathcal{H}_b(x)\delta_{,a}(x,x') - (ax \leftrightarrow bx'), \tag{8.1}$$

$$[\mathcal{H}_a(x),\mathcal{H}(x')] = \mathcal{H}(x)\delta_{,a}(x,x') , \tag{8.2}$$

$$[\mathcal{H}(x),\mathcal{H}(x')] = \mathcal{H}^a(x)\delta_{,a}(x,x') - (x \leftrightarrow x') . \tag{8.3}$$

These relations remain the same for the kinematic constraints (5.4), the constraints (6.2) of a parametrized field theory on a given background, and for the gravitational constraints (2.7), (2.9). The universal form of these relations suggests that they reflect the geometrical law of composition of hypersurface deformations in a Riemannian space-time. As the hypersurface is pushed through space-time from a given initial hypersurface to a given final hypersurface along two different families of interpolated hypersurfaces, the observed field which started in a given configuration on the initial hypersurface must always end in the same configuration on the final hypersurface irrespective of the family along which it was developed. The universal character of the relations (8.1)–(8.3) expresses this principle of path independence of the dynamical evolution. The Poisson brackets among the kinematic constraints thus dictate the form of the Poisson brackets among the constraints of a parametrized field theory or geometrodynamics. The hypersurfaces march on playing their tune like the Pied Piper; the fields follow like rats pursuing the tune.

Equation (8.1) is identical to the Lie algebra of the generators $\mathcal{H}_a$ of the group of diffeomorphisms in a three-dimensional space. Equation (8.2) says that $\mathcal{H}(x)$ transforms under such diffeomorphisms as a scalar density of weight 1. Equation (8.3) means that two infinitesimal normal deformations performed in an arbitrary order end on the same final hypersurface, but not at the same point of that hypersurface. It is the only equation which contains an explicit reference to the metric, due to the fact that $\mathcal{H}^a = g^{ab}\mathcal{H}_b$.

9. INTERCONNECTION THEOREMS

Because the constraints satisfy the Lie algebra (8.1)–(8.3), they are intimately linked together and with the dynamical evolution which they generate. The relations (8.1)–(8.3) enter into the proof of numerous interconnection theorems:

1. The constraints are preserved in the dynamical evolution: i.e., if they hold on an initial hypersurface and the canonical data are developed by the Hamilton equations to another hypersurface, the developed data satisfy the constraints on that new hypersurface.
2. If the constraints hold on an arbitrary hypersurface, the canonical data on any two hypersurfaces are necessarily connected by the Hamilton equations.
3. If the Jacobi principal functional S satisfies the super-Hamiltonian equation at a given space point $\underset{\circ}{x}$ and the supermomentum equation everywhere, it also satisfies the super-Hamiltonian equation everywhere:

$$\mathcal{H}(\underset{\circ}{x})[S] = 0, \quad H_a(x)[S] = 0 \Rightarrow H(x)[S] = 0 . \tag{9.1}$$

4. If the Jacobi principal functional S satisfies the super-Hamiltonian equation everywhere, it must satisfy the supermomentum equation everywhere:

$$\mathcal{H}(x)[S] = 0 \Rightarrow \mathcal{H}_a(x)[S] = 0 . \tag{9.2}$$

5. The geometrodynamical constraints (2.7), (2.9) are essentially the only expressions which can be constructed solely from the space metric $g_{ab}(x)$ and a conjugate momentum $p^{ab}(x)$ so that they satisfy the algebra (8.1)–(8.3).

Theorems 3 and 4 have counterparts in the quantum theory provided that the factor ordering problem has been satisfactorily resolved.

10. SUPERMETRIC

The super-Hamiltonian (3.4) of a non-relativistic particle is a parabolic function of the momenta p_α, being linear in p_0 and quadratic in p_a. On the other hand, the super-Hamiltonian of a relativistic particle (3.7) is hyperbolic in the momenta p_α, with coefficients $1/2m\ g^{\alpha\beta}$ proportional to the space-time metric.

There is a similar difference between the super-Hamiltonian of a field theory on a given background and the gravitational super-Hamiltonian. The field super-Hamiltonian (6.2), (5.4), (4.3) is parabolic in the momenta $\{P_\alpha, \pi\}$, being linear in P_α and quadratic in π. The gravitational super-Hamiltonian is hyperbolic in the momenta p^{ab}, with coefficients $G_{ab\ cd}(x)[g_{mm}]$ given by eqn. (2.8). At a given point x, these coefficients are functions of the metric $g_{ab}(x)$ at that point. When I consider the coefficients in this way, I shall call $G_{ab\ cd}(x)[g_{mn}]$ the (local) supermetric.

The analogy with a relativistic particle indicates that $G_{ab\ cd}(x)[g_{mn}]$ is to be interpreted as a contravariant supermetric. The covariant supermetric $G^{ab\ cd}(x)[g_{mn}]$ is its inverse,

$$G^{ab\ cd}(x)G_{cd\ ef}(x) = \delta^{ab}_{ef} \equiv \frac{1}{2}\left(\delta^a_e\delta^b_f + \delta^a_f\delta^b_e\right). \tag{10.1}$$

Explicitly,

$$G^{ab\ cd} = \frac{1}{2}\, g^{\frac{1}{2}}(g^{ac}g^{bd} + g^{ad}g^{bc} - 2g^{ab}g^{cd}) . \tag{10.2}$$

The expression (10.2) is a spatial tensor density of weight 1. Integrating over all space, I can then define the 'interval' between the original metric $g_{ab}(x)$ and a neighbouring metric $g_{ab}(x) + \delta g_{ab}(x)$:

$$\delta S^2 \equiv (\delta g, \delta g) \equiv \int d^3x\ G^{ab\ cd}(x)[g_{mn}]\delta g_{ab}(x)\delta g_{cd}(x). \tag{10.3}$$

When two metrics differ only by a diffeomorphism, they define the same geometry. From the interval (10.3) between two metrics DeWitt was able to define an interval between two geometries. In the following, I shall work only with the local metric (2.8) and its inverse (10.2).

11. GEOMETRODYNAMICS WITH SOURCES

Does the supermetric preserve its form when gravitation is coupled to sources? When the coupling is non-derivative, like that of a scalar or an electromagnetic field, the answer is yes. The canonical action of the coupled fields has the super-Hamiltonian and supermomentum which are simply the sums of the gravitational and the field parts. Thus, for the scalar field,

$$\mathcal{H} = \mathcal{H}^{G} + H^{\phi}, \qquad \mathcal{H}_a = \mathcal{H}^{G}_a + H^{\phi}_a \,. \tag{11.1}$$

The total super-Hamiltonian is again hyperbolic in the momenta $\{p^{ab}, \pi\}$, with the coefficients

$$\begin{array}{c|c|c} & p^{cd} & \pi \\ \hline p^{ab} & G_{ab\;cd}[g_{mn}] & 0 \\ \hline \pi & 0 & \frac{1}{2}\, g^{\frac{1}{2}} \end{array} \tag{11.2}$$

The supermetric in the extended momentum space thus has the block-diagonal form, in which the p^{ab} - p^{cd} block is the gravitational supermetric and the π - π block depends on g_{mn}, but not on the field ϕ.

The derivative coupling changes the whole picture. The supermetric in the extended momentum space does not decompose into blocks, its components depend on the source field, and even the p^{ab} - p^{cd} part of the supermetric is in general modified by the presence of the source field variables. Moreover, the super-Hamiltonian acquires terms which are linear in the momenta.

Spinor fields can be incorporated into the formalism only at an additional price of replacing the metric g_{ab} by the triad $h^{(m)}_a$ as a basic variable. In the following, I shall talk only about the gravitational field decoupled from sources.

12. CONSTRAINT QUANTIZATION

Hamiltonian geometrodynamics looked special because its Hamiltonian was a linear combination of terms which were constrained to vanish. By now, I have cast the particle dynamics and field theory on a background into the same mould. I am thus ready to discuss the quantization of canonical theories subject to such constraints. For simplicity, I shall first state the rules for systems with a finite number of degrees of freedom. Schematically, the constraint quantization proceeds in four steps:

1. The canonical variables are turned into operators satisfying the standard commutation rules on some vector space,

$$x^\alpha, p_\beta \mapsto \hat{x}^\alpha, \hat{p}_\beta : [\hat{x}^\alpha, \hat{p}_\beta] = i\delta^\alpha_\beta \ . \tag{12.1}$$

2. These operators are substituted into the super-Hamiltonian which is thereby also turned into an operator:

$$\mathcal{H}(x^\alpha, p_\beta) \longmapsto \hat{\mathcal{H}}(\hat{x}^\alpha, \hat{p}_\beta) \ . \tag{12.2}$$

The super-Hamiltonian constraint is then imposed as a restriction which selects from all states ψ those states which are physically permissible,

$$\hat{\mathcal{H}}\psi = 0 \ . \tag{12.3}$$

3. Special representation of the operators can be chosen, such that

$$\hat{x}^\alpha = x^\alpha \cdot , \qquad \hat{p}_\alpha = -i\partial_\alpha \tag{12.4}$$

and the states ψ become some functions $\psi(x^\alpha)$ of x^α.

4. The space of solutions of eqn. (12.3) must be turned into a Hilbert space and the physically significant operators identified with self-adjoint operators on this space.

It is the last step which presents the most serious difficulties in geometrodynamics. I shall approach it in steps. First, I shall discuss how to build the Hilbert space for theories with parabolic super-Hamiltonians, namely, for parametrized non-relativistic particle dynamics and parametrized field theory. Then I shall pass to the same problem in theories with hyperbolic super-Hamiltonians, namely, in parametrized relativistic particle dynamics and in geometrodynamics.

There are numerous other problems associated with the steps 1–4. Let me briefly mention at least two of them. The first one is the factor ordering problem. When the super-Hamiltonian is turned into an operator, various orders of the non-commuting variables $\hat{x}^\alpha$, $\hat{p}_\alpha$ can be chosen. The proper ordering is guided by the covariance requirements and, for geometrodynamics, by the desire to preserve the algebra of constraints (8.1)–(8.3). The factor ordering problem is connected with the Hilbert space problem, because the choice of the detailed form of the inner product depends on the factor ordering. The second problem is the positivity problem. The physical nature of some variables, like the energies $-p_0$, energy densities $-n^\alpha P_\alpha$, or spatial metrics g_{ab}, requires that these variables be represented by operators which have a positive spectrum. Obviously, this problem is again related to the choice of an appropriate inner product in the Hilbert space. In this lecture, I shall largely concentrate on the problem of constructing the Hilbert space.

13. NON-RELATIVISTIC PARTICLE DYNAMICS

My starting point is the super-Hamiltonian (3.4), (3.2). I want to write the quantum constraint so that it is covariant under diffeomorphisms in the x^a-space. One way of doing it is to choose $\psi(x^a)$ as a scalar density of weight $\frac{1}{2}$ and order $\hat{\mathcal{H}}$ so that the kinetic energy term yields the Laplacian when acting on $\psi(x^a)$:

$$\hat{\mathcal{H}} = \hat{p}_0 + \frac{1}{2}\, g^{-\frac{1}{4}}\hat{p}_a g^{\frac{1}{2}} g^{ab}\hat{p}_b g^{-\frac{1}{4}} + V\ . \qquad (13.1)$$

As a consequence, when $\underset{1}{\psi}$ and $\underset{2}{\psi}$ satisfy the constraint equation $\mathcal{H}\underset{1}{\psi} = 0 = \mathcal{H}\underset{2}{\psi}$, the expressions

$$\underset{12}{\rho} \equiv \underset{1}{\psi}{}^*\underset{2}{\psi}, \qquad \underset{12}{s}{}^a \equiv \frac{1}{2}\, i g^{ab}(\underset{1}{\psi}{}^*\overleftrightarrow{\partial}_b\underset{2}{\psi}) \qquad (13.2)$$

satisfy the continuity equation

$$\dot{\rho}_{12} + S^{a}_{12,a} = 0 . \qquad (13.3)$$

This enables me to introduce the inner product

$$\langle \psi_1, \psi_2 \rangle \equiv \int_{t=\text{const},} d^3x \; \psi_1^*(x,t)\psi_2(x,t) \qquad (13.4)$$

in the space of solutions of the constraint equation (12.3) and prove that this product does not depend on the time slice on which it is evaluated. The solutions of the constraint equation (12.3) with the finite norm (13.4) then form the Hilbert space. Note that the integration in eqn. (13.4) is not taken over the time variable t. Of course, the constraint (12.3) is nothing else but the Schrödinger equation and the inner product is the standard Schrödinger product between two wave functions.

14. FIELD THEORY ON A BACKGROUND

The super-Hamiltonian of the parameterized scalar field is parabolic in the momenta $\{P_\alpha, \pi\}$. The canonical momenta P_α which enter into it linearly have the meaning of the momentum and energy densities, while the canonically conjugate co-ordinates X^α specify the hypersurface, i.e., an instant of time. The canonical pair $\{\phi, \pi\}$ describes the field on the hypersurface $X^\alpha(x^a)$.

I shall follow steps 1–4 which are straightforward generalizations of the constraint quantization of a particle system.

1. The canonical variables are turned into operators satisfying the commutation rules

$$[\hat{X}^\alpha(x), \hat{P}_\beta(x')] = i\delta^\alpha_\beta \delta(x,x') ,$$

$$[\hat{\phi}(x), \hat{\pi}(x')] = i\delta(x,x') , \qquad (14.1)$$

all other commutators being equal to zero.

2. These operators are substituted into the super-Hamiltonian and supermomentum (6.2), (5.4), (4.3), (4.4). No factor ordering problem arises in the field Hamiltonian (4.3), because all operators in the kinetic term and all operators in the potentia

energy term commute. I order the operators $\hat{X}^{\alpha}(x)$ and $\hat{P}_{\alpha}(x)$ in the kinematic terms (5.4) so that $\hat{X}^{\alpha}$ stands to the left of $\hat{P}_{\alpha}$. With this choice, the kinematic terms acting on a functional of X^{α} in the X^{α}, ϕ-representation induce the change of this functional from the original to the deformed hypersurface. Similarly, I order the operators $\hat{\phi}$ and $\hat{\pi}$ in the momentum density (4.4) so that $\hat{\phi}$ stands to the left of $\hat{\pi}$.

3. I choose the functional representation of the operators,

$$\hat{X}^{\alpha}(x) = X^{\alpha}(x). \quad , \quad \hat{P}_{\alpha}(x) = -\, i\delta/\delta X^{\alpha}(x) \ ,$$

$$\hat{\phi}(x) = \phi(x). \quad , \quad \hat{\pi}(x) = -\, i\delta/\delta\phi(x) \ , \qquad (14.2)$$

and let them act on the functionals $\Psi[X^{\alpha},\phi]$ of the variables $X^{\alpha}(x)$ and $\phi(x)$. The constraints take the form of functional differential equations

$$\begin{aligned}\hat{\mathcal{H}}(x)\Psi \equiv &-\, in^{\alpha}[X]\frac{\delta\Psi}{\delta X^{\alpha}(x)} - \frac{1}{2}\, g^{-\frac{1}{2}}(x)[X]\frac{\delta^2\Psi}{\delta\phi(x)^2}\\ &+ \frac{1}{2}\, g^{\frac{1}{2}}(x)[X]g^{ab}(x)[X]\phi_{,a}(x)\phi_{,b}(x)\Psi\\ &+ \frac{1}{2}\, m^2 g^{\frac{1}{2}}(x)[X]\phi^2(x)\Psi = 0 \qquad (14.3)\end{aligned}$$

and

$$\hat{\mathcal{H}}_{a}(x)\Psi \equiv -\, i\left(X^{\alpha}_{a}(x)\frac{\delta\Psi}{\delta X^{\alpha}(x)} + \phi(x)\frac{\delta\Psi}{\delta\phi(x)}\right) = 0\ . \qquad (14.4)$$

The notation $n^{\alpha}[X]$, etc., emphasizes that the expressions are to be considered as appropriate functionals of the hypersurface variables $X^{\alpha}(x^{a})$.

I now face the problem of turning the space of solutions $\Psi[X^{\alpha},\phi]$ of the functional differential equations (14.3), (14.4) into a Hilbert space. Heuristically, I follow the procedure suggested by the analogy with the Schrödinger equation of a single non-relativistic particle. If the functionals Ψ_1 and Ψ_2 satisfy the constraints (14.3) and (14.4), the expressions $\rho_{12} \equiv \Psi_1^{*}\Psi_2$ and

$$S_{\alpha}(x)[X^{\beta},\phi] \equiv \frac{1}{2}\, ig^{-\frac{1}{2}}[X]n_{\alpha}[X]\Psi_1^{*}\frac{\overleftrightarrow{\delta}}{\delta\phi(x)}\Psi_2 + X^{a}_{\alpha}(x)\phi_{,a}(x)\Psi_1^{*}\Psi_2 \tag{14.5}$$

satisfy the first-order functional differential equation

$$\frac{\delta\rho_{12}}{\delta X^{\alpha}(x)} + \frac{\delta S_{\alpha}(x)[X^{\beta},\phi]}{\delta\phi(x)} = 0. \tag{14.6}$$

This is not a single equation, like the continuity equation (13.3), but an ∞^3 of equations, one for each point x^{a}. I can obtain from eqns. (14.6) a single equation by asking how $\rho_{12}[X]$ changes along a one-parameter family of hypersurfaces $X^{\alpha}(x^{a},t)$:

$$\frac{\partial\rho_{12}}{\partial t} = \int d^3x\, \dot{X}^{\alpha}(x)\frac{\delta\rho_{12}[X^{\beta}]}{\delta X^{\alpha}(x)} = -\int d^3x\, \frac{\delta}{\delta\phi(x)}\, \dot{X}^{\alpha}(x)S_{\alpha}(x)[X^{\beta},\phi]\ . \tag{14.7}$$

I integrate this equation over all field configurations $\phi(x^{a})$ on the $X^{\alpha}(x)$ hypersurface. Heuristically, the functional divergence on the right-hand side of the equation vanishes by the application of a functional version of the Gauss theorem, so that

$$\int d^3x\, \dot{X}^{\alpha}(x)\frac{\delta}{\delta X^{\alpha}(x)}\int D\phi\ \rho_{12}[X^{\beta},\phi] = 0\ . \tag{14.8}$$

Because the deformation vector $\dot{X}^{\alpha}(x)$ is arbitrary. I conclude that

$$\frac{\delta}{\delta X^{\alpha}(x)}\int_{X^{\alpha}(x^{a})\ \text{prescribed}} D\phi\ \Psi_1^{*}[X^{\beta},\phi]\Psi_2[X^{\beta},\phi] = 0\ . \tag{14.9}$$

This indicates that the functional integral in eqn. (14.9) is independent of the choice of the hypersurface $X^{\alpha}(x)$ and defines an inner product

$$\langle \Psi_1, \Psi_2 \rangle \equiv \int\limits_{X^\alpha(x^a)\ \text{prescribed}} D\phi\ \Psi_1^*[X^\beta,\phi]\Psi[X^\beta,\phi] \tag{14.10}$$

which turns the space of solutions to the constraint equations (14.3), (14.4) into a Hilbert space.

Unfortunately, this heuristic argument has serious flaws. It is difficult to give a good mathematical meaning to the terms like $\delta^2\Psi/\delta\phi(x)\,\delta\phi(x)$ in eqn. (14.3) and it is known that a Lebesgue measure $D\phi$ in an infinitely dimensional function space $\phi(x)$ does not exist. Investigations by Kay and Ashtekars indicate that the operators $\hat{\mathcal{H}}(x)$ and $\hat{\mathcal{H}}_a(x)$, or rather the smeared operators $\hat{\mathcal{H}}_N \leftrightarrow \int d^3X\ N(x)\hat{\mathcal{H}}(x)$ and $\hat{\mathcal{H}}_{\vec{N}} \leftrightarrow \int d^3x\ N^a(x)\hat{\mathcal{H}}_a(x)$, cannot all be interpreted as well-defined operators acting on a single Hilbert space. It rather seems that for each choice of the lapse and shift functions it is necessary to use a different Hilbert space to give the meaning to the operators $\hat{\mathcal{H}}_N$ and $\hat{\mathcal{H}}_{\vec{N}}$ smeared by these functions. Unfortunately, this circumstance vastly limits the heuristic power of the canonical techniques and casts serious doubt on whether such techniques can be successfully applied to quantum geometrodynamics. Still, I want to follow the discussion at the same heuristic level, comparing the relativistic particle dynamics with canonical geometrodynamics. I shall show that even at this level there are further obstacles, in addition to those of a rigorous definition of the constraint operators and of the measure, which prevent me from constructing a satisfactory Hilbert space.

15. RELATIVISTIC PARTICLE IN CURVED SPACE-TIME

I shall pass now to the constraint quantization of a relativistic particle described by the super-Hamiltonian (3.7). I want the quantum constraint to be covariant under space-time diffeomorphisms. As in the non-relativistic case, it is quite possible to work with state functions which are $\frac{1}{2}$-densities. The standard picture, however, is to choose $\phi(x^\alpha)$ to be a scalar and order the super-Hamiltonian (3.7) so that it yields the D'Alembert operator when acting on scalars,

$$\hat{\mathcal{H}} = |{}^4g|^{-\frac{1}{2}}\,\hat{p}_\alpha |{}^4g|^{\frac{1}{2}}\, g^{\alpha\beta}\hat{p}_\beta + m^2 \tag{15.1}$$

The constraint $\hat{\mathcal{H}}\phi(x^\alpha) = 0$ in the x^α-representation then takes the form of the Klein–Gordon equation. The real-valued solutions of this equation which vanish sufficiently fast at spatial infinity form the space which I want to turn into a Hilbert space.

From two solutions, $\underset{1}{\phi}$ and $\underset{2}{\phi}$, of the constraint equation I can construct the four-vector

$$\underset{12}{S}^\alpha = \frac{1}{2}\, g^{\alpha\beta}\underset{1}{\phi}\overleftrightarrow{\partial}_\beta\underset{2}{\phi} \tag{15.2}$$

which satisfies the continuity equation

$$\nabla_\alpha \underset{12}{S}^\alpha = 0\ . \tag{15.3}$$

The value of the functional

$$\Omega[\underset{1}{\phi},\underset{2}{\phi}] \equiv \frac{1}{2}\int_\sigma d\sigma_\alpha\, g^{\alpha\beta}\underset{1}{\phi}\overleftrightarrow{\partial}_\beta\underset{2}{\phi} \tag{15.4}$$

is thus independent of the space-like hypersurface $\sigma\!:\!X^\alpha(x^a)$ on which it is evaluated. Here,

$$d\sigma_\alpha = \varepsilon_{\alpha\beta\gamma\delta}X^\beta_b X^\gamma_c X^\delta_d\, dx^b\, dx^c\, dx^d \tag{15.5}$$

is the invariant surface element of the hypersurface. However, Ω is antisymmetric in its arguments, so that $\Omega[\phi,\phi] = 0$ cannot serve as a norm in the space of solutions of the constraint equation. In fact, Ω is the sympletic form in this space.

In general, the solution space does not possess a conserved inner product at all! This is just another way of saying that the generic metric field $g_{\alpha\beta}(x^\gamma)$ will produce particles so that the one-particle theory does not necessarily make good sense. The solution space can be turned into a complex Hilbert space only in stationary space-times. The consequences for geometrodynamics, if we believe in the fundamental similarity of the super-Hamiltonian (2.7), (2.8) with the relativistic particle super-Hamiltonian (3.7), are far reaching. I shall thus spend some time reviewing the familiar process by which the Hilbert space is constructed from solutions of the Klein–Gordon equation in stationary space-times.

For future analogy with quantum geometrodynamics, I can restrict myself to the discussion of static space-times. To turn the space of *real* solutions of the Klein–Gordon equation into a *complex* Hilbert space, I must construct both the complex structure and the inner product. Both tasks are achieved by separating the positive and negative frequency solutions of the Klein–Gordon equation.

Static space-times are characterized by the presence of a time-like hypersurface orthogonal Killing vector field t^α,

$$t^\alpha = N^2 g^{\alpha\beta} t_{,\beta} \ . \tag{15.6}$$

I shall choose the privileged co-ordinates $x^\alpha = \{t, x^a\}$, where t is the scalar function entering into eqn. (15.6) and the x^a co-ordinates are constant along the flowlines of the vector field t^α. The scalar field $N(x^a)$ is the lapse function between the space-like hypersurfaces t = const.; the shift vector N^a vanishes, due to the hypersurface orthogonality condition.

In the $\{t, x^a\}$ co-ordinates, the Klein–Gordon equation takes the form

$$N^2 \hat{\mathcal{H}} \phi \equiv \ddot{\phi} + \hat{H}_N \phi = 0 \ , \tag{15.7}$$

where the operator $\hat{H}_N$ is defined by the prescription

$$\hat{H}_N \phi \equiv N\left[-g^{-\frac{1}{2}}\left(N g^{\frac{1}{2}} g^{ab} \phi_{,b}\right)_{,a} + N m^2 \phi\right] . \tag{15.8}$$

I turn the space of functions $\phi(x^a)$ into an auxiliary real Hilbert space by prescribing in it the inner product

$$(\underset{1}{\phi}, \underset{2}{\phi}) \equiv \int d^3x \, g^{\frac{1}{2}} N^{-1} \underset{1}{\phi} \, \underset{2}{\phi} \ . \tag{15.9}$$

Under this inner product, the operator $\hat{H}_N$ is symmetric and positive definite and it can be extended into a self-adjoint operator. Consequently, it possesses a complete set of generalized eigenfunctions $u_E(x^a)$,

$$\hat{H}_N u_E(x^a) = E^2 u_E(x^a), \quad E \geqslant E_0 > 0 . \tag{15.10}$$

The general solution of the Klein–Gordon equation (15.7) can be expressed through the eigenfunctions $u_E(x)$ by separating time from the spatial co-ordinates,

$$\phi(x^a, t) = \phi^+(x^a, t) + \phi^-(x^a, t) , \tag{15.11}$$

where $\phi^\pm(x^a,t)$ are the positive (negative) frequency solutions

$$\phi^\pm(x^a, t) = \int_{E_0}^{\infty} \mathrm{d}E \; u_E(x^a) \exp(\mp\, \mathrm{i}Et) . \tag{15.12}$$

The complex structure J is defined so that it maps any real solution ϕ of the Klein–Gordon equation into another real solution $J\phi$ by the formula

$$J\phi \equiv \mathrm{i}\phi^+ + (-\mathrm{i})\phi^- . \tag{15.13}$$

By virtue of its construction, J does not depend on t. It is a complex structure, because

$$J^2 = -1 . \tag{15.14}$$

One says that J is compatible with Ω, because it meets the following conditions:

$$\Omega[\phi_1, J\phi_2] = \Omega[\phi_2, J\phi_1] , \tag{15.15}$$

$$\Omega[\phi, J\phi] \geqslant 0 \text{ is a positive definite form of } \phi . \tag{15.16}$$

The complex structure J gives the solution space the structure of a complex vector space. One can then define in this space the inner product

$$\langle \phi_1, \phi_2 \rangle \equiv \Omega[\phi_1, J\phi_2] + \mathrm{i}\Omega[\phi_1, \phi_2] \tag{15.17}$$

which turns it into a (complex) Hilbert space. In particular,

$$\langle \phi, \phi \rangle = \Omega[\phi, J\phi] \tag{15.18}$$

is positive definite by the condition (15.16), and it has the required linearity—antilinearity behaviour

$$\langle \phi_1, J\phi_2 \rangle = i\langle \phi_1,\phi_2 \rangle, \quad \langle J\phi_1,\phi_2 \rangle = -i\langle \phi_1,\phi_2 \rangle \tag{15.19}$$

by the conditions (15.14), (15.15).

Alternatively, I can characterize the solution $\phi(x^a,t)$ of the Klein—Gordon equation by the initial 'co-ordinate' $\phi(x^a)$ and 'momentum' $\pi(x^a)$ on a fixed hypersurface $t = t_0$ = const. The sympletic structure Ω and the complex structure J can be specified by their action on these initial data:

$$\Omega\left[\{\phi_1,\pi_1\},\{\phi_2,\pi_2\}\right] = \frac{1}{2}\int_{\sigma_{t_0}} d^3x(\phi_1\pi_2 - \phi_2\pi_1) \ , \tag{15.20}$$

$$J\{\phi,\pi\} = \{\hat{H}_N^{-\frac{1}{2}}(Ng^{-\frac{1}{2}}\pi)\, , \; - N^{-1}g^{\frac{1}{2}}\hat{H}_N^{\frac{1}{2}}\phi\}. \tag{15.21}$$

Ω is independent of the choice of σ, and thus of the choice of t. By virtue of its original construction, J is also independent of the choice of t. These properties are essential for maintaining the one-particle interpretation of the scheme, i.e., for the construction of the t-independent inner product (15.17). If σ_t is not a static foliation, the method of construction breaks down.

The second property which is vital for constructing J is the positivity of the operator $\hat{H}_N$. This is especially clear from eqn. (15.21) which characterizes J by its action on the initial data. If $\hat{H}_N$ were not a positive definite operator, I could not have defined its square roots $H_N^{\frac{1}{2}}$ and $H_N^{-\frac{1}{2}}$ by spectral decomposition. Also, the inner product (15.17) would not be positive definite.

16. SCHRÖDINGER QUANTIZATION OF A RELATIVISTIC PARTICLE

The explained construction of a complex Hilbert space can be generalized from static to stationary space-times, but not to generic space-times. The few remarks which I made indicate why such a type of construction is likely to fail in these more general cases. Is there then any way to keep the one-particle interpretation in an arbitrary space-time?

One possibility is to return to the Schrödinger quantization. This approach is not likely to be the one chosen by Nature in the case of relativistic particles. There we have experimental reasons to believe that particles are produced by fields varying with time and the metric field is not supposed to be an exception. Further, there are invariance reasons to believe that the Schrödinger approach is not correct. However, I am using the particle theory only as a model on which I can illustrate the logical possibilities which are open in canonical geometrodynamics. There, I have neither experimental nor compelling theoretical reasons to exclude the Schrödinger quantization from the very beginning. I thus continue by explaining how the Schrödinger quantization works for a finite-dimensional 'relativistic' particle model.

I take the super-Hamiltonian (3.7) of a relativistic particle moving in an arbitrary, possibly time-dependent, space-time. I choose an arbitrary foliation of space-like hypersurfaces σ_t by hand. The transition from one hypersurface in the foliation to a neighbouring one is characterized by the lapse function $N(x^a,t)$ and the shift vector $N^a(x^b,t)$. I then split the super-Hamiltonian (3.7) into the product of two factors,

$$\mathcal{H} = -\frac{1}{2mN^2}\left[p_0 - N^a p_a - N\sqrt{g^{ab}p_a p_b + m^2}\,\right]$$

$$\cdot\left[p_0 - N^a p_a + N\sqrt{g^{ab}p_a p_b + m^2}\,\right], \tag{16.1}$$

and replace the Hamiltonian constraint by the statement that the second one of these factors is equal to zero,

$$\mathcal{H}_S \equiv p_0 - N^a p_a + N\sqrt{g^{ab}p_a p_b + m^2} = 0\ . \tag{16.2}$$

Next, I try to convert $\mathcal{H}_S$ into an operator. To succeed, I must define the square root of an operator version of $N^2(g^{ab}p_a p_b + m^2)$. This I can do by spectral analysis provided the operator I start from is a positive-definite self-adjoint operator on some Hilbert space. A possible way to achieve this is to base the Hilbert space product on the volume element corresponding to the space metric $\bar{g}_{ab} = N^{-2} g_{ab}$,

$$\langle \phi_1, \phi_2 \rangle \equiv \int d^3x \ \bar{g}^{\frac{1}{2}} \phi_1^* \phi_2 \qquad (16.3)$$

and order the factors in $N^2 g^{ab} p_a p_b$ so that it becomes the Laplacian with respect to that metric,

$$N^2 (g^{ab} p_a p_b + m^2) \mapsto \bar{g}^{-\frac{1}{2}} \hat{p}_a \bar{g}^{\frac{1}{2}} \bar{g}^{ab} \hat{p}_b + m^2 N^2 \ . \qquad (16.4)$$

The operator (16.4) can be extended into a positive definite self-adjoint operator under the inner product (16.3) and its square root can thus be defined by spectral analysis. Similarly, I order the term $N^a p_a$ so that it becomes self-adjoint under the inner product (16.3),

$$N^a p_a \mapsto \frac{1}{2}(N^a \hat{p}_a + \bar{g}^{-\frac{1}{2}} \hat{p}_a \bar{g}^{\frac{1}{2}} N^a) \ , \qquad (16.5)$$

and put

$$p_0 \mapsto \bar{g}^{-\frac{1}{4}} p_0 \bar{g}^{\frac{1}{4}} \qquad (16.6)$$

which takes care of the time dependence of the volume element in the inner product (16.3). With this ordering, the inner product between two state functions which satisfy the constraint $\hat{\mathcal{H}}_S \phi = 0$ does not depend on t. Because p_0 enters into $\mathcal{H}_S$ linearly, the constraint equation is a first order differential equation in t and the inner product (16.3) has the 'Schrödinger' structure.

In the static case, I can choose the Hilbert space structure and the factor ordering of $\hat{\mathcal{H}}_S$ (different from those described above) so that the scheme is unitarily equivalent to the Klein–Gordon quantization.

The Schrödinger scheme allows me to treat hyperbolic constraints in non-stationary space-times by replacing them by the 'factored' $\mathcal{H}_S$ constraints. However, there is no obvious way in which the Schrödinger schemes based on different foliations are related. Because there is no privileged foliation in a generic space-time, the choice of the Schrödinger quantization is arbitrary. This, of course, makes it highly contrived on theoretical grounds.

17. PARTICLE MODELS WITH 'VARIABLE MASS'

I shall generalize the relativistic particle model one step further, to bring it into an even closer correspondence with geometrodynamics: I replace the mass term m^2 by an arbitrary potential $V(x^\alpha)$,

$$m^2 \mapsto V(x^\alpha) \ . \tag{17.1}$$

For a positive V, the super-Hamiltonian (3.7) describes the motion of a particle whose rest mass depends on location; a feature which occurs, e.g., in the Nordström theory of gravitation. I shall, however, not restrict myself to positive potentials; the classical trajectory of the particle does not need then to obey the light cone structure.

I quantize the motion by imposing the hyperbolic super-Hamiltonian constraint (15.1), (17.1) on the state function $\phi(x^\alpha)$. I then ask the old question: when is it possible to maintain the one-particle interpretation?

If the space-time is static and $V(x^\alpha)$ is positive and constant along the flowlines of t^α, $L_{t^\alpha}V = 0$, the complex Hilbert space with a t-independent inner product may be introduced as before. However, if V is time-dependent ($L_{t^\alpha}V \neq 0$) or such that the operator $\hat{H}_N$ introduced by eqns. (15.8), (17.1) is not positive definite, the construction breaks down and one expects particle production.

For the 'Schrödinger quantization', it does not matter whether V depends on time or not. However, if V becomes negative, the operator (16.4), (17.1) may lose its positivity and the Schrödinger quantization fails on the extraction of the square root.

The replacement of m^2 by a potential, however, opens a new route in the Klein–Gordon quantization. Because it is no longer physically necessary to preserve the term m^2 in the super-Hamiltonian as a constant, I can scale the constraint first by a factor $\Lambda^{-2}(x^\alpha)$ and then 'quantize' the scaled constraint in place of the old constraint. Now, it can happen that the scaled metric $\tilde{g}_{\alpha\beta} = \Lambda^2 g_{\alpha\beta}$ and the scaled potential $\tilde{V} = \Lambda^{-2} V$ are static,

$$L_{t\gamma}\, \tilde{g}^{\alpha\beta} = 0, \qquad L_{t\alpha}\, \tilde{V} = 0 \tag{17.2}$$

while the original metric $g_{\alpha\beta}$ and the original potential V were not. The construction of the one-particle Hilbert space can be then carried through by using the scaled variables $\tilde{g}_{\alpha\beta}$ and $\tilde{V}$. The positivity of $\hat{\tilde{\mathcal{H}}}_N$, of course, is still vital.

The conditions (17.2) expressed in terms of the old variables read

$$L_{t\gamma}\, g^{\alpha\beta} = \lambda g^{\alpha\beta}, \quad L_{t\gamma}\, V = \lambda V. \tag{17.3}$$

Therefore, if the space-time in which the particle with variable mass moves has a *conformal* Killing vector and the potential V is 'conformally constant' along its flowlines, eqn. (17.3), the one-particle interpretation of the theory can be maintained.

18. QUANTUM GEOMETRODYNAMICS: KLEIN—GORDON APPROACH

After all preparatory considerations, I turn finally to the real problem which is quantum gravity itself. The most conspicuous feature which distinguishes canonical geometrodynamics from parametrized field theories is the hyperbolic nature of its super-Hamiltonian. It is thus natural to try a Klein—Gordon quantization of this infinitely dimensional system.

The first three steps of the familiar algorithm have straightforward counterparts in canonical geometrodynamics:

1. I fix the manifold structure of the x^a-space (quantum geometrodynamics does not easily accommodate the idea of 'quantum foam'). I turn then the canonical variables g_{ab}, p^{ab} into operators satisfying the canonical commutation relations

$$[\hat{g}_{ab}(x), \hat{p}^{cd}(x')] = i\delta^{cd}_{ab}\delta(x,x') \; . \tag{18.1}$$

All other commutators are supposed to vanish.

2. I turn the super-Hamiltonian (2.7) and the supermomentum (2.9) into operators $\hat{\mathcal{H}}(x)$, $\hat{\mathcal{H}}_a(x)$ and impose the constraints

$$\hat{\mathcal{H}}(x)\Psi = 0, \quad \hat{\mathcal{H}}_a(x)\Psi = 0 \tag{18.2}$$

on the state vector Ψ. At this point, the factor ordering problem arises which I shall not attempt to resolve.

3. I choose the 'metric representation' in which Ψ becomes a functional $\Psi[g_{ab}]$ of $g_{ab}(x)$ and g_{ab}, p^{ab} are represented by the operators

$$\hat{g}_{ab}(x) = g_{ab}(x) \cdot , \quad \hat{p}^{ab}(x) = -\,i\delta/\delta g_{ab}(x) \,. \tag{18.3}$$

As in the parametrized field theory, this raises questions about the measure in the $g_{ab}(x)$-space and about different inequivalent representations of the canonical commutation relations. At this heuristic level, I shall side-step such questions.

When I write the constraints (18.2) in the g_{ab}-representation explicitly, they assume the familiar form

$$\hat{\mathcal{H}}(x)\Psi \equiv \text{"}G_{ab\ cd}(x)[g_{mn}]\frac{\delta^2\Psi}{\delta g_{ab}(x)\delta g_{cd}(x)}\text{"} - g^{\frac{1}{2}}R(x)\Psi = 0 \tag{18.4}$$

and

$$i\hat{\mathcal{H}}^a(x)\Psi \equiv \left[\frac{\delta\Psi}{\delta g_{ab}(x)}\right]_{|b} = 0 \,. \tag{18.5}$$

The quotation marks in eqn. (18.4) are inserted as a reminder that the factor ordering of the kinetic term is left unresolved On the other hand, I have decided to put the metric terms entirely on the left of the momenta in the supermomentum constraint (18.5). This is analogous to what I have done in the parametrized field theory, eqn. (14.4). With this factor ordering, the supermomentum constraint (18.5) has a simple meaning: the functional $\Psi[g_{ab}]$ should depend only on the 3-geometry $\mathcal{G}$, not on its particular representation by the metric components $g_{ab}(x^c)$ in some system of co-ordinates x^a.

The space of all positive definite 3-metrics is denoted by the symbol Riem (^{3}M). The space which is obtained by identifying all the metrics which are connected by a 3-diffeomorphism, i.e., $\mathcal{S}$ = Riem(^{3}M)/Diff(^{3}M), is called superspace. The supermomentum constraint means that the range of the state functiona

Ψ is superspace.

In order that the four constraints (18.2) do not give rise to additional constraints by commutation, the operators $\hat{\mathcal{H}}$, $\hat{\mathcal{H}}_a$ are expected to satisfy the original Dirac algebra (with Poisson brackets replaced by $\frac{1}{i}\times$ commutators), in which $\hat{\mathcal{H}}^a$ in eqn. (8.3) must come out with g^{ab} standing on the left of the supermomentum $\hat{\mathcal{H}}_b$, $\hat{\mathcal{H}}^a = g^{ab}\hat{\mathcal{H}}_b$. This imposes a strict condition on the factor ordering. Other restrictions on the factor ordering follow if one wants to preserve the constraints under transformations of 'co-ordinates' in Riem(^{3}M).

If the factor ordering is fixed so that it preserves the Dirac algebra, with $\hat{\mathcal{H}}^a = g^{ab}\hat{\mathcal{H}}_b$, the quantum constraints satisfy two interconnection theorems which are direct counterparts of the theorems 3 and 4 for the Hamilton–Jacobi function:

1. If the state vector Ψ satisfies the super-Hamiltonian constraint at a given space point $\underset{\circ}{x}$ and the supermomentum constraint everywhere, it also satisfies the super-Hamiltonian constraint everywhere:

$$\hat{\mathcal{H}}(\underset{\circ}{x})\Psi = 0, \quad \hat{\mathcal{H}}_a(x)\Psi = 0 \Rightarrow \hat{\mathcal{H}}(x)\Psi = 0 .$$

2. If the state vector Ψ satisfies the super Hamiltonian constraint everywhere, it must satisfy the supermomentum constraint everywhere:

$$\hat{\mathcal{H}}(x)\Psi = 0 \Rightarrow \hat{\mathcal{H}}_a(x)\Psi = 0 .$$

The quantum constraints thus form a tightly knit structure. The super-Hamiltonian constraint (18.4) is often called the Wheeler–DeWitt equation. As in the parametrized field theory, it is really a variational equation, i.e., '∞^3' equations, one per each space point x^a.

I now face the really difficult step in the quantization programme, namely, the question whether the space of functionals $\Psi[\mathcal{G}]$ which satisfy the Wheeler–DeWitt equation can be turned, at least heuristically, into a Hilber space. The negative answer has far more serious consequences than in the relativistic particle theory. There, if I fail to construct a one-particle Hilbert space, I can blame the failure on 'particle production' and reinterpret the Klein–Gordon equation in the context of the many-particle ('second-quantized'), Klein–Gordon field theory. However, the Wheeler–DeWitt equation already stands for a 'second-quantized' theory of massless particles of spin 2, similarly as quantum electrodynamics stands for a 'second-quantized' theory of massless particles of spin 1. Gravitons, whatever this term may mean, ought to be already produced and annihilated according to eqn. (18.4). Any attempt to resolve the Hilbert space problem by interpreting the state functional $\Psi[\mathcal{G}]$ as an *operator* would amount to something resembling a 'third quantization' and would represent a radical departure from the conventional wisdom of ordinary quantum field theory. This is a good reason for viewing the failure in constructing a Hilbert space from the solutions $\Psi[\mathcal{G}]$ of eqn. (18.4) as the failure of eqn. (18.4) itself.

How far does the procedure used for constructing a complex Hilbert space out of solutions of the Klein–Gordon equation work in geometrodynamics? At the beginning, I can certainly view the solutions $\Psi[g_{ab}]$ to the linear eqns. (18.4), (18.5) as a real vector space. Can I define on this vector space an analogue of the 'symplectic form' (15.4)? From two solutions, $\underset{1}{\Psi}$ and $\underset{2}{\Psi}$ of eqns. (18.4), (18.5) I can construct the analogue of the current vector (15.2), namely,

$$\underset{12}{S}{}_{ab}(x)[\underset{1}{\Psi},\underset{2}{\Psi}] = \frac{1}{2}\, G_{ab\;cd}(x)\underset{1}{\Psi}[\Phi]\frac{\overleftrightarrow{\delta}}{\delta g_{cd}(x)}\underset{2}{\Psi}[\Phi] \ . \tag{18.6}$$

I can argue that, modulo the factor ordering difficulties, this current equation satisfies an 'equation of continuity'

$$\frac{\nabla}{\delta g_{ab}(x)}\underset{12}{S}{}_{ab}(x) = 0 \ .$$

The symbol $\nabla/\delta g_{ab}(x)$ denotes the covariant variational derivative in the $g_{ab}(x)$-space with respect to the supermetric $G_{ab\;cd}(x)[g_{mn}]$. I then choose a 'spacelike hypersurface in Riem(^{3}M)', by putting

$$g_{ab}(x) = g_{ab}(x)[g_A(y)] \ , \tag{18.7}$$

where $g_A(y^c)$, $A = 1, \ldots, 5$, are arbitrary function 'co-ordinat on that hypersurface, and define the surface area $D\Sigma^{ab}(x)$ of the hypersurface by an analogue of eqn. (15.5). Finally, by reasoning similar to that used in the 'proof' of eqn. (14.9), I can argue that the functional integral

$$\Omega[\underset{1}{\Psi},\underset{2}{\Psi}] \equiv \prod_{x} \int_{\Sigma} D\Sigma^{ab}(x) \underset{12}{S}{}_{ab}(x)[\underset{1}{\Psi},\underset{2}{\Psi}] \tag{18.8}$$

does not depend on the choice of the hypersurface.

The next question is whether the supermetric $G_{ab\ cd}(x)[g_{mn}]$ has a 'conformal time-like Killing vector'. I shall look for such a 'supervector' belonging to the supermetric at a fixed point x^a. The 'supervector' is represented by a symmetric tensor $t_{mn}(x)[g_{pq}(x)]$ constructed from the metric at that point. It must satisfy the conformal Killing equation

$$L_{t_{mn}} G_{ab\ cd} = \frac{\partial G_{ab\ cd}}{\partial g_{mn}} t_{mn} - G_{ab\ mn} \frac{\partial t_{cd}}{\partial g_{mn}}$$

$$- G_{mn\ cd} \frac{\partial t_{ab}}{\partial g_{mn}} = \lambda G_{ab\ cd} \,. \tag{18.9}$$

One can easily see that this equation is fulfilled, with $\lambda = -3/2$, for $t_{mn} \sim g_{mn}$. This t_{mn} is the Killing vector of the scaled metric

$$\begin{aligned} \tilde{G}_{ab\ cd} &\equiv g^{\frac{1}{2}} G_{ab\ cd} \,, \\ L_{t_{mn}} \tilde{G}_{ab\ cd} &= 0 \,. \end{aligned} \tag{18.10}$$

The scaled metric belongs to the scaled super-Hamiltonian

$$\tilde{\mathcal{H}} = g^{\frac{1}{2}} \mathcal{H} \tag{18.11}$$

which is a scalar density of weight 2. Because

$$\tilde{G}^{ab\ cd} g_{ab} g_{cd} = -6$$

the Killing vector $t_{ab} = g_{ab}$ is 'time-like'. It is handy to normalize t_{ab} so that the normalized vector

$$\tilde{t}_{mn} = \frac{1}{\sqrt{6}} g_{mn} \tag{18.12}$$

has the norm -1. The Killing vector (18.12) is hypersurface orthogonal, because

$$\tilde{t}^{ab} \equiv \tilde{G}^{ab\ cd} \tilde{t}_{cd} = -\frac{\partial T}{\partial g_{ab}} , \tag{18.13}$$

with $T \equiv 2\sqrt{2/3}\ \ell g\ g^{\frac{1}{2}}$. I shall introduce T and the conformal metric $\gamma_{ab} \equiv g^{-1/3} g_{ab}$ as privileged co-ordinates in Riem(^{3}M) (any function of γ_{ab} would also serve the same purpose). From the condition $\gamma \equiv \det\|\gamma_{ab}\| = 1$ I get

$$\tilde{t}^{mn} \delta\gamma_{mn} = 0 , \tag{18.14}$$

which means that the γ_{ab} 'co-ordinates' are preserved along the 'flowlines' of $\tilde{t}^{mn}$. On the other hand, keeping γ_{ab} fixed and varying T yields

$$\tilde{t}_{mn} = \partial g_{mn}(T, \gamma_{ab})/\partial T . \tag{18.15}$$

The momenta P, π^{ab} conjugate to the new co-ordinates T, γ_{ab} are

$$P = \frac{1}{\sqrt{6}} p, \quad \pi^{ab} = g^{1/3}(p^{ab} - \frac{1}{3}\, p g^{ab}) . \tag{18.16}$$

The kinetic term of rescaled super-Hamiltonian becomes

$$\tilde{G}_{ab\ cd} p^{ab} p^{cd} = -P^2 + \pi^{mn}\pi_{mn} , \tag{18.17}$$

where π_{mn} has its indices lowered by γ_{mn}. Because $\pi^{mn}\pi_{mn}$ is positive definite, eqn. (18.17) reveals that the supermetric is indeed hyperbolic, with signature (-,+,+,+,+,+).

The second condition needed for the construction of J is the existence of a positive time-independent rescaled potential. It is here that the method fails. Using the familiar formula for the conformal transformation of the curvature scalar,

$$g_{ab} = \Phi^4 \gamma_{ab} \Rightarrow R[g_{mn}] = \Phi^{-4} R[\gamma_{mn}] - 8\Phi^{-5}\Delta_\gamma \Phi , \tag{18.18}$$

I see that the rescaled potential term

$$-gR[g_{mn}] = -\exp(\sqrt{2/3}T)R[\gamma_{mn}] + 8 \exp\left(\frac{7}{4\sqrt{6}}T\right)\Delta_{\gamma} \exp\left(\frac{1}{4\sqrt{6}}T\right) \tag{18.19}$$

depends on the variable T. Moreover, there is no reason why the potential (18.19) could not be negative.

One can argue that looking for a conformal Killing vector of *local* supermetric is to take an oversimplified view of the analogy between geometrodynamics and the Klein–Gordon theory. Because the gravitational super-Hamiltonian is a function of x^a, it can be smeared by different x^a-dependent functionals of g_{mn} and the search for conformal Killing vectors carried on a non-local level. I can only conclude that a straightforward search for a local conformal Killing vector of the supermetric does not lead to the desired result and that nobody has as yet succeeded in turning the solution space of the Wheeler–DeWitt equation into a complex Hilbert space by another method.

19. QUANTUM GEOMETRODYNAMICS: SCHRÖDINGER'S APPROACH

Because of the difficulties in constructing a Klein–Gordon type inner product in geometrodynamics, many people have adopted the Schrödinger type approach when dealing with specific mini-superspace models. While the explicit time dependence is no objection to this method, the indefinite character of the gravitational potential $\sim R[g_{mn}]$ prevents us from pursuing the Schrödinger quantization of generic space-times. The Schrödinger approach is not the universal remedy to the Hilbert space problem.

20. QUANTUM GRAVITY: EXTRINSIC TIME APPROACH

The trouble with both the Klein–Gordon and the Schrödinger approach to quantum geometrodynamics stems from taking the hyperbolic structure of the super-Hamiltonian too seriously. The hyperbolic structure suggests that $P(x)$ may be interpreted as an 'energy-type' variable and the conjugate quantity T as a 'time-type' variable. Comparison with the parametrized field theories points, however, in an entirely different direction: the energy-type variables enter into the constraints linearly. This suggests that the 'energy-type' variables are likely to be constructed from the metric and hidden in the $g^{\frac{1}{2}}R$ term, rather than being associated with the momenta p^{ab}. Conversely, the 'time-type' variable, being conjugate to the 'energy-type' variable, must be constructed from p^{ab}. Because the momentum p^{ab} is essentially the extrinsic curvature, I shall call this kind of the time variable 'the extrinsic time'.

The extrinsic time approach works extremely well in the linearized Einstein theory and for special wave-type solutions of the Einstein equations. The general features of the procedure may be described as follows:

1. One performs a canonical transformation

$$g_{ab}(x),\ p^{ab}(x) \mapsto X^{\alpha}(x),\ P_{\alpha}(x);\ g_{A}(x),\ p^{A}(x)\ ,$$

$$\alpha = 0,1,2,3;\quad A = 1,2 \tag{20.1}$$

from the old variables g_{ab}, p^{ab} to a suitably chosen set of $2.4\ \infty^3 + 2.2\ \infty^3$ new variables.

2. In the new variables, the constraints can be naturally written in the form

$$P_{\alpha} + H_{\alpha}\left[X^{\beta}, g_{A}, p^{A}\right] = 0\ . \tag{20.2}$$

without the use of '$\surd$' or similar operations.

3. The new conjugate variables are turned into operators satisfying the Dirac commutation relations. These operators are substituted into the new constraints (20.2) which are then imposed as restrictions on the state vector. The choice of the variables is geared to the use of the X^{α}, g_A-representation.

The criterion for the split (20.1) is the nebulous idea of 'simplicity'. One desired feature is the positivity of the 'energy density' $-P_0$. Another desideratum, probably much too good to be realizable, is the requirement that if I impose the co-ordinate conditions $X^a = x^a$, $X^0 = t$, H_{α} shall not depend on x^a and t.

The only sufficiently analysed general split of canonical

variables along the lines (20.1) was introduced by York. I shall discuss only its simplified version, when $X^0(x) \equiv \tau(x)$ is split from the rest of the variables, but X^a and g_A are left unseparated. The proposal then reduces essentially to the reversal of the role of the T and P variables which I have introduced in eqns. (18.13) and (18.16). In particular, I put

$$\tau = \frac{2}{3}\, g^{-\frac{1}{2}} p\,, \qquad -\pi_\tau = g^{\frac{1}{2}} > 0\ , \tag{20.3}$$

leaving the rest of the canonical variables γ_{ab}, π^{ab} unchanged. The energy density type variable is positive. Using eqn. (18.18), with $\Phi = g^{1/12} = (-\pi_\tau)^{1/6}$, I can write the super-Hamiltonian constraint in the form

$$g^{-1/12}\mathcal{H} \equiv \Delta_\gamma \Phi - \frac{1}{8}\, R[\gamma]\Phi + \frac{1}{8}\, \pi^{mn}\pi_{mn}\Phi^{-7} - \frac{3}{64}\tau^2\Phi^5 = 0\ . \tag{20.4}$$

One virtue of eqn. (20.4) is that it is a quasi-linear elliptic equation for Φ and the general theory of such equations is well developed. Roughly speaking, in a closed space, if one stays away from maximal slices $\tau = 0$ and has at least some non-vanishing $\pi^{ab}\pi_{ab}$, there is always a solution Φ of this equation with no curvature restrictions on the metric. This allows me to replace the super-Hamiltonian constraint by the statement

$$-\, g^{\frac{1}{2}}(x) + g^{\frac{1}{2}}(x)[\tau;\gamma_{ab},\pi^{ab}] = 0\ , \tag{20.5}$$

where $g^{\frac{1}{2}}(x)[\tau;\gamma_{ab},\pi^{ab}]$ is defined implicitly by the solution Φ of eqn. (20.4). The quantum constraint then takes the form of the Schrödinger equation

$$i\,\frac{\delta\Psi[\tau;\gamma_{ab}]}{\delta\tau(x)} = \hat{g}^{\frac{1}{2}}(x)[\tau;\hat{\gamma}_{ab},\hat{\pi}^{ab}]\Psi\ . \tag{20.6}$$

The explained version does not do full justice to York's theory, which handles the supermomentum constraint together with the super-Hamiltonian constraint and, on τ = const. hypersurfaces, decouples the solutions of these problems. The main disadvantage from the point of view of the quantum theory remains, unfortunately, the same. It is not known how to handle the Hamiltonian density $\hat{g}^{\frac{1}{2}}(x)[\tau;\hat{\gamma}_{ab},\hat{\pi}^{ab}]$ which is defined only implicitly through the solution of a complicated partial differential equation. In particular, it is difficult to say how to order the $\hat{\gamma}_{ab}$ and $\hat{\pi}^{ab}$ operators so that $\hat{g}^{\frac{1}{2}}(x)$ becomes self-adjoint. For such reasons, the extrinsic time approach to quantum gravity remains incomplete, like the other approaches which I discussed before.

21. CONCLUSIONS

I described some basic difficulties encountered in quantizing Einstein's theory of gravitation by canonical methods. First, I cast the theory into a canonical form. However, it was not quite an ordinary canonical form. The Hamiltonian of the theory was constrained to vanish. I showed then that ordinary systems – particles and fields – can be discussed in a similar language when time and energy, and possibly space co-ordinates and field momentum, are adjoined to the rest of the canonical variables. I explained how to quantize these simpler theories by what is known as the Dirac constraint method. However, when I applied the same method to canonical geometrodynamics, I did not entirely succeed in making the scheme work. What is the underlying reason?

Of course, I can blame the ultimate failure on the complexity of the gravitational field equations and excuse it by saying that, after all, no other method has succeeded in quantizing gravity. However, this does not meet the question. Why should one then try different methods at all? When success is not the criterion, one must find the justification in failure. Different methods fail for different reasons; by studying the reasons, we learn what problems should be given special attention.

I believe that the main reason why Dirac's constraint quantization cannot be carried too far in canonical geometrodynamics is that it does not include a general prescription for separating kinematic variables (time variables in mechanics, hypersurface variables in field theory) from the dynamical variables ('true degrees of freedom') and, based on this split, construct a conserved inner product. Where we have a privileged time, as in classical mechanics, relativistic particle theory in flat space-time, or free field theory in flat space-time, we know how to build the Hilbert space. When we are not sure what to use as time, as in relativistic particle theory and field theory in curved space-time, or in geometrodynamics, we lose our way. In the relativistic particle theory, we can shift the problem. Being unable to define the one-particle Hilbert space, we blame the problem on the particle production. This brings us to a field theory which incorporates particle production; however, the problem reappears in a more sophisticated form. The field Hilbert space seems to depend on the choice of the hypersurface and, still worse, on the choice of the lapse and shift functions.

The problem thus seems to stem from the fundamentally different manner in which relativity and quantum theory view the concept of time. Relativity insists that the choice of time is essentially arbitrary and should not influence the physical predictions of the theory. Quantum mechanics, on the other hand, knows how to make unique predictions only when time is separated from the rest of the variables and treated as a *c*-number. This discrepancy was clearly perceived by the founders of the modern quantum mechanics. I want to quote Schrödinger and von Neumann in this context.

Schrödinger frequently returned to the problem of time in quantum mechananics in his writings. The passage which I selected is from his article 'Die gegenwärtige Situation in der Quantenmechanik' in *Die Naturwissenschaften* from the year 1935.

> Daß die zeitlich scharfe Voraussage ein Mißgriff ist, ist auch aus anderen Gründen Wahrscheinlich. Die Maßzahl der Zeit ist wie jede andere das Resultat einer Beobachtung. Darf man gerade der Messung an einer Uhr eine Ausnahmestellung einräumen? Soll sie sich nicht wie jede andere auf eine Variable Beziehen, die im allgemeinen keinen scharfen Wert hat und ihn jedenfalls nicht zugleich mit *jeder* anderen Variablen haben kann? Wenn man den Wert einer *anderen* für einen bestimmten *Zeitpunkt* voraussagt, muß man nicht befürchten, daß beide zugleich gar nicht scharf bekannt sein können? Innerhalb der heutigen Q.M. läßt sich der Befürchtung kaum recht nachgeben. Denn die Zeit wird a priori als dauernd genau bekannt angesehen, obwohl man sich sagen müßte, daß jedes Auf-die-Uhr-Sehen den

Fortschritt der Uhr in unkontrollierbarer Weise stört. Ich möchte wiederholen, daß wir eine Q.M., deren Aussagen *nicht* für scharf bestimmte Zeitpunkte gelten sollen, nicht besitzen. Mir scheint, daß dieser Mangel sich gerade in jenen Antinomien kundgibt. Womit ich nicht sagen will, daß es der einzige Mangel ist, der sich in ihnen kundgibt.

Those of you who are inclined to dismiss Schrödinger as a perpetually dissatisfied critic of the conceptual foundations of quantum mechanics, I want to remind that von Neumann (1955) makes essentially the same point in the *Mathematical Foundations of Quantum Mechanics*:

Second, it is to be noted ... that we have repeatedly shown that a measurement ... must be instantaneous.... This is now questionable in principle, because it is well known that there is a quantity which, in classical mechanics, is canonically conjugate with the time: the energy. Therefore, it is to be expected that for canonically conjugate pair time-energy, there must exist indeterminacy relations similar to those of the pair cartesian coordinate-momentum.

After elaborating on the nature of the time-energy uncertainty relation, von Neumann continues:

First of all we must admit that this objection points at an essential weakness which is, in fact, the chief weakness of quantum mechanics: its non-relativistic character, which distinguishes the time t from the three space coordinates x, y, z, and presupposes an objective simultaneity concept. In fact, while all other quantities (especially those x, y, z closely connected with t by the Lorentz transformation) are represented by operators, there corresponds to the time an ordinary number-parameter t, just as in classical mechanics.... It may be connected with this non-relativistic character of quantum mechanics that we can ignore the natural law of minimum duration of the measurements. This might be a clarification, but not a happy one!

Von Neumann ascribed the difficulties to the non-relativistic character of quantum mechanics. However, as anyone familiar with the extensive literature on position and time operators in relativistic quantum mechanics realizes, the relativistic theory really did not escape these problems. If anything, they became more puzzling than before. Moreover, as I have remarked earlier, the transition from one-particle quantum mechanics to field theory does not help either. The problem of time merely reappears in a different guise. The ultimate irony, of course, is that general relativity, supposedly the most 'relativistic' theory in our arsenal, encounters the very same difficulty when subjected to the canonical quantization.

I thus believe that the main problem which we face in an attempt to build quantum geometrodynamics is not a technical problem, but a conceptual one. It consists in reconciling the diametrically opposite ways in which relativity and quantum mechanics view the concept of time. What are the options we have in this dilemma? I shall present you with three suggestions without stating my preferences.

1. Find a privileged structure in the constraints which would allow a natural separation of time from the dynamical variables. In this lecture, I reviewed several unsuccessful or at least incomplete attempts in that direction.

2. Accept the fact that the prominent features of classical geometrodynamics, in particular, the clear separation of the multipliers from the intrinsic and extrinsic geometry of the slices, and the fundamental mingling of an internal time with the dynamical variables, are an illusion which cannot be carried over into the quantum theory. Try to get used to the idea that different choices of time are possible, but that they lead to inequivalent quantum theories.

3. Blame the present day quantum mechanics for the troubles. Try to put the dynamical variables and time on an equal footing in quantum theory, broadening simultaneously von Neumann's theory of measurement.

22. A GUIDE TO LITERATURE

Canonical geometrodynamics

General canonical theory of systems with constraints has been introduced by Dirac (1950). In the late sixties, Dirac (1958, 1959) applied it to the gravitational field. At the same time, Arnowitt, Deser, and Misner (ADM) developed the theory in great detail. They discussed the geometrical meaning of the lapse-shift decomposition and introduced the concepts of gravitational energy and momentum through 'deparametrization'. A summary of their classical work with references to the original papers is to be found in ADM (1962). The formalism which I use emphasizes the idea of embedding independent of the choice of special coordinates. For details, see Kuchař (1972, 1976, 1977). Alternative reviews of the generalized canonical theory of gravitation may be found in Wheeler (1964), Dirac (1965), Kundt (1966), DeWitt (1967), Brill and Gowdy (1970), Misner, Thorne, and Wheeler (1973), MacCallum (1975), Hanson, Regge, and Teitelboim (1976), Fischer and Marsden (1979), Isenberg and Nester (1979), Bergman and Komar (1980), Teitelboim (1980).

The boundary terms in eqn. (2.5) are written explicitly in DeWitt (1967) or Kuchař (1977). The canonical action (2.6)–(2.9) holds in compact space manifolds. The role of boundary terms and the appropriate form of the action in asymptotically flat space-times is discussed in Regge and Teitelboim (1974).

Parametrized particle dynamics

Parametrized particle dynamics is discussed, e.g., by Synge (1960), or Lanczos (1970). Comparison with general relativity is drawn in Kuchař (1973) and in Hanson, Regge, and Teitelboim (1976).

Field theory on a background

The space-time action of general tensor fields is cast into the ADM canonical form in Kuchař (1976*c*).

Kinematics of deformations

This discussion of deformations is based on Kuchař (1978).

Parametrized field theory

Parametrized field theories with non-derivative coupling are discussed in Dirac (1965). The heuristic treatment which I give follows Kuchar (1978). A consistent derivation of the action (6.1) by 'parametrization' is given in Kuchař (1976), Section 7.

The Hamilton–Jacobi equation

Careful discussion of the Hamilton–Jacobi theory is given, e.g. in Lanczos (1970). The Hamilton–Jacobi equation for geometrodynamics was introduced by Peres (1962). Its role in the WKB approximation to quantum geometrodynamics was discussed by Gerlach (1969) and DeWitt (1967).

The algebra of constraints

The Poisson brackets among the geometrodynamical constraints were derived by Dirac (1965); those among the kinematical constraints or constraints in a parametrized field theory by Dirac (1965). The underlying geometrical composition of hypersurface deformations was noticed by Teitelboim (1973). The algebra (8.1)–(8.3) was interpreted as the Lie algebra of a non-holonomic basis in hyperspace in Kuchař (1976*a*). The Lie algebra of the generators $\mathcal{H}_a$ of the group of diffeomorphisms was analysed by DeWitt (1964).

Interconnection theorems

Theorem 1 was proved in Dirac (1965). An easy way to prove 2 is through the connection of $G_{\perp\perp}$ and $G_{\perp a}$ with the constraints and of G_{ab} with the dynamical equations; Kuchař (1976*b*), Section 8. Theorems 3 and 4 and their quantum counterparts were proved by Moncrief (1972), Teitelboim (1973), Moncrief and Teitelboim (1972). Theorem 5, in versions of increasing generality, was proved by Hojman, Kuchař, and Teitelboim (1973, 1976) and by Kuchař (1973).

Supermetric

The local supermetric was introduced by DeWitt (1968) and its 'nested' structure analysed by means of geodesic equations. Later, DeWitt (1970) passed from the local supermetric to the supermetric in superspace. This concept was further elaborated on by Gowdy (1970). Misner (1972) used a conformally scaled supermetric extensively in the study of homogeneous cosmologies. Parabolic super-Hamiltonians were cast into hyperbolic super-Hamiltonians for several models by Kuchař (1978).

Geometrodynamics with sources

A general theory of tensor fields with non-derivative gravitational coupling was developed by Kuchař (1977). Here, the mixing of the field and metric variables in the extended supermetric is analysed. For the use of tetrad formalism in canonical geometrodynamics, see, e.g., Deser and Isham (1976). Nelson and Teitelboim (1977), Geheniau and Henneaux (1977), and

Henneaux (1978) show how to accommodate spinor fields in the generalized canonical formalism.

Constraint quantization

The general strategy of the constraint quantization was developed by Dirac (1965). For the approach to quantum geometrodynamics using the analogy with parametrized theories, see Kuchař (1973). For constraint quantization of other models, see Hanson, Regge, and Teitelboim (1976). The factor ordering problem in quantum geometrodynamics is discussed by Dirac (1965), Anderson (1959, 1963), Schwinger (1963), Dirac (1969), DeWitt (1967), Komar (1979). The positivity problem was pointed out by Klauder (1970) and analysed by him on simplified models. Klauder's method was applied to the super-local version of geometrodynamics by Pilati (1980). Misner (1972) and his collaborators often handled the positivity problem in mini-superspace models by an 'exponential representation'; for a survey, see, e.g., Ryan (1972). Isham (1976) suggested that the positivity problem can be removed by the use of triads as basic variables.

Non-relativistic particle dynamics

Schrödinger's equation in curvilinear co-ordinates and the factor ordering appropriate for scalar wave functions is discussed by Pauli (1958). The advantage of ½-densities over scalars is that the volume element does not appear in the inner product (13.4) and $p_0 \mapsto \hat{p}_0$ in eqn. (13.1). Compare with the scalar wave function ordering in eqns. (16.3)–(16.6) which was chosen for correspondence with the Klein–Gordon quantization.

Field theory on a background

The construction of the variational continuity equation (14.6) and of the inner product (14.10) follows Kuchař (1973), where it is carried out for the 1 + 1 dimensional Minkowskian background. The necessity of using a different Hilbert space for each hypersurface and each choice of the lapse and shift follows from the work of Kay (1977) and Ashtekar and Magnon-Ashtekar (1975).

Relativistic particles in curved space-time

My discussion of how to turn the space of real solutions of the Klein–Gordon constraint into a complex Hilbert space follows the method developed by Magnon-Ashtekar (1976). The generalization to stationary background and the discussion of difficulties posed by non-stationary space-times is to be found in the same reference.

Schrödinger quantization of a relativistic particle

The discussion follows unpublished 1975 notes by the author.

Particle models with 'variable mass'

The difficulty in maintaining the one-particle interpretation even in a static space-time when the operator $\hat{H}_N$ becomes indefinite was analysed for a conformally coupled scalar particle by Ashtekar and Magnon-Ashtekar (1978). The idea that the constraints can be scaled and attention paid to their conformal structure comes from Misner (1972).

Quantum geometrodynamics: Klein–Gordon approach

The Klein–Gordon approach to quantum geometrodynamics based on the superspace representation was advocated by Wheeler (1962, 1964, 1968) and vigorously pursued by his students and associates. For historical details, see DeWitt (1967, 1970), Misner (1972). Quantum geometrodynamics is reviewed in DeWitt (1967), Brill and Gowdy (1970), Kuchař (1973), Isham (1972, 1975, 1976). The testing ground of quantum geometrodynamics was provided by mini-superspace models; this development is reviewed, with extensive references to original articles, in Ryan (1972) and MacCallum (1975).

On 'quantum foam', see Wheeler (1964, 1967). On factor ordering, see the papers quoted under *Field theory on a background*. On inequivalent representations of the canonical commutation relations (18.1) and the measure problem, see Isham (1976). The relation between co-ordinatization independence and the supermomentum constraint was discovered by Misner (1957) and Higgs (1959). Wheeler's concept of superspace was put on a formal footing by Fischer (1970). The interconnection theorems 1 and 2 between quantum constraints were discovered by Moncrief (1972), Teitelboim (1973), Moncrief and Teitelboim (1972). The 'symplectic form' (18.8) and the argument that it does not depend on the choice of hypersurface in Riem(^{3}M) was given by DeWitt (1967). The existence of the conformal Killing vector t_{ab} and of the associated time variable T also follows from the 'canonical nested form' of the supermetric given by DeWitt (1967).

Quantum geometrodynamics: Schrödinger's approach

Schrödinger quantization was tried on several mini-superspace models. Blyth and Isham (1975) are especially clear in discussing its relation to other approaches.

Quantum gravity: extrinsic time approach

Extrinsic time was first used in the linearized theory by Arnowitt, Deser, and Misner (1962). Its advantages over the intrinsic time were analysed by Kuchař (1970). Its use enabled Kuchar (1971, 1973) to quantize cylindrical gravitational waves in a midi-superspace fashion. General splitting of the hypersurface variables and momenta from the dynamical variables is discussed in Kuchař (1972). York and his collaborators designed a powerful algorithm for solving the initial value constraints on hypersurfaces of a constant extrinsic time. A recent summary of these techniques with references to original papers can be found in Choquet-Bruhat and York (1980).

REFERENCES

Anderson, J. (1959). *Phys. Rev.* **114**, 1182.
—— (1963). In *Proc. First East. Theor. Phys. Conf., Univ. Virginia, 1962*, ed M.E. Rose. Gordon and Breach, New York.
Arnowitt, R., Deser, S., and Misner, C.W. (1962). In *Gravitation: an introduction to current research*, ed L. Witten. John Wiley, New York.
Ashtekar, A. and Magnon, A. (1975). *Proc. R. Soc. (Lond.)*, **A346**, 375.
—— and Magnon-Ashtekar, A. (1975). *C.R. Acad. Sci. (Paris)*, **281**, 875.
—— —— (1978). *C.R. Acad. Sci. (Paris)*, **286**, 531.
Bergman, P. and Komar, A. (1980). In *Einstein centenary volume*, ed P. Bergman, J. Goldberg, and A. Held. Plenum, New York.
Blyth, W. and Isham, C. (1975). *Phys. Rev.*, **D11**, 768.
Brill, D. and Gowdy, R.H. (1970). *Rep. Prog. Phys.*, **33**, 413.
Choquet-Bruhat, Y. and York, J.W. (1980). In *Einstein centenary volume*, ed P. Bergman, J. Goldberg, and A. Held. Plenum, New York.
Deser, S. and Isham, C.J. (1976). *Phys. Rev.*, **D14**, 2505.

DeWitt, B.S. (1964). In *Relativity, groups and topology*, ed C. DeWitt and B. DeWitt. Gordon and Breach, New York and London.
—— (1968). *Phys. Rev.*, **160**, 1113.
—— (1970). In *Relativity*, ed M. Carmeli, S. Fickler, and L. Witten. Plenum, New York.
—— (1970). *GRG*, **1**, 181.
Dirac, P.A.M. (1950). *Can. J. Math.*, **2**, 129.
—— (1958). *Proc. R. Soc. (Lond.)*, **A246**, 326; **A246**, 333.
—— (1959). *Phys. Rev.*, **114**, 924.
—— (1965). *Lectures on quantum mechanics*. Academic Press, New York.
—— (1969). In *Contemporary Physics: Trieste Symposium 1968*. IAEA, Vienna.
Fischer, A.E. (1970). In *Relativity*, ed M. Carmeli, S. Fickler, and L. Witten. Plenum, New York.
—— and Marsden, J.E. (1979). In *General relativity – an Einstein centenary survey*, ed S.W. Hawking and W. Israel. Cambridge University Press, Cambridge.
Geheniau, J. and Henneaux, M. (1977). *GRG*, **8**, 611.
Gerlach, U.H. (1969). *Phys. Rev.*, **117**, 1929.
Gowdy, R.M. (1970). *Phys. Rev.*, **D2**, 2774.
Hanson, A.J., Regge, T., and Teitelboim, C. (1976). *Constrained Hamiltonian systems*. Academia Nationale dei Lincei, Roma.
Henneaux, M. (1978). *GRG*, **9**, 1031.
Higgs, P.W. (1959). *Phys. Rev.* Lett. **1**, 373.
Isenberg, J. and Nester, J. (1980). In *Einstein centenary volume*, ed P. Bergman, J. Goldberg, and A. Held. Plenum, New York.
Isham, C. (1972). ICTP/72/8 Imperial College Preprint of the lecture notes of the 1972 Boston Conference.
—— (1975). In *Quantum gravity: an Oxford symposium*, ed C.J. Isham, R. Penrose, and D.W. Sciama. Clarendon Press, Oxford.
—— (1976). *Proc. R. Soc. (Lond.)*, **A351**, 209.
Kay, B. (1977). Ph.D. Thesis, London. The relevant results are only partially covered in (1978). *Comm. Math. Phys.*, **62**, 55.
Klauder, J. (1970). In *Relativity*, ed M. Carmeli, S. Fickler, and L. Witten. Plenum, New York.
Komar, A. (1979). *Phys. Rev.*, **D20**, 830.
Kuchař, K. (1970). *J. Math. Phys.*, **11**, 3322.
—— (1972). *J. Math. Phys.*, **13**, 768.
—— (1973). In *Relativity, astrophysics and cosmology*, ed W. Israel. Reidel, Dordrecht.
—— (1976). *J. Math. Phys.*, **17**, 177; **17**, 792; **17**, 801.
—— (1977). *J. Math. Phys.*, **18**, 1589.
—— (1978). *J. Math. Phys.*, **19**, 390.
Kundt, W. (1966). *Canonical quantization of gauge invariant field theories*. Springer, Berlin, Heidelberg and New York.
Lanczos, C. (1970). *The variational principles of mechanics*. University of Toronto Press, Toronto.
MacCallum, M.A.H. (1975). In *Quantum gravity: an Oxford symposium*, ed C.J. Isham, R. Penrose, and D.W. Sciama. Clarendon Press, Oxford.
Magnon-Ashtekar, A. (1976). Thesis. Départment de Mathématiques Pures. Université de Clarmont; Preprint.
Misner, C.W. (1957). *Rev. mod. Phys.*, **29**, 497.
—— (1972). In *Magic without magic* (J.A. Wheeler 60th Birthday Volume), ed J.R. Klauder. Freeman, San Francisco.
——, Thorne, K.S., and Wheeler, J.A. (1973). *Gravitation*. Freeman, San Francisco.
Moncrief, V. (1972). *Phys. Rev.*, **D5**, 277.
—— and Teitelboim, C. (1972). *Phys. Rev.*, **D6**, 966.
Nelson, J.E. and Teitelboim, C. (1977). *Phys. Lett.*, **69B**, 81.

von Neumann, J. (1955). *Mathematical foundations of quantum mechanics.* Transl. from German by R.T. Beyer. Princeton University Press, Princeton.
Pauli, W. (1958). In *Handbuch der Physik*, Vol. 5, ed S. Flügge. Springer, Berlin, Göttingen and Heidelberg.
Peres, A. (1962). *Nuovo Cim.*, **26**, 53.
Pilati, M. (1980). Ph.D. Thesis. Princeton Preprint.
Regge, T. and Teitelboim, C. (1974). *Ann. Phys. (N.Y.)*, **88**, 286.
Ryan, M. (1972). *Hamiltonian cosmology*. Springer, Berlin, Heidelberg and New York.
Schrödinger, E. (1935). *Naturwissenschaften*, **23**, 807.
Schwinger, J. (1963). *Phys. Rev.*, **132**, 1317.
Synge, J.L. (1960). In *Encyclopedia of physics*, Vol. III/1, ed S. Flügge, Springer, Berlin.
Teitelboim, C. (1973). *The Hamiltonian structure of space-time*, Ph.D. Thesis, Princeton (unpublished); *Ann. Phys. (N.Y.)*, **79**, 542.
—— (1973). *Phys. Rev.* **D8**, 3266.
—— (1980). In *Einstein Centenary Volume*, ed P. Bergman, A. Goldberg, and A. Held. Plenum, New York.
Wheeler, J.A. (1962). In *Topics of modern physics* (Italian Physical Society). Academic Press, New York.
—— (1964). In *Relativity, groups and topology*, ed C. DeWitt and B.S. DeWitt. Gordon and Breach, New York and London.
—— (1968). In *Batelle Rencontres 1967*, ed C. DeWitt and J.A. Wheeler. Benjamin, New York.

16

THE ROLE OF INSTANTONS IN QUANTUM GRAVITY

C.N. Pope

Department of Applied Mathematics and Theoretical Physics, Cambridge University

1. INTRODUCTION

In recent years a great deal of effort has been directed towards understanding the role of topology in both elementary particle physics and quantum gravity. In non-Abelian gauge theories it has been found that the gauge field can adopt topologically non-trivial configurations, and that these could give rise to physically significant effects, such as the helicity non-conserving fermion Green's functions first studied by 't Hooft. The most appropriate framework within which to study the topological properties of the theory seems to be the functional integral approach, in which the vacuum to vacuum amplitude or the expectation value of an operator is expressed as a sum over all field configurations, weighted by the classical action. The dominant contributions to the functional integral come from the stationary points of the action, i.e. the classical solutions of the field equations. In Yang–Mills theory these are known as instantons. The quadratic fluctuations around these stationary points give the one loop quantum corrections.

In quantum gravity the possibility arises that space-time itself may adopt configurations with non-trivial topology, and once again the functional integral approach seems to be the most appropriate. Certain difficulties arise which do not occur in flat space Yang–Mills theory, associated with the fact that the gravitational vacuum is not scale invariant and is not positive definite in the Euclidean regime. There are also difficulties connected with passing to the Euclidean regime, which is necessary in order to improve the convergence of the functional integral, because not all Lorentzian metrics admit Euclidean sections. These will be discussed in greater detail below.

The stationary points of the action for quantum gravity are

just the classical solutions of the Euclidean Einstein equations, and by analogy with the case of Yang–Mills theory they have become known as instantons. A gravitational instanton may thus be defined as a positive definite metric $g_{\mu\nu}$ on a complete and non-singular manifold, which satisfies the equations $R_{\mu\nu} = 0$. A variety of gravitational instantons are known, corresponding to certain types of boundary conditions imposed on the metrics in the functional integral. These boundary conditions are determined by the physical problem under consideration, and in quantum gravity fall into three basic categories. These are (i) metrics which approach flat Euclidean space at infinity, corresponding to the vacuum state for gravity, (ii) metrics which approach flat space identified periodically in imaginary time, which correspond to the thermal Gibbs state, and (iii) compact metrics, which arise for example in Hawking's space-time foam picture of the gravitational vacuum. Instantons of all three types are described below, preceded by a discussion of the functional integral in quantum gravity.

2. THE FUNCTIONAL INTEGRAL

(a) *Quantum fields in flat space*

In flat space-time quantum field theory the functional integral may be written as

$$Z = \int_{\Phi} d[\phi]\exp(iI[\phi]) , \tag{2.1}$$

where $d[\phi]$ is a measure on the space of all field configurations, $I[\phi]$ is the classical action of the field ϕ and Φ is the class of fields over which the integration is to be performed. If ϕ represents a gauge field then gauge fixing and ghost terms must be included in $I[\phi]$ in order to remove the unphysical degrees of freedom. The integral (2.1) is not absolutely convergent, but the convergence properties can be improved by Wick rotating the time axis in the complex t-plane to $\tau = it$, giving

$$Z = \int_{\Phi} d[\phi]\exp(-\hat{I}[\phi]) , \tag{2.2}$$

where $\hat{I}[\phi]$ is the 'Euclidean' classical action, which is bounded below and so may be chosen to satisfy the inequality $\hat{I}[\phi] \geqslant 0$. Z may be estimated by locating the stationary points of $\hat{I}$ subject to the boundary conditions implied by Φ, and looking at the quadratic fluctuations around these classical solutions or instantons. The resulting infinite dimensional Gaussian integrals still require a further regularization, since formally the result involves the determinant of the operator giving the second variation of $\hat{I}$, and the eigenvalues of this operator are unbounded above. Many regularization schemes are available, but one of the most convenient in the context of the functional integral approach is the generalized zeta function (Critchley and Dowker 1976; Hawking 1977*a*). In this scheme the determinant of an operator Δ, with eigenvalues λ, is defined to be $\exp(-\zeta'(0))$, where $\zeta(s)$ is the meromorphic function analytic in the neighbourhood of $s = 0$ which is defined for $s > 2$ by the expression $\zeta(s) = \sum_{\lambda} \lambda^{-s}$.

In Yang–Mills theory the well known instantons arise as the stationary points of the action in the functional integral over all fields which die away sufficiently rapidly at large distances in all directions in Euclidean space. The instantons are all self-dual or anti-self-dual ($F = \pm * F$). The boundary conditions imply that the potential A must be a pure gauge at infinity ($A \to g^{-1}dg$, $g \in SU(2)$), but because the field configuration is topologically non-trivial g cannot be continuously deformed to the identity. In fact the instantons are characterized by the integer k, given by

$$k = -\frac{1}{8\pi^2}\,\mathrm{tr}\int F \wedge F\,, \tag{2.3}$$

the degree of the map from the boundary three sphere at infinity into the gauge group SU(2). The integer k, the Pontrjagin number, may be thought of as the instanton number of the solution. The action of a k instanton solution is a local minimum, and takes the value $2\pi^2|k|$. The minimally coupled Dirac operator has k positive helicty zero modes (if $k > 0$), and these give rise to helicity non-conserving Green's functions for fermions, by the 't Hooft mechanism ('t Hooft 1976).

(b) *Quantum gravity*

The functional integral for quantum gravity is given by (2.1) with ϕ representing the metric tensor $g_{\mu\nu}$. The analogue of the Wick rotation procedure in flat space is to perform the integration over metrics of positive definite signature rather than the physical Lorentzian signature. However, unlike the case of flat space quantum theory, this 'Euclideanization' procedure is rather more than just a mathematical device for giving a meaning to the functional integral. The reason for this is that not all Lorentzian metrics have Euclidean sections, and vice versa. Therefore if the integral is performed over all Euclidean ized metrics then some Lorentzian metrics will have been omitted whilst if the integration is taken over all Euclidean metrics then this will include some which do not admit Lorentzian sections, and are thus not 'physical'. Hawking has proposed a radical solution to this problem; that the integral should correctly be performed over all Euclidean metrics, and that this can be regarded as being rather like a contour integral in the space of all complex metrics, in which the contour has been rotated from the Lorentzian to the Euclidean section (Hawking 1979*a*). Thus individual metrics in the functional integral need not necessarily have a direct physical interpretation, and only Z itself, or Green's functions, etc. calculated from Z, should be analytically continued back to the Lorentzian regime at the end of the calculation. Ultimately one would hope to be able to test this hypothesis by comparing the predictions of the theory with experiment.

The functional integral is therefore given by (2.2) with $\phi = g_{\mu\nu}$, and

$$\hat{I}[g] = -\frac{1}{16\pi}\int_M R\sqrt{g}\,d^4x - \frac{1}{8\pi}\int_{\partial M}(K - K^0)\sqrt{h}\,d^3x\ , \qquad (2.4)$$

where R is the Ricci scalar of the metric $g_{\mu\nu}$ in the manifold M, K is the trace of the second fundamental form of the boundary ∂M and $h_{\mu\nu}$ is the induced metric on ∂M (Gibbons and Hawking 1977). The term involving K^0 is a constant term which can be added in cases where ∂M can be embedded in flat space, so that the action of flat space is normalized to zero; K^0 is the trace of the second fundamental form of ∂M embedded in flat space. Of course in the functional integral one also has to add a gauge fixing term to (2.4), and introduce ghost fields in the standard manner.

Another difficulty that arises for gravity is that (2.4) is not bounded below; by performing a conformal transformation on a given metric $g_{\mu\nu}$ the action can be made arbitrarily negative. Gibbons, Hawking, and Perry (1978) have shown how this problem can be circumvented in the case of the functional integral over all metrics which approach flat Euclidean space at infinity (Asymptotically Euclidean metrics). The integral can be split up into an integral over all such metrics with Ricci scalar $R = 0$, and an integral over the conformal deformations of these metrics. It can be shown that the action of an asymptotically Euclidean metric with $R = 0$ is positive or zero, and vanishes only if the metric is flat (Schoen and Yau 1979; Hawking 1979*b*). The troublesome behaviour has therefore been isolated in the conformal degrees of freedom, and by rotating to an integration over complex conformal factors this can be rendered convergent also.

The asymptotically Euclidean boundary condition arises when considering the vacuum state in quantum gravity, and the calculation of scattering matrix elements. It is therefore natural to look for instanton solutions which would dominate in this functional integral for the gravitational vacuum; these would be the gravitational analogues of the Yang–Mills instantons. In fact, as will be discussed in the next section, such non-trivial solutions cannot occur, and so some other approximation scheme has to be adopted in order to compute the effects of gravity on the S-matrix (e.g. Hawking, Page, and Pope, 1979, 1980). If the boundary condition is relaxed somewhat, so that the metric approaches flat space identified under a discrete subgroup of SO(4) at infinity, then instanton solutions can occur. It is not clear what the physical interpretation of this boundary condition is.

Another type of boundary condition which is of interest is where the metrics in the functional integral tend to the flat metric in three directions but are periodic in the fourth dimension, which may be thought of as the imaginary time dimension. In other words the functional integral is evaluated over all metrics which 'fill in' an $S^1 \times S^2$ boundary, where the S^1 has a given periodicity β. $Z[\beta]$ is then the partition function for states of the gravitational field in equilibrium at a temperature $T = \beta^{-1}$. Such metrics are known as 'asymptotically flat'. The Schwarzschild and Kerr solutions are instantons of this type; the periodicity of these solutions in imaginary time on the Euclidean section leads directly to the thermal properties of black holes. There is also a 'twisted' version of the thermal boundary condition, in which the $S^1 \times S^2$ boundary is replaced by a non-trivial S^1 bundle over S^2. This is analogous to the local version of the asymptotically Euclidean metrics mentioned above; instanton solutions are known in this case also. These are discussed in section 4.

Both the asymptotically Euclidean and asymptotically flat instantons discussed above are non-compact. There is another situation of physical interest in quantum gravity, in which compact instantons arise; i.e. instantons of finite volume. In order to investigate the topological structure of the gravitational vacuum Hawking has tried to estimate quantities such as the Euler number per unit volume of space-time (Hawking 1978). This can be conveniently done by first of all normalizing all the metrics in the functional integral to have a given 4-volume V, and then estimating the instanton contributions to Z as a function of their topological complexity. At the end of the calculation the volume V is then sent to infinity. The constraining of the metrics in the functional integral to have volume V may be achieved by adding a term $\frac{\Lambda}{8\pi} V$ to the action, where Λ is a Lagrange multiplier for the 4-volume, and taking the functional integral over all metrics of finite volume. The stationary points of this modified action are then solutions of the Einstein equations with cosmological constant Λ,

$$R_{\mu\nu} = \Lambda g_{\mu\nu} \, . \tag{2.5}$$

Some examples of such compact instantons are described in section 5.

In practice one cannot hope to find explicitly all the compact instantons, most of which will have extremely complicated topologies, so that one has to make estimates of the action of the instantons as a function of χ and Λ. This is discussed in Hawking (1978) (but see also Christensen and Duff (1979)). It would seem that space-time has a 'foamlike' nature at the Planck length scale or less, while on 'everyday' length scales it appears to be more-or-less smooth and only slightly curved — as one would expect in the classical limit.

There are two topological invariants which may be expressed as integrals of the curvature over the manifold which are useful for classifying gravitational instantons. These are the Euler number χ and the Hirzebruch signature τ. They are given by the expressions

$$\chi = \frac{1}{32\pi^2}\int_M R_{\mu\nu\alpha\beta}\,{}^*R^{*\mu\nu\alpha\beta}\sqrt{g}\;d^4x + \text{surface terms} \qquad (2.6)$$

and

$$\tau = \frac{1}{48\pi^2}\int_M R_{\mu\nu\alpha\beta}\,{}^*R^{\mu\nu\alpha\beta}\sqrt{g}\;d^4x + \text{surface terms}\;. \qquad (2.7)$$

The surface terms arise in the case of non-compact instantons, where one introduces a boundary at some large distance and then lets it recede to infinity. The surface terms which arise for the known gravitational instantons are discussed in Gibbons, Pope, and Römer (1979).

3. ASYMPTOTICALLY EUCLIDEAN INSTANTONS

By analogy with the Yang–Mills instantons one might hope to find asymptotically Euclidean gravitational instantons, and that these would give rise to the occurrence of zero modes of the Dirac operator, and consequently lead to helicity violating Green's functions by the 't Hooft mechanism ('t Hooft 1976). However, as a direct consequence of the Positive Action Theorem, the only asymptotically Euclidean instanton is flat space (Hawking 1979*a*; Schoen and Yau 1979). The theorem states that the action of an asymptotically Euclidean metric with $R = 0$ is greater than or equal to zero, and vanishes if and only if the metric is flat. Now if such a metric were to be a solution of the Einstein equations ($R_{\mu\nu} = 0$) its action would, in particular, have to be stationary under constant conformal scalings of the metric. But under the scaling $g_{\mu\nu} \rightarrow k^2 g_{\mu\nu}$, the action $\hat{I} \rightarrow k^2\hat{I}$, and so it could only be stationary (and finite) if $\hat{I} = 0$, which, by the positive action theorem, means that the metric must be flat.

There is another way to prove this result in the case of a metric whose Riemann tensor is self-dual ($R_{\mu\nu\alpha\beta} = {}^*R_{\mu\nu\alpha\beta}$), based on the Atiyah–Singer index theorem. This states that the difference between the number of regular L^2 solutions of the Dirac equation of positive helicity and negative helicity is

$$n_+ - n_- = -\frac{1}{384\pi^2}\int_M R_{\mu\nu\alpha\beta}{}^*R^{\mu\nu\alpha\beta}\sqrt{g}\; d^4x \; . \tag{3.1}$$

In general there are extra terms if the manifold has a boundary, but these are absent in the case of asymptotically Euclidean metrics. Now a straightforward generalization of Lichnerowicz's theorem to non-compact manifolds shows that there can be no normalizable zero modes, and so $n_+ = n_- = 0$. Therefore because of the self-duality of the curvature, it follows from (3.1) that the positive definite expression $\int (R_{\mu\nu\alpha\beta})^2\sqrt{g}\; d^4x$ is zero, and hence the metric is flat (Gibbons and Pope 1979).

There is a local form of the asymptotically Euclidean boundary condition for which instanton solutions can occur, in which the boundary at infinity has the topology S^3/Γ rather than simply S^3, where Γ is a discrete subgroup of the local tetrad rotation group SO(4). The simplest of these was discovered by Eguchi and Hanson, and corresponds to $\Gamma = Z_2$, i.e. $\partial M = \mathbb{R}P^3$ (Eguchi and Hanson 1978; Belinskii, Gibbons, Page, and Pope 1978). In order to describe this instanton it is convenient to introduce a set of three left-invariant one forms $\{\sigma_i\}$ on the three sphere, which satisfy the SU(2) algebra $d\sigma_i = -\frac{1}{2}\epsilon_{ijk}\sigma_j \wedge \sigma_k$, and may be parametrized by Euler angles (θ, ϕ, ψ), thus:

$$\sigma_1 = \cos\psi\; d\theta + \sin\psi\; \sin\theta\; d\phi \; ,$$

$$\sigma_2 = -\sin\psi\; d\theta + \cos\psi\; \sin\theta\; d\phi \; ,$$

$$\sigma_3 = d\psi + \cos\theta\; d\phi \; , \tag{3.2}$$

where $0 \leqslant \theta \leqslant \pi$ and $0 \leqslant \phi \leqslant 2\pi$. For an unidentified three sphere the angle ψ is identified modulo 4π. In terms of these one forms the Eguchi–Hanson instanton may be written in the Bianchi IX form

$$ds^2 = \left(1 - \frac{a^4}{r^4}\right)^{-1} dr^2 + \frac{1}{4}\, r^2\left[\sigma_1{}^2 + \sigma_2{}^2 + \left(1 - \frac{a^4}{r^4}\right)\sigma_3{}^2\right], \quad (3.3)$$

where $a \leqslant r < \infty$. The metric appears to be singular at $r = a$, but this is only a coordinate singularity. In terms of a new radial coordinate ρ defined by $4\frac{\rho^2}{a^2} = 1 - \frac{a^4}{r^4}$, the metric near $r = a$ is approximately

$$ds^2 \simeq d\rho^2 + \rho^2(d\psi + \cos\theta\ d\phi)^2 + \frac{1}{4}\, a^2(d\theta^2 + \sin^2\theta\ d\phi^2)\ , \quad (3.4)$$

which is regular at $\rho = 0$ provided that ψ is identified with period 2π. Because the period has to be 2π rather than 4π, the level surfaces r = constant have the topology $\mathbb{R}P^3$ rather than S^3. At $r = a$ the metric reduces to that on a 2-sphere of radius $\frac{1}{2}a$; in the terminology of Gibbons and Hawking (1979), $r = a$ is a 'bolt', where the action of the Killing vector $\partial/\partial\psi$ has a two dimensional fixed point set. The Euler number χ of the Eguchi–Hanson instanton is 2, and the Hirzebruch signature $\tau = 1$. The curvature tensor $R_{\mu\nu\alpha\beta}$ is self-dual.

There is a family of multi-instanton solutions, in which the group Γ is Z_k, and so $\partial M = L(k,1)$, the cyclic lens space of order k (Gibbons and Hawking 1978). They all have self-dual curvature, and may be written in the form

$$ds^2 = V^{-1}(d\tau + \underline{\omega}.d\underline{x})^2 + Vd\underline{x}.\ \underline{x}\ , \quad (3.5)$$

where V and $\underline{\omega}$ are functions only of the three dimensional coordinates $\underline{x}$ on an auxiliary flat three space with metric $d\underline{x}.d\underline{x}$. The metric (3.5) satisfies the Einstein vacuum equations if grad V = curl $\underline{\omega}$, which implies

$$\underline{\nabla}^2 V = 0\ . \quad (3.6)$$

The asymptotically locally Euclidean multi-instantons are obtained by taking

$$V = \sum_{i=1}^{k} \frac{1}{|\underline{x} - \underline{x}_i|} \, . \tag{3.7}$$

When $k = 1$, (3.5) reduces to flat space, and when $k = 2$ it is just the Eguchi–Hanson solution. For general k, (3.5) has Euler number $\chi = k$, and Hirzebruch signature $\tau = k - 1$. For $k > 2$ there are $3k - 6$ arbitrary parameters, corresponding to the freedom to choose the positions $\underline{x}_i$ of the singularities in V. These singularities correspond to coordinate singularities in (3.5), and are removable by a suitable coordinate transformation.

Hitchin has given an implicit construction for instantons in which Γ is any discrete subgroup of one of the SU(2) factors of SO(4), containing the L(k,1) instantons as a special case (Hitchin 1979). He has recently obtained explicit metrics in the case where Γ is the binary dihedral group of order n. Once again the curvature is self-dual; in fact Hawking has made a Generalized Positive Action Conjecture, which implies that any asymptotically locally Euclidean instanton must be self-dual (Hawking 1979*a*).

4. ASYMPTOTICALLY FLAT INSTANTONS

The asymptotically flat (or thermal) boundary condition corresponds to metrics which approach flatness in three directions at infinity but which are periodic in the fourth, which may be interpreted as the imaginary time coordinate. The simplest, and perhaps the most important, non-trivial example is provided by the Euclidean Schwarzschild solution, which can be obtained from its Lorentzian analogue by rotating the time coordinate to lie along the imaginary axis. In the standard coordinate system it takes the form

$$ds^2 = \left(1 - \frac{2M}{r}\right)d\tau^2 + \left(1 - \frac{2M}{r}\right)^{-1}dr^2 + r^2(d\theta^2 + \sin^2\theta \, d\phi^2). \tag{4.1}$$

The radial coordinate r lies in the range $2M \leqslant r < \infty$. There is an apparent singularity at $r = 2M$, but just as in the Eguchi–Hanson solution, this may be removed by a redefinition of the radial coordinate. The metric is then regular at $r = 2M$ provided τ is identified with period $8\pi M$ (Hawking 1977*b*), and

$r = 2M$ is then a bolt of area $16\pi M^2$, on which the Killing vector $\frac{\partial}{\partial\tau}$ vanishes. This periodicity on the Euclidean section leads to the interpretation of the Schwarzschild solution as describing a black hole in thermal equilibrium with gravitons at a temperature $(8\pi M)^{-1}$. The fact that any matter field Green's functions on this background will also be periodic in imaginary time leads directly to some of the well known thermal emission properties of black holes (Hartle and Hawking 1976). There is a generalization of (4.1), the Kerr solution, which describes a rotating black hole. These instantons have Euler number $\chi = 2$ and Hirzebruch signature $\tau = 0$.

If the metric (4.1) is cut off at some large radius, the boundary ∂M has the topology $S^1 \times S^2$. There is a local version of the asymptotically flat boundary condition in which ∂M has the topology of a non-trivial S^1 bundle over S^2, i.e. S^3/Γ, where Γ is a discrete subgroup of SO(4). Unlike the asymptotically Euclidean boundary condition, here the S^3 is distorted and expands with increasing radius in only two directions rather than three (Gibbons, Pope, and Römer 1979). The simplest example of an asymptotically locally flat instanton is the self-dual Taub–NUT solution, which arises as a special case of the two parameter Taub–NUT metrics (Hawking 1977*b*)

$$ds^2 = \left(\frac{r+M}{r-M}\right)dr^2 + 4M^2\left(\frac{r-M}{r+M}\right)\sigma_3{}^2 + (r^2 - M^2)(\sigma_1{}^2 + \sigma_2{}^2), \tag{4.2}$$

where $\{\sigma_i\}$ are the left-invariant 1-forms introduced in the previous section. The radial coordinate lies in the range $M \leqslant r < \infty$, and the apparent singularity at $r = M$ is removable by a coordinate redefinition, provided ψ defined in (3.2) is identified modulo 4π, which shows that the level surfaces r = constant have the topology S^3. In fact $r = M$ is actually a point, at which the isometry generated by the Killing vector $\frac{\partial}{\partial\psi}$ has a zero-dimensional fixed point set; a 'nut' in the terminology of Gibbons and Hawking (1979). Topologically the Taub–NUT instanton is $\mathbb{R}^4$, and so $\chi = 1$ and $\tau = 0$.

There is a family of asymptotically locally flat multi-Taub–NUT instantons, analogous to the asymptotically locally Euclidean multi-instantons described in the previous section. They are given by the same expression for the metric, (3.5), but now V is taken to be (Hawking 1977*b*):

$$V = 1 + \sum_{i=1}^{k} \frac{2M}{|\underline{x} - \underline{x}_i|} \, . \tag{4.3}$$

As before, the apparent singularities at $\underline{x} = \underline{x}_i$ are removable, and the topology of the boundary at infinity is L(k,1). The instantons are all self-dual, with $\chi = k$ and $\tau = k - 1$.

Unlike the case of the asymptotically locally Euclidean instantons, which are all self-dual, there is in addition a non-self-dual asymptotically locally flat instanton, which, like the self-dual solution (4.2), has a boundary which is topologically S^3, and is a special case of the two parameter Taub–NUT metrics (Page 1978*a*):

$$ds^2 = \frac{r^2-N^2}{r^2-\frac{5}{2}Nr+N^2}dr^2 + 4N^2\left(\frac{r^2-\frac{5}{2}Nr+N^2}{r^2-N^2}\right)\sigma_3{}^2 + (r^2-N^2)(\sigma_1{}^2+\sigma_2{}^2) \, . \tag{4.4}$$

The radial coordinate lies in the range $2N \leqslant r < \infty$; and in this case $r = 2N$ is a bolt; for this reason (4.4) has become known as the Taub–BOLT solution. It can be regarded as a 'twisted' version of the Schwarzschild solution. There is also a generalization which has a rotation parameter; a twisted analogue of the Kerr solution (Gibbons and Perry 1980). These instantons have $\chi = 2$ and $\tau = 1$.

5. COMPACT INSTANTONS

Very few compact gravitational instantons are known explicitly. Apart from the trivial flat metric on the 4-torus, there are only four. These are the 4-sphere, S^4, which is the analytic continuation to the Euclidean regime of de Sitter space; the complex projective plane, $\mathbb{C}P^2$; the product manifold $S^2 \times S^2$; and the non-trivial S^2 bundle over S^2. The Einstein metrics on all four of these manifolds may be cast into the Bianchi IX form.

The metric on S^4, with cosmological constant Λ set equal to 3, may be written as

$$ds^2 = d\beta^2 + \frac{1}{4}\sin^2\beta[\sigma_1{}^2 + \sigma_2{}^2 + \sigma_3{}^2] \, , \tag{5.1}$$

where the angle β lies in the range $0 \leqslant \beta \leqslant \pi$. The apparent singularities at $\beta = 0$ and π are regular nuts, provided the Euler angle ψ defined in (3.2) is identified modulo 4π. The surfaces β = constant are, therefore, topologically S^3. Because the coefficients of the 1-forms $\{\sigma_i\}$ are all the same the surfaces are not only left invariant but also right invariant. The isometry group of the metric (5.1) is SO(5), which acts transitively. S^4 has Euler number $\chi = 2$ and Hirzebruch signature $\tau = 0$. The metric is conformally flat.

$\mathbb{C}P^2$ may be defined as the space obtained by identifying the points (z_1, z_2, z_3) and $(\lambda z_1, \lambda z_2, \lambda z_3)$ in $\mathbb{C}^3$ for all non-zero complex λ. This two-dimensional complex space may be given a real four-dimensional metric, which satisfies the Einstein equations with cosmological constant Λ (Gibbons and Pope 1978). In a convenient coordinate system, with Λ set equal to 6, it takes the form

$$ds^2 = d\beta^2 + \frac{1}{4}\sin^2\beta[\sigma_1{}^2 + \sigma_2{}^2 + \cos^2\beta\,\sigma_3{}^2] , \qquad (5.2)$$

where $0 \leqslant \beta \leqslant \frac{\pi}{2}$. Near $\beta = 0$ this looks like (5.1), and has the canonical form near a nut. There is a bolt at $\beta = \frac{\pi}{2}$, where $\frac{\partial}{\partial\psi}$ has a two-dimensional fixed point set. The isometry group of this metric is locally SU(3), which has a U(2) subgroup acting on the 3-spheres β = constant. $\mathbb{C}P^2$ has $\chi = 3$ and $\tau = 1$. The Weyl tensor $C_{\mu\nu\alpha\beta}$ is self-dual.

The conventional Einstein metric on $S^2 \times S^2$ is obtained simply as the direct sum of the metrics on two 2-spheres,

$$ds^2 = \frac{1}{\Lambda}\sum_{i=1}^{2}(d\theta_i{}^2 + \sin^2\theta_i\, d\phi_i{}^2) .$$

This is manifestly invariant under the SO(3) × SO(3) isometry group of $S^2 \times S^2$, but is not of Bianchi IX type. A straightforward coordinate transformation shows that the above metric may be expressed in Bianchi IX form as

$$ds^2 = d\beta^2 + \cos^2\beta\,\sigma_1{}^2 + \sin^2\beta\,\sigma_2{}^2 + \sigma_3{}^2 , \qquad (5.3)$$

where $\Lambda = 2$ and $0 \leqslant \beta \leqslant \frac{\pi}{2}$. The metric is regular at $\beta = 0$ and $\beta = \frac{\pi}{2}$ provided that ψ is identified modulo 2π, which means that the surfaces β = constant have the topology $\mathbb{R}P^3$. This is the only known regular Bianchi IX Einstein metric in which the coefficients of σ_1, σ_2 and σ_3 are all different. The topological invariants χ and τ for $S^2 \times S^2$ take the values 4 and 0 respectively.

The final example of a compact instanton whose Einstein metric is explicitly known is the non-trivial S^2 bundle over S^2, which is a sort of 'twisted' version of $S^2 \times S^2$ (Page 1978*b*). Like $S^2 \times S^2$, it has $\chi = 4$, $\tau = 0$, and its metric may be case in the form

$$ds^2 = (1+\nu^2)\left\{\frac{1-\nu^2x^2}{3-\nu^2-\nu^2(1+\nu^2)x^2}\,\frac{dx^2}{(1-x^2)} + \frac{1-\nu^2x^2}{3+6\nu^2-\nu^4}(\sigma_1{}^2+\sigma_2{}^2) + \right.$$
$$\left. + \frac{3-\nu^2-\nu^2(1+\nu^2)x^2}{(3-\nu^2)(1-\nu^2x^2)}(1-x^2)\sigma_3{}^2\right\} \qquad (5.4)$$

when $\Lambda = 3$. The coordinate x lies in the range $0 \leqslant x \leqslant 1$, and ν is the positive root of $\nu^4 + 4\nu^3 - 6\nu^2 + 12\nu - 3 = 0$. The isometry group is U(2).

There is one other compact instanton which has been studied in quantum gravity, even though an explicit Einstein metric has not yet been found. This is K3, which is diffeomorphic to any non-singular quartic surface in $\mathbb{C}P^3$. It is the unique simply connected compact instanton to admit an Einstein metric which has a self-dual Riemann tensor, and therefore satisfies $R_{\mu\nu} = 0$ (Hitchin 1975); its topological invariants take the values $\chi = 24$, $\tau = 16$. Because it is self-dual a great deal of information can be deduced by means of the various index theorems; some examples of this are discussed in Hawking and Pope (1978).

6. CONCLUSION

In the functional integral approach to quantum gravity three basic types of boundary condition for the metrics arise; asymptotically Euclidean, asymptotically flat, and compact. For the first two kinds there are also versions which hold only locally – though whether these have any physical significance is not clear. Thus although the self-dual asymptotically locally Euclidean and locally flat multi-instantons are of great interest mathematically, it is not known what, if any, physical question they provide the answer to. Indeed it would seem that any non-compact self-dual instanton necessarily cannot satisfy a physical boundary condition; the argument presented in section 3 based on the Atiyah–Singer index theorem can be generalized to show that any non-compact self-dual instanton must have a non-physical 'twisted' boundary at infinity.

For the asymptotically flat boundary condition, the non-trivial one-loop contributions come from fluctuations around the Schwarzschild and Kerr black hole metrics, giving rise to the well known thermal properties of black holes as a consequence of the time periodicity on the Euclidean section.

For asymptotically Euclidean metrics the only contribution at the one-loop level comes from fluctuations around flat space. In order to determine the effects of topologically non-trivial configurations, such as are implied by Hawking's space-time foam calculations, it is necessary to adopt some other approximation scheme, such as a finite-dimensional approximation to asymptotically Euclidean metrics. One possibility is to consider a class of such metrics (which would not of course be vacuum solutions) containing a large number of arbitrary parameters which could be varied in the functional integral and thus would at least fill some region in the space of all asymptotically Euclidean metrics. Yuille (1980) has considered one possible class of metrics.

A much more satisfactory approach might be to adopt a lattice type approximation, such as in the simplicial approximations of Regge calculus (Regge 1961). In this approach a space-time manifold is triangulated with 4-simplices, and the infinite dimensional functional integral over metrics is replaced by a finite dimensional integral over the edge of lengths of the simplices. The approximation can be made arbitrarily accurate by taking sufficiently complicated triangulations. Such a scheme should also help with the problem of understanding the measure on the space of all metrics in quantum gravity.

Finally, there are the compact instantons, which satisfy Einstein's equations with a cosmological constant. These arise in the space-time foam picture of the gravitational vacuum, and may also be of interest in extended supergravity theories.

Many compact instantons are known to exist, but only a few are known explicitly. In the foam picture it is the instantons with very complicated topology which are the most important, and these are likely to have metrics which have very complicated functional forms. Fortunately it is possible to deduce a lot about the properties of these instantons on general grounds without needing to know the metrics explicitly; see for example Hawking (1978). In the future, it will be of interest to see how the picture changes in the case of supergravity.

REFERENCES

Belinskii, V.A., Gibbons, G.W., Page, D.N., and Pope, C.N. (1978). *Phys. Lett.*, **B76**, 433.
Christensen, S.M. and Duff, M.J. (1980). *Nucl. Phys.*, **B170**, 480.
Critchley, R. and Dowker, J.S. (1976). *Phys. Rev.*, **D13**, 3224.
Eguchi, T. and Hanson, A.J. (1978). *Phys. Lett.*, **B74**, 249.
Gibbons, G.W. and Hawking, S.W. (1977). *Phys. Rev.*, **D15**, 2752.
—— —— (1978). *Phys. Lett.*, **B78**, 430.
—— —— (1979). *Comm. Math. Phys.*, **66**, 291.
—— and Perry, M.J. (1980). *Phys. Rev.*, **D22**, 313.
—— and Pope, C.N. (1978). *Comm. Math. Phys.*, **61**, 239.
—— —— (1979). *Comm. Math. Phys.*, **66**, 267.
——, Hawking, S.W., and Perry, M.J. (1978). *Nucl. Phys.*, **B138**, 141.
——, Pope, C.N. and Romer, H. (1979). *Nucl. Phys.*, **B157**, 377.
Hartle, J.B. and Hawking, S.W. (1976). *Phys. Rev.*, **D13**, 2118.
Hawking, S.W. (1977*a*). *Comm. Math. Phys.*, **55**, 133.
—— (1977*b*). *Phys. Lett.*, **A60**, 81.
—— (1978). *Nucl. Phys.*, **B144**, 349.
—— (1979*a*). Euclidean Quantum Gravity. In *Cargese Lectures in Physics, 1978*. Plenum Press, New York.
—— (1979*b*). The Path Integral Approach to Quantum Gravity. In *General relativity: an Einstein centenary survey*. Cambridge University Press.
—— and Pope, C.N. (1978). *Nucl. Phys.*, **B146**, 381.
——, Page, D.N., and Pope, C.N. (1979). *Phys. Lett.*, **B86**, 175.
—— —— —— (1980). *Nucl. Phys.*, **B170**, 283.
Hitchin, N.J. (1975). *J. diff. Geom.* **9**, 435.
—— (1979). *Math. proc. Camb. phil. Soc.* **85**, 465.
't Hooft, G. (1976). *Phys. Rev.*, **D14**, 3432.
Page, D.N. (1978*a*). *Phys. Lett.*, **B78**, 249.
—— (1978*b*). *Phys. Lett.*, **B79**, 235.
Regge, T. (1961). *Nuovo Cim.*, **19**, 558.
Schoen, R.M. and Yau, S.T. (1979). *Phys. Rev. Lett.*, **42**, 547.
Yuille, A.L. (1980). Israel-Wilson Metrics in Quantum Gravity. Department of Applied Mathematics and Theoretical Physics, Cambridge. Preprint.

17

ACAUSAL PROPAGATION IN QUANTUM GRAVITY

S.W. Hawking

Department of Applied Mathematics and Theoretical Physics, University of Cambridge

1. INTRODUCTION

One of the properties of gravity which distinguishes it from other field theories is that it determines the causal structure of space-time. Causality has been extensively investigated within the context of classical General Relativity. (See, for instance, Kronheimer and Penrose (1967); Hawking and Ellis (1973).) It is usually assumed that causality is not violated on a classical level, i.e. that there are no closed or almost closed non-spacelike curves. If this were not the case, one could obtain a number of paradoxes such as going back and shooting one's grandfather before one's father was conceived. Such causality conditions play an important role in many, but not all, of the singularity theorems.

In ordinary quantum theory on a flat space-time background, the assumption of causality again plays an important role: it restricts the singularities of the n-point Green's functions to lie within the future tube. This means that the Green's functions propagate positive frequencies purely forward in time and negative frequencies purely backwards. However the situation is very different when one quantizes the gravitational field or considers other quantum fields on a fixed curved space background. The reason is that the concept of positive frequency requires for its definition a complexification of the real Lorentzian manifold. If the topology of this complexified manifold is non-trivial, there may be an obstruction to the Wick rotation in the complex time coordinate that one performs in order analytically to continue the Green's functions from the Euclidean regime where they are unambiguously defined to the Lorentzian regime. This will result in the Green's functions having additional singularities which are acausal, i.e. they lie outside the future tube.

This kind of acausal behaviour was first noticed in the case of quantum fields on a black hole background which I discussed at the first Oxford Conference six years ago (Hawking (1975)). At first sight this seemed to be just an application of the standard formalism of quantum fields in an external potential, the gravitational field of the black hole. However there were two unusual features; first, there was a steady rate of particle creation and emission, even though the black hole was essentially stationary; second, the emitted radiation was random and thermal and was described by a density matrix rather than by a pure quantum state (Wald 1975; Hawking 1976). The reason for these properties was eventually traced to the fact that the Green's functions for the fields were regular on a section (the Euclidean section) of the complexified black hole manifold on which the metric is real and positive definite. The Euclidean section of a black hole metric has non-trivial topology with Euler number 2. This means that there is an obstruction to the Wick rotation of the time axis with the result that when the Green's functions are analtyically continued to the Lorentzian regime, they contain acausal singularities periodically distributed in the imaginary time coordinate. The Green's functions can in fact be interpreted as the expectation values of time-ordered products of field operators where the expectation value is taken, not in the vacuum state, but in a density matrix corresponding to thermal radiation. They can therefore give rise to processes in which an initial pure quantum state at past infinity evolves into a density matrix with a finite entropy at future infinity. This loss of quantum coherence has no parallel in other field theories and it can be directly attributed to the fact that gravity can give rise to acausal singularities in the Green's functions.

If this kind of behaviour can occur in quantum theory on a fixed background space, it should also arise in a complete theory in which the gravitational field itself was quantized. One expects there to be large quantum fluctuations of the metric and of the topology of space-time on scales of the Planck length or less. The best way to handle these seems to be to use a path integral approach evaluated over positive definite metrics (the Euclidean approach). For example, if one was interested in scattering experiments, one would integrate over all asymptotically Euclidean metrics, that is, all positive definite metrics which, outside some compact set, approach the standard flat metric on $\mathbb{R}^4$. Inside the compact set, however, the topology may differ from that of $\mathbb{R}^4$. If it does, there will be an obstruction to the Wick rotation and the Green's function will contain acausal singularities when analytically continued to asymptotically flat Lorentzian space at infinity. These acausal singularities will cause initially pure quantum states to evolve to density matrices with a loss of quantum coherence. They will also violate all other conservation laws except those like charge, energy-momentum and angular momentum which are connected with the existence of a long range field. They will lead to proton decay in about 10^{50} years. This is much longer than both the predictions of some Grand Unified Theories and of any hope of measurement. They might also lead to the decay of muons into $e\gamma$ with a lifetime of order 10^{10} years. However this would depend on the muon and the electron being truly point particles and on a certain matrix element not being zero. The effect of the acausal singularities would become larger at high energies. Thus they would be very important in the early universe and might explain why the universe started off in thermal equilibrium.

Section 2 contains a brief reminder about density matrices. In section 3 positive frequency is defined in terms of regularity in the Euclidean regime and the analytic continuation from Euclidean to Lorentzian space is discussed. It is shown that the ordering of field operators in expectation values corresponds to their displacements in imaginary time. Annihilation and creation operators at infinity are introduced in section 4, and are used to define a new kind of perturbation diagram. In section 5 it is shown how to use these diagrams to derive the emission from black holes in the presence of interactions and to describe the gravitational decay of particles such as the proton or muon.

2. DENSITY MATRICES

Density matrices are something everyone learns about as an

undergraduate and most people forget. So, just to refresh your memories, a system in a pure quantum state can be represented by a vector λ in some Hilbert space $\mathcal{H}$. In the case of asymptotic 'in' or 'out' states, this Hilbert space will be the Fock space of many particle states. If $\{|A\rangle\}$ are an orthonormal basis for $\mathcal{H}$, one can express the state vector in terms of components

$$\lambda = \sum_A \lambda_A |A\rangle \quad .$$

A system in a *mixed* state is described by a density operator or matrix, $\rho \in \mathcal{H} \otimes \bar{\mathcal{H}}$. In other words ρ is a 2-index tensor. In terms of components

$$\rho = \sum_A \sum_B \rho_A{}^B |A\rangle\langle B|; \quad \rho_A{}^B = \bar{\rho}_B{}^A; \quad \mathrm{tr}\,\rho = 1 \ .$$

The expectation value of an observable θ is

$$\langle \theta \rangle = \mathrm{tr}(\theta\rho) \ .$$

In particular, the two-point function for a complex scalar field is

$$G(x,y) = \mathrm{tr}(T\bar{\phi}(x)\phi(y)\rho) \ .$$

In the case of a massless free field and the vacuum density matrix

$$\rho = |0\rangle\langle 0|$$

this will be the usual Feynmann propagator with Fourier transform

$$\frac{1}{k^2 + i\varepsilon} \ .$$

This is purely causal, i.e. it propagates positive frequencies forward in time and negative frequencies backwards. However if ρ is not the vacuum density matrix, the propagator will contain an acausal part. For example, ρ might be the thermal density matrix at temperature $T = \beta^{-1}$, i.e.

$$\rho = \exp(-\beta H) = \sum_{A} |A\rangle \exp(-\beta E_A) \langle A|$$

where H is the Hamiltonian operator and E_A is the energy of the state $|A\rangle$. In this case, the propagator would be

$$\frac{\exp(\beta w)}{\exp(\beta w) - 1} \cdot \frac{1}{k^2 + i\varepsilon} + \frac{1}{\exp(\beta w) - 1} \cdot \frac{1}{k^2 - i\varepsilon}$$

where $w = k^0$ is the energy. The second term with the $- i\varepsilon$ is acausal: it propagates positive frequencies backwards.

3. WICK ROTATION

A function $f(t)$ of the form $\exp(-i\omega t)$, $\omega > 0$ is said to be positive frequency (I have never understood why it is $\exp(-i\omega t)$ and not $\exp(i\omega t)$). More generally $f(t)$ is said to be positive frequency if it can be extended to a function which is analytic in the lower complex t-plane (Fig. 1). Thus a positive frequency function $f(t,\underset{\sim}{x})$ on Minkowski space can be analytically continued to a function on the Euclidean space defined by $(\tau,\underset{\sim}{x})$ where $\tau = it$ such that $f(\tau,\underset{\sim}{x})$ is regular in the lower half space $\tau < 0$. Let f be a positive frequency solution of the massless wave equation:

$$\Box f = 0 \ .$$

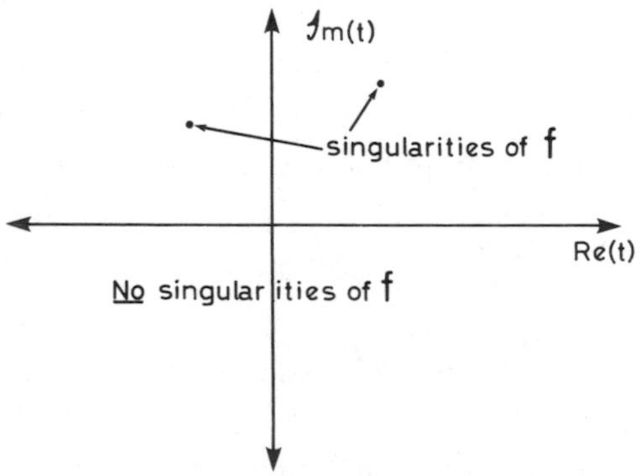

FIG. 1.

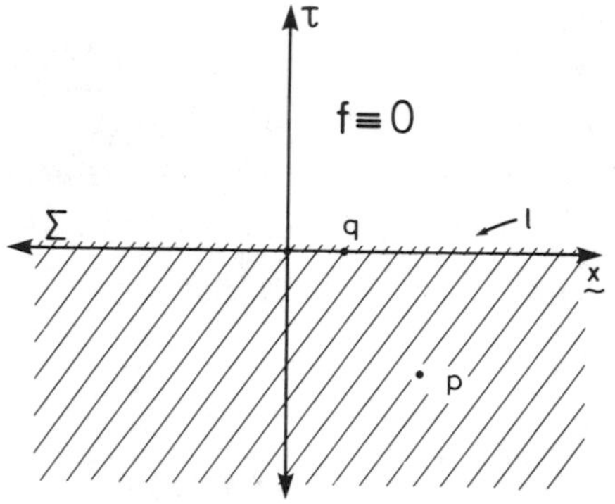

FIG. 2.

Suppose that we are given the value of f and its normal derivative on the hypersurface $\tau = 0$ in Euclidean space. Then we can propagate to a general point $(\tau,\underset{\sim}{x})$ using the Green's function arising from the vacuum density matrix:

$$G(p,q) = \frac{4\pi}{(p - q)^2} = \langle 0|\bar{\phi}(p)\phi(q)|0\rangle$$

$$f(p) = \int_\Sigma G(p,q)\overleftrightarrow{\partial}_\mu f(q)\,d\Sigma^\mu(q)$$

where the integral is taken over the hypersurface, Σ, $\tau = 0$. This will give the correct value of f if p is the lower half space. However, it will give zero if p is in the upper half space. In other words, the vacuum Green's function propagates positive frequencies purely downwards in the Euclidean space and negative frequencies purely upwards. From this one can see that one will obtain the correct Feynman causal Green's function

$$\langle 0|T\bar{\phi}(x)\phi(y)|0\rangle$$

in Minkowski space if one analytically continues the Euclidean Green's function in the complex t-plane to a line which almost coincides with the real t axis, but which slopes downwards

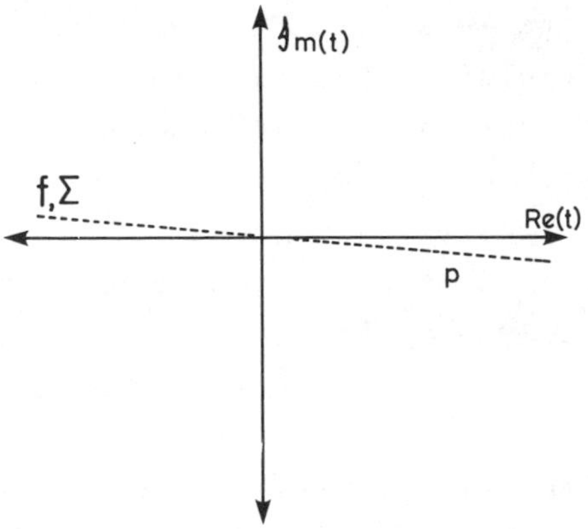

FIG. 3.

slightly from left to right (Fig. 3). To see that this is the case, consider the propagation formula

$$f(p) = \int_{\Sigma} G(p,q)\overleftrightarrow{\partial}_{\mu} f(q)\,d\Sigma^{\mu}(q) \ .$$

Suppose that the point p lies to the future of the surface Σ. Then the propagation from Σ to p will be in the direction of decreasing τ and so will propagate the positive frequency function f to its value at p.

In general when one is dealing with the vacuum expectation value of a number of field operators

$$\langle 0|\bar{\phi}(p)\phi(q)\bar{\phi}(r)\phi(s)\ \ldots\ |0\rangle$$

the rule is that one obtains the correct operator ordering if one analytically continues the n-point Euclidean Green's function to Minkowski space but slightly displaces the points p, q, r, s, ... from the real t-axis in that order in imaginary time, i.e. $\mathrm{Im}(p) < \mathrm{Im}(q) < \ldots$. In cases like Euclidean space where there is no obstruction to rotating the Euclidean section to the Lorentzian one, this will give a Green's function which is purely causal. However, if there is an obstruction, they will be acausal. A simple example is provided by the thermal case in flat space. Here the Euclidean Green's functions are defined not on Euclidean space $\mathbb{R}^4$, but on $\mathbb{R}^3 \times S^1$, a Euclidean space periodically identified in the τ direction with period β. In this case the Euclidean Green's function will consist of the ordinary vacuum Green's function plus a contribution from a series of image charges periodically distributed in τ (Fig. 4). Because of the images, the Green's function will propagate positive frequencies upwards as well as downwards. This means that when the Green's function is analytically continued to Minkowskian space according to the previous prescription, it will propagate positive frequencies backwards as well as forwards in time.

Obstructions to the Wick rotation of the time axis also arise when the effect of gravity is included, either as a fixed background metric or as a fully quantized field. The Green's functions for the fields are defined by a path integral in the Euclidean regime, i.e. on a manifold or manifolds with a positive definite metric with an integral over all such metrics in the case that the gravitational field is quantized. In general the complexification of a manifold with a positive definite metric will not contain a section on which the metric is real and Lorentzian. However, if one restricts oneself to metrics which are asymptotically flat or asymptotically Euclidean, there will be such a section at infinity. One can analytically continue the Euclidean Green's functions to this section at infinity. If the Euler number of the positive definite metric is different from one, the value for $\mathbb{R}^4$, one

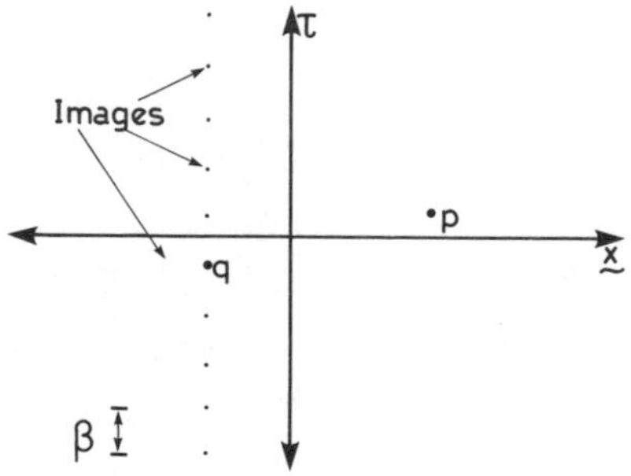

FIG. 4.

cannot introduce a time function on the positive definite metric which has non-zero gradient everywhere and which is similar to the time function on $\mathbb{R}^4$ in the asymptotic region near infinity. This means that there will be an obstruction to the Wick rotation of the time axis and the analytically continued Green's functions will be acausal. Such Green's functions cannot be represented as the time-ordered products of field operators in any pure quantum state, but they can be represented as expectation values in a mixed state.

4. ANNIHILATION AND CREATION OPERATORS

I shall consider situations in which there is an asymptotically flat 'in' region in the infinite past and an asymptotically flat 'out' region in the infinite future. In the asymptotic regions the interactions between the fields will be neglected and the metric will be treated as essentially flat. The field operators $\phi(x)$ and $\bar{\phi}(x)$ will obey the wave equation

$$\Box \phi = 0 \ .$$

They can be decomposed into annihilation and creation operators:

$$\phi_{\text{in}}(x) = \sum_i (a_{i\,\text{in}} f^i(x) + b^{+i}_{\text{in}} \bar{f}_i(x))$$

$$\bar{\phi}_{\text{in}}(x) = \sum_i (a^{+i}_{\text{in}} \bar{f}_i(x) + b_{\text{in}\,i} f^i(x))$$

where $\{f^i(x)\}$ is a complete orthonormal basis of positive frequency solutions of the wave equation and $a_{i\,\text{in}}$ and a^{+i}_{in} are annihilation and creation operators for a particle at past infinity in the mode f^i. Similarly $b_{i\,\text{in}}$ and b^{+i}_{in} are the annihilation and creation operators for an incoming anti-particle.

Using these creation and annihilation operators, one can change the density matrix at past infinity. For example, if ρ is the vacuum matrix at past infinity

$$\rho = |0_{\text{in}}\rangle\langle 0_{\text{in}}|$$

then

$$a^{+i}_{\text{in}} \, \rho \, a_{i\ \text{in}} = |i,\text{in}\rangle\langle i,\text{in}|$$

is the density matrix corresponding to an incoming particle in the mode f^i. One can operate with annihilation and creation operators at future infinity and take the trace to obtain the probability that one can annihilate a particle in the mode f^i at future infinity

$$P(j|i) = \text{tr}\left\{a_{j\ \text{out}} \, a^{+i}_{\text{in}} \, \rho \, a_{i\ \text{in}} \, a^{+j}_{\text{out}}\right\} .$$

The in and out annihilation and creation operators can be expressed as integrals of the field operators ϕ and $\bar{\phi}$ with basis functions f^i and $\bar{f}_i$ at past and future infinity, e.g.

$$a_{i\ \text{in}} = i \int \bar{f}_i(x) \overleftrightarrow{\nabla}_\mu \phi(x) d\Sigma^\mu(x)$$

where the integral is taken over a spacelike surface at past infinity. Thus

$$P(j|i) = \iiiint \bar{f}_j(u) f^i(v) \bar{f}_i(x) f^j(y) \overleftrightarrow{\nabla}^u_\mu \overleftrightarrow{\nabla}^v_\nu \overleftrightarrow{\nabla}^x_\lambda \overleftrightarrow{\nabla}^y_\delta$$

$$\text{tr}\left\{\phi(u)\bar{\phi}(v)\rho\phi(x)\bar{\phi}(y)\right\} \, d\Sigma^\mu(u) d\Sigma^\nu(v) d\Sigma^\lambda(x) d\Sigma^\delta(y)$$

where the integrals over v and x are taken at past infinity and the integrals over u and y are at future infinity. By the cyclic property of the trace

$$\text{tr}\left\{\phi(u)\bar{\phi}(v)\rho\phi(x)\bar{\phi}(y)\right\} = \text{tr}\left\{\phi(x)\bar{\phi}(y)\phi(u)\bar{\phi}(v)\rho\right\} .$$

Note that the $\phi(u)$ and $\bar{\phi}(v)$ are time ordered but $\phi(x)$ and $\bar{\phi}(y)$ are anti-time ordered. Thus the above expectation value is not that of the time-ordered product of the field operators at the four points x, y, u and v. However, one can obtain the expectation of the operators in the prescribed order from the Euclidean four-point Green's function if one follows the procedure described above, i.e. one arranges the points x, y, u and v in that order in imaginary time (Fig. 5).

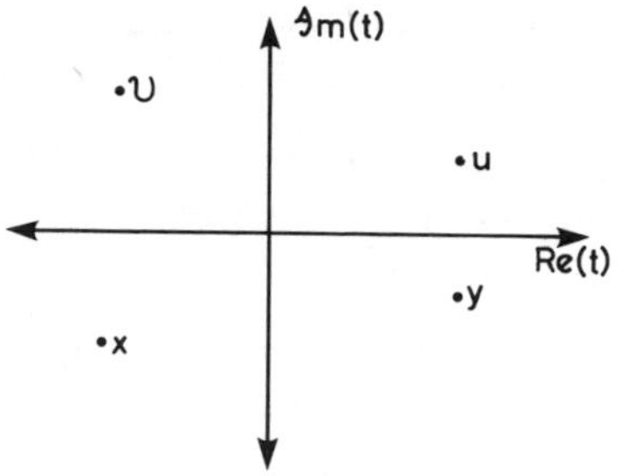

Im(t)
•v
•u
Re(t)
•y
•x

FIG. 5.

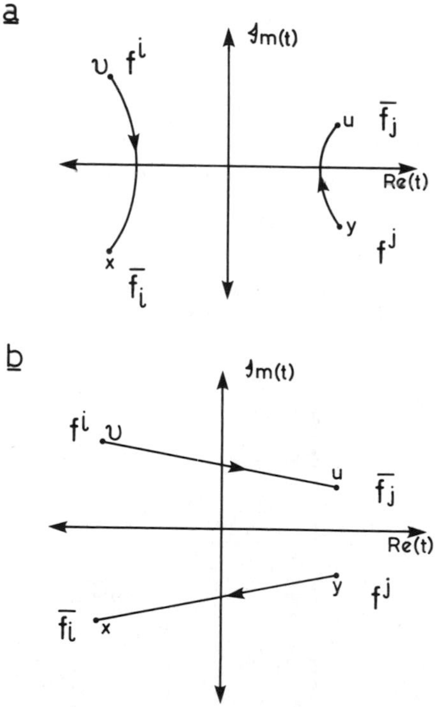

a
Im(t)
v
f^i
u
$\bar{f}_j$
Re(t)
y
f^j
x
$\bar{f}_i$
b
Im(t)
f^i
v
u
$\bar{f}_j$
Re(t)
y
f^j
$\bar{f}_i$
x

FIG. 6.

If the field is non-interacting, the four-point Green's function will be the sum of products of two-point Green's functions. There will be two possible diagrams (Fig. 6a, Fig. 6b). If one is working in flat space, at zero temperature, ρ will be the vacuum density matrix and the propagators will be causal. In this case the diagram 6a will be zero because although the positive frequency functions f^i can propagate downwards from v to x, the function f^j cannot propagate upwards from y to u. Diagram 6b consists of two disconnected paths. The upper one is the ordinary Feynman diagram for the amplitude to propagate from the state $|i\rangle$ at v to $\langle j|$ at u. It is multiplied by the complex conjugate diagram from x to y to yield the probability of a particle in the mode i at past infinity propagating to a mode j at future infinity.

In the case that there are interactions one can use perturbation theory to express the four-point Green's function in terms of diagrams in which the lines correspond to the free-field, two-point Green's function and the vertices correspond to interactions. If the propagation is causal, one can have diagrams like that in Fig. 7 which are separated into two disconnected parts by the real time axis (this diagram represents the creation of several particles at past infinity and the annihilation of several particles at future infinity. Creation operators at past infinity commute with each other and can therefore all be put at the same distance above the real t axis. Similarly

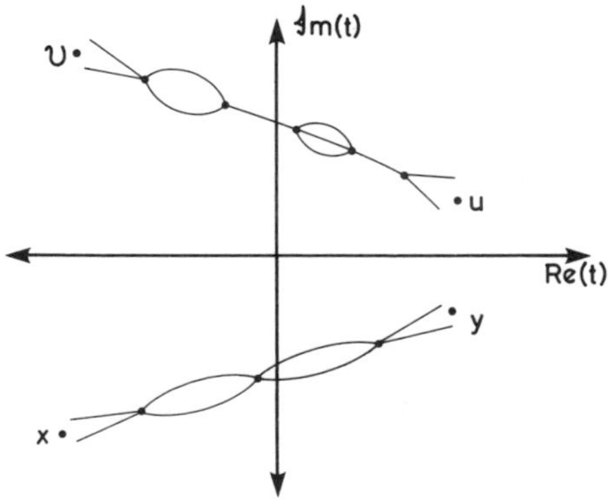

FIG. 7.

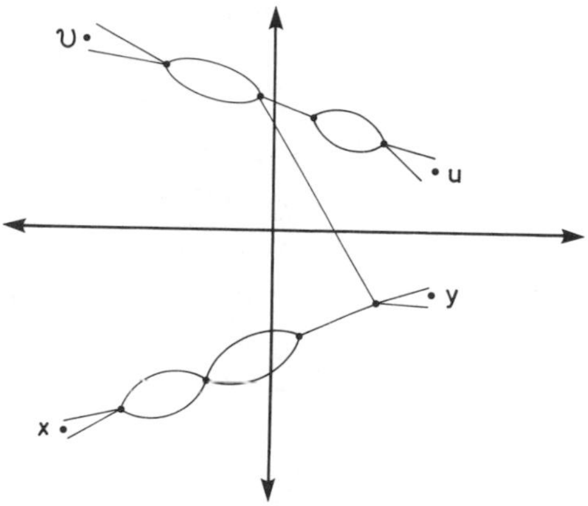

FIG. 8.

for annihilation operators.) The upper part of Fig. 7 can be interpreted as an ordinary Feynman diagram for an amplitude and it is multiplied by a complex conjugate Feynman diagram corresponding to the lower part of Fig. 7. Note that the upper and lower parts need not be complex conjugates of each other. There are also diagrams like that in Fig. 8 which are not separated by the real time axis. A line which crosses the real time axis must carry purely positive frequencies downwards. It therefore can be broken by introducing another annihilation operator at u and a creation operator at y. One can thus interpret it as an additional particle that is present at future infinity but which is not measured by the given number of annihilation and creation operations. In performing a prescribed number of annihilations and creations at future infinity one is averaging over all the unobserved particles.

In the case that the propagators are acausal, for instance in flat space at a finite temperature, one can get non-zero diagrams like that in Figs. 9 and 10, in which a particle is detected at future infinity in mode i without anything having been created at past infinity. This diagram measures the moment of the density matrix ρ_{out} at future infinity:

$$\sum_A \rho_{\text{out}}{}_A{}^A \, N(A,i)$$

where $N(A,i)$ is the number of particles of mode i in the state A. Similarly by performing several annihilation and creation operations at future infinity, one can measure the higher moments of the density matrix. The set of all moments determines the density matrix uniquely though the solution is very complicated in general. It is however simple in the thermal case for non-interacting fields. The thermal case with interactions can be solved by perturbation theory where the density matrix to a given order is determined by the finite number of diagrams to that order.

In a similar manner one can determine the ingoing density

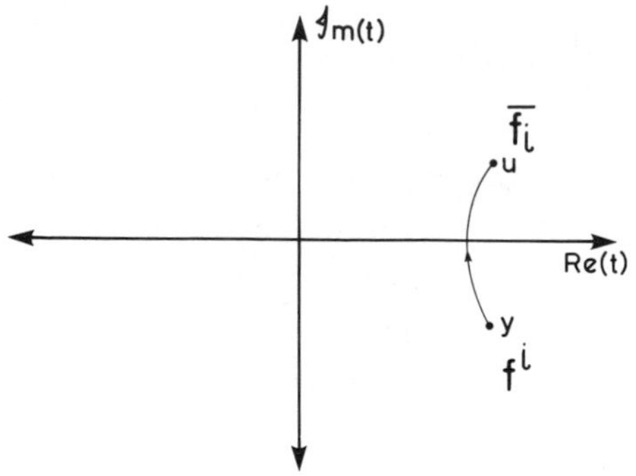

FIG. 9.

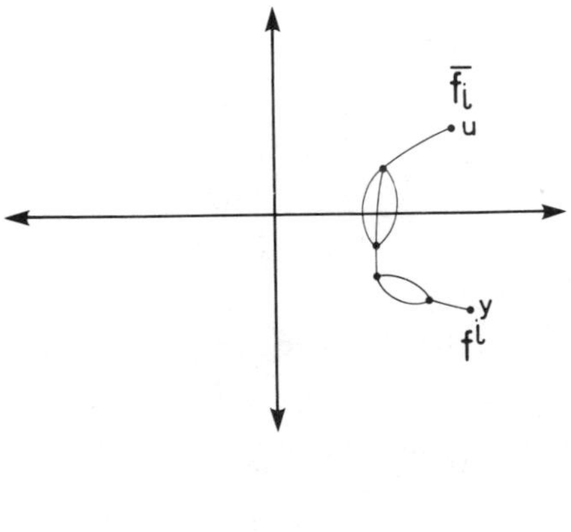

FIG. 10.

matrix ρ by performing annihilation and creation operations at past infinity. In most cases this will be the same as the outgoing density matrix. One can then change the ingoing density matrix by a certain number of annihilations and creations at past infinity and see what effect this has on the outgoing density matrix. By the superposition principal of quantum mechanics, the outgoing density matrix will be related to the ingoing density matrix by a linear operator \$, the superscattering operator (Hawking 1976)

$$\rho_{\text{out}\,A}{}^{B} = \$_{A}{}^{BC}{}_{D}\, \rho_{\text{in}\,C}{}^{D} .$$

In the case of causal propagation, the four-index \$ operator will be the product of two index S matrix operators

$$\$_{A}{}^{BC}{}_{D} = S_{A}{}^{C}\bar{S}^{B}{}_{D} .$$

If the propagation is acausal however, the superscattering operator cannot be decomposed in this way and it will convert a density matrix corresponding to a pure state into one representing a mixed state. It will thus increase entropy in general.

5. APPLICATIONS

A. *Interacting Fields around Black Holes*

The behaviour of non-interacting fields around black holes is now well understood. The black hole creates and emits particles, as if it were a hot body with temperature $T = \frac{1}{8\pi M}$. One can regard this emission as a consequence of pair creations near the horizon of the black hole with one particle falling into the hole and the other escaping to infinity. In this view the initial pure vacuum state evolves into a pure quantum state containing particles going out to infinity and other particles falling into the hole. It is however a somewhat metaphysical picture because an observer at infinity cannot measure what goes into the hole. He therefore has to sum over all possible horizon states. This means that the outgoing radiation is described by a density matrix rather than a pure state.

For non-interacting rields on a black hole background, the horizon states do not cause much problem. This is not the case, however, if there are interactions. If one tries to deal with these by writing down Feynman diagrams for the amplitude, one has to include diagrams with lines coming out of and going into the black hole. One does not know what to put at the end of these lines. Moreover they seem to lead to unphysical divergences near the horizon.

One can treat interacting fields on a black hole background quite satisfactorily however, using the formalism described above. A complexified black hole metric such as the Schwarzschild solution admits a section (the Euclidean section) on which the metric is real and positive definite (Hartle and Hawking 1976). The imaginary time coordinate τ is periodic on this section with period $\beta = 8\pi M$ (in the non-rotating case). The Green's functions for the non-interacting fields, which can be defined on the Euclidean section by standard perturbation theory, will thus all be periodic in τ and hence will be the Green's functions corresponding to thermal equilibrium at a temperature $T = \beta^{-1}$ (Gibbons and Perry 1976, 1978).

One can perform annihilation and creation operations at future infinity to determine the density matrix of the outgoing radiation. In particular, the expectation value of the number of outgoing particles in the mode i is given by the sum of all diagrams like that in Fig. 10 which have two outgoing lines, one going to u where there is a function $\bar{f}_i$ and the other going to y where there is f^i. Unlike the non-interacting case, there will be correlations between different modes so that the density matrix will not simply be given by a product of factors for each mode.

Because the system is in thermal equilibrium there will be a similar density matrix of ingoing radiation. However it is not often that we find a black hole in a box in thermal equilibrium. One is more interested in the case of a black hole radiating into empty space with no ingoing radiation. One can calculate the density matrix of outgoing radiation in this case also by first performing a suitable number of annihilations at past infinity and then measuring moments at future infinity. In the case of non-interacting fields, one can treat each mode

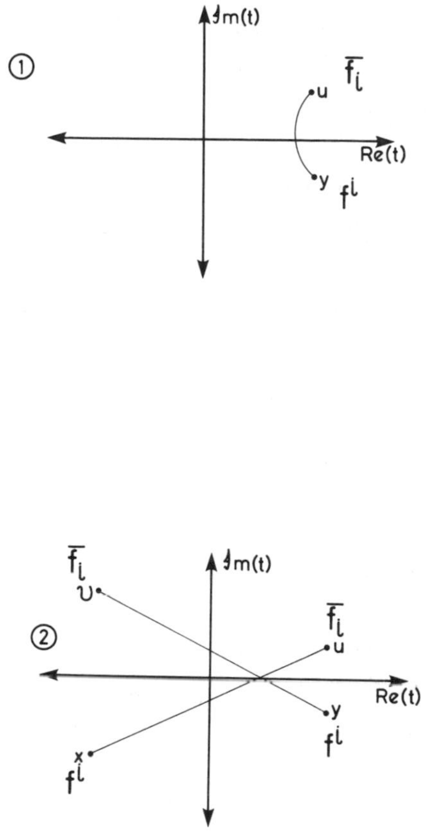

FIG. 11.

separately. There are only two diagrams (Fig. 11). Diagram (1) is what one would have for the expectation value of the number of particles of mode i in the equilibrium situation. It is equal to

$$(1) = x(1 \mp x)^{-1}$$

where $x = \exp(-\beta\omega)$ is the Boltzmann factor for the mode and the - sign is for bosons, the + sign for fermions. Diagram (2) involves propagation from past infinity to future infinity and is equal to

$$(2) = x^2(1 \mp x)^{-2}R$$

where R is theprobability that a particle in the mode i which is sent in from past infinity will be reflected back by the gravitational field of the black hole and go out to future infinity. The expectation value of the number of outgoing particles when there are no ingoing particles is then given by

$$(1) - \frac{(2)}{(1)} = x(1 \mp x)^{-1}T$$

where $T = 1 - R$ is the probability that an ingoing particle in mode i will fall into the black hole.

In the interacting case the solution is more complicated. To get the expectation value for the number of outgoing particles in mode i one starts with the graphs of the form of Fig. 10 which give the answer in the equilibrium situation. One then breaks n lines by introducing n annihilations and creations at past infinity with mode functions $\bar{f}_{j_1}, \bar{f}_{j_2}, \ldots$ (Fig. 12 is the diagram in Fig. 10 with two lines broken). One then multiplies

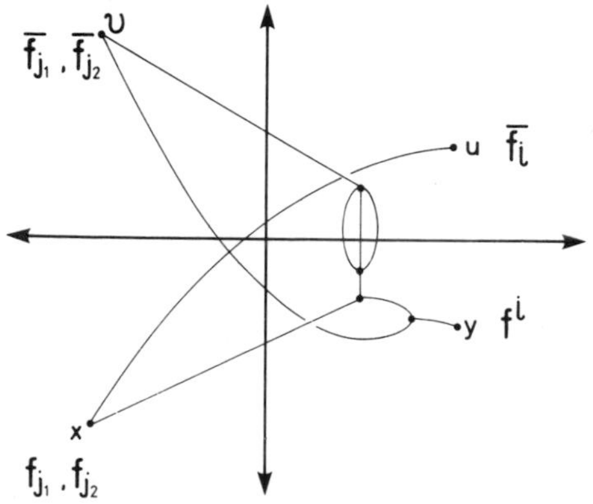

FIG. 12.

these diagrams by

$$(-1)^n \frac{\prod_{j=1}^{n} (1 \mp x_j)}{\prod_{j=1}^{n} x_j} .$$

One sums over all such graphs and then divides by a normalization factor C which is given by a sum of all graphs of the form of Fig. 12 but without the two external lines going to u and y.

With this formalism one can investigate such questions as the effect of radiative corrections on the emission of electrons and positrons and whether symmetry is restored near the black hole in theories with spontaneous symmetry breaking. One can also study whether CP non-invariant interactions can lead to an excess of baryons over antibaryons in theories in which baryon number is locally conserved. This is probably not of much practical importance in the case of black holes, but it would demonstrate that quantum gravitational effects in the early universe could give rise to the observed baryon asymmetry without resource to theories like GUTS in which baryon number is not locally conserved.

B. The super-scattering operator

A system of particles in an initial pure state can collapse to produce a black hole which then evaporates and disappears leaving radiation in a mixed state. If such a loss of quantum coherence and violation of baryon conservation can occur on a macroscopic scale, one would also expect that they could take place on a microscopic scale, because of virtual black holes produced by quantum fluctuations of the metric. To study this one should consider path integrals over asymptotically Euclidean positive definite metrics. As explained above, the fact that some of these manifolds have non-trivial topology means that the Green's functions will be acausal when analytically continued to asymptotically flat Lorentzian space at infinity. They will therefore give rise to a superscattering operator which is not the product of S-matrix operators and which increases entropy and can violate baryon conservation.

We do not know how to evaluate a path integral over all metrics on all manifolds but one can make some order of magnitude estimates of the acausal effects by considering finite dimensional approximations (Hawking, Page, and Pope 1980). These indicate that at energies, low compared to the Planck mass, the acausal effects can be represented by effective interactions which are proportional to k^s for each particle line where s is the spin and k is a typically centre of mass momentum or energy. Suppose first one considers diagrams like that in Fig. 13, in which one tries to annihilate particles at future infinity without having created anything at past infinity. Such a diagram should be zero for the following reasons. Let A be the value diagram in a fixed background metric g. By functionally differentiating A with respect to g, we obtain an object like an energy–momentum tensor

$$\frac{\delta \log A}{\delta g_{\mu\nu}} = \frac{1}{2} \sqrt{g}\, T^{\mu\nu} \;.$$

This will be locally conserved in the background metric g, i.e.

$$T^{\mu\nu}{}_{;\nu} = 0 \;.$$

However in general the flux of outgoing energy at future infinity will be greater than the ingoing flux at past infinity. This means that there can be no metric g such that the classical field equation is satisfied

$$R^{\mu\nu} - \frac{1}{2} R g^{\mu\nu} = 8\pi T^{\mu\nu} \;.$$

This in turn implies that there can be no stationary phase point in the path integral which probably means that the diagram averages to zero. Thus the outgoing density matrix will be the vacuum if no particle has been created at past infinity.

Suppose now we create some particles at past infinity and then annihilate some other particles at future infinity (Fig. 14). In this case the energy fluxes of the ingoing particles can balance those of the outgoing particles, so that it is possible to find a stationary phase metric. Thus the diagram

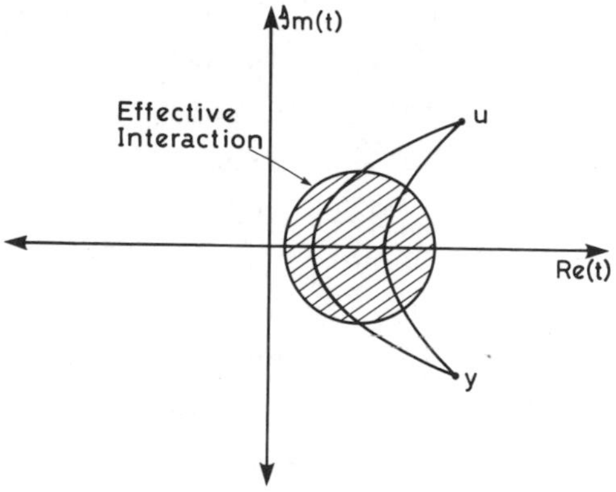

FIG. 13.

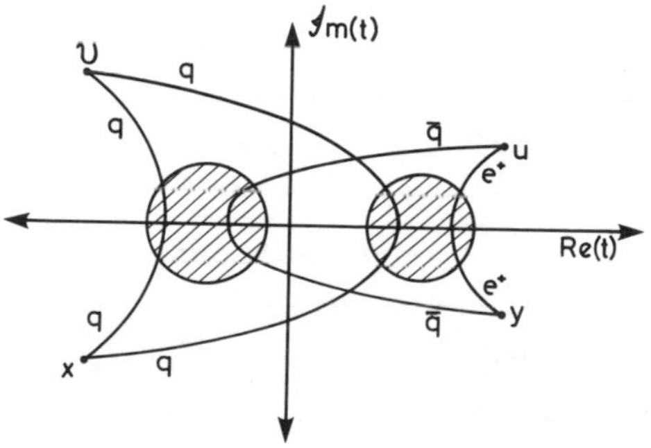

FIG. 14.

can be non-zero.

One can estimate the order of magnitude from the finite dimensional bubble calculations. In the case of proton decay one would have a diagram like Fig. 14, where the lines going to v and x correspond to two ingoing quarks and the lines going to u and y correspond to an outgoing antiquark and a positron. This diagram would be of order of magnitude

$$\left(\frac{m_{\mathrm{b}}}{m_{\mathrm{p}}}\right)^4$$

where $m_b = 1$ GeV is the mass of a baryon and $m_p = 10^{19}$ GeV is the Planck mass. This leads to a lifetime of order of 10^{50} years which is way beyond experimental test as well as being much larger than that predicted by GUTS. Such processes might be important, however, in the early universe. Another process is the decay of a muon into $e\gamma$. In this case, one would have a diagram like Fig. 15. The line from v to x represents the muon. The solid line from u to y represents the electron which interacts with the photon denoted by the wavy line at the two points. In order to get a non-zero value for this diagram, the momenta of the muon states at v and x have to be different. It is not clear to me whether such off-diagonal diagrams may not average to zero but if they do not, they will be of order

$$e^2\left(\frac{m_\mu}{m_p}\right)^2 .$$

This would give a lifetime of the order of 10^{10} years, which is way beyond observational test. The lifetime would be even longer if the muon and the electron were not point particles but were composed of some number of smaller 'preons'. On the other hand, scalar particles would decay or lose quantum coherence without any suppression by factors involving the Planck mass. The fact that we do not observe gross violation of causality or quantum coherence indicates that there cannot be any elementary scalar particles; any observed scalar particles must be bound states.

Further details will be published elsewhere.

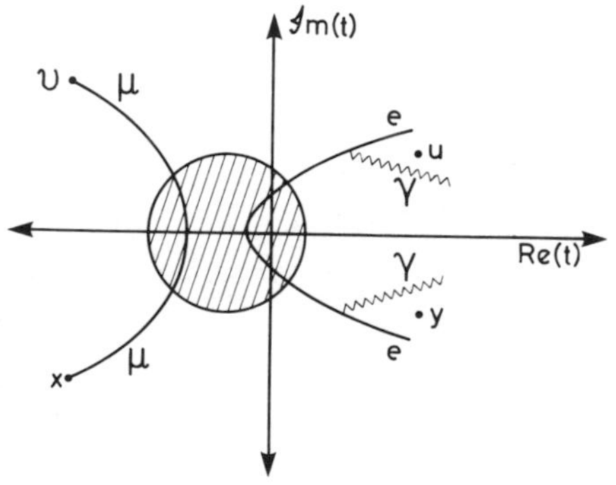

FIG. 15.

The author is grateful to I.G. Moss and N.P. Warner for their help in preparing this paper.

REFERENCES

Gibbons, G.W. and Perry, M.J. (1976). *Phys. Rev. Lett.*, **36**, 985.
—— —— (1978). *Proc. R. Soc. Lond.*, **A358**, 467.
Hartle, J.B. and Hawking, S.W. (1976). *Phys. Rev.*, **D13**, 2188.
Hawking, S.W. (1975). In *Quantum gravity – An Oxford symposium*, ed Isham, C.J., Penrose, R., Sciama, D.W. Oxford University Press; *Commun. Math. Phys.*, **43**, 199.
—— (1976). *Phys. Rev.* **D14**, 2460.
—— and Ellis, G.F.R. (1973). *The Large Scale Structure of Space-time.* Cambridge University Press.
——, Page, D.N., and Pope, C.N. (1980). *Nuc. Phys.* **B170**, 283-306.
Kronheimer, E.H. and Penrose, R. (1967). *Proc. Camb. phil. Soc.* **63**, 481.
Wald, R. (1975). *Commun. Math. Phys.*, **45**, 9.

18

QUANTIZATION OF THE RADIATIVE MODES OF THE GRAVITATIONAL FIELD

A. Ashtekar

Université de Clermont-Fd., 63170 Aubière, France

1. INTRODUCTION

Traditional attempts at obtaining a quantum theory of gravity may be divided into two broad categories: 'covariant' approaches and 'canonical' approaches. In the first method, one treats gravity in thespirit of other field theories and focuses on scattering processes involving gravitons, while in the second, one takes into account the geometrical aspect of the gravitational field and uses symplectic (i.e. Hamiltonian) methods to obtain the quantum description. While both avenues have led to a number of insights, they also have obvious limitations. In the covariant approach, for example, one begins by fixing a background metric, usually chosen to be flat, which serves as a 'kinematical arena': notions such as microcausality of field operators, asymptotic regions for incoming and outgoing states, spin and mass of the particles involves, etc., refer to this metric. Aesthetically, introduction of such a background is very unsatisfactory: it violates the very spirit of general relativity. Even if such considerations are ignored, one is still faced with severe practical limitations: since the global structure of the background space-time is fixed once and for all, one cannot hope to encompass, within such frameworks, processes involving 'topology changes' such as black-hole formation and evaporation. And, presumably, it is precisely via such qualitatively new effects that quantum gravity will make its impact felt. Indeed, detailed numerical predictions for scattering amplitudes, which have played a crucial role in the development of other field theories seem, at the moment, uninteresting in the gravitational case, given the weakness of the coupling constant and the current feasibilities in experimental physics. The issues of immediate interest are, rather, the conceptual ones. To some extent, canonical approaches have an edge over the covariant ones in this respect: canonical methods acknowledge the geometrical role of the gravitational field and, already at the classical level, try to understand ways in which general relativity *differs* from other field theories. Unfortunately, the usual Hamiltonian formulation leads to non-trivial constraints and, over the years, relatively little progress has been made in the incorporation of these at the quantum level. For example, the precise structure of the resulting Hilbert space of quantum states (or a substitute thereof) is still far from being clear (see, e.g. Kuchar 1980). Consequently, one does not yet have a sufficiently rich framework in the quantum domain in terms of which one can analyse the various conceptual issues or even formulate questions of physical interest. Thus, one has the uneasy feeling that the covariant approaches are pragmatic but not sufficiently deep while the canonical approaches are broader in their goals but not sufficiently supple to manoeuvre. The question naturally arises: can one place oneself 'in between' the two?

The purpose of this article is to present one possible way.

The new framework that we shall introduce here has the following features. As in covariant approaches, the aim is to obtain a S-matrix description, while, as in canonical methods, the passage to quantum theory is via symplectic techniques. However, *at no stage in the analysis do we introduce a background metric or linearize Einstein's equation*; gravitons, for example, arise as 'asymptotic entities' in the exact theory rather than spin-2 quanta in Minkowski space. At the same time, the resulting quantum description is quite rich in structure; it incorporates, in particular, notions – normally associated with the Poincaré invariance – such as energy–momentum, angular momentum, spin and mass, in terms of which one can hope to analyse issues of direct physical interest. Of course, by the very fact that it places itself in between the two traditional methods, the framework inherits some of their limitations. For example, due to its orientation towards the S-matrix theory, the approach is unlikely to be of direct use in the analysis of quantum fluctuations at the Planck length. On the other hand, at least in principle, it is capable of predicting macroscopic* effects (which hopefully exist) of these fluctuations; in particular, it can handle processes such as black hole formation and evaporation.

The basic idea underlying the new approach is the following. Consider, to begin with, (source-free) Maxwell fields in Minkowski space. In the usual quantization procedure one begins by isolating the 'true' degrees of freedom of this field. This is achieved via Fourier transforms: every (sufficiently regular) solution $F_{ab}(x)$ to field equations can be completely characterized by the equivalence class $\{A_a(k)\}$ of vector fields on the light cone in the momentum space, satisfying $A_a(k)k^a = 0$, where two vector fields are considered as equivalent if they differ by a multiple of k^a. (We have: $\Box A_a(x) = 0 \Leftrightarrow A_a(k)$ has support on the light cone; $\nabla^a A_a(x) = 0 \Leftrightarrow A_a(k)k^a = 0$; and, the restricted gauge freedom in $A_a(x) \Leftrightarrow$ the equivalence relation on $A_a(k)$.) Thus, $\{A_a(k)\}$ can be considered as the freely specifiable data and represents the radiative modes associated with $F_{ab}(x)$. In quantum theory, each $\{A_a(k)\}$ leads to a creation and an annihilation operator, or, equivalently, to a one photon state. One would like to proceed along similar lines in the gravitational case. Unfortunately, the above procedure cannot be extended unless one introduces a suitable background metric: in general relativity, a natural momentum space is simply not available! Can one somehow isolate the radiative degrees of freedom of $F_{ab}(x)$ *without* any reference to Fourier transforms? It turns out that the answer is in fact 'yes': Penrose's (1963) null infinity, I, provides a way. For, every regular, source-free solution F_{ab} to Maxwell's equations in Minkowski space can also be characterized (Penrose 1965) by its null datum ε_a on I, the pull back to I of $F_{ab}n^b$, where n^b is the null normal to I. The two freely specifiable components of ε_a represent the radiative degrees of freedom associated with F_{ab}. Using the presence of the BMS group (Bondi, Metzner, and Van der Burg 1961; Sachs 1962) on I, one can quantize ε_a directly, without *any* reference to the interior of space-time. Via field equations, the resulting quantum description turns out to be equivalent to the usual

---

* By 'macroscopic' we only mean length scales much larger than the Planck length.

one. The idea now is to carry out a similar 'asymptotic quantization' in the gravitational case.*

There is a fair amount of diversity in the discussion that follows: one needs notions from the gravitational radiation theory which may be unfamiliar to most quantum theorists and also some technical material from quantum field theory to which most relativists may not be exposed. In order to make the article accessible to both sets of readers, I will focus only on main ideas. I apologize in advance to the experts who may be disappointed by the lack of detailed proofs; these will be published elsewhere (Ashtekar 1980*b*).

2. ISOLATION OF RADIATIVE MODES OF THE GRAVITATIONAL FIELD

In order to investigate the properties of the gravitational radiation in a coordinate free manner, Penrose (1963, 1965) introduced the notion of conformal completion: one brings infinity to a finite distance via a conformal transformation and considers it as a boundary of space-time. The asymptotic properties of the gravitational field can then be investigated using techniques from local differential geometry at the points on the boundary. Let $(\hat{M},\hat{g}_{ab})$ denote the physical space-time, and (M,g_{ab}) a Penrose completion thereof. Then, we have: $M = \hat{M} \cup \partial M$, and, on $\hat{M}$, $g_{ab} = \Omega^2\hat{g}_{ab}$ for some function Ω. ∂M, the newly attached boundary representing infinity, is denoted by I. The function Ω vanishes on I, reflecting the fact that I is at infinite 'distance' w.r.t. the physical metric $\hat{g}_{ab}$ from points in the physical space-time $\hat{M}$. If Einstein's vacuum equation is satisfied by $\hat{g}_{ab}$ in a neighbourhood of I (*as we shall always assume*) one can show, using the regularity of g_{ab} on M, that I is a null surface: it may be thought of as a null-cone at infinity. In general I has two disconnected pieces, I^+ and I^-, which serve, respectively, as the future and the past boundary of $\hat{M}$ in M. They provide a natural arena for the S-matrix theory of zero rest mass fields.

---

* An analysis of Yang–Mills fields along these lines has recently been carried out by Newman (1978, 1980) and by Hawking and Pope (1979). The approach to quantum gravity using null infinity was suggested by Sachs as early as 1962!

It turns out that if a space-time admits a conformal completion with the required properties, then, one can always choose a conformal factor Ω which makes I divergence-free; i.e., which satisfies $\nabla^a\nabla_a\Omega = 0$ on I, where ∇ is the derivative operator on (M,g_{ab}). Throughout our discussion, we assume that such a choice has been made. What is the structure available on I? First, because it is a null surface its normal, $\nabla^a\Omega =: n^a$, is also tangential to I. It is easy to show that the field n^a is geodetic; I is ruled by null geodesics. The orbits of n^a are called *generators* of I. The space S of these generators is diffeomorphic to S^2; I has a topology of $S^2 \times R$. The pull-back to I of the metric g_{ab} gives us a tensor field q_{ab} on I. This q_{ab} is the degenerate intrinsic metric of the null surface I: $q_{ab}V^b = 0$ if and only if V^b is proportional to n^b. Using the fact that I is divergence-free, one can show that $L_n\, q_{ab} = 0$. Thus, q_{ab} is the lift to I of a (positive definite) metric on the space S of generators. The pair (q_{ab},n^a) is the 'first-order' structure on I.

The 'second-order' structure is induced by the derivative operator ∇ on (M,g_{ab}). This ∇ gives rise to a unique (torsion-free) derivative operator D *within* I which satisfies*

$$D_a q_{bc} = 0 \qquad D_a n^b = 0 \qquad \text{and,} \tag{2.1a}$$

$$D_a V_b = 2D_{[a}V_{b]} + 2L_V\, q_{ab}, \text{ if } V_a n^a = 0 \; ; \tag{2.1b}$$

where, V^a is any vector field on I such that $V^a q_{ab} = V_b$, the given co-vector field on I. (Note: since I is a null surface, we must distinguish between vectors and co-vectors when we discuss fields which are intrinsic to I.) Equation (2.1a) is not surprising. Equation (2.1b) gives the action of D on all co-vectors V_b orthogonal to n^a in terms of exterior and Lie derivatives, neither of which requires a derivative operator for its definition. Thus, *any* derivative operator D, induced on I by *some* ∇ has a prescribed action on V_b if $V.n = 0$: such a D can be completely specified by giving its action on a co-vector field 1_a with $1_a n^a = 1$ on I.

---

* For details, see Geroch (1977).

Next, we note that there is a certain amount of 'conformal freedom' in all these fields. For, if (M, g_{ab}) is a permissible completion of the physical space-time $(\hat{M}, \hat{g}_{ab})$, so is $(M, g'_{ab} = \omega^2 g_{ab})$ where ω is a smooth nowhere vanishing function on M. The requirement that I be divergence-free restricts this freedom somewhat: ω must satisfy $L_n \omega = 0$ on I. Under this rescaling, we have:

$$q_{ab} \to q'_{ab} = \omega^2 q_{ab}; \qquad n^a \to n'^a = \omega^{-1} n^a \tag{2.2a}$$

$$D_a K_b \to D'_a K_b = D_a K_b - 2\omega^{-1} K_{(a} D_{b)} \omega + (\underline{\omega}^m K_m) \omega^{-1} q_{ab} \tag{2.2b}$$

where $\underline{\omega}^m$ is the restriction to I of $\nabla^m \omega$. A curious thing happens if we require $\omega = 1$ *on* I (but not necessarily on M). We have: $q'_{ab} = q_{ab}$; $n'^a = n^a$; but $D' \neq D$! Indeed, $D'_a K_b = D_a K_b + f K_c n^c q_{ab}$; where $\nabla^m \omega = f n^m$ on I. (Since ω is constant on I, $D_m \omega = 0$ and $\nabla^m \omega$ is parallel to n^a, the null normal to I.) Thus, fixing a 'conformal frame' (q_{ab}, n^a) on I does not fix the derivative operator D. This merely reflects the fact that D contains second-order structure while (q_{ab}, n^a) contains only the first-order structure; although defined intrinsically on I, D has a better memory of the particular conformal metric g_{ab} that has been used in the Penrose completion.

The asymptotic properties of the physical metric $\hat{g}_{ab}$ are thus coded (to first and second order) in the equivalence classes of triplets (q, n, D) satisfying eqn. (2.1) where two triplets, (q, n, D) and (q', n', D') are considered as equivalent if they are related by eqn. (2.2). In practice, it is often easier to fix a conformal frame (q_{ab}, n^a) on I and work with an equivalence class $\{D\}$ of derivatives, satisfying eqn. (2.1), where D and $\tilde{D}$ are considered as equivalent if and only if there exists a function f on I such that

$$(\tilde{D}_a - D_a) K_b = f K_c n^c q_{ab} \tag{2.3}$$

for all K_b on I. In what follows, we shall work with these $\{D\}$.

It turns out that the equivalence class $\{D\}$ represents, precisely, the 'radiative degrees of freedom' associated with the physical metric $\hat{g}_{ab}$! To see this, let us fix a D in $\{D\}$ and compute its curvature tensor. From eqn. (2.1), it follows that there exists on I a tensor field $S_a{}^b$ such that:

$$2D_{[a}D_{b]}K_c =: \mathcal{R}_{abc}{}^d K_d = (q_{c[a}S_{b]}{}^d + S_{c[a}\delta_{b]}{}^d). \qquad (2.4)$$

where, $S_{ab} = S_a{}^c q_{ab}$ (which turns out to be symmetric). Unfortunately, $S_a{}^b$ does not have an invariant significance: it has a complicated behaviour under conformal rescalings of eqn. (2.2). One can, however, extract the 'invariant part' of $S_a{}^b$ by subtracting from it a tensor field which absorbs this complicated structure: there exists on I a tensor field (for details see Geroch 1977) ρ_{ab}, independent of our choice of D, such that

$$N_{ab} := S_{ab} - \rho_{ab} \qquad (2.5)$$

is *invariant* under conformal rescalings. This N_{ab} — the 'invariant' part of the curvature of D — is called the News tensor. It contains information about the energy—momentum carried away by gravitational waves. In analogy with gauge theories, one can regard {D} as the 'potential' and N_{ab} as the 'field'.*

To summarize, null infinity, I, is equipped with triplets (q_{ab}, n^a, D_a). The pairs (q_{ab}, n^a) have only 'kinematical' information; we obtain the same collection of these pairs on I irrespective of which particular space-time is being analysed. The connections D, on the other hand, differ from one space-time to another; they contain the information about 'dynamical' or 'radiative' degrees of freedom of the gravitational field.

3. THE SYMPLECTIC DESCRIPTION

We now wish to construct the phase space of *all* radiative modes

---

* One can also recover the components containing radiative information of the asymptotic Weyl curvature directly from {D}. Set $*K^{ab} = (1/4)\varepsilon^{amn}D_m S_n{}^b$. Then, one can show that $*K^{ab}$ is symmetric and trace-free: its five independent components represent $\psi_o{}^4, \psi_o{}^3$ and $\mathrm{Im}\psi_o{}^2$ in the Newman-Penrose notation. (The real part of $\psi_o{}^2$, which contains information about longitudinal modes, on the other hand, cannot be recovered from {D}; {D} contains only 'radiative' information!) In what follows we shall say that {D} has *trivial curvature* if $N_{ab} = 0$ and $*K^{ab} = 0$.

For this purpose, we must first introduce a 'kinematical arena' on which the action is to take place. If we were dealing with any field other than gravitation, the underlying space-time itself would have served as this arena. In the present case, however, we have no background geometry to refer to. Rather, we must deal with the collection of all possible space-times! The discussion in the previous section suggests that we use the 3-manifold I (together with its universal structure, i.e., the structure it inherits irrespective of which asymptotically flat space-time is being considered) as the required arena. This will enable us to handle all space-times which are asymptotically flat irrespective of their topologies in the 'interior'.

Fix a 3-manifold I equipped with a collection of pairs (q_{ab}, n^a) of nowhere vanishing fields satisfying: (i) $q_{ab}V^b = 0$ if and only if V^b is proportional to n^b; (ii) $L_n q_{ab} = 0$; (iii) pairs (q_{ab}, n^a) and (q'_{ab}, n'^b) are both in the collection if and only if there exists a function ω, with $L_n\omega = 0$, such that $q'_{ab} = \omega^2 q_{ab}$; and $n'^a = \omega^{-1} n^a$; and, (iv) the vector field n^a is complete and manifold of its orbits is diffeomorphic to S^2. (The last condition ensures that I has the correct global structure.) This is the kinematical arena on which various fields of interest are to live. Note that I is considered here as an *abstract* 3-manifold; there is *no* background space-time!*

Using the discussion of the previous section as a guide, we proceed as follows. Fix a pair (q_{ab}, n^a) on I. Denote by C the collection of derivative operators D which satisfy eqn. (2.1) on I. Two elements D and $\tilde{\mathrm{D}}$ will be said to be equivalent if there exists a function f relating them via eqn. (2.3). Denote each equivalence class by {D} and the collection of all equivalence classes by Γ. This Γ is the required phase space. It has the structure of an affine space: there is no natural 'origin' in Γ.

Let us first 'count' the degrees of freedom available. Using eqn. (2.1) it is easy to show that any two derivative operators D and D′ in C are related by: $(\mathrm{D}'_a - \mathrm{D}_a)K_b = \Sigma_{ab} n^c K_c$, for some symmetric tensor field Σ_{ab} satisfying $\Sigma_{ab} n^b = 0$. Hence, the difference between the corresponding equivalence classes, {D}

---

* The Cheshire cat disappears and only the smile remains!

and {D′} can be characterized by $\gamma_{ab} = \Sigma_{ab} - 1/2 q_{ab}\Sigma_{mn}q^{mn}$, the trace-free part of Σ_{ab}. (Here, q^{ab} is any 'inverse' of q_{ab} satisfying $q_{am}q^{ab}q_{bn} = q_{mn}$. Since q_{ab} is degenerate, q^{ab} is not unique: it is determined only up to addition of the terms of the type $V^{(a}n^{b)}$ where V^a is any vector field on I.) Therefore, by fixing an arbitrary element {D°} as the origin, one can coordinatize Γ by γ_{ab}. How many components does γ_{ab} have? Precisely two! This suggests that Γ has the 'right size'.

However, to establish rigorously that Γ is the correct phase space, one must solve the initial value problem on I. That is, one must show that every {D} in Γ leads to a unique asymptotically flat solution of Einstein's equation: the discussion of the previous section shows only that *given* an asymptotically flat space-time, the information about its radiative aspects are coded in the equivalence class {D}. This initial value problem is obviously very hard. Some results in this direction were obtained by Newman and Unti as early as 1962. These do indicate that a given {D} whose news vanishes in the future (respectively, past) of a cross-section of I will lead, upon integration of Einstein's equation, to a unique solution (up to isometries) which is asymptotically flat at future (past) null infinity and *flat* in the future (past) of a retarded (advanced) instant of time. (The restriction to space-times which are flat in the future (past) of a retarded (advanced) instant ensures that there are no bound states in the future (past); the uniqueness claim would otherwise be false trivially!) However, certain issues of mathematical rigour – such as convergence of the power series appearing in the solution – were ignored in this analysis, and, surprisingly, a complete resolution of this issue is still not in sight.

It is desirable, therefore, to have some independent checks that Γ is indeed the correct phase space. A powerful check is provided by the action of the BMS group on I which is the group of asymptotic symmetries in presence of gravitational radiation.*

* Just as the Poincaré group is a semi-direct product of the Translation group with the Lorentz, the BMS group is a semi-direct product of the super-translation group (which is infinite dimensional) with the Lorentz. The BMS group does admit a preferred 4-dimensional translation subgroup. It does not, however, admit a preferred Poincaré subgroup. For details, see Sachs (1962).

To see this, we must first introduce a symplectic structure Ω on Γ which will also lead to the canonical commutation relations in the next section. Ω is to be a tensor field on Γ. The tangent vectors at any point {D} of Γ may be interpreted as the linearized radiative modes in the background space-time corresponding to {D}. Hence, what we need is a symplectic tensor for linearized gravity, but off arbitrary radiative backgrounds. An extension of the results of Palmer (1978) on these tensors using the information provided by Geroch and Xanthopoulos (1978) on the asymptotic behaviour of gravitational perturbations yields the desired result:

$$\Omega_{\{D\}}(\gamma,\tilde{\gamma}) = \int_{I} (\gamma_{ab}L_n\tilde{\gamma}_{cd} - \tilde{\gamma}_{ab}L_n\gamma_{cd})q^{ac}q^{bd}dI \tag{3.1}$$

where γ and $\tilde{\gamma}$ (satisfying $\gamma_{ab} = \gamma_{(ab)}$; $\gamma_{ab}n^b = 0$; $\gamma_{ab}q^{ab} = 0$ on I) represent any two tangent vectors at the point {D} on Γ, and, q^{ab}, as before, is any 'inverse' of q_{ab}. (Note: since we have $\gamma_{ab}n^b = 0$, the expression is independent of which 'inverse' q^{ab} one uses.) It is easy to verify that Ω is conformally invariant (i.e. remains unchanged under transformations induced by eqn. (2.2)) and has, as one might expect, the dimensions of action. Finally, by inspection, Ω is a constant tensor field (w.r.t. the affine structure) on Γ.

The action of the BMS group on I provides a step by step check on our construction of (Γ,Ω). First, the action leaves the collection C of our preferred derivative operators invariant. (While eqn. (2.1a) is trivially preserved, the fact that eqn. (2.1b) is also preserved is non-trivial.) Next, the action respects our equivalence relation in eqn. (2.3): if D and $\tilde{D}$ are equivalent, so are their images under any BMS transformation. Finally, the induced action on Γ preserves the symplectic structure Ω. Thus, the 'kinematical' symmetry group associated with I automatically gives rise to canonical transformations (i.e. symmetries) in the phase-space description. One can compute the corresponding generating functions, i.e. Hamiltonians. For the special case of BMS translations, αn^a, one obtains:

$$H_{\alpha n}(\{D\}) = \int_{I} \alpha N_{ab}N_{cd}q^{ac}q^{bd}dI \tag{3.2}$$

which is precisely the 4-momentum flux formula obtained by Bondi *et al.* (1961); Sachs (1962); Penrose (1963); Winicour (1968), using entirely different considerations!

Finally, let us consider the collection of 'classical vacua'. These are the points $\{\underset{o}{D}\}$ in Γ with trivial curvature. (Since the news N_{ab} of $\{\underset{\bullet}{D}\}$ vanishes, so does the associated energy in any 'rest-frame' (eqn. (3.2)). Hence the terms 'vacua'.) A detailed examination of the BMS action on Γ shows that although the entire group maps the collection of these vacua to itself, it is only the (4-dimensional) translation subgroup that leaves each individual $\{\underset{o}{D}\}$ invariant. Furthermore, given *any* two vacua, there exists a BMS supertranslation connecting them. Thus, the quotient, ST/T, of the supertranslation subgroup by the translation subgroup, acts *effectively and transitively* on the collection of classical vacua. This fact plays an important role in the quantum theory.

4. QUANTIZATION

4.1 *The spin-2 gravitons*

The first step in quantization will be the construction of the algebra of quantum operators using the classical symplectic description. Recall that, due to factor-ordering problems, not all classical observables can admit unambiguous classical analogues. One must first single out a preferred collection of classical observables (closed under Poisson brackets) which are to be promoted directly to quantum operators. How is this selection to be made? Quantization schemes applicable to simpler systems provide a hint. Consider for example non-relativistic systems. Not only is their phase-space a symplectic manifold, but it also admits a natural cotangent bundle structure. It is *precisely* those classical observables which generate canonical transformations preserving this additional structure (i.e. observables independent or linear in momenta) that admit natural quantum analogues! (See, e.g. Ashtekar 1980*a*.) The situation is similar also for relativistic free fields. In this case, the phase space is, naturally, a vector space, and, it is again *precisely* the generators of canonical transformations preserving this additional linear structure (i.e., the linear observables) that lead, unambiguously, to field operators. (See, e.g. Ashtekar and Magnon-Ashtekar 1980.) In the present case, the additional structure is that of an affine space. What are the generators of canonical transformations which preserve the affine character of Γ? These are (constants and) the functions on Γ of the type

$$N(f) := \int_I N_{ab} f_{cd} q^{ac} q^{bd} dI \quad ,$$

where, f_{ab}, satisfying $f_{ab} = f_{(ab)}$; $f_{ab}n^b = 0$; and $f_{ab}q^{ab} = 0$ is a test field* on I. (The Hamiltonian vector-field generated by $N(f)$ is the constant vector field on Γ represented by the tensor field f_{ab} on I.) These observables have the following Poisson brackets: $\{N(f), N(f')\} = \Omega(f,f')$. To construct the algebra A of quantum operators, therefore, we proceed as follows. Introduce on I an operator-valued-distribution $\underline{N}_{ab}(x)$ (which will be called the news operator) subject to the canonical commutation relations (CCR):

$$[\underline{N}(f), \underline{N}(f')] = \hbar/i \cdot \Omega(f,f')I \tag{3.3}$$

where $N(f) = \int_I \underline{N}_{ab}(x) f_{cd}(x) q^{ac} q^{bd} dI$. The required A is the *-algebra (over complexes) generated, as usual, by the smeared out news operators $\underline{N}(f)$.

Next, we construct the Fock space of states. Note first that the affine parameter u, along the integral curves of n^a (i.e. a function u on I satisfying $L_n u = 1$) provides a natural decomposition of fields into positive and negative frequency parts: $f_{ab} = f^+_{ab} + f^-_{ab}$. (The decomposition is invariant under conformal rescalings $n^a \to \omega^{-1} n^a$ since $L_n \omega = 0$.) Let H denote the (complex) Hilbert space of positive frequency fields f^+_{ab} (having the same algebraic properties as the test fields f_{ab}), equipped with the inner-product $\langle f^+, f'^+ \rangle = (i/\hbar)\Omega(\overline{f^+}, f'^+)$, where the bar denotes complex-conjugation, and F the symmetric Fock space over H. Then, H is the Hilbert space of one-graviton states and one can

---

* For definiteness, we shall use the space S(I) of C^∞ test fields which, together with all their derivatives, fall off faster than any power of u as u tends to $\pm\infty$ on I.

represent, as usual, the operators $\underline{N}(f)$ by sums of creation and annihilation operators on F. The choice of the inner-product ensures that these sums satisfy the CCR.

The BMS group on I has a natural action of the space H of 1-graviton states since the notion of positive frequency fields is BMS invariant. This action is in fact *unitary*: recall that the BMS group preserves the symplectic structure Ω! This representation of the BMS group is, however, reducible: it can be decomposed into two irreducible parts corresponding to the two possible helicities of gravitons. More precisely, we have the following. A graviton state f^{+}_{ab} will be said to be *right-handed* if $\varepsilon^{mnp} l_p q_{nb} f^{+}_{am} = i f^{+}_{ab}$, and, *left-handed* if $\varepsilon^{mnp} l_p q_{nb} f^{+}_{am} = - i f^{+}_{ab}$, where, ε^{mnp} is the natural alternating tensor on (I, q_{ab}) and l_p, *any* co-vector field on I such that $l_p n^p = 1$. Then, we have: $H = H_R \oplus H_L$, where H_R contains only the right-handed gravitons and H_L only the left-handed. Each of H_R and H_L provides us with an irreducible representation of the BMS group as well as that of *any* of its Poincaré subgroups. One can therefore compute the values of the Poincaré Casimir operators: these are, $m = 0$ and $s = \pm 2$, positive sign for H_R and negative for H_L.

Finally, the action of the BMS group also provides us with supermomentum—angular momentum operators on F.

4.2 *New graviton states with internal charge*

It turns out that the CCR admit other, unitarily inequivalent representations. These owe their existence to the enlargement of the four dimensional translation group to the infinite dimensional supertranslation group that occurs in the presence of gravitational radiation. Although this enlargement is an essential feature of general relativity, one cannot see it in perturbative treatments off flat (or, more generally, stationary) backgrounds. Hence, it is likely that the new representations contain *physical* information which cannot be captured in the Minkowskian spin-2 frameworks.

The emergence of these representations is analogous to that of the 'topologically charged' ones in the case of scalar field Φ satisfying, in 2-dimensional Minkowski space, $\Box \Phi = 0$. Let me therefore briefly recall the 2-dimensional model. (For details, see Streater and Wilde (1970) or Streater (1974).) In the classical description of this model, there appears a (trivial) vacuum degeneracy: the field $\Phi = k$, being of zero energy, represents a ground state for all constants k. This degeneracy is characteristic of the zero rest-mass fields and does not depend on the dimensionality of the underlying space-time. The fact that the dimension is two has another consequence: using the alternating tensor ε^{ab} one can now construct a conserved current $J^a := \varepsilon^{ab}\nabla_b\Phi$. The corresponding 'topological charge' Q is given by $Q = \Phi(x = +\infty, t) - \Phi(x = -\infty, t)$. Note that the charge associated with any one classical vacuum is zero; *all* classical vacua belong to the trivial sector. It is the topological charge (rather than the vacuum degeneracy itself) that leads to interesting structure in quantum theory. For, due to technical reasons associated with the fact that the space-time is 2-dimensional, it is only the operator-valued distribution $d\underline{\Phi}/dx$ (rather than $\underline{\Phi}(x)$ itself) that can be considered as a *local* Weightman field. (More precisely, one can smear out $\underline{\Phi}(x)$ only by test fields $f(x)$ for which $\int_0^x f(y)dy \in S(R)$, rather than by arbitrary elements of $S(R)$.) Consequently, the quantum fluctuations off *all* classical vacua lead to the *same* (Fock) representation of the CCR; the 1-particle states in this representation correspond to classical fields with zero topological charge. (Note that this is in sharp contrast with the situation in the 4-dimensional space-times where each classical vacuum leads, individually, to a quantum vacuum (and hence to a new representation of the CCR) due to spontaneous breaking of the classical symmetry $\Phi \to \Phi + k$ (Streater 1965; Ashtekar and Sen 1980). In the 2-dimensional model, the symmetry is *not* broken! New representations arise due to quantum fluctuations of classical configurations with non-zero topological charges; i.e. configurations which 'connect' two distinct vacua $\Phi = k := \Phi(-\infty, t)$ and $\Phi = k' := (+\infty, t)$. This situation is different from the more familiar one in $\lambda\Phi^4$ or Yang–Mills theories where each non-trivial classical vacuum itself (rather than the configuration connecting two such vacua) leads to a new representation of the CCR.

Let us now return to the gravitational field. Consider the classical observable $Q_{ab}(\theta, \phi)$ defined on the phase space Γ by

$Q_{ab}(\theta,\phi) = \int_{-\infty}^{\infty} N_{ab}(u,\theta,\phi)du$ where u is an affine parameter along the generators of I and where θ and ϕ label the space of generators.* What is the interpretation of these observables? Fix a point {D} of Γ whose curvature is trivial outside a compact region of I. Then, in the future of this region, {D} coincides with a classical vacuum $\{\underset{o}{D}\}$ and in the past, with another classical vacuum $\{\underset{o}{D}'\}$: one can regard {D} as connecting two vacua. When does $\{\underset{o}{D}\}$ coincide with $\{\underset{o}{D}'\}$? Precisely when $Q_{ab}(\theta,\phi) = 0$! In general, Q_{ab} is a measure of the 'difference' between $\{\underset{o}{D}\}$ and $\{\underset{o}{D}'\}$. Recall, however, that this difference determines a unique element of ST/T. Hence, there is a natural isomorphism between the collection (or, more precisely, the group under addition) of all possible Q_{ab} and the group ST/T. There is a simple geometric meaning to this isomorphism: Q_{ab} represents the difference between the asymptotic 'shears' associated with {D} while the corresponding element of ST/T gives the amount by which the shear-free cross-sections (i.e. 'good cuts') of I in the future are supertranslated w.r.t. those in the past. The observable $Q_{ab}(\theta,\phi)$ has an interesting relation with observables $N(f)$ introduced before: the Poisson brackets $\{Q_{ab}(\theta,\phi),N(f)\}$ vanish for all test fields f_{ab}! This suggests that in the quantum description, there would exist several inequivalent irreducible representations of the algebra A, labelled by values of $Q_{ab}(\theta,\phi)$. This is indeed what happens: quantum analogue of $Q_{ab}(\theta,\phi)$ exists and commute with all smeared-out field operators, thereby acting as Casimir operators for classifying the irreducible representations of the CCR. The eigenvalues of this operator can be thought of as 'internal charges' which take values in ST/T. The standard Fock representation has, of course, zero charge.

Starting from the Fock representation, one can explicitly construct the 'internally-charged' representations as follows. Consider the automorphism Λ generated on A by the transformation

---

* It would have been more elegant to avoid charts on I and consider instead observables of the type $Q(k) = \int N_{ab}k_{cd}q^{ac}q^{bd}dI$; where k_{ab} has the same algebraic properties as the test fields f_{ab} but also satisfies $L_n k_{ab} = 0$. The observables $Q(k)$ obtained by considering all possible k_{ab} have the same information as $Q_{ab}(\theta,\phi)$. It just turns out that the ideas will be more transparent with $Q_{ab}(\theta,\phi)$.

$$\underline{N}_{ab}(x) \to \Lambda \cdot \underline{N}_{ab}(x) = \underline{N}_{ab}(x) + N^{\circ}_{ab}(x)\underline{I} \tag{3.4}$$

where $N^{\circ}_{ab}(x)$ is the news tensor of a {D} in Γ whose curvature is trivial outside a compact region. This automorphism can be shown to be unitarily implementable on the Fock representation if and only if $Q^{\circ}_{ab} \equiv \int_{-\infty}^{\infty} N^{\circ}_{ab} du = 0$! Now, any automorphism Λ which fails to be unitarily implementable on F naturally leads to inequivalent representations of A. For, one can just define a new vacuum expectation value functional V_Λ on A by $V_\Lambda(A) = V(\Lambda \cdot A)$ for all A in A where V is the Fock vacuum expectation value functional, and construct a representation of A via the gelfand-Naimark (1943)—Segal (1947) construction; Unitary inequivalence with the Fock representation is ensured by the fact that Λ fails to be unitarily implementable on F. Thus, by choosing an appropriate $N^{\bullet}_{ab}(x)$, we can construct an irreducible representation of the CCR with any prescribed internal charge. We wish to emphasize that, since $Q_{ab}(\theta,\phi)$ vanishes at *all* classical vacua, quantum fluctuations off *any* of these vacua define states in the Fock representation; the internally charged quantum states correspond to fluctuations of classical configurations {D} at which $Q_{ab}(\theta,\phi)$ fails to vanish. Thus, the situation resembles the 2 dimensional model discussed above ($\underline{N}_{ab}$ playing the role of $d\underline{\Phi}/d\underline{x}$ and $Q_{ab}(\theta,\phi)$, of Q) and differs from that in $\lambda\Phi^4$ and Yang-Mills theories.

It turns out that the new, internally charged graviton states have a well-defined mass (which is again zero) but *not* spin. This occurs because although the translation sub-group T leaves each charged sector (i.e. irreducible representation of the CCR) invariant, *no* charged sector is left invariant by *any* Poincaré subgroup of the BMS group. Therefore, at least in the conventional sense of the terms, spin is just not a good quantum number for these 'particles'. An alternate (and perhaps better) viewpoint would be the following. One may consider the direct sum of all charged representations of the CCR and investigate the action of the Poincaré subgroups of the BMS group on it. Each irreducible representation of the Poincaré group will carry a definite mass as well as spin. However, any one of these representations would contain admixtures of states with different internal charges. The viewpoint would be that the particle interpretation should be associated with these irreducible representations of the Poincaré or the BMS group, rather than with the irreducible representations of the CCR. One would then be led to (massless) gravitons with well-defined spins, which, however, need not be confined to the value 2. (Compare: Friedman and Sorkin 1980.) According to this stand, internal charge would not be a good quantum number; gravitons with a given value of spin could emit a variety of internal charges. Note that both viewpoints agree as far as the Fock representation is concerned: since the Fock space is invariant under the action of the algebra A *and* since the 1-particle Hilbert space is Poincaré invariant, these states carry a well-defined mass ($m = 0$) and spin ($|s| = 2$) *as well as* the internal charge ($Q_{ab} = 0$). The ambiguity pertains only to the states which do *not* belong to the Fock representation.

In the classical description, all states (i.e. elements of Γ) were treated on equal footing; the internal charge $Q_{ab}(\theta,\phi)$ played no essential role. Why is then the situation so different in quantum theory? How did it come about that the Fock representation admits only those states which have zero charge? The essential reason is that new input was fed in the transition between the classical to the quantum description: we required that only those fields f_{ab} which have finite norm $\langle f,f\rangle < \infty$, be admitted in the 1-particle Hilbert space. More precisely, the situation is the following. Each test-field f_{ab} defines a constant vector field on Γ. Since Γ has an affine structure, each f_{ab} inherits an internal charge: the difference between the values of $Q_{ab}(\theta,\phi)$ at any two points of Γ 'connected' by the constant vector field f_{ab}. It is easy to verify that the norm $\langle f,f\rangle$ of f_{ab} is finite only if its charge vanishes; i.e., only if the charge of any two classical states connected by f_{ab} is the same! In the classical limit, this requirement just disappears and all constant vector fields appear on the same footing. Thus, the distinction between the charged and the uncharged states emerges as a genuinely quantum phenomenon.

5. OUTLOOK

Using the notion of null infinity, we have obtained a new kinematic setup for quantum gravity. The starting point is the framework describing gravitational radiation in *exact* general relativity. That one should take this framework seriously in quantum physics was stressed by several authors (particularly McCarthy 1972) in the early seventies. However, most of these discussions pertained to issues such as the classification of elementary particles using the BMS group, involving the effects of the boundary conditions provided by the gravitational radiation theory on the rest of physics. The present standpoint is much more conservative in that we focus on the gravitational field itself: here, the idea is simply to take these boundary conditions seriously in the problem of quantization of the gravitational radiative modes! Put differently, we have simply applied the established techniques from quantum theory to the established, exact, classical theory of gravitational radiation in a way that does not violate the spirit of either general relativity or quantum mechanics. The result is a framework which is more geometric than the one used in the covariant approaches and, in the quantum domain, more complete than the one provided by the canonical approaches.

The framework does, none the less, share a number of features with the two approaches – as well as with the programmes based on Newman's H-spaces (see e.g. Newman, Ko, and Todd 1977) and Penrose's (1976) non-linear graviton – so that an exchange of techniques and ideas is feasible. For example, as in the covariant approaches, the emphasis is on the S-matrix description. Thus, the mathematical machinery includes operator valued distributions, Fock spaces of ingoing and outcoming states and the notion of zero rest mass, spin-2 gravitons. This may enable one to reformulate the S-matrix theory in a way which avoids the introduction of a background geometry and of linearization of Einstein's equation, reformulation which may perhaps be free of some of the difficulties faced in the standard theory. As in the canonical approaches, we have a symplectic description at the classical level and the passage to quantum theory is via the canonical commutation relations. The only significant difference is that the present analysis uses $\mathcal{I}$ in place of Cauchy surfaces. One may therefore gain new insight into the constraint and the Hilbert space problem faced in the canonical approaches by analysing in detail the relation between the two mathematical frameworks. Next, consider the H-space programme. Since the construction of the H-space itself is based on the $\mathcal{I}$-formalism, the programme shares a number of mathematical structures with the present approach. For example, the internal-charge observable $Q_{ab}(\theta,\phi)$ plays an important role also in the H-space theory. It would be certainly fruitful to investigate the detailed relation. Finally, it appears that the equivalence classes {D} of connections which constitute the phase space Γ in our description arise naturally also in the twistor framework (Penrose 1980). Thus, not only does the present approach have qualitative ideas in common with a number of apparently disjoint approaches, but it also appears to be closely related to these approaches in its detailed mathematical structure. Hence, it may serve as a common platform from where one can compare and contrast the ideas arising from different sources.

Although the emphasis of the present programme is on the S-matrix theory, its primary goal is to shed light on conceptual issues rather than to obtain numerical predictions. The situation one has in mind is the following: states come in from $\mathcal{I}^-$, interact and scatter off to $\mathcal{I}^+$. At the classical level, the incoming states would be just gravitational radiation. Typically, news would have compact support in the future of a cross-section of $\mathcal{I}^-$ and the space-time would be flat in the past of a retarded instant of time. If the radiation is properly focused, one would be able to form a bound state — say a black-hole. Such an occurrence could be deduced *directly* from the classical S-matrix, without any information on the details of the space-time: known relations (Ashtekar and Magnon-Ashtekar 1979) between the ADM and Bondi 4-momenta imply that a bound state would occur if and only if the energy contained in the incoming waves is greater than that in the outgoing waves; i.e. if and only if

$$\left|\sum_{0}^{3} \varepsilon_i \left|\int_{\mathcal{I}^-} \alpha_i N_{ab} N_{cd} q^{ac} q^{bd} d\mathcal{I}^-\right|^2\right| > \left|\sum_{0}^{3} \varepsilon_i \left|\int_{\mathcal{I}^+} \alpha_i N_{ab} N_{cd} q^{ac} q^{bd} d\mathcal{I}^+\right|^2\right|,$$

where the vector fields $\underset{i}{\alpha} n^a$ on $I^{\pm}$ represent orthonormal basis vectors in the space of BMS translations, $\varepsilon_0 = -1$, and, $\varepsilon_1 = \varepsilon_2 = \varepsilon_3 = 1$. Similarly, in the quantum domain, one wishes to deduce directly from the S-matrix elements, the occurrence (or absence) of processes of a certain type. For instance, a non-zero matrix element between an incoming state whose (radiative) energy is less than that of a given outgoing state of a certain type (time reverse of the situation considered above) will signal the occurrence of a black-hole evaporation. Thus, even in the absence of a detailed S-matrix, one can play around and get insight into the way things would finally fit together.

The framework also provides a natural arena for discussing the role of discrete symmetries in quantum gravity. For example, there has been a considerable interest over the past year or so in the issue of CPT-violation by the gravitational interaction. (See, e.g. Page (1980) and Wald (1980).) This discussion invariably assumes that one can introduce operators on the Hilbert space of graviton-states corresponding to time and parity reversal. *A priori*, the assumption seems to be grounded in the Minkowskian physics: it is unclear that the operators can be meaningfully introduced in absence of a flat background geometry. The present framework is well-suited for checking this issue. If the required operators do exist on the Hilbert spaces constructed in Section 4, one would be able to make the existing discussion more concrete and rigorous. If, on the other hand, the operators cannot be introduced unambiguously in spite of the rich structure made available by asymptotic flatness, in my view, the assumption would be a suspect. In either case, a detailed investigation is bound to provide new insight.

Finally, since it provides quantum kinematics, the framework is, in a certain sense, complementary to the Feynman path integral approach. (I do not have the Euclidean programme in mind; the term 'path integral' is used in a wider and a more loose sense.) In the case of non-relativistic systems, for example, the path integral provides only the amplitude $\langle x,t|x',t'\rangle$ for a particle at the point x' at instant t' to appear at the point x at time t. To obtain predictions of physical interest (e.g. transition amplitude between two energy levels) one needs to know quantum kinematics; in this case, the

Hilbert space of states $\Psi(x)$, and the energy eigen-vectors, $\Psi_E(x)$ and $\Psi_{E'}(x)$. It is only when one knows this that one can compute the desired amplitude $\langle \Psi_E(x,t), \int dx' \langle x,t|x',t'\rangle \Psi_{E'}(x')\rangle$. A similar situation exists in field theory as well, where the Fock space provides the required kinematics. The present framework serves the same purpose for quantum gravity. Stated differently, it provides the appropriate boundary conditions for the path integral formalism. (Note that, from this viewpoint, the issue of whether or not the initial-value problem on I is really well-posed at the classical level has, at least in principle, little relevance. One would just specify the boundary conditions – i.e. the incoming quantum states on I^- and the outgoing one on I^+ as suitable (holomorphic) functionals on the classical phase space – and consider all kinematically possible classical paths connecting the relevant points of the phase space (i.e. points in the support of the wave functions). If classical dynamical trajectories connecting these points exist, they would make significant contributions to the integral. If they do not exist, the amplitude in question would be very small. The framework would be meaningful in any case.)

The most surprising feature of the framework is the prediction that, in addition to the spin-2 gravitons, there exist graviton-states with an internal charge. One would like to understand the physical significance of these states better: their occurrence is tied down to properties of gravitational radiation in the fully relativistic regime for which one has a very limited intuition. One may, none the less, make certain speculations. Since the value of the internal charge, represents, in a certain sense, the obstruction to the reduction of the BMS group to the Poincaré, one might imagine that the charged gravitons will play an important role in processes involving space-time topologies which admit no flat metrics. That this may not be a far-fetched idea is suggested by the following considerations. In the 2-dimensional model mentioned in Section 4.2, the direct sum of representations with all possible topological charges admits certain 'condensed' states which have fermionic character. (Streater and Wilde 1970.) The close similarity that our framework shares with this model suggests that the situation may be similar in the case of charged graviton states. Furthermore, since we have a natural action of the BMS group on *I*, one would presumably be able to show that the states of fermionic character do actually have half integral spin. If this turns out to be the case, these states would almost certainly be closely related to those discovered by Friedman and Sorkin (1980) using the canonical approach, which owe their existence to non-trivial topologies on space-like Cauchy surfaces. Thus, although their origin, in the present framework, lies in the technical aspects of gravitational radiation which refer only to the asymptotic structure of the space-time, the internally-charged graviton-states may well provide insight into the role of topological issues in quantum gravity.

ACKNOWLEDGEMENTS

I thank Michael Streubel for discussions on the phase space formulation of the radiative modes. Thanks are also due to the members of the Mathematical Physics Group at Clermont-Fd. and colleagues at the Oxford Conference for their comments and suggestions.

REFERENCES

Ashtekar, A. (1980*a*). *Commun. Math. Phys.*, **71**, 59.
—— (1980*b*). *Isolation and quantization of the radiative degrees of freedom of the gravitational field. J. Math. Phys.* (to appear).
—— and Magnon-Ashtekar, A. (1979). *Phys. Rev. Lett.*, **43**, 181.
—— —— (1980). *J. Gen. rel. Grav.*, 12, 205.
—— and Sen, A. (1980). *J. Math. Phys.*, **21**, 526.
Bondi, H., Metzner, A.W.K., and Van der Burg, M.J.G. (1961). *Proc. R. Soc.*, **A269**, 21.
Friedman, J.L. and Sorkin, R. (1980). *Phys. Rev. Lett.* **44**, 1100.
Gelfand, I.M. and Naimark, M.A. (1943). *Mat. Sobernik* **12**, 197.
Geroch, R. (1977). In *Asymptotic structure of space-time*, ed P. Esposito and L. Witten. Plenum, New York.
—— and Xanthopoulos, B.C. (1978). *J. Math. Phys.*, **19**, 714.
Hawking, S.W. and Pope, C.N. (1979). *Nucl. Phys.*, **B161**, 93.
Kuchar, K. (1980). Article in this volume.
McCarthy, P.J. (1972). *Proc. R. Soc.*, **A330**, 517.
Newman, E.T. (1978). *Phys. Rev.*, **D18**, 2901.
—— (1980). *Self-dual gauge fields*. Preprint.
——, Ko, M., and Todd, P. (1977). In *Asymptotic structure of space-time*, ed P. Esposito and L. Witten. Plenum, New York.
—— and Unti, T. (1962). *J. Math. Phys.*, **3**, 891.
Page, D. (1980). *Phys. Rev. Lett.*, **44**, 301.

Palmer, T.N. (1978). *J. Math. Phys.*, **19**, 2324.
Penrose, R. (1963). *Phys. Rev. Lett.*, **10**, 66.
—— (1965). *Proc. R. Soc.*, **A284**, 159.
—— (1976). *J. Gen. Rel. Grav.*, **7**, 31.
—— (1980). Private communication.
Sachs, R.K. (1962). *Phys. Rev.*, **128**, 2851.
Segal, I.E. (1947). *Bull. Am. Math. Soc.*, **53**, 73.
Streater, R.F. (1965). *Proc. R. Soc.*, **A287**, 510.
—— (1974). In *Physical reality and mathematical description*, ed J. Mehra and P. Enz. D. Reidel, Dordrecht.
—— and Wilde, I.F. (1970). *Nucl. Phys.*, **B24**, 561.
Wald, R. (1980). Article in this volume.
Winicour, J. (1968). *J. Math. Phys.*, **9**, 861.

Note added in proof: Recent investigations have shown that the new representations of the CCR are the analogues of the infrared sectors which arise in QED and QCD. In QED, where the situation is clearer, one has the following result: even after renormalization (which handles the ultraviolet divergences the S-matrix fails to be well-defined (in the sense of perturbation theory) unless the infrared sectors are included in the construction of the spaces of asymptotic states; S-matrix elements between the incoming and the outgoing Fock space states vanish identically. One expects that the situation would be analogous in the gravitational case. The special feature in this case is the geometrical interpretation of the new representations in terms of the enlargement of the symmetry group from the Poincaré to the BMS. Becaus of this, the supertranslation freedom in the radiation theory can now be seen as the imprint left by the infrared behaviour of the quantum gravitational field on the classical general relativity. Thus, the geometrical nature of the gravitational field makes the impact of the infrared issues more profound unlike in QED and QCD, the impact is felt already on the classical level. (See: Ashtekar (1980*b*) and *Phys. Rev. Lett.* **46**, 573 (1981).)

19

IS QUANTUM GRAVITY FINITE?

M.R. Brown
Department of Astrophysics, University of Oxford

1. INTRODUCTION

We shall present a formulation of quantum gravity that is finite to all orders in the Planck length. Moreover we shall argue that the inclusion of quantum gravitational effects makes other flat space quantum field theories similarly finite. The underlying philosophy of the approach is as follows.

Implicit in the usual formulation of any flat space quantum field theory, Q.E.D. for example, is the assumption that the distance between any two points, x and y, in the background space-time is described by the Minkowski metric, η, no matter how close x is to y. This assumption is directly responsible for the ultraviolet divergence of Q.E.D. We know that this assumption is probably incorrect: we have yet to take into account quantum gravity. Thus we should more correctly say that we can approximate the distance between x and y by using the Minkowski metric only when this distance is greater than the Planck length. We simply do not know how to describe distances less than the Planck length until we have quantized the gravitational field. How then do we do this?

The conventional approach, where one treats General Relativity as a spin-two field propagating in Minkowski space with a complicated self-interaction, is doomed to failure: one is making the same assumption concerning distance in the background space-time and the theory appears to be non-renormalizable. An extremely interesting, but highly ambiguous, alternative is provided by the non-polynomial Lagrangian work of Isham, Salam, and Strathdee (1971, 1972). We shall return to this in a moment. For the present we shall sketch another alternative, one that specifically addresses itself to the problem of finding a metric, let us call it η', that, in some appropriate sense, represents the quantum gravitational correction to the Minkowski metric. It might be called the effective action approach.

One uses the background field method of DeWitt (1967) to define an effective action, $W(\eta',\hbar)$, that incorporates all quantum gravitational interactions when the Einstein action, $G^{-1}\int(-\det g_{ab})^{\frac{1}{2}}R(g)\mathrm{d}^4x$, is quantized in a general background space-time having a metric η'. To find the particular metric that represents the quantum correction to η one solves the effective action field equations, $\delta W/\delta\eta'_{ab} = 0$, subject to suitable boundary conditions. In the past this approach has also met with disaster for the following reason: it is possible to develop $W(\eta',\hbar)$ as a power series in $\hbar$; the lowest order term in the series being the original Einstein action. Attempts have been made (Weinberg 1979) to compute W to first order, the one-loop approximation. These calculations are all based on the assumption that the background metric η', in suitable co-ordinates, has the local Taylor series expansion,

$$\eta'_{ab}(y) = \eta_{ab}(x) - \tfrac{1}{4}R_{acbd}(\eta'(x))(x-y)^{c}(x-y)^{d} + 0(x-y)^3 \,. \tag{1.1}$$

It must come as no surprise that W, so defined, contains all sorts of divergent integrals. By eqn. (1.1) one is yet again assuming that the local structure of the background space-time is given by the Minkowski metric; there is simply some additional weak curvature to be taken into account. This way of computing W is directly at odds with what one hopes will emerge from the field equations, namely a metric η' that describes small distances in such a way that potentially divergent integrals in the theory are rendered finite. We shall clarify this discussion by taking a particular example.

A typical integral occurring in the computation of either $W(\eta',\hbar)$ or the effective action for any matter fields propagating in a background space-time having a metric η' is given by $I(\eta')$, where

$$I(\eta') \equiv \int \mathrm{d}^4x\ \mathrm{d}^4y\,(-\det\eta'_{ab}(x))^{\frac{1}{2}}(-\det\eta'_{ab}\ (y))^{\frac{1}{2}}V(x)V(y)\sigma^{-2}(x,y,\eta') \tag{1.2}$$

$\sigma(x,y,\eta')$ is the square of the distance along the geodesic joining the points x and y in the background space-time. It is sufficient for our purposes that V is some arbitrary vertex function of the theory in question. Let us briefly record some relevant properties of I:

$$I(\eta) = \int d^4x\, d^4y V(x)V(y)(x - y)^{-4} \tag{1.3}$$

is divergent. Dimensional regularization would replace $I(\eta)$ by $I_n(\eta)$, where

$$I_n(\eta) \equiv \int d^nx\, d^ny V(x)V(y)(x - y)^{2(2-n)}. \tag{1.4}$$

$I_n(\eta)$ can be written in terms of a Fourier transform,

$$I_n(\eta) = \int d^nx\, d^np V(x)\overline{V}(p)\exp(ip.x) \int d^nu\, \exp(ip.u)u^{2(2-n)}, \tag{1.5}$$

and exhibited as the Laurent series,

$$I_n(\eta) = \int d^nx\, d^np V(x)\overline{V}(p)\exp(ip.x)\pi^2\Big\{3\psi(1)+2-\ln(\tfrac{1}{4}\pi p^2) + \\ + \frac{2}{(4-n)} + O(n-4)\Big\}, \tag{1.6}$$

$$\psi(z) \equiv \frac{1}{\Gamma(z)} \cdot \frac{d}{dz}\Gamma(z). \tag{1.7}$$

The divergent part of $I(\eta)$ can be defined to be the coefficient of the pole at $n = 4$ in eqn. (1.6).

There are two cases we wish to discuss; let us call them (A) and (B):

(A) Suppose that we are trying to compute the effective action for Q.E.D. in flat space. In so doing we encounter the divergent integral $I(\eta)$ as given by eqn. (1.3). V will be some function of the background photon and electron fields. We can proceed in at least three ways:

(a) We keep Minkowski space as the background space, dimensionally regularize the theory and renormalize the coefficient of the term $\int d^4xV^2(x)$ in the original action so as to absorb the pole term of eqn. (1.6). This is conventional.

(b) We can again keep Minkowski space as the background space-time and follow the non-polynomial Lagrangian method to incorporate an infinite number of graviton interactions. In our example this will have the effect of replacing the divergent integral $I(\eta)$ by the convergent, gravity modified integral, $I_N(\eta)$.

$$I_N \equiv \int d^4x\, d^4y V(x)V(y)(x - y)^{-4}\exp-(G\hbar/(x - y)^2) \quad . \qquad (1.8)$$

It is not clear how to proceed any further with this method. We only include it here because the form of the integrand in eqn. (1.8) will be significant later on.

(c) We can argue that Minkowski space is only an approximation to the true background space-time and compute the quantum gravity corrected metric η'. We are then led to consider the effective action for Q.E.D. in this new background space. This causes the integral $I(\eta)$ to be replaced by $I(\eta')$. $I(\eta')$ does not necessarily diverge. It does so or not depending on the local properties of the background space-time. We have no *a priori* knowledge of these properties. Thus we are led to discuss B.

(B) Suppose that we attempt to compute the pure gravitational effective action, $W(\eta',\hbar)$. We shall encounter an integral like $I(\eta')$, where now V is a function of η'; it might be the Ricci scalar, $R(\eta')$, for example. If we assume that η' has the local behaviour described by eqn. (1.1) and substitute this into eqn. (1.2) we shall immediately recover the divergent integral $I(\eta)$. At this point we should ask why we are making this assumption. Is it necessary? The answer is that it is not. We must compute the gravitational effective action, as far as is possible, without making assumptions concerning the local properties of the space-time that it is designed to predict!

It should now be apparent that the crucial factor in overcoming the ultraviolet divergence problems in cases (A) and (B), and others not mentioned, is the structure of the effective metric η' – if, indeed, it is a metric. We shall now describe a way of computing this object that is finite to any finite order in the Planck length.

2. THE QUANTUM GRAVITATIONAL CORRECTION TO THE MINKOWSKI METRIC

Consider the one-loop approximation to the gravitational effective action. This approximation is valid when $(G\hbar)/(\text{distance})^2 \ll 1$. When we are looking for a solution to the field equations that is to be a first order approximation to the Minkowski metric, the only distance that can enter the analysis is $\sigma(x,y,\eta)$. In this context the one-loop approximation is valid and provides a measure of the distance between points in the space-time only when this distance is greater than the Planck length. In other words, we are forced to consider, not the expansion of equation (1.1), but instead the expansion

$$\eta'_{ab}(y) = \eta_{ab}(x) + \frac{G\hbar}{(x-y)^4}\left\{\lambda_1 (x-y)^2\eta_{ab} + \mu_1 (x-y)_a (x-y)_b\right\} + \\ + O(G\hbar/(x-y)^2) \quad . \tag{2.1}$$

Equation (2.1) is valid in the region $(x - y)^2 > G\hbar$. The convergence of the integral (1.2) is determined by the behaviour of η' in the region $(x - y)^2 < G\hbar$. We can discover this behaviour from eqn. (2.1) only by computing all the terms in the series and analytically the sum to inside a Planck radius of the point x. Consequently we must consider the complete solution, to all orders in $G\hbar$, to the effective action field equations. This has the form

$$\eta'_{ab}(y) = \sum_{r=0}^{\infty} \frac{(G\hbar)^r}{(x-y)^{2(r+1)}}\left\{\lambda_r (x-y)^2\eta_{ab} + \mu_r (x-y)_a (x-y)_b\right\} , \tag{2.2}$$

where $\lambda_0 = 1$, $\mu_0 = 0$. If the theory is to make sense at all, we must assume that the sum in eqn. (2.2) converges in a region $(G\hbar)/(x - y)^2 < \varepsilon; \varepsilon > 0$. The radius of convergence ε is a function of the coefficients (λ_r, μ_r).

It is meaningless to substitute eqn. (2.2) into eqn. (1.2) and integrate term by term; all integrals diverge. However we should recognize that our inability to do this is merely a consequence of our representation of η' as the series (2.2). We are free to rearrange this convergent series as best suits our purposes. Consider the following argument.

For $(x - y)^2 > \varepsilon G\hbar$ eqn. (2.2) implies that we can write the integrand of $I(\eta')$ in the form

$$\frac{\eta'^{\frac{1}{2}}(x)\eta'^{\frac{1}{2}}(y)V(x)V(y)}{\sigma^2(x,y,\eta')} = \frac{V(x)V(y)}{(x-y)^4}\sum_{r=0}^{\infty} a_r \frac{(G\hbar)^r}{(x-y)^{2r}} , \tag{2.3}$$

where $\eta' \equiv -\det(\eta'_{ab})$. In this same region we can write

$$\frac{\eta'^{\frac{1}{2}}(x)\eta'^{\frac{1}{2}}(y)V(x)V(y)}{\sigma^2(x,y,\eta')} =$$

$$= \frac{V(x)V(y)}{(x-y)^4}\exp-(G\hbar/(x-y)^2) \sum_{r=0}^{\infty} b_r \frac{(G\hbar)^r}{(x-y)^{2r}} , \tag{2.4}$$

where

$$b_r = \sum_{s=0}^{r} a_{r-s}(s!)^{-1} . \tag{2.5}$$

Equation (2.4) has the advantage that the integrals of individual terms in the series all exist. The integration in eqn. (1.2) can now be performed giving the answer in the form of a sum over the coefficients (b_r), or, equivalently, (a_r). $I(\eta')$ will be finite or not depending upon the properties of this sum. These properties are determined from the requirement that eqn. (2.2) is a solution to the field equations. The important point is that by proceeding in this way we give quantum gravity the opportunity of being finite. We shall know whether it is or not when the coefficients (λ_r, μ_r) have been computed and we can answer the question: Does the theory possess a self-consistent solution, η', having the convergent series expansion (2.2)?

It is of interest to examine more closely the consequences of substituting eqn. (2.4) into eqn. (1.2) and performing the integrations. This is relatively straightforward. We make use of the easily established formula

$$J(p) \equiv \int d^4u\ u^{-4}\exp(ip.u)\exp(-(G\hbar/u^2))\ , \tag{2.6}$$

$$= \pi^2 \sum_{s=0}^{\infty} \frac{(\frac{1}{4}G\hbar p^2)^s}{(s+1)(s!)^3}\left\{3\psi(s+1) + \frac{1}{(s+1)} - \ln(\tfrac{1}{4}G\hbar p^2)\right\}\ . \tag{2.7}$$

Whence $I(\eta')$ can be written

$$I(\eta') = \int d^4x\ d^4p V(x)\overline{V}(p)\exp(ip.x) \sum_{r=0}^{\infty} b_r \frac{d^r}{dG^r} J(p)\ , \tag{2.8}$$

$$= \int d^4x\ d^4p V(x)\overline{V}(p)\exp(ip.x)\pi^2\left\{3\psi(1) + 1 - \ln(\tfrac{1}{4}G\hbar p^2) + \right.$$

$$\left. + \sum_{r=0}^{\infty} b_r(r-1)! + 0(G\hbar p^2)\right\}\ . \tag{2.9}$$

Here there are several points worth mentioning:

1. Notice that if we set $\eta' = \eta$, then $b_r = (r!)^{-1}$ and we recover the infinite part of the integral (1.3), now in the form of the divergent sum $\sum_{r=0}^{\infty} r^{-1}$.

2. The expressions for the dimensionally regularized integral $I_n(\eta)$, eqn. (1.6), and $I(\eta')$, eqn. (2.9), are strikingly similar. If eqn. (1.6) led to an infinite renormalization of the coefficient of $\int d^4xV^2(x)$, eqn. (2.9) now gives a (potentially) finite renormalization of the same term. Alternatively, if eqn. (1.6) meant that the coefficient of $\int d^4xV^2(x)$ was unknown, eqn. (2.9) now gives it a value.

3. In eqn. (2.4) we extracted the exponential convergence factor, $\exp-(G\hbar/(x - y)^2)$, as an allowable mathematical device permitting integration of the series. We could have chosen other functions having similar properties; it makes little difference to the argument. Of course, one wants to choose a function that best approximates the final answer. In this spirit the non-polynomial approach, eqn. (1.8), is suggestive. It may be that this exponential behaviour really is a characteristic of the behaviour of the distance between x and y when x is close to y. One has only to compare the functions

$$\sigma(x,y,\eta) = (x - y)^2$$

and

$$\sigma(x,y,\eta') \cong (x-y)^2 \exp-(G\hbar/(x-y)^2)$$

to see that it is at least a plausible behaviour. The discussion as to the physical significance of gravity providing an ultraviolet cut-off of this form can be found in Isham *et al.* (1971, 1972) and we shall not repeat it here. There are many extremely interesting possibilities, not least the finite determination of renormalization constant.

4. We should add, by way of a footnote, that where there is any difficulty in interpreting the above formulae when x and y can be both space-like and time-like separated, one can Euclideanize the Minkowski metric in the usual way. In this sense, it can be seen that the value of the imaginary part for the lowest order term of eqn. (2.9) is in agreement with that obtained from eqn. (1.6), as, indeed, it must be.

All higher loop contributions to W can be treated in a similar fashion. One has only to assume that the eventual solution to the field equations has the form of eqn. (2.2) and require that the theory converges. This must be a more profitable way to proceed than has been the case in the past where one assumes from the outset, by eqn. (1.1), that the theory is divergent. Notice that eqn. (2.2) and the subsequent analysis does not preclude this unpleasant possibility we are only saying that it need not necessarily be the case.

3. CONCLUSION

We hope to have shown that, by making only necessary assumptions, it is possible, at least in principle, for quantum field theories that either are or include quantum gravity to be finite to any given order in the Planck length. Moreover, that it is possible, again in principle, to systematically calculate whether they are or not to all orders. In practice, these calculations are probably sufficiently complicated for them never to be done. There is some hope in that one only really needs to prove convergence of a numerical sum. What is needed is a useful, valid approximation of the local structure of space-time when quantum gravitational interactions are included. In this paper, we have given a framework in which one might exist.

Finally, it is interesting to ask how our picture of quantum field theory would be altered if quantum gravity did indeed act as a universal regulator. It has been argued (correctly, it seems) that the criterion of renormalizability is useful in distinguishing those theories that might be said to describe the physical world (Salam–Weinberg, for example) from those that do not (the Fermi theory of weak interactions, for example). If, by including gravity, we remove the divergencies of quantum field theory, then this criterion is also removed. It might then (perversely) be argued that gravity should not, or is unlikely to, fulfil this role and that we should content ourselves in describing nature with a search for renormalizable theories.

We wish to make the point that renormalizable field theories retain their privileged position even if the inclusion of gravity superficially makes them comparable with previously non-renormalizable theories.

Suppose that we could successfully quantize gravity. We might then have many candidates for a unified field theory of weak, electromagnetic and gravitational interactions. Given the present experimental capabilities, it would be impossible to observe much, if any, of the Planck length structure of these theories. We would then want to find a theory that dealt exclusively with physics on the weak and electromagnetic length scales. Theories that were renormalizable, previous to the inclusion of gravity, have the property that they successfully make no explicit reference to the Planck length. When gravity is included, it is precisely these theories that are again wanted. It may even be that the more fundamental Planck length structure determines the coupling constants of these derived, or effective, theories. In the foreseeable future, this is likely to be the only test of quantum gravity. It would be a supreme achievement.

ACKNOWLEDGEMENT

The author wishes to thank professors Roger Penrose and Dennis Sciama for their encouragement and conversation during the completion of this work.

REFERENCES

DeWitt, B.S. (1967). *Phys. Rev.*, **162**, 1195.
Isham, C.J., Salam, A., and Strathdee, J. (1971). *Phys. Rev.*, **D3**, 867.
—— —— —— (1972), *Phys. Rev.*, **D5**, 2548.
Weinberg, S. (1979). In *General relativity – an Einstein centenary survey*, ed S.W. Hawking and W. Israel. Cambridge University Press.

20

A GAUGE INVARIANT EFFECTIVE ACTION

B.S. DeWitt

Department of Physics, University of Texas, Austin, Texas 78712

1. INTRODUCTION

The background field method was originally introduced (DeWitt 1965) in order to maintain manifest gauge covariance in quantum field theory despite the introduction of special gauges to secure well-defined propagators. In its earliest formulation it worked well only for single-loop processes. In order to exploit the advantages of the method also in renormalization calculations involving multi-loop graphs 't Hooft (1976) proposed a refinement in which gauge covariant and background dependent sources are introduced and the background itself is not required to satisfy the classical field equations. 't Hooft's proposal has unfortunately not yet been implemented in actual calculations,* partly because the diagrammatic rules to which it leads are unfamiliar and partly because 't Hooft left the details implicit and his reasoning schematic. It is the purpose of this paper (1) to present an alternative (although probably equivalent) proposal in which the rules are fully spelled out, (2) to construct, in the process of formulating these rules, a gauge invariant effective action for the theory, and (3) to discuss the utility as well as the broader theoretical significance of this effective action.

We begin by recalling some standard notation and results. All fields are collectively denoted by φ. The symbol φ^i denotes a particular component of a particular field at a particular space-time point, the index i being a shorthand for a set of both discrete and continuous labels. All the φ^i will be assumed real. Functional differentiation will often be abbreviated by

$$F_{,ij\cdots}[\varphi] \stackrel{\text{def}}{=} \cdots \frac{\delta}{\delta\varphi^j}\frac{\delta}{\delta\varphi^i}F[\varphi] \qquad (1.1)$$

* Since this paper was written two calculations have been performed, using the new methods in gauge theories: L.F. Abbott (CERN preprint), C.F. Hart (Ph.D. thesis, University of Texas).

where $F[\varphi]$ is any functional of the field(s) φ. The classical field equations are

$$S_{,i}[\varphi] = 0 \,, \tag{1.2}$$

where $S[\varphi]$ is the classical action functional. $S[\varphi]$ is invariant under a set of gauge transformations which, infinitesimally, take the form

$$\delta\varphi^{i} = Q^{i}{}_{\alpha}[\varphi]\delta\xi^{\alpha} \,, \tag{1.3}$$

the $\delta\xi$s being infinitesimal gauge parameters and the $Q^{i}{}_{\alpha}$ typically being components of a φ-dependent local distribution (sum of delta function and its derivatives) which vanish when the space-time points associated with the indices i and α are distinct. (Here and in what follows summation-integration is implied over repeated indices.) The Qs are equivalent to local differential operators acting on the $\delta\xi$s.

If some of the fields that φ denotes are of fermion type special notations must be introduced to distinguish right and left functional derivatives, right and left components, etc. We shall confine our attention here to boson fields, for which these distinctions are irrelevant. This limitation is not imposed merely for convenience. The most interesting cases in which fermion fields are present are those for which $S[\varphi]$ is invariant under a set of local supersymmetry transformations. How to extend 't Hooft's ideas (or those of the present paper) to these cases is as yet unknown. This is because both 't Hooft and the author make two enormously simplifying assumptions, neither of which holds in conventional formulations of locally supersymmetric theories. The first assumption is that the transformations (1.3) form a *group*, which implies (arguments suppressed)

$$Q^{i}{}_{\alpha,j}Q^{j}{}_{\beta} - Q^{i}{}_{\beta,j}Q^{j}{}_{\alpha} \equiv Q^{i}{}_{\gamma}c^{\gamma}{}_{\alpha\beta} \,, \tag{1.4}$$

where the $c^{\gamma}{}_{\alpha\beta}$ are φ-independent distributions (the *'structure constants'*) satisfying

$$c^{\delta}{}_{\alpha\varepsilon}c^{\varepsilon}{}_{\beta\gamma} + c^{\delta}{}_{\beta\varepsilon}c^{\varepsilon}{}_{\gamma\alpha} + c^{\delta}{}_{\gamma\varepsilon}c^{\varepsilon}{}_{\alpha\beta} \equiv 0 \ . \tag{1.5}$$

The second assumption is that these transformations are *linear* (homogeneous or inhomogeneous) in φ, which implies that the $Q^{i}{}_{\alpha,j}$ are φ-independent distributions and that

$$Q^{i}{}_{\alpha,jk} \equiv 0 \ . \tag{1.6}$$

These two assumptions make manifest covariance almost trivial. They permit one to infer the transformation character of the various quantities appearing in the theory by simply noting the positions and types of the indices that they bear. Thus, from the gauge invariance of the classical action, which may be expressed by the formula

$$S_{,i}Q^{i}{}_{\alpha} \equiv 0 \ , \tag{1.7}$$

one obtains immediately, by repeated differentiation, the transformation law for the field equations as well as the higher functional derivatives of S (i.e., the *bare vertex functions*):

$$\delta S_{,i} \equiv S_{,ij}Q^{j}{}_{\alpha}\delta\xi^{\alpha} \equiv - S_{,j}Q^{j}{}_{\alpha,i}\delta\xi^{\alpha} \ , \tag{1.8}$$

$$\delta S_{,ij} \equiv S_{,ijk}Q^{k}{}_{\alpha}\delta\xi^{\alpha} \equiv - (S_{,kj}Q^{k}{}_{\alpha,i} + S_{,ik}Q^{k}{}_{\alpha,j})\delta\xi^{\alpha}, \text{ etc.} \tag{1.9}$$

In a similar manner one obtains from eqn. (1.4) the transformation law for the Qs:

$$\delta Q^{i}{}_{\alpha} \equiv Q^{i}{}_{\alpha,j}Q^{j}{}_{\beta}\delta\xi^{\beta} \equiv (Q^{i}{}_{\beta,j}Q^{j}{}_{\alpha} - Q^{i}{}_{\gamma}c^{\gamma}{}_{\beta\alpha})\delta\xi^{\beta} \ . \tag{1.10}$$

We note that, with the $\delta\xi$s removed, the series of equations beginning with (1.9) are the *bare Ward identities* of the theory, which relate vertex functions of adjacent order.

2. THE VACUUM TO VACUUM AMPLITUDE

If space-time is asymptotically flat and diffeomorphic to $\mathbf{R}^4$

then it is known that the complete content of the quantum theory of the system with classical action $S[\varphi]$ is derivable (in principle) from a knowledge of the vacuum to vacuum amplitude

$$\exp(iW[\varphi]) \overset{\text{def}}{=\!=} N\int \exp(iS[\varphi+\phi])d\phi, \quad d\phi \overset{\text{def}}{=\!=} \prod_i d\phi^i \ . \qquad (2.1)$$

Here φ is a *background field*, which is assumed to be infinitesimally close to the classical vacuum at infinity and to satisfy the classical field equations there, but to be *arbitrary* elsewhere. N is a normalizing constant, and the functional integral is assumed to be defined as an asymptotic series of Gaussian integrals satisfying Feynman boundary conditions, with regularizations being performed and counter terms being added to the exponent $iS[\varphi+\phi]$ as needed.*

The amplitude (2.1) may be regarded as reflecting a division of the quantum *operator* field (here written in boldface) into the classical background and an operator remainder:

$$\boldsymbol{\varphi}^i = \varphi^i + \boldsymbol{\phi}^i \ . \qquad (2.2)$$

Because of the asymptotic boundary condition both $\boldsymbol{\varphi}^i$ and $\boldsymbol{\phi}^i$ may be regarded as satisfying the linearized field equations (around the classical vacuum) at infinity, and hence $\boldsymbol{\phi}^i$ may be decomposed into standard creation and annihilation operators there.† These operators define 'in' and 'out' Fock spaces, the vacua of which are *relative* vacua, i.e., relative to the background φ^i. Expression (2.1) gives the probability amplitude

---

* Although this integral is widely believed to have a meaning that transcends perturbation theory, every definition of it to date (in the context of field theory) ultimately makes use of a regularized asymptotic expansion around one or more chosen backgrounds, which are usually stationary points of $S[\varphi]$. This is true whether the integral is computed directly in the Lorentzian sector or by analytic continuation from the Euclidean domain. Fortunately this limitation is not a crippling one because backgrounds can be chosen that are not themselves obtainable by perturbation expansions starting from the classical vacuum.

† That is, the operator $\boldsymbol{\phi}^i$ may be regarded as 'infinitesimally weak' at infinity. What is meant by this is that its matrix elements, between all relevant normalized physical states, have infinitesimal magnitude there. Because of the gauge freedom $\boldsymbol{\phi}^i$ is determined at infinity only *modulo* a linearized gauge transformation. The creation and annihilation operators for asymptotic physical states are nevertheless well-defined. (See for example, DeWitt 1965.

for the state of the field to be the relative vacuum in the remote future if it was known to be the relative vacuum in the remote past. We stress that this amplitude depends only on our choice of the background field at infinity. Because φ is totally arbitrary elsewhere, the amplitude cannot and does not depend on φ at finite points. This may be shown formally by functionally differentiating eqn. (2.1) and making use of the fact that the integrand on the right depends on φ only in the combination $\varphi + \phi$:

$$\frac{\delta W[\varphi]}{\delta\varphi^i} = -iN\,\exp(-iW[\varphi]) \int \frac{\delta}{\delta\phi^i}\exp(iS[\varphi+\phi])\,d\phi = 0\,. \qquad (2.3)$$

Because of the gauge invariance of the integrand the integration in (2.1) is redundant. Gauge transformations can be viewed in two guises, both useful. One can either make the background field φ and the remainder ϕ transform separately, as in

$$\left.\begin{aligned} \delta\varphi^i &= Q^i{}_\alpha[\varphi]\,\delta\xi^\alpha, \\ \delta\phi^i &= Q^i{}_{\alpha,j}\phi^j\,\delta\xi^\alpha \end{aligned}\right\} \qquad (2.4a)$$

or else one can hold the background fixed and make the field ϕ carry the whole burden of transformation:

$$\left.\begin{aligned} \delta\varphi^i &= 0, \\ \delta\phi^i &= Q^i{}_\alpha[\varphi+\phi]\,\delta\xi\,. \end{aligned}\right\} \qquad (2.4b)$$

It is the latter version that is relevant in analysing and removing the redundancy in the integral (2.1).

A convenient way to remove the redundancy is to factor out the infinite dimensional gauge group (Fadde'ev and Popov 1967) and then perform a Gaussian average over selected linear gauges. To use this procedure one must treat the gauge group formally as if it were compact. This requires the adoption of a regularization procedure that yields*

---

* When one attempts to evaluate expressions (2.5) explicitly in specific cases (e.g., for Yang–Mills or gravitational fields) one obtains delta

$$Q^i{}_{\alpha,i} = 0, \qquad c^\beta{}_{\alpha\beta} = 0 , \tag{2.5}$$

which incidentally guarantees the gauge invariance of the functional volume element $d\phi$. In principle any independent set of linear gauge functionals can be chosen for the Gaussian average, but for the purpose of maintaining manifest covariance it is convenient to choose them in the form

$$P^\alpha[\varphi,\phi] \equiv P^\alpha{}_i[\varphi]\phi^i , \tag{2.6}$$

where the $P^\alpha{}_i$ are components of a φ-dependent local distribution obeying the gauge transformation law

$$\delta P^\alpha{}_i \equiv P^\alpha{}_{i,j}Q^j{}_\beta\delta\xi^\beta \equiv (c^\alpha{}_{\beta\gamma}P^\gamma{}_i - P^\alpha{}_j Q^j{}_{\beta,i})\delta\xi^\beta , \tag{2.7}$$

and such that the (effective) local differential operator

$$\mathfrak{F}[\varphi,\phi] \stackrel{\text{def}}{=\!=} P^\alpha{}_i[\varphi]Q^i{}_\beta[\varphi+\phi] \tag{2.8}$$

is non-singular.

One then defines

$$\Delta[\varphi,\phi,\zeta] \stackrel{\text{def}}{=\!=} \int \delta[P[\varphi]\phi^\xi - \zeta]\mu_R[\xi]d\xi , \tag{2.9}$$

where the integration is carried out over the entire gauge group, $\mu_R[\xi]$ is a right-invariant measure on the group, $\delta[\ ]$ is the delta *functional*, and ϕ^ξ denotes a finite gauge transform of ϕ generated by the set of infinitesimal transformations (2.4b). Since we are confining our attention to perturbation theory (albeit about an arbitrary background) we may ignore any potential Gribov phenomenon (Gribov 1977) and assume that for each ϕ and ζ there exists exactly one point, denoted by ϕ_ζ, on the group orbit through ϕ, such that

---

functions and their derivatives with coincident arguments. All such quantities automatically vanish in the dimensional regularization procedure when the associated fields are massless. Since the dimensional procedure also leaves gauge invariance inviolate it is almost uniquely the procedure of choice in gauge theories.

$$P^{\alpha}{}_{i}[\varphi]\phi^{\zeta}{}_{i} = \zeta^{\alpha} . \tag{2.10}$$

To evaluate the integral (2.9) it is then not necessary to know anything about $\mu_R[\xi]$ beyond the fact that it may be taken equal to unity when ξ is the identity. Since expression (2.9) is obviously invariant under the gauge transformations (2.4b) one may evaluate it at $\phi = \phi_\zeta$, obtaining by well known arguments,

$$\Delta[\varphi,\phi,\zeta] = \det \hat{\mathfrak{G}}[\varphi,\phi_\zeta] , \tag{2.11}$$

where $\hat{\mathfrak{G}}^{\alpha}{}_{\beta}[\varphi,\phi]$ is the Feynman Green's function for the operator $\mathfrak{F}^{\alpha}{}_{\beta}[\varphi,\phi]$.

One next inserts unity, in the guise of

$$\Delta^{-1}[\varphi,\phi,\zeta] \int \delta[P[\varphi]\phi^{\xi} - \zeta]\mu_R[\xi]d\xi ,$$

into the integrand of (2.1), changes the order of integrations, and makes use of (2.11) as well as the gauge invariance of various factors, obtaining

$$\exp(iW[\varphi]) = N' \int \exp(iS[\varphi+\phi])(\det \hat{\mathfrak{G}}[\varphi,\phi])^{-1}\delta[P[\varphi]\phi-\zeta]d\phi , \tag{2.12}$$

$$N' \stackrel{\text{def}}{=} N \int \mu_R[\xi]d\xi . \tag{2.13}$$

Finally, one performs a Gaussian average over the ζs, and uses the fact that $W[\varphi]$ is independent of the ζs. The result is

$$\exp(iW[\varphi]) = N''(\det\eta[\varphi])^{\frac{1}{2}}$$
$$\times \int \exp(i(S[\varphi+\phi]+\tfrac{1}{2}\eta_{\alpha\beta}[\varphi]P^{\alpha}_{i}[\varphi]P^{\beta}_{j}[\varphi]\phi^{i}\phi^{j}))(\det\hat{\mathfrak{G}}[\varphi,\phi])^{-1}d\phi , \tag{2.14}$$

$$N'' = N'/C, \quad \int \exp\left(\frac{i}{2}\eta_{\alpha\beta}\zeta^{\alpha}\zeta^{\beta}\right)d\zeta = C(\det \eta)^{-\frac{1}{2}} . \tag{2.15}$$

$\eta_{\alpha\beta}$ may in principle be any non-singular symmetric continuous matrix, but for the sake of manifest covariance we choose it to be a φ-dependent local distribution obeying the gauge transformation law

$$\delta\eta_{\alpha\beta} \equiv \eta_{\alpha\beta,i}Q^i{}_\gamma\delta\xi^\gamma \equiv -(\eta_{\delta\beta}c^\delta{}_{\gamma\alpha}+\eta_{\alpha\delta}c^\delta{}_{\gamma\beta})\delta\xi^\gamma . \tag{2.16}$$

$\eta_{\alpha\beta}$ is usually an undifferentiated delta function, and hence its inverse, denoted by $\eta^{\alpha\beta}$, is also a delta function.

In the following sections an important role will be played by the local differential operator

$$\hat{F}_{ij}[\varphi,\phi] \stackrel{\text{def}}{=} S_{,ij}[\varphi+\phi] + \eta_{\alpha\beta}[\varphi]P^\alpha{}_i[\varphi]P^\beta{}_j[\varphi] . \tag{2.17}$$

This operator can be shown to be non-singular, possessing a Feynman Green's function $\hat{G}^{ij}[\varphi,\phi]$, whenever the Ps are chosen so as to make the operator $\mathfrak{F}^\alpha{}_\beta$ non-singular. It is also an easy exercise to show that when eqns (2.7) and (2.16) hold, $\hat{F}$, $\hat{G}$, $\hat{\mathfrak{F}}$, $\hat{\mathfrak{G}}$ transform as follows under (2.4a):

$$\left.\begin{aligned}
\delta\hat{F}_{ij} &= -(\hat{F}_{kj}Q^k{}_{\alpha,i} + \hat{F}_{ik}Q^k{}_{\alpha,j})\delta\xi^\alpha ,\\
\delta\hat{G}^{ij} &= (Q^i{}_{\alpha,k}\hat{G}^{kj} + Q^j{}_{\alpha,k}\hat{G}^{ik})\delta\xi^\alpha ,\\
\delta\hat{\mathfrak{F}}^\alpha{}_\beta &= (c^\alpha{}_{\gamma\beta}\hat{\mathfrak{F}}^\delta{}_\beta - \hat{\mathfrak{F}}^\alpha{}_\delta c^\delta{}_{\gamma\beta})\delta\xi^\gamma ,\\
\delta\hat{\mathfrak{G}}^\alpha{}_\beta &= (c^\alpha{}_{\gamma\delta}\hat{\mathfrak{G}}^\delta{}_\beta - \hat{\mathfrak{G}}^\alpha{}_\delta c^\delta{}_{\gamma\beta})\delta\xi^\gamma .
\end{aligned}\right\} \tag{2.18}$$

3. *P*- AND η-INDEPENDENCE OF THE FUNCTIONAL INTEGRAL

The functional integral (2.14) provides the point of departure for a straightforward perturbation theory based on a well known diagrammatic expansion involving propagators (G and $\hat{\mathfrak{G}}$) for 'regular' and 'ghost' quanta. Since it was obtained by purely formal manipulations one would like, in order to gain confidence in the validity of such manipulations, to verify independently that although the propagators G and $\hat{\mathfrak{G}}$ depend on the choice of $P^\alpha{}_i$ and $\eta_{\alpha\beta}$, the integral itself does not. This is most easily done by making infinitesimal changes, $\delta P^\alpha{}_i$ and $\delta\eta_{\alpha\beta}$, in $P^\alpha{}_i$ and $\eta_{\alpha\beta}$, and showing that the resulting change in the integrand is equivalent to making the change

$$\left.\begin{aligned} \phi^i &\to \phi^i + Q^i{}_\alpha[\varphi+\phi]\,\delta\xi^\alpha[\varphi,\phi] \ , \\ \delta\xi^\alpha[\varphi,\phi] &\overset{\text{def}}{=\!=} -\hat{\mathfrak{G}}^\alpha{}_\beta[\varphi,\phi](\delta P^\beta{}_j[\varphi] + \\ &\quad + \tfrac{1}{2}\eta^{\beta\gamma}[\varphi]\,\delta\eta_{\gamma\delta}[\varphi]P^\delta{}_j[\varphi])\phi^j \ , \end{aligned}\right\} \tag{3.1}$$

wherever ϕ appears, namely in both the integrand and in the functional volume element. Since ϕ is just a dummy variable of integration the integral remains unaffected. In the demonstration (for the details of which see DeWitt (1967)) it is necessary to use eqns (2.5).

The proof of P- and η-independence can also be carried out order by order in perturbation theory provided the background field, which is arbitrary except at infinity, is chosen to satisfy the classical field equations everywhere so that the term linear in ϕ, in the expansion of $S[\varphi+\phi]$, is absent. For every positive integer n, the sum of all n-loop graphs is then, by itself, P- and η-independent. For each n the proof requires the use of the easily verified identity (DeWitt 1965)

$$Q^i{}_\alpha \mathfrak{G}^\alpha{}_\beta = G^{ij}P^\alpha{}_j\eta_{\alpha\beta} \qquad (S_{,i}[\varphi] = 0) \ , \tag{3.2}$$

where

$$\mathfrak{G}^\alpha{}_\beta[\varphi] \overset{\text{def}}{=\!=} \hat{\mathfrak{G}}^\alpha{}_\beta[\varphi,0] \ , \tag{3.3}$$

$$G^{ij}[\varphi] \overset{\text{def}}{=\!=} \hat{G}^{ij}[\varphi,0] \ . \tag{3.4}$$

When $n \geqslant 2$ the proof also requires the use of eqns (2.5) as well as the Ward identities (1.9). (For the case $n = 2$ see DeWitt 1965.)

The functionals $\mathfrak{G}[\varphi]$ and $G[\varphi]$ are the Feynman Green's functions for the operators

$$\mathfrak{F}^\alpha{}_\beta[\varphi] \overset{\text{def}}{=\!=} \hat{\mathfrak{F}}^\alpha{}_\beta[\varphi,0] = P^\alpha{}_i[\varphi]Q^i{}_\beta[\varphi] \ , \tag{3.5}$$

$$F_{ij}[\varphi] \overset{\text{def}}{=\!=} \hat{F}_{ij}[\varphi,0] = S_{,ij}[\varphi] + \eta_{\alpha\beta}[\varphi]P^\alpha{}_i[\varphi]P^\beta{}_j[\varphi] \ . \tag{3.6}$$

Under (1.3) (or 2.4a)) $F[\varphi]$, $G[\varphi]$, $\mathfrak{F}[\varphi]$, $\mathfrak{G}[\varphi]$ obey transformation laws identical in form to those of eqns (2.18). Under (2.4b) the functionals $F[\varphi+\phi]$, $G[\varphi+\phi]$, $\mathfrak{F}[\varphi+\phi]$, $\mathfrak{G}[\varphi+\phi]$ transform similarly. Note that all these transformation laws hold whether φ satisfies the classical field equations or not.

4. CHRONOLOGICAL AVERAGES: THE S-MATRIX

Let $A[\varphi,\phi]$ be any functional of φ and ϕ that remains invariant under the transformation (2.4b). Define

$$\langle A[\varphi,\boldsymbol{\phi}]\rangle \overset{\text{def}}{=\!=} N\exp(-iW[\varphi])\int A[\varphi,\phi]\exp(iS[\varphi+\phi])d\phi\ . \quad (4.1)$$

This quantity is $\exp(-iW[\varphi])$ times the matrix element, between the 'in' and 'out' relative vacua, of the associated quantum operator $A[\varphi,\boldsymbol{\phi}]$ arranged in chronologically ordered form. Expression (4.1) may be called the *chronological average* of $A[\varphi,\boldsymbol{\phi}]$. By steps identical with those used in passing from eqn. (2.1) to eqn. (2.4) one may recast eqn. (4.1) into the form

$$\langle A[\varphi,\boldsymbol{\phi}]\rangle \overset{\text{def}}{=\!=} N''\exp(-iW[\varphi])(\det\eta[\varphi])^{\frac{1}{2}}\int A[\varphi,\phi]$$
$$\times\exp(i(S[\varphi+\phi]+\eta_{\alpha\beta}[\varphi]P^{\alpha}{}_{i}[\varphi]P^{\beta}{}_{j}[\varphi]\phi^{i}\phi^{j}))(\det\hat{\mathfrak{G}}[\varphi,\phi])^{-1}d\phi \quad (4.2)$$

This integral, like (2.14) is P- and η-independent.

If $A[\varphi,\phi]$ is not invariant under the transformation (2.4b) the integral (4.1) is ambiguous (like $\int_{-\infty}^{\infty}x\,dx$). The integral (4.2), however, is unambiguous (at least in perturbation theory) and can be used to define the symbol $\langle A[\varphi,\boldsymbol{\phi}]\rangle$ in all cases, although it is no longer generally independent of the choice of $P^{\alpha}{}_{i}$ and $\eta_{\alpha\beta}$. The quantity $\langle A[\varphi,\boldsymbol{\phi}]\rangle$ then becomes $\exp(-iW[\varphi])$ times the matrix element, between the 'in' and 'out' relative vacua, of the following operator:

$$T(A[\varphi,\boldsymbol{\phi}]) \overset{\text{def}}{=\!=} C^{-1}(\det\eta[\varphi])^{\frac{1}{2}}\int T(A[\varphi,\boldsymbol{\phi}_{\zeta}])\exp(\tfrac{i}{2}\eta_{\alpha\beta}[\varphi]\zeta^{\alpha}\zeta^{\beta})d\zeta\ , \quad (4.3)$$

where

$$T(A[\varphi,\boldsymbol{\phi}_\zeta]) \overset{\text{def}}{=\!=} T\Big((\det \hat{\mathfrak{G}}[\varphi,\boldsymbol{\phi}_\zeta])^{-1} \int A[\varphi,\boldsymbol{\phi}^\xi]\delta[P[\varphi]\phi^\xi-\zeta]\mu_R[\xi]d\xi\Big) \,. \tag{4.4}$$

The 'T' on the right of eqn. (4.4) denotes the ordinary chronological ordering operation; its use allows the non-commutativity of $A[\varphi,\boldsymbol{\phi}^\xi]$ with the inverse determinant and the delta functional to be ignored. Note that because the gauge group acts linearly on ϕ there is no ordering ambiguity about the symbol ϕ^ξ. Note also that in the special case $A[\varphi,\phi] = P^\alpha{}_i[\varphi]\phi^i$ we have

$$T(P^\alpha{}_i[\varphi]\boldsymbol{\phi}^\zeta{}_i) = \zeta^\alpha, \qquad P^\alpha{}_i[\varphi]\langle\boldsymbol{\phi}^i\rangle = \langle P^\alpha{}_i[\varphi]\boldsymbol{\phi}^i\rangle = 0 \,, \tag{4.5}$$

a result that will prove useful later.

Equations (4.3) and (4.4) describe a procedure for constructing out of an arbitrary operator $A[\varphi,\boldsymbol{\phi}]$ one that is gauge invariant (under (2.4b)) although P- and η-dependent. We obtain an operator that is P- and η-independent only if the original operator is gauge invariant, in which case (4.4) reduces to the ordinary chronological ordering operation and (4.3) becomes an identity.

The following gauge-invariant operators are of fundamental importance:

$$\boldsymbol{\phi}_A \overset{\text{def}}{=\!=} u^i{}_A \vec{S}{}^{\circ}{}_{,ij}\boldsymbol{\phi}^j \,. \tag{4.6}$$

Here $S^\circ{}_{,ij}$ is $S_{,ij}$ evaluated at the classical vacuum, and the arrow indicates the direction in which, as a differential operator, it is to be understood as acting. The $u^i{}_A$ are appropriately normalized positive-frequency wave functions satisfying

$$\vec{S}{}^{\circ}{}_{,ij}\, u^j{}_A = 0 \,, \tag{4.7}$$

and the index A labels the asymptotic particle states of the theory. The gauge invariance of the $\boldsymbol{\phi}_A$ may be seen as follows. Under (2.4b) we have

$$\delta\boldsymbol{\phi}_A = u^i{}_A \vec{S}{}^{\circ}{}_{,ij}\, Q^j{}_\alpha[\varphi+\boldsymbol{\phi}]\delta\xi^\alpha \,. \tag{4.8}$$

The group of gauge transformations is always to be understood as restricted to transformations that reduce to the identity at infinity. The gauge parameters $\delta\xi^{\alpha}$ are, in fact, required to fall off at infinity rapidly enough so that the arrow in eqn. (4.8) may, through an integration by parts, be reversed. In virtue of eqn. (4.7), therefore, we obtain*

$$\delta\boldsymbol{\phi}_{\mathrm{A}} = 0 \,. \tag{4.9}$$

The importance of the operators $\boldsymbol{\phi}_{\mathrm{A}}$ lies in the fact that the S-matrix of the theory can be constructed from the chronological averages of their products (DeWitt 1964). If we define

$$\begin{aligned}
&(\mathrm{out},\, \mathrm{B}_1,\, \ldots\, \mathrm{B}_n|\mathrm{in},\, \mathrm{A}_1,\, \ldots\, \mathrm{A}_m) \\
&\stackrel{\mathrm{def}}{=} (-\mathrm{i})^{\mathrm{m+n}}\exp(\mathrm{i}W[\varphi])\langle \boldsymbol{\phi}_{\mathrm{B}_1}{}^{*}\, \ldots\, \boldsymbol{\phi}_{\mathrm{B}_\mathrm{n}}{}^{*}\boldsymbol{\phi}_{\mathrm{A}_1}\, \ldots\, \boldsymbol{\phi}_{\mathrm{A}_\mathrm{m}}\rangle \\
&= (-\mathrm{i})^{\mathrm{m+n}} u^{j_1}{}_{\mathrm{B}_1}{}^{*} \,\ldots\, u^{j_\mathrm{n}}{}_{\mathrm{B}_\mathrm{n}}{}^{*}\, \vec{S}^{\mathrm{o}}{}_{,j_1 l_1} \,\ldots\, \vec{S}^{\mathrm{o}}{}_{,j_\mathrm{n} l_\mathrm{n}} \Big\{ N''(\det \eta[\varphi])^{\frac{1}{2}} \\
&\times \int \phi^{l_1} \,\ldots\, \phi^{l_\mathrm{n}} \phi^{k_1} \,\ldots\, \phi^{k_\mathrm{m}} \exp(\mathrm{i}(S[\varphi+\boldsymbol{\phi}] + \tfrac{1}{2}\eta_{\alpha\beta}[\varphi]P^{\alpha}{}_{i}[\varphi]P^{\beta}{}_{j}[\varphi]\phi^{i}\phi^{j}) \\
&\times (\det\hat{\boldsymbol{\mathfrak{G}}}[\varphi,\phi])^{-1}\mathrm{d}\phi\Big\} \overleftarrow{S}^{\mathrm{o}}{}_{,k_1 i_1} \,\ldots\, \overleftarrow{S}^{\mathrm{o}}{}_{,k_m i_m}\, u^{i_1}{}_{\mathrm{A}_1} \,\ldots\, u^{i_\mathrm{m}}{}_{\mathrm{A}_\mathrm{m}} \,,
\end{aligned} \tag{4.10}$$

then the elements of the S-matrix are given by (DeWitt 1965)

$$\begin{aligned}
&\langle \mathrm{out},\, \mathrm{B}_1,\, \ldots\, \mathrm{B}_\mathrm{n}|\mathrm{in},\, \mathrm{A}_1,\, \ldots\, \mathrm{A}_\mathrm{m}\rangle \\
&= \sum_{l=0}^{\infty} P_{(m,n;l)}\, \delta_{\mathrm{B}_1\mathrm{A}_1} \,\ldots\, \delta_{\mathrm{B}_l\mathrm{A}_l}(\mathrm{out},\, \mathrm{B}_{l+1},\, \ldots\, \mathrm{B}_n|\mathrm{in},\, \mathrm{A}_{l+1},\, \ldots\, \mathrm{A}_m
\end{aligned} \tag{4.11}$$

where the symbol P in the summand indicates that the expression following it is to be summed over all distinct permutations of the As and Bs, the subscript

$$(m,n;l) \stackrel{\mathrm{def}}{=} \frac{m!\,n!}{(m-l)!\,(n-l)!\,l!} \tag{4.12}$$

---

* In Yang–Mills theory it is well known that there are gauge transformations that reduce to the identity at infinity but which cannot be continuously deformed to the identity everywhere, and which therefore cannot be generated by the infinitesimal transformations (2.4b). The operators $\boldsymbol{\phi}_{\mathrm{A}}$ are also invariant under these so-called 'large' gauge transformations.

giving the number of permutations required.

Since the S-matrix describes physical processes it must be P- and η-independent. This follows from the gauge invariance of the operators ϕ_A (and hence of their products) and can also be verified directly from expression (4.10). By repeating the dummy-variable arguments used in proving the P- and η-independence of (2.4) one can show that making infinitesimal changes in $P^\alpha{}_i$ and $\eta_{\alpha\beta}$ is equivalent to subjecting each of the factors $\phi^{l_1} \ldots \phi^{l_n}\phi^{k_1} \ldots \phi^{k_n}$ in the integrand of (4.10) to the inverse of the transformation (2.9). One can then use the fact that the S-matrix is implicitly understood to be defined in terms of asymptotic wave packets that become infinitely broad only in a formal limit, to argue that $\delta\xi^\alpha[\varphi,\phi]$ effectively falls off at infinity sufficiently rapidly for the arrows on the $S^{\text{o}}{}_{,ij}$ terms to be reversed. Or else one can use the well known Lehman–Symanzik–Zimmermann reduction arguments (DeWitt 1965) in reverse to work back from expression (4.10) to an expression involving hypersurface integrals at infinity, where $Q^i{}_\alpha[\varphi+\phi]$ reduces effectively to $Q^i{}_\alpha{}^{\text{o}}$ and use the fact that $S^{\text{o}}{}_{,ij}Q^j{}_\alpha{}^{\text{o}} \equiv -S^{\text{o}}{}_{,j}Q^j{}_{\alpha,i} = 0$ (see eqn. (1.8)).

5. THE GENERATING FUNCTIONAL FOR CHRONOLOGICAL AVERAGES

Since chronological averages play such a fundamental role in the theory it is useful to have a device for dealing with them systematically. One introduces an external source J_i and replaces eqns (2.14) and (4.2) by

$$\exp(i\hat{W}[\varphi,J]) \overset{\text{def}}{=\!=} N''(\det \eta[\varphi])^{\frac{1}{2}}$$

$$\times \exp(i(S[\varphi+\phi]+\tfrac{1}{2}\eta_{\alpha\beta}[\varphi]P^\alpha{}_i[\varphi]P^\beta{}_j[\varphi]\phi^i\phi^j+J_i\phi^i))$$

$$\times (\det \hat{\mathfrak{G}}[\varphi,\phi])^{-1}d\phi \quad , \tag{5.1}$$

$$\langle A[\varphi,\phi]\rangle_J \overset{\text{def}}{=\!=} N''\exp(-i\hat{W}[\varphi,J])(\det \eta[\varphi])^{\frac{1}{2}} \int A[\omega,\phi]$$

$$\times \exp(i(S[\varphi+\phi]+\tfrac{1}{2}\eta_{\alpha\beta}[\varphi]P^\alpha{}_i[\varphi]P^\beta{}_j[\varphi]\phi^i\phi^j+J_i\phi^i)$$

$$\times (\det \hat{\mathfrak{G}}[\varphi,\phi])^{-1}d\phi \; . \tag{5.2}$$

Evidently $W[\varphi] = \hat{W}[\varphi,0]$ and $\langle A[\varphi,\boldsymbol{\phi}]\rangle = \langle A[\varphi,\boldsymbol{\phi}]\rangle_{J=0}$. When $J_i \neq 0$ the functional $\hat{W}[\varphi,J]$, unlike $W[\varphi]$, depends on the background field φ at finite points as well as at infinity. This is because J_i is coupled directly to $\boldsymbol{\phi}^i$, and when the background is changed $\boldsymbol{\phi}^i$ becomes a different operator. In his use of the background field method 't Hooft (1976) makes the source itself depend on the background and constructs it in a particular way by an iterative procedure. In lowest order he sets it equal to $-S_{,i}[\varphi]$ so that it cancels the term linear in ϕ in the expansion of $S[\varphi+\phi]$. We shall adopt a different approach in this paper.

Let ΔJ_i be a finite increment in the source. It is easy to see that

$$\sum_{n=0}^{\infty} \frac{i^n}{n!}\Delta J_{i_1}\ldots\Delta J_{i_n}\langle\boldsymbol{\phi}^{i_1}\ldots\boldsymbol{\phi}^{i_n}\rangle_J = \exp(\mathrm{i}(\hat{W}[\varphi,J+\Delta J]-\hat{W}[\varphi,J]))$$

$$= \exp\Big(i\Delta J_i\bar{\phi}^i + i\sum_{n=0}^{\infty}\frac{1}{n!}\Delta J_{i_1}\ldots\Delta J_{i_n}\hat{\Gamma}^{i_1\ldots i_n}[\varphi,\bar{\phi}]\Big)\;, \tag{5.3}$$

where

$$\bar{\phi}^i \;\underline{\underline{\mathrm{def}}}\; \frac{\delta}{\delta J_i}\hat{W}[\varphi,J]\;, \tag{5.4}$$

$$\hat{\Gamma}^{i_1\ldots i_n}[\varphi,\bar{\phi}] \;\underline{\underline{\mathrm{def}}}\; \frac{\delta}{\delta J_{i_1}}\cdots\frac{\delta}{\delta J_{i_n}}\hat{W}[\varphi,J]\;. \tag{5.5}$$

The $\hat{\Gamma}$s are known as *many-particle propagators*. In all that follows they are to be regarded as functionals of φ and $\bar{\phi}$, obtained by solving eqn. (5.4) for the J_i in terms of the φ^i and $\bar{\phi}^i$ and substituting in the right hand side of eqn. (5.5). This procedure depends on the Jacobian matrix

$$\frac{\delta\bar{\phi}^i}{\delta J_i} = \frac{\delta}{\delta J_i}\frac{\delta}{\delta J_j}\hat{W}[\varphi,J] = \hat{\Gamma}^{ij}[\varphi,\phi] \tag{5.6}$$

being non-singular, a condition that can be verified *a posteriori* in perturbation theory. We note that the $\hat{\Gamma}$s are symmetric in their indices and hence that

$$\delta\bar{\phi}^i/\delta J_j = \delta\bar{\phi}^j/\delta J_i\;. \tag{5.7}$$

We also note that in virtue of eqn. (4.5) we have

$$P^{\alpha}{}_{i}[\varphi]\bar{\phi}^{i} \underset{J\to 0}{\longrightarrow} 0 \tag{5.8}$$

and hence that the field $\bar{\phi}^{i}$ satisfies a linear gauge condition in the limit of vanishing source.

By picking out terms of equal degree in the ΔJs on both sides of eqn. (5.3) we obtain the following sequence of equations:

$$\left.\begin{aligned}
\langle \phi^{i} \rangle_{J} &= \bar{\phi}^{i} , \\
\langle \phi^{i}\phi^{j} \rangle_{J} &= \bar{\phi}^{i}\bar{\phi}^{j} - i\hat{\Gamma}^{ij} , \\
\langle \phi^{i}\phi^{j}\phi^{k} \rangle_{J} &= \bar{\phi}^{i}\bar{\phi}^{j}\bar{\phi}^{k} - iP_{3}\bar{\phi}^{i}\hat{\Gamma}^{jk} + (-i)^{2}\hat{\Gamma}^{ijk} , \\
\langle \phi^{i}\phi^{j}\phi^{k}\phi^{l} \rangle_{J} &= \bar{\phi}^{i}\bar{\phi}^{j}\bar{\phi}^{k}\bar{\phi}^{l} - iP_{6}\bar{\phi}^{i}\bar{\phi}^{j}\hat{\Gamma}^{kl} + (-i)^{2}P_{4}\bar{\phi}^{i}\hat{\Gamma}^{jkl} \\
&\quad + (-i)^{2}P_{3}\hat{\Gamma}^{ij}\hat{\Gamma}^{kl} + (-i)^{3}\hat{\Gamma}^{ijkl} , \text{ etc.,}
\end{aligned}\right\} \tag{5.9}$$

where P indicates that a summation is to be performed over all distinct permutations of the indices appearing in the term to which it is affixed, and the subscript on P indicates the number of permutations required in each case. This sequence of equations shows that the many-particle propagators may also be viewed as *correlation functions* for the chronological average.

Equation (5.3) yields a useful series expansion for the chronological average of an arbitrary operator:

$$\langle A[\varphi,\phi]\rangle_J = \sum_{n=0}^{\infty}\frac{1}{n!}\langle\phi^{i_1}\cdots\phi^{i_n}\rangle_J\left(\frac{\delta}{\delta\phi^{i_1}}\cdots\frac{\delta}{\delta\phi^{i}}A[\varphi,\phi]\right)_{\phi=0}$$

$$= \left[\exp\left(\bar{\phi}^i\frac{\delta}{\delta\phi^i}+i\sum_{n=2}^{\infty}\frac{(-i)^n}{n!}\hat{\Gamma}^{i_1\cdots i_n}[\varphi,\bar{\phi}]\frac{\delta}{\delta\phi^{i_1}}\cdots\frac{\delta}{\delta\phi^{i_n}}\right)A[\varphi,\phi]\right]_{\phi}$$

$$= :\exp\left(i\sum_{n=2}^{\infty}\frac{(-i)^n}{n!}\hat{\Gamma}^{i_1\cdots i_n}[\varphi,\bar{\phi}]\frac{\delta}{\delta\bar{\phi}^{i_1}}\cdots\frac{\delta}{\delta\bar{\phi}^{i_n}}\right):A[\varphi,\bar{\phi}]$$

$$= \left\{1+\frac{(-i)}{2!}\hat{\Gamma}^{ij}[\varphi,\bar{\phi}]\frac{\delta}{\delta\bar{\phi}^i}\frac{\delta}{\delta\bar{\phi}^j}+\frac{(-i)^2}{3!}\hat{\Gamma}^{ijk}[\varphi,\bar{\phi}]\frac{\delta}{\delta\bar{\phi}^i}\frac{\delta}{\delta\bar{\phi}^j}\frac{\delta}{\delta\bar{\phi}^k}\right.$$

$$+\frac{(-i)^3}{4!}\left(\hat{\Gamma}^{ijkl}[\varphi,\bar{\phi}]+3i\hat{\Gamma}^{ij}[\varphi,\bar{\phi}]\hat{\Gamma}^{kl}[\varphi,\bar{\phi}]\right)\frac{\delta}{\delta\bar{\phi}^i}\frac{\delta}{\delta\bar{\phi}^j}\frac{\delta}{\delta\bar{\phi}^k}\frac{\delta}{\delta\bar{\phi}}$$

$$\left.+\ \cdots\right\}A[\varphi,\bar{\phi}]\ . \qquad (5.10)$$

The :s in the third line indicate that the terms in the expansion of the exponential are to be ordered with all the $\hat{\Gamma}$s standing to the left and all the $\delta/\delta\bar{\phi}$s standing to the right.

6. THE EFFECTIVE ACTION

Consider the formal equation

$$-\ iN''\exp(-i\hat{W}[\varphi,J])(\det\eta[\varphi])^{\frac{1}{2}}$$

$$\times\int\frac{\delta}{\delta\phi^i}\left\{\exp\left(i(S[\varphi+\phi]+\tfrac{1}{2}\eta_{\alpha\beta}[\varphi]P^{\alpha}{}_j[\varphi]P^{\beta}{}_k[\varphi]\phi^j\phi^k+J_j\phi^j)\right)\right.$$

$$\left.\times\ (\det\hat{\mathfrak{G}}[\varphi,\phi])^{-1}\right\}d\phi = 0. \qquad (6.1)$$

Carrying out the indicated differentiation in the integrand and making use of eqn. (5.2) and the law for differentiating determinants, one easily verifies that this equation is equivalent to

$$\langle S_{,i}[\varphi+\boldsymbol{\phi}] + i\hat{\mathfrak{G}}^{\alpha}{}_{\beta}[\varphi,\boldsymbol{\phi}]V^{\beta}{}_{\alpha i}[\varphi]\rangle_J + \eta_{\alpha\beta}[\varphi]P^{\alpha}{}_{i}[\varphi]P^{\beta}{}_{j}[\varphi]\bar{\phi}^{j} = -J_i \ , \tag{6.2}$$

where

$$V^{\beta}{}_{\alpha i}[\varphi] \ \underline{\underline{\text{def}}} \ \frac{\delta}{\delta\phi^{i}}\hat{\mathfrak{F}}^{\alpha}{}_{\beta}[\varphi,\phi] = P^{\beta}{}_{j}[\varphi]Q^{j}{}_{\alpha,i} \ . \tag{6.3}$$

When no gauge group is present eqn. (6.2) may be recognized as the chronological average of the operator field equations $S_{,i}[\varphi+\boldsymbol{\phi}] = -J_i$. In this case the space-time points associated with the factors appearing in every term of the field equations (which are assumed to be local) may be consistently imagined as separated by infinitesimal space-like intervals so that the chronological ordering operation leaves these equations unaffected even if they are non-linear. When a gauge group is present this is no longer true for two reasons. First, separation by a space-like interval, even an infinitesimal one, is not a gauge invariant operation. Second, $S_{,i}[\varphi+\boldsymbol{\phi}]$ itself is not gauge invariant, and $\langle S_{,i}[\varphi+\boldsymbol{\phi}]\rangle$ is not the chronological average of $S_{,i}[\varphi+\boldsymbol{\phi}]$ but rather of a gauge invariant operator constructed from $S_{,i}[\varphi+\boldsymbol{\phi}]$ by the operation embodied in eqns. (4.3) and (4.4). The result is that even when the source vanishes eqn. (6.2) reduces not to $\langle S_{,i}[\varphi+\boldsymbol{\phi}]\rangle = 0$ but, in virtue of (5.8), to

$$\langle S_{,i}[\varphi+\boldsymbol{\phi}] + i\,\hat{\mathfrak{G}}^{\alpha}{}_{\beta}[\varphi,\boldsymbol{\phi}]V^{\beta}{}_{\alpha i}[\varphi]\rangle = 0 \ . \tag{6.4}$$

Let us now differentiate both sides of eqn. (6.2) with respect to J_j. Making use of eqn. (5.6) we find

$$\left\{\frac{\delta}{\delta\bar{\phi}^{k}}\langle S_{,i}[\varphi+\boldsymbol{\phi}]+i\,\hat{\mathfrak{G}}^{\alpha}{}_{\beta}[\varphi,\boldsymbol{\phi}]V^{\beta}{}_{\alpha i}[\varphi]\rangle_J + \eta_{\alpha\beta}[\varphi]P^{\alpha}{}_{i}[\varphi]P^{\beta}{}_{k}[\varphi]\right\} \times \hat{\Gamma}^{kj}[\varphi,\bar{\phi}] = -\,\delta_i{}^{j} \ . \tag{6.5}$$

The quantity inside the brackets { } is seen to be the negative inverse of the propagator $\hat{\Gamma}^{kj}$. Since $\hat{\Gamma}^{kj}$ is symmetric in its indices its inverse must also be symmetric (i.e., in i and k). But this implies the existence of a functional $\hat{\Gamma}[\varphi,\bar{\phi}]$ such that

$$\langle S_{,i}[\varphi+\phi] + i\,\hat{\mathfrak{G}}^{\alpha}{}_{\beta}[\varphi,\phi]V^{\beta}{}_{\alpha i}[\varphi]\rangle_J + \eta_{\alpha\beta}[\varphi]P^{\alpha}{}_{i}[\varphi]P^{\beta}{}_{j}[\varphi]\bar{\phi}^j$$
$$= \frac{\delta}{\delta\bar{\phi}^i}\hat{\Gamma}[\varphi,\bar{\phi}] \ . \tag{6.6}$$

$\hat{\Gamma}[\varphi,\bar{\phi}]$ is called the *effective action.*

In terms of the effective action the basic equations of the theory become (arguments suppressed)

$$\frac{\delta\hat{\Gamma}}{\delta\bar{\phi}^i} = -J_i \ , \tag{6.7}$$

$$\frac{\delta^2\hat{\Gamma}}{\delta\bar{\phi}^i\delta\bar{\phi}^k}\hat{\Gamma}^{kj} = -\delta_i{}^j \ . \tag{6.8}$$

The many-particle propagators may be expressed in terms of $\hat{\Gamma}^{ij}$ and the derivatives of $\hat{\Gamma}$ by repeated use of the identity

$$\frac{\delta\hat{\Gamma}^{ij}}{\delta\bar{\phi}^m} = \hat{\Gamma}^{ik}\frac{\delta^3\hat{\Gamma}}{\delta\bar{\phi}^k\delta\bar{\phi}^l\delta\bar{\phi}^m}\hat{\Gamma}^{lj} \ . \tag{6.9}$$

For example

$$\hat{\Gamma}^{ijk} = \frac{\delta}{\delta J_k}\hat{\Gamma}^{ij} = \hat{\Gamma}^{kc}\frac{\delta\hat{\Gamma}^{ij}}{\delta\bar{\phi}^c} = \hat{\Gamma}^{ia}\hat{\Gamma}^{jb}\hat{\Gamma}^{kc}\frac{\delta^3\hat{\Gamma}}{\delta\bar{\phi}^a\delta\bar{\phi}^b\delta\bar{\phi}^c} \ . \tag{6.10}$$

Equation (6.10) and the equations that follow from it by repeated differentiation with respect to the source all have simple graphical representations. The propagators $\hat{\Gamma}^{ij}$ are represented by heavy solid lines and the derivatives of $\hat{\Gamma}$ by vertices or forks having prongs equal in number to the number of functional differentiations. The many-particle propagators are represented by diagrams in which the heavy lines are joined together at vertices in the same ways that the $\hat{\Gamma}^{ij}$ in the explicit expressions are coupled to the derivatives of $\hat{\Gamma}$ by dummy indices. It is easy to see that differentiation with respect to the source corresponds to the insertion of an external heavy line in all possible ways into a given diagram (including at pre-existing vertices). Iteration of this rule leads to the theorem that each many-particle propagator is expressible as the sum of all *simply connected* or *tree* diagrams having a fixed number of external lines, the indices attached to the latter being permuted just sufficiently to yield complete symmetry.

The functional derivatives (with respect to $\bar{\phi}$) of $\hat{\Gamma}$ of order three and higher are the full radiatively corrected vertex functions of the theory, and $\hat{\Gamma}^{ij}$ is the full radiatively corrected 1-particle propagator. In virtue of eqns. (4.10), (4.11), and (5.9) we see that the elements of the S-matrix are all expressible as sums of products of tree diagrams in which the lines and vertices are full 1-particle propagators and full vertex functions respectively and every external line terminates in a wave function u_A or $u_A{}^*$.

Although eqn. (6.5) merely allows us to *infer* the existence of the functional $\hat{\Gamma}$, eqn. (6.6), combined with eqns. (5.10) and (6.8), allows us to compute it. We begin by writing

$$\frac{\delta}{\delta\bar{\phi}^i}\hat{\Gamma}[\varphi,\bar{\phi}]$$

$$= :\exp\left(i\sum_{n=2}^{\infty}\frac{(-i)^n}{n!}\hat{\Gamma}^{i\cdots i_n}[\varphi,\bar{\phi}]\frac{\delta}{\delta\bar{\phi}^i}\cdots\frac{\delta}{\delta\bar{\phi}^{i_n}}\right):$$

$$\times\left(S_{,i}[\varphi+\bar{\phi}] + i\hat{\mathfrak{G}}^{\alpha}{}_{\beta}[\varphi,\bar{\phi}]V^{\beta}{}_{\alpha i}[\varphi]\right)$$

$$+\ \eta_{\alpha\beta}[\varphi]P^{\alpha}{}_{i}[\varphi]P^{\beta}{}_{j}[\varphi]\bar{\phi}^{j}\ , \qquad (6.11)$$

which is a hybrid equation involving both bare vertex functions $S_{,ijk}\cdots[\varphi+\bar{\phi}]$ and, through the $\hat{\Gamma}^{ijk\cdots}$, full vertex functions. By differentiating this equation we obtain hybrid expressions for the full vertex functions, and by inverting the expression for $\delta^2\hat{\Gamma}/\delta\bar{\phi}^i\delta\bar{\phi}^j$ we obtain an expansion of the full propagator $\hat{\Gamma}^{ij}$ in terms of the *bare* propagators $\hat{G}^{ij}[\varphi,\bar{\phi}]$ and $\hat{\mathfrak{G}}^{\alpha}{}_{\beta}[\varphi,\bar{\phi}]$ that were introduced earlier (but with ϕ now replaced by $\bar{\phi}$) In this way an iteration procedure may be developed to express $\delta\hat{\Gamma}/\delta\bar{\phi}^i$ completely in terms of bare propagators and bare vertex functions.

The result may be expressed as an infinite sum of diagrams built out of dotted lines and light solid lines. The dotted lines represent propagators $\hat{\mathfrak{G}}^{\alpha}{}_{\beta}[\varphi,\bar{\phi}]$ and bear arrows to remind us that owing to the lack of symmetry in its indices $\hat{\mathfrak{G}}$ carries an orientation. The light solid lines represent propagators $\hat{G}^{ij}[\varphi,\bar{\phi}]$. It turns out that a given solid line can meet two or more others in a bare vertex $S_{,ijk}\ldots[\varphi+\bar{\phi}]$, or it can meet two dotted lines in a vertex having the value $V^{\beta}{}_{\alpha i}[\varphi]$ (see eqn. (6.3)). The dotted lines always appear in closed oriented loops. They couple to the rest of a diagram only through the vertices $V^{\beta}{}_{\alpha i}[\varphi]$. The diagrams for $\delta\hat{\Gamma}/\delta\bar{\phi}^i$ all bear a dangling prong, representing the differentiation with respect to $\bar{\phi}^i$, and can be obtained by inserting a single prong in all possible ways into a more basic set of prongless diagrams. The latter diagrams are those of the effective action itself. The first few of these are depicted in Fig. 1.

The first two terms on the right in the figure, when differentiated twice with respect to $\bar{\phi}$, yield the operator $\hat{F}_{ij}[\varphi,\bar{\phi}]$ inverse to the bare propagator $\hat{G}^{ij}[\varphi,\bar{\phi}]$. The third term is independent of $\bar{\phi}$ and is a constant integration for eqn. (6.11), introduced in order to secure agreement with eqn. (7.3) below. The solid circle stands for ln det $\hat{G}[\varphi,\bar{\phi}]$ and the dotted circle stands for ln det $\hat{\mathfrak{G}}[\varphi,\bar{\phi}]$. It will be observed that the figure contains no 1-particle irreducible diagrams, i.e., diagrams that can be separated into two parts by cutting a single line. This is because all propagators and vertex functions have been evaluated at $\phi = \bar{\phi}$ rather than $\phi = 0$.

7. INVARIANCE PROPERTIES OF $\hat{\Gamma}$

$\hat{\Gamma}$ has a number of important invariance properties that can be most easily derived by first relating it to the functional $\hat{W}$. Observe that the equation

$$\bar{\phi}^i = \frac{\delta\hat{W}}{\delta J_i} = \hat{\Gamma}^{ij}\,\frac{\delta\hat{W}}{\delta\bar{\phi}^i} \tag{7.1}$$

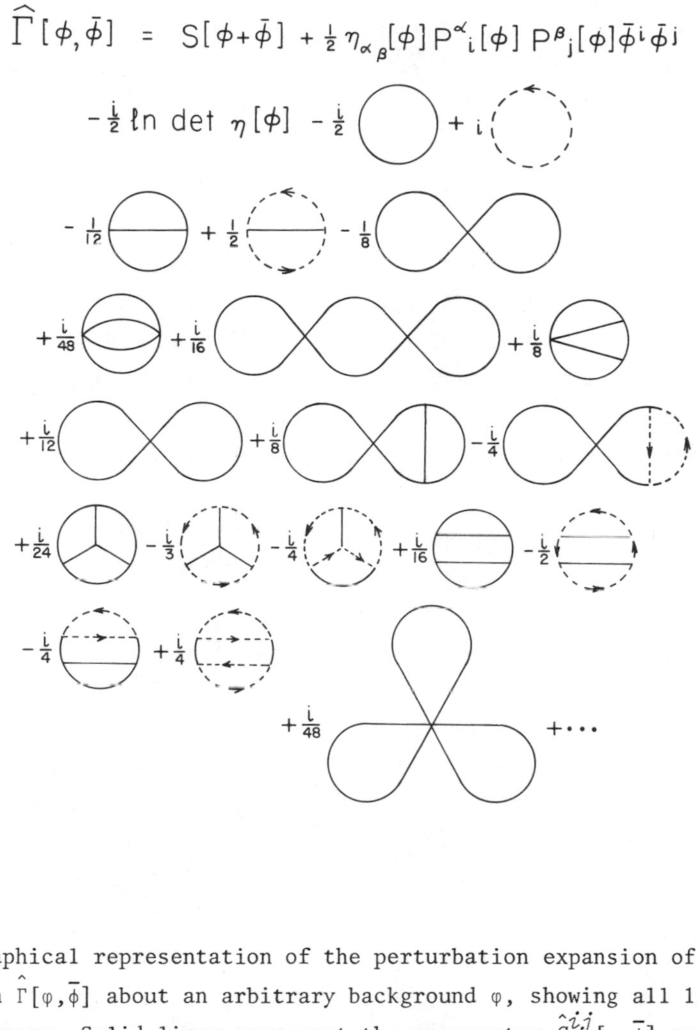

FIG. 1. Graphical representation of the perturbation expansion of the effective action $\hat{\Gamma}[\varphi,\bar{\phi}]$ about an arbitrary background φ, showing all 1, 2 and 3-loop diagrams. Solid lines represent the propagator $\hat{G}^{ij}[\varphi,\bar{\phi}]$ and dotted lines the propagator $\hat{\mathfrak{G}}^{\alpha}{}_{\beta}[\varphi,\bar{\phi}]$. Solid lines meet each other in the vertices $S_{,ijk\ldots}[\varphi+\bar{\phi}]$. They meet dotted lines in the vertex $V^{\alpha}{}_{\beta i}[\varphi]$.

implies

$$\frac{\delta\hat{W}}{\delta\bar{\phi}^{i}} = -\frac{\delta^{2}\hat{\Gamma}}{\delta\bar{\phi}^{i}\delta\bar{\phi}^{j}}\,\bar{\phi}^{j} = \frac{\delta}{\delta\bar{\phi}^{i}}\left(\hat{\Gamma} - \frac{\delta\hat{\Gamma}}{\delta\bar{\phi}^{j}}\,\bar{\phi}^{j}\right) , \tag{7.2}$$

whence, up to an arbitrary functional of φ,

$$\hat{W} = \hat{\Gamma} - \frac{\delta\hat{\Gamma}}{\delta\bar{\phi}^i}\,\bar{\phi}^i + \text{const.} = \hat{\Gamma} + J_i\bar{\phi}^i + \text{const.} \tag{7.3}$$

With $\hat{\Gamma}$ given as in Fig. 1, with the term $-\frac{i}{2}\,ln \det \eta[\varphi]$ included, the 'const.' in eqn. (7.3) is a true constant, independent not only of φ and $\bar{\phi}$ but also of J and of the choice of P and η. We note that eqn. (7.3) provides the basis for an alternative construction of $\hat{\Gamma}$. One starts from eqn. (5.1) and evaluates both $\delta\hat{W}/\delta J_i$ $(= \bar{\phi})$ and $\hat{W} - J_i\delta\hat{W}/\delta J_i$ $(= \hat{\Gamma})$ as asymptotic series of Gaussian integrals, computed in terms of $\hat{G}^{ij}[\varphi,0]$, $\hat{\mathfrak{G}}^{\alpha}{}_{\beta}[\varphi,0]$, $S_{,ijk\ldots}[\varphi]$, $V^{\alpha}{}_{\beta i}[\varphi]$, and J_i. One then shows that the result for $\hat{\Gamma}$, which graphically involves many 1-particle reducible diagrams, is equivalent to Fig. 1, in which only the 1-particle irreducible diagrams are retained but the propagators and vertex functions are $\hat{G}^{ij}[\varphi,\bar{\phi}]$, $\hat{\mathfrak{G}}^{\alpha}{}_{\beta}[\varphi,\bar{\phi}]$, $S_{,ijk\ldots}[\varphi+\bar{\phi}]$ and $V^{\alpha}{}_{\beta i}[\varphi]$.

Consider now the equation

$$\begin{aligned}
\frac{\delta\hat{W}[\varphi,J]}{\delta\varphi^i}Q^i{}_{\alpha}[\varphi] &= -i\,\exp(-i\hat{W}[\varphi,J])\left(\frac{\delta}{\delta\varphi^i}\exp(i\hat{W}[\varphi,J])\right)Q^i{}_{\alpha}[\varphi] \\
&= N''\exp(-i\hat{W}[\varphi,J])(\det\eta[\varphi])^{\frac{1}{2}}\int\Big\{-\tfrac{i}{2}\eta^{\beta\gamma}[\varphi]\eta_{\beta\gamma,i}[\varphi] + \\
&\quad + S_{,i}[\varphi+\phi] + \tfrac{1}{2}(\eta_{\beta\gamma}[\varphi]P^{\beta}{}_{j}[\varphi]P^{\gamma}{}_{k}[\varphi])_{,i}\phi^j\phi^k - \\
&\quad - i\hat{\mathfrak{G}}^{\beta}{}_{\gamma}[\varphi,\phi](P^{\gamma}{}_{j,i}[\varphi]Q^j{}_{\beta}[\varphi+\phi] + P^{\gamma}{}_{j}[\varphi]Q^j{}_{\beta,i})\Big\}Q^i{}_{\alpha}[\varphi] \\
&\quad \times \exp(i(S[\varphi+\phi] + \tfrac{1}{2}\eta_{\delta\varepsilon}[\varphi]P^{\delta}{}_{l}[\varphi]P^{\varepsilon}{}_{m}[\varphi]\phi^l\phi^m + \\
&\quad + J_l\phi^l))\,(\det\hat{\mathfrak{G}}[\varphi,\phi])^{-1}d\phi\,.
\end{aligned} \tag{7.4}$$

Making use of the transformation laws (2.7) and (2.16) together with eqns (1.4) and (2.5) and the corollary

$$S_{,i}[\varphi+\phi]Q^i{}_\alpha[\varphi] \equiv - S_{,i}[\varphi+\phi]Q^i{}_{\alpha,j}\phi^j$$

of eqn. (1.7), one easily converts expression (7.4) to

$$\frac{\delta\hat{W}[\varphi,J]}{\delta\varphi^i}Q^i{}_\alpha[\varphi]$$

$$= N''\exp(-i\hat{W}[\varphi,J])(\det\eta[\varphi])^{\frac{1}{2}}\int(S_{,i}[\varphi+\phi]+\eta_{\beta\gamma}[\varphi]P^\beta{}_i[\varphi]P^\gamma{}_k[\varphi]\phi^k$$

$$+ i\hat{\mathfrak{G}}^\beta{}_\gamma[\varphi,\phi]V^\gamma{}_{\beta i}[\varphi])Q^i{}_{\alpha,j}\phi^j$$

$$\times \exp(i(S[\varphi+\phi]+\tfrac{1}{2}\eta_{\delta\varepsilon}[\varphi]P^\delta{}_l[\varphi]P^\varepsilon{}_m[\varphi]\phi^l\phi^m+J_l\phi^l))(\det\hat{\mathfrak{G}}[\varphi,\phi])^{-1}d\phi$$

$$= N''\exp(-i\hat{W}[\varphi,J])(\det\eta[\varphi])^{\frac{1}{2}}\int\left(J_i + i\frac{\delta}{\delta\phi^i}\right)\Big\{Q^i{}_{\alpha,j}\phi^j$$

$$\times \exp(i(S[\varphi+\phi]+\tfrac{1}{2}\eta_{\beta\gamma}[\varphi]P^\beta{}_k[\varphi]P^\gamma{}_l[\varphi]\phi^k\phi^l+J_k\phi^k))(\det\hat{\mathfrak{G}}[\varphi,\phi])^{-1}\Big\}d\phi$$

$$= J_iQ^i{}_{\alpha,j}\bar{\phi}^j\ , \qquad (7.5)$$

or, alternatively,

$$\frac{\delta\hat{W}[\varphi,J]}{\delta\varphi^i}Q^i{}_\alpha[\varphi] - J_iQ^i{}_{\alpha,j}\frac{\delta\hat{W}[\varphi,J]}{\delta J_i} \equiv 0\ . \qquad (7.6)$$

Evidently $\hat{W}$ is invariant under the set of extended gauge transformations

$$\left.\begin{aligned}\delta\varphi^i &= Q^i{}_\alpha[\varphi]\delta\xi^\alpha\ ,\\ \delta J_i &= -J_jQ^j{}_{\alpha,i}\delta\xi^\alpha\ .\end{aligned}\right\} \qquad (7.7)$$

From eqn. (7.6) one can obtain a corresponding invariance statement about $\hat{\Gamma}$. Using eqn. (7.3) together with eqn. (5.4) or eqn. (6.7) one easily verifies that

$$\frac{\delta \hat{W}[\varphi, J]}{\delta \varphi^i} = \frac{\delta \hat{\Gamma}[\varphi, \bar{\phi}]}{\delta \varphi^i} . \tag{7.8}$$

Therefore

$$\frac{\delta \hat{\Gamma}[\varphi, \phi]}{\delta \varphi^i} Q^i{}_\alpha[\varphi] + \frac{\delta \hat{\Gamma}[\varphi, \bar{\phi}]}{\delta \bar{\phi}^i} Q^i{}_{\alpha, j} \bar{\phi}^j \equiv 0 . \tag{7.9}$$

$\hat{\Gamma}$ is seen to be invariant under the transformation (2.4a) with ϕ replaced by $\bar{\phi}$. This invariance, which also holds individually for every term of the diagrammatic expansion in Fig. 1, is a consequence of our having chosen *P*s and ηs that transform covariantly (eqns (2.7) and (2.16)) under gauge transformations of the background field.

$\hat{\Gamma}$ is *not* invariant under changes in the *P*s and ηs. Using the dummy-variable arguments invoked in proving the *P*- and η-invariance of $W[\varphi]$, we find the infinitesimal variation law

$$\delta \hat{W}[\varphi, J] = - \mathrm{i} \exp(-\mathrm{i}\hat{W}[\varphi, J]) \delta \exp(\mathrm{i}\hat{W}[\varphi, J])$$

$$= - J_i \langle Q^i{}_\alpha[\varphi + \boldsymbol{\phi}] \delta \xi^\alpha[\varphi, \boldsymbol{\phi}] \rangle_J , \tag{7.10}$$

where $\delta\xi^\alpha[\varphi, \boldsymbol{\phi}]$ is the quantum operator corresponding to the $\delta\xi^\alpha[\varphi, \phi]$ of eqn. (3.1). If the chronological average on the right is expressed in terms of φ and J, then (7.10) gives the change in the functional form of $\hat{W}$ with J held fixed. To get the corresponding change in the functional form of $\hat{\Gamma}$ one must hold $\bar{\phi}$ fixed. (The background is held fixed in both cases.) This requires shifting J_i by an appropriate infinitesimal amount δJ_i, which adds a term $\bar{\phi}^i \delta J_i$ to the right of (7.10). Therefore

$$\delta \hat{\Gamma}[\varphi, \bar{\phi}] = \delta(\hat{W}[\varphi, J] - J_i \bar{\phi}^i)$$

$$= \frac{\delta \hat{\Gamma}[\varphi, \phi]}{\delta \bar{\phi}^i} X^i[\varphi, \bar{\phi}] \tag{7.11}$$

where X^i is the chronological average appearing in (7.10), re-expressed in terms of φ and $\bar{\phi}$, e.g., by use of eqn. (5.10):

$$X^i[\varphi,\bar{\phi}] = :\exp\left(i \sum_{n=2}^{\infty} \frac{(-i)^n}{n!} \hat{\Gamma}^{i_1\ldots i_n}[\varphi,\bar{\phi}] \frac{\delta}{\delta\bar{\phi}^{i_1}} \cdots \frac{\delta}{\delta\bar{\phi}^{i_n}}\right): (Q^i{}_\alpha[\varphi+\bar{\phi}]\delta\xi^\alpha[\varphi,\bar{\phi}]) \quad (7.12)$$

We stress that it is the functional form and not the value of a given expression that is of interest to us here. $\bar{\phi}$ is to be kept completely unconstrained at this stage. In principle, for every $\bar{\phi}$ for which $\hat{\Gamma}^{ij}$ is non-singular it is possible to choose a source J that yields that $\bar{\phi}$. Note that $\bar{\phi}$, being $\exp(-i\hat{W})$ times an 'in-out' matrix element rather than an expectation value of $\boldsymbol{\phi}$, is generally complex even when J is real.

Even with $\bar{\phi}$ unconstrained there are relations between some of the functionals of φ and $\bar{\phi}$ that have been introduced thus far. One that will be used presently is the following (cf. eqns (4.5) and (5.8)):

$$\begin{aligned} P^\alpha{}_i[\varphi]\bar{\phi}^i &= \langle P^\alpha{}_i[\varphi]\boldsymbol{\phi}^i \rangle_J \\ &= \langle P^\alpha{}_i[\varphi]\boldsymbol{\phi}^i(1-\exp(-iJ_j\boldsymbol{\phi}^j)) \rangle_J \\ &= iJ_i\langle \boldsymbol{\phi}^i P^\alpha{}_j[\varphi]\boldsymbol{\phi}^j f(iJ_k\boldsymbol{\phi}^k) \rangle_J \end{aligned} \quad (7.13)$$

where

$$f(x) \overset{\text{def}}{=\!=} \frac{1 - e^{-x}}{x} = 1 - \frac{1}{2!}x + \frac{1}{3!}x^2 - \ldots. \quad (7.14)$$

Let $Y^{i\alpha}$ be the chronological average appearing in the last line of eqn. (7.13), expressed in terms of φ and $\bar{\phi}$. Then we may also write

$$P^\alpha{}_i[\varphi]\bar{\phi}^i \equiv -i \frac{\delta\hat{\Gamma}[\varphi,\bar{\phi}]}{\delta\bar{\phi}^i} Y^{i\alpha}[\varphi,\bar{\phi}] . \quad (7.15)$$

Since φ and $\bar{\phi}$ are arbitrary this is necessarily an identity.

8. THE GAUGE INVARIANT EFFECTIVE ACTION

Suppose, in equations (5.1) and (5.2), we replace $\eta_{\alpha\beta}[\varphi]$ and $P^{\alpha}{}_{i}[\varphi]$ respectively by $\eta_{\alpha\beta}[\varphi+\bar{\phi}]$ and $P^{\alpha}{}_{\hat{i}}[\varphi+\bar{\phi}]$ everywhere they occur, including in the factors $(\det \mathfrak{G}[\varphi,\phi])^{-1}$. There will then be a functional dependence on $\bar{\phi}$ *inside* the integrands even before $\bar{\phi}$ (defined by (5.4) or (5.9)) is computed. Of course, if the source vanishes and $A[\varphi,\phi]$ is invariant under (2.4b), these replacements have no effect on the integrals because of their P- and η-independence. However, the value of $\bar{\phi}$ is altered whether the source vanishes or not, and we have here a self-consistency problem. In principle this problem can be solved by an iterative procedure (cf. 't Hooft 1976) but it turns out, in what follows, to be unnecessary to go through the actual steps.

Let us examine what these replacements do to some of the basic functionals of the theory. Observe, first of all, that $\hat{\mathfrak{F}}^{\alpha}{}_{\beta}[\varphi,\bar{\phi}]$ and $\hat{F}_{ij}[\varphi,\bar{\phi}]$ are changed to $\mathfrak{F}^{\alpha}{}_{\beta}[\varphi+\bar{\phi}]$ and $F_{ij}[\varphi+\bar{\phi}]$ respectively, where $\mathfrak{F}^{\alpha}{}_{\beta}$ and F_{ij} are defined by (3.5) and (3.6). $\hat{\mathfrak{G}}^{\alpha}{}_{\beta}[\varphi,\bar{\phi}]$ and $\hat{G}^{ij}[\varphi,\bar{\phi}]$ are accordingly changed to $\mathfrak{G}^{\alpha}{}_{\beta}[\varphi+\bar{\phi}]$ and $G^{ij}[\varphi+\bar{\phi}]$. Next consider $\hat{\Gamma}[\varphi,\bar{\phi}]$, which can be written in the form

$$\hat{\Gamma}[\varphi,\bar{\phi}] = S[\varphi+\bar{\phi}] + \tfrac{1}{2}\eta_{\alpha\beta}[\varphi]P^{\alpha}{}_{i}[\varphi]P^{\beta}{}_{j}[\varphi]\bar{\phi}^{i}\bar{\phi}^{j} + \hat{\Sigma}[\varphi,\bar{\phi}] \ , \tag{8.1}$$

where $\hat{\Sigma}$ denotes everything below the first line in Fig. 1. The first term on the right of eqn. (8.1) remains unchanged and the second term changes in an obvious manner. To get the new $\hat{\Sigma}$ from the old one, replace $-\frac{1}{2}\ ln \det \eta[\varphi]$ by $-\frac{1}{2}\ ln \det \eta[\varphi+\bar{\phi}]$ and leave all the diagrams of Fig. 1 exactly as they are but reinterpret them as follows. Solid lines now denote the propagator $G^{ij}[\varphi+\bar{\phi}]$, dotted lines the propagator $\mathfrak{G}^{\alpha}{}_{\beta}[\varphi+\bar{\phi}]$. The vertices at which solid lines meet are $S_{,ijk\ldots}[\varphi+\bar{\phi}]$ (as before), and the vertex at which a solid line meets a dotted line becomes $V^{\alpha}{}_{\beta i}[\varphi+\bar{\phi}]$.

Observe that $\hat{\Sigma}$ now depends on φ and $\bar{\phi}$ only in the combination $\varphi + \bar{\phi}$. It is therefore useful to introduce a new symbol,

$$\bar{\varphi}^{i} \overset{\text{def}}{=\!=} \varphi^{i} + \bar{\phi}^{i} \tag{8.2}$$

and to define

$$\Sigma[\bar{\varphi}] \overset{\text{def}}{=} (\hat{\Sigma}[\varphi,\bar{\phi}])_{\rightarrow} \tag{8.3}$$

where '$(\ )_{\rightarrow}$' applied to any quantity indicates that the quantity is to be computed with the new Ps and ηs. Because of the covariance conditions (2.7) and (2.16) Σ satisfies the gauge-invariance identity

$$\Sigma_{,i}[\bar{\varphi}]Q^{i}{}_{\alpha}[\bar{\phi}] \equiv 0 \ , \tag{8.4}$$

which is, in fact, a corollary of eqn. (7.9). Note that, when applied to $\bar{\varphi}$, the gauge transformation laws (2.4a) and (2.46), with ϕ replaced by $\bar{\phi}$, become identical. From now on we shall call $\bar{\varphi}$ the *effective field*.

Define

$$\Gamma[\bar{\varphi}] \overset{\text{def}}{=} S[\bar{\varphi}] + \Sigma[\bar{\varphi}]$$

$$\equiv (\hat{\Gamma}[\varphi,\bar{\phi}])_{\rightarrow} - \tfrac{1}{2}\eta_{\alpha\beta}[\bar{\varphi}]P^{\alpha}{}_{i}[\bar{\varphi}]P^{\beta}{}_{j}[\bar{\varphi}]\bar{\phi}^{i}\bar{\phi}^{j}$$

$$\equiv (\hat{\Gamma}[\varphi,\bar{\phi}])_{\rightarrow} + \frac{1}{2}\left(M^{ij}[\varphi,\bar{\phi}]\ \frac{\delta\hat{\Gamma}[\varphi,\bar{\phi}]}{\delta\bar{\phi}^{i}}\ \frac{\delta\hat{\Gamma}[\varphi,\bar{\phi}]}{\delta\bar{\phi}^{i}}\right)_{\rightarrow} \tag{8.5}$$

where (see eqn. (7.15))

$$M^{ij}[\varphi,\bar{\phi}] \overset{\text{def}}{=} \eta_{\alpha\beta}[\varphi]Y^{i\alpha}[\varphi,\bar{\phi}]Y^{j\beta}[\varphi,\bar{\phi}] \ . \tag{8.6}$$

Evidently

$$\Gamma_{,i}[\bar{\varphi}]Q^{i}{}_{\alpha}[\bar{\varphi}] \equiv 0 \ . \tag{8.7}$$

$\Gamma[\bar{\varphi}]$ is the *gauge invariant effective action* promised at the beginning. Our chief remaining task is to show how it can be used in place of $\hat{\Gamma}$ to compute the tree amplitudes and hence the S-matrix elements of the theory.

9. THE EFFECTIVE FIELD EQUATIONS: TREE THEOREMS

Consider the functional derivative $\delta(\hat{\Gamma}[\varphi,\bar{\phi}])_{\rightarrow}/\delta\bar{\phi}^{i}$. It consists of two parts: (i) a part $(\delta\hat{\Gamma}[\varphi,\bar{\phi}]/\delta\bar{\phi}^{i})_{\rightarrow}$ which is computed by differentiating the old $\hat{\Gamma}$ and then making the shift to the new Ps and ηs; (ii) a part in which the new Ps and ηs on which $(\hat{\Gamma})_{\rightarrow}$ depends get differentiated. Since differentiating the Ps and ηs is similar to varying them, the second part can be computed by steps analogous to those followed in obtaining eqns. (7.10) and (7.11). The total result is

$$\frac{\delta}{\delta\bar{\phi}^{i}}(\hat{\Gamma}[\varphi,\bar{\phi}])_{\rightarrow} = \left(\frac{\delta\hat{\Gamma}[\varphi,\bar{\phi}]}{\delta\bar{\phi}^{i}} + \frac{\delta\hat{\Gamma}[\varphi,\bar{\phi}]}{\delta\bar{\phi}^{j}}\, X^{j}{}_{i}[\varphi,\bar{\phi}]\right)_{\rightarrow} \tag{9.1}$$

where

$$X^{j}{}_{i}[\varphi,\bar{\phi}] = -\langle Q^{j}{}_{\alpha}[\varphi+\boldsymbol{\phi}]\hat{\boldsymbol{\mathfrak{G}}}^{\alpha}{}_{\beta}[\varphi,\boldsymbol{\phi}](P^{\beta}{}_{k,i}[\varphi] + \tfrac{1}{2}\eta^{\beta\gamma}[\varphi]\eta_{\gamma\delta,i}[\varphi]P^{\delta}{}_{k}[\varphi])\boldsymbol{\phi}^{k}\rangle_{J}. \tag{9.2}$$

Equations (8.5) and (9.1) together yield

$$\begin{aligned}\Gamma_{,i}[\bar{\varphi}] = {} & \left(\frac{\delta\hat{\Gamma}[\varphi,\bar{\phi}]}{\delta\bar{\phi}^{i}} + \frac{\delta\hat{\Gamma}[\varphi,\bar{\phi}]}{\delta\bar{\phi}^{j}}\, X^{j}{}_{i}[\varphi,\bar{\phi}]\right)_{\rightarrow} \\ & + \frac{1}{2}\left\{\frac{\delta}{\delta\bar{\phi}^{i}}(M^{jk}[\varphi,\bar{\delta}])_{\rightarrow}\right\}\left(\frac{\delta\hat{\Gamma}[\varphi,\bar{\phi}]}{\delta\bar{\phi}^{j}}\,\frac{\delta\hat{\Gamma}[\varphi,\bar{\phi}]}{\delta\bar{\phi}^{k}}\right)_{\rightarrow} \\ & + \left\{\frac{\delta}{\delta\bar{\phi}^{i}}\left(\frac{\delta\hat{\Gamma}[\varphi,\bar{\phi}]}{\delta\bar{\phi}^{j}}\right)_{\rightarrow}\right\}\left(M^{jk}[\varphi,\bar{\phi}]\,\frac{\delta\hat{\Gamma}[\varphi,\bar{\phi}]}{\delta\bar{\phi}^{k}}\right)_{\rightarrow}.\end{aligned} \tag{9.3}$$

Since $(\delta\hat{\Gamma}[\varphi,\bar{\phi}]/\delta\bar{\phi}^{i})_{\rightarrow} = -J_{i}$ it follows that when the source vanishes we have

$$\Gamma_{,i}[\bar{\varphi}] = 0 \qquad (J_{i} = 0)\ . \tag{9.4}$$

The source can be viewed as a device for permitting $\bar{\phi}$ to assume all possible values and for constructing an effective action having a definite functional form. The physics becomes real only when the source vanishes. $\bar{\phi}$ is then, of course, no longer unconstrained but becomes a definite functional of φ. Here, however, the background field comes into its own. It not only combines linearly with $\bar{\phi}$ to yield the total effective field $\bar{\varphi}$ but it can be chosen to be an arbitrary classical field at infinity. This has the consequence that *all* solutions of eqns (9.4) are of physical relevance. We shall call these equations the *equations for the effective field.*

Let us assume that there is no dynamical symmetry breaking in the theory. Then the classical vacuum, which may be denoted by φ_o, will be a solution not only of the classical field eqns (1.2) but also of eqns (9.4) and will correspond to the unique non-degenerate ground state of the system. Furthermore, there will exist an important class of solutions of eqns (9.4) that tend arbitrarily closely to the classical vacuum at infinity. Let $\bar{\varphi}$ be one of these solutions. Then we can obtain another solution, $\bar{\varphi} + \Delta\bar{\varphi}$, which differs finitely from $\bar{\varphi}$ at finite points, by the following construction. Let $\Gamma^{ij}[\bar{\varphi}]$ be the Feynman Green's function for the integro-differential operator $\Gamma_{,ij}[\bar{\varphi}] + \eta_{\alpha\beta}[\bar{\varphi}]P^{\alpha}{}_{i}[\bar{\varphi}]P^{\beta}{}_{j}[\bar{\varphi}]$:*

$$(\Gamma_{,ik} + \eta_{\alpha\beta}P^{\alpha}{}_{i}P^{\beta}{}_{k})\Gamma^{kj} = -\delta_{i}{}^{j} \text{ (arguments suppressed)}. \tag{9.5}$$

Define

$$\Delta\bar{\varphi}_o^i = \Gamma^{ij}[\bar{\varphi}]\overleftarrow{S}^o{}_{,jk}(u^k{}_A a_A + u^k{}_A{}^* \; a_A{}^*) \tag{9.6}$$

where the us are the asymptotic physical-particle wave functions introduced in eqn. (4.6) and the as and a^*s are arbitrary coefficients, chosen, however, so as to give the quantity in parentheses a wave packet structure. Then solve the non-linear integral equation

* This is an integro-differential operator because the effective field equations, unlike the classical ones, are *non-local*. One often calls the effective action itself non-local.

$$\Delta\bar{\varphi}^i = \Delta\bar{\varphi}_0^i + \Gamma^{ij}[\bar{\varphi}](\Gamma_{,j}(\bar{\varphi}+\Delta\bar{\varphi}) - \Gamma_{,jk}[\bar{\varphi}]\Delta\bar{\varphi}^k)$$

$$= \Delta\bar{\varphi}_0^i + \Gamma^{ij}[\bar{\varphi}] \sum_{n=2}^{\infty} \frac{1}{n!} \Gamma_{,ji_1\ldots i_n}[\bar{\varphi}]\Delta\bar{\varphi}^{i_1}\ldots\Delta\bar{\varphi}^{i_n} \qquad (9.7)$$

by iteration.

The proof that this construction yields a solution of the effective field equations involves several steps. First observe that differentiation of eqn. (8.7) and use of (9.4) leads to

$$\Gamma_{,ij}[\bar{\varphi}]Q^j{}_\alpha[\bar{\varphi}] = 0 , \qquad (9.8)$$

and hence (suppressing arguments)

$$(\Gamma_{,ij} + \eta_{\alpha\beta}P^\alpha{}_i P^\beta{}_j)Q^j{}_\gamma = \eta_{\alpha\beta}P^\alpha{}_i \mathfrak{F}^\beta{}_\gamma . \qquad (9.9)$$

This implies (cf. eqn. (3.2))

$$Q^i{}_\alpha \mathfrak{G}^\alpha{}_\beta = \Gamma^{ij}P^\alpha{}_j\eta_{\alpha\beta} \text{ or } P^\alpha{}_i\Gamma^{ij} = \eta^{\alpha\beta}\mathfrak{G}^\gamma{}_\beta Q^j{}_\gamma \qquad (9.10)$$

and therefore

$$P^\alpha{}_i[\bar{\varphi}]\Delta\bar{\varphi}_0^i = \eta^{\alpha\beta}[\bar{\varphi}]\mathfrak{G}^\gamma{}_\beta[\bar{\varphi}]Q^j{}_\gamma[\bar{\varphi}]\overleftarrow{S}^0{}_{,jk}(u^k{}_A a_A + u^k{}_A{}^* a_A{}^*) . \qquad (9.11)$$

The $Q^j{}_\gamma[\bar{\varphi}]$ on the right of this equation can be split into two terms:

$$Q^j{}_\gamma[\bar{\varphi}] = Q^j{}_\gamma[\varphi_0] + Q^j{}_{\gamma,l}(\bar{\varphi}^l - \varphi_0^l) , \qquad (9.12)$$

where φ_0 is the classical vacuum. The operator $\overleftarrow{S}^0{}_{,jk}$ annihilates the first term and, when juxtaposed to the second term, can have its arrow reversed because of the fall-off at infinity of the product of $\bar{\varphi}^l - \bar{\varphi}_0^l$, the Green's function $\mathfrak{G}^\gamma{}_\beta$ and the wave packet $u^k{}_A a_A + u^k{}_A{}^* a_A{}^*$. Therefore

$$P^\alpha{}_i[\bar{\varphi}]\Delta\bar{\varphi}_0^i = 0 . \qquad (9.13)$$

Next observe that if the operator $\Gamma_{,ij} + \eta_{\alpha\beta}P^\alpha{}_i P^\beta{}_j$ is applied to the right of eqn. (9.6) the arrow on $S^0{}_{,jk}$ can again be reversed, leading to

$$(\Gamma_{,ij}[\bar{\varphi}] + \eta_{\alpha\beta}[\bar{\varphi}]P^{\alpha}{}_{i}[\bar{\varphi}]P^{\beta}{}_{j}[\bar{\varphi}])\Delta\bar{\varphi}_0^j = 0 \ , \tag{9.14}$$

which, in view of (9.13), implies also

$$\Gamma_{,ij}[\bar{\varphi}]\Delta\bar{\varphi}_0^j = 0 \ . \tag{9.15}$$

Finally, apply the operator $\Gamma_{,ij}$ to eqn. (9.7) and make use of (9.8) and (9.15) together with the following corollaries of eqns (8.7) and (9.10):

$$\begin{aligned} 0 &\equiv \Gamma_{,i}[\bar{\varphi}+\Delta\bar{\varphi}]Q^{i}{}_{\alpha}[\bar{\varphi}+\Delta\bar{\varphi}] \\ &\equiv \Gamma_{,i}[\bar{\varphi}+\Delta\bar{\varphi}](Q^{i}{}_{\alpha}[\bar{\varphi}]+Q^{i}{}_{\alpha,j}\Delta\bar{\varphi}^k) \ , \end{aligned} \tag{9.16}$$

$$\begin{aligned} \Gamma_{,ik}[\bar{\varphi}]\Gamma^{kj}[\bar{\varphi}] &= -\ \delta_{i}{}^{j} - \eta_{\alpha\beta}[\bar{\varphi}]P^{\alpha}{}_{i}[\bar{\varphi}]P^{\beta}{}_{k}[\bar{\varphi}]\Gamma^{kj}[\bar{\varphi}] \\ &= -\ (\delta_{i}{}^{j}+Q^{j}{}_{\alpha}[\bar{\varphi}]\mathfrak{G}^{\alpha}{}_{\beta}[\bar{\varphi}]P^{\beta}{}_{i}[\bar{\varphi}]) \ . \end{aligned} \tag{9.17}$$

Obtain

$$\begin{aligned} 0 &= \Gamma_{,ij}[\bar{\varphi}]\Big\{\Delta\bar{\varphi}^j - \Delta\bar{\varphi}_0^j - \Gamma^{jk}[\bar{\varphi}](\Gamma_{,k}[\bar{\varphi}+\Lambda\bar{\varphi}] - \Gamma_{,kl}[\bar{\varphi}]\Delta\bar{\varphi}^l)\Big\} \\ &= \Gamma_{,ij}[\bar{\varphi}]\Delta\bar{\varphi}^j + (\delta_{i}{}^{j} + Q^{j}{}_{\alpha}[\bar{\varphi}]\mathfrak{G}^{\alpha}{}_{\beta}[\bar{\varphi}]P^{\beta}{}_{i}[\bar{\varphi}]) \\ &\qquad\qquad \times (\Gamma_{,j}[\bar{\varphi}+\Delta\bar{\varphi}] - \Gamma_{,jk}[\bar{\varphi}]\Delta\bar{\varphi}^k) \\ &= (\delta_{i}{}^{j} - P^{\beta}{}_{i}[\bar{\varphi}]\mathfrak{G}^{\alpha}{}_{\beta}[\bar{\varphi}]Q^{i}{}_{\alpha,k}\Delta\bar{\varphi}^k)\Gamma_{,j}[\bar{\varphi}+\Delta\bar{\varphi}] \end{aligned} \tag{9.18}$$

The operator in the final parentheses in (9.18) is generally non-singular. Hence it may be removed, leaving

$$\Gamma_{,i}[\bar{\varphi}+\Delta\bar{\varphi}] = 0 \ . \tag{9.19}$$

Consider now the iterated solution of eqn. (9.7):

$$\Delta\bar{\varphi}^i = \Delta\bar{\varphi}_0^i + \Gamma^{ij}[\bar{\varphi}] \sum_{n=2}^{\infty} \frac{1}{n!} T_{ji_1\ldots i_n}[\bar{\varphi}]\Delta\bar{\varphi}_0^{i_1} \ldots \Delta\bar{\varphi}_0^{i_n} \ . \tag{9.20}$$

It is not difficult to see that the coefficients $T_{ji_1\ldots i_n}$ can be represented as the sum of all tree diagrams of a given order, with their external lines removed. These diagrams differ from the tree diagrams mentioned earlier. Here the vertices are the derivatives of $\Gamma[\bar{\varphi}]$ with respect to $\bar{\varphi}$ and not the derivatives of $\hat{\Gamma}[\varphi,\bar{\phi}]$ with respect to $\bar{\phi}$. Moreover, the lines connecting the vertices represent the propagator $\Gamma^{ij}[\bar{\varphi}]$, not $\hat{\Gamma}^{ij}[\varphi,\bar{\phi}]$. These diagrams nevertheless yield the same renormalized S-matrix elements as the earlier ones, at least when $\bar{\varphi} = \varphi_o$. This may be shown by the following arguments.

From eqns (9.7), (9.19) and (9.20) we have

$$\sum_{n=3}^{\infty} \frac{1}{(n-1)!} T_{i_1\ldots i_n}[\bar{\varphi}]\Delta\bar{\varphi}_o^{i_1}\ldots\Delta\bar{\varphi}_o^{i_n}$$
$$= \Delta\bar{\varphi}_o^{i}(\Gamma_{,i}[\bar{\varphi}+\Delta\bar{\varphi}] - \Gamma_{,ij}[\bar{\varphi}]\Delta\bar{\varphi}^{j})$$
$$= -\Delta\bar{\varphi}_o^{i}\overrightarrow{\Gamma}_{,ij}[\bar{\varphi}]\Delta\bar{\varphi}^{j} \,. \tag{9.21}$$

In virtue of eqn. (9.13), $\Gamma_{,ij}$ may be replaced by $\Gamma_{,ij} + \eta_{\alpha\beta}P^{\alpha}{}_{i}P^{\beta}{}_{j}$ in this equation. Substituting expression (9.6) for $\Delta\bar{\varphi}_o^{i}$, therefore, we find

$$\sum_{n=3}^{\infty} \frac{1}{(n-1)!} T_{i_1\ldots i_n}[\bar{\varphi}]\Delta\bar{\varphi}_o^{i_1}\ldots\Delta\bar{\varphi}_o^{i_n}$$
$$= (u^{i}{}_{A}a_{A} + u^{i}{}_{A}{}^{*}a_{A}{}^{*})\overrightarrow{S}^{o}{}_{,ij}\Delta\bar{\varphi}^{j} \,. \tag{9.22}$$

By the same reasoning as was used to infer the gauge invariance of the operators $\boldsymbol{\phi}_A$ of eqn. (4.6) we may infer the invariance of expression (9.22) under gauge transformations of $\Delta\bar{\varphi}$ of the form

$$\delta\Delta\bar{\varphi}^{i} = Q^{i}{}_{\alpha}[\bar{\varphi}+\Delta\bar{\varphi}]\delta\xi^{\alpha} \,. \tag{9.23}$$

This is important because the transformed $\Delta\bar{\varphi}$ still corresponds to the same *physical* solution of eqn. (9.19), and therefore expression (9.22) has the same value for all equivalent physical solutions.

Now, it is possible to introduce the same construction using the old effective action $\hat{\Gamma}$. Thus we may define

$$\Delta\bar{\phi}_0^i \stackrel{\text{def}}{=} \hat{\Gamma}^{ij}[\varphi,\bar{\phi}]\overleftarrow{S}^0{}_{,jk}(u^k{}_A a_A + u^k{}_A{}^* a_A{}^*) \tag{9.24}$$

and solve the equation

$$\Delta\bar{\phi}^i = \Delta\bar{\phi}_0^i + \hat{\Gamma}^{ij}[\varphi,\bar{\phi}]\left(\frac{\delta\hat{\Gamma}[\varphi,\bar{\phi}+\Delta\bar{\phi}]}{\delta\bar{\phi}^i} - \frac{\delta^2\hat{\Gamma}[\varphi,\bar{\phi}]}{\delta\bar{\phi}^i\delta\bar{\phi}^j}\,\Delta\bar{\phi}^j\right)$$

$$\Delta\bar{\phi}_0^i + \hat{\Gamma}^{ij}[\varphi,\bar{\phi}]\sum_{n=2}^{\infty}\frac{1}{n!}\left(\frac{\delta}{\delta\bar{\phi}^{i_1}}\cdots\frac{\delta}{\delta\bar{\phi}^{i_n}}\,\hat{\Gamma}[\varphi,\bar{\phi}]\right)\Delta\bar{\phi}^{i_1}\cdots\Delta\bar{\phi}^{i_n} \tag{9.25}$$

by iteration. The result will satisfy the equation

$$\frac{\delta\hat{\Gamma}[\varphi,\bar{\phi}+\Delta\bar{\phi}]}{\delta\bar{\phi}^i} = 0\,, \tag{9.26}$$

and, moreover, the coefficients of the iterated solution will correspond exactly to the sums of the tree diagrams that we originally introduced, with their external lines removed. If we denote these coefficients by $\hat{T}_{i_1\ldots i_n}$ then we have

$$\sum_{n=3}^{\infty}\frac{1}{(n-1)!}\,\hat{T}_{i_1\ldots i_n}[\varphi,\bar{\phi}]\Delta\bar{\phi}_0^{i_1}\cdots\Delta\bar{\phi}_0^{i_n}$$

$$= (u^i{}_A a_A + u^i{}_A{}^* a_A{}^*)\overrightarrow{S}^0{}_{,ij}\Delta\bar{\phi}^j\,. \tag{9.27}$$

The structural elements of the S-matrix are obtained by repeatedly differentiating expression (9.27) with respect to as and a^*s and then setting these coefficients equal to zero. This procedure yields the so-called *tree amplitudes*. These amplitudes, like the S-matrix itself are P- and η-independent.

Hence they are not changed if we replace the coefficients in (9.27) by the shifted coefficients $(\hat{T}_{i_1 \ldots i_n})_{\rightarrow}$ and redefine $\Delta\bar{\phi}^i_0$ and $\Delta\bar{\phi}^i$ accordingly.

$$\Delta\bar{\phi}^i_0 \overset{\text{def}}{=\!=} (\hat{\Gamma}^{ij}[\varphi,\bar{\phi}])_{\rightarrow} \overleftarrow{S}^{\circ}{}_{,jk}(u^k{}_{\mathrm{A}}a_{\mathrm{A}} + u^k{}_{\mathrm{A}}{}^{*}a_{\mathrm{A}}{}^{*}) , \tag{9.28}$$

$$\Delta\bar{\phi}^i = \Delta\bar{\phi}^i_0 + (\hat{\Gamma}^{ij}[\varphi,\bar{\phi}])_{\rightarrow} \sum_{n=2}^{\infty} \frac{1}{n!}(\hat{T}^{i_1 \ldots i_n}[\varphi,\bar{\phi}])_{\rightarrow}\Delta\bar{\phi}^{i_1}_0 \ldots \Delta\bar{\phi}^{i_n}_0 \tag{9.29}$$

$\Delta\bar{\phi}$ then satisfies the equation

$$\left(\frac{\delta\hat{\Gamma}[\varphi,\bar{\phi} + \Delta\bar{\phi}]}{\delta\bar{\phi}^i}\right)_{\rightarrow} = 0 \tag{9.30}$$

and also (see eqn. (9.9))

$$\Gamma_{,i}[\bar{\varphi} + \Delta\bar{\phi}] = 0 . \tag{9.31}$$

Now let $\bar{\varphi} = \varphi = \varphi_0$, $\bar{\phi} = 0$, and let the labels A in eqns (9.6) and (9.28) refer to momentum and helicity states. Because of manifest covariance and the fact that the classical vacuum feeds no external momentum or helicity into the wave packet $u^i{}_{\mathrm{A}}a_{\mathrm{A}} + u^i{}_{\mathrm{A}}{}^{*}a^{*}_{\mathrm{A}}$, it follows that any divergences contained in the full propagators $\Gamma^{ij}[\varphi_0]$ and $(\hat{\Gamma}^{ij}[\varphi_0,0])_{\rightarrow}$ must be momentum and helicity independent and that $\Delta\bar{\varphi}^i_0$ and $\Delta\bar{\phi}^i_0$ must reduce to

$$\left.\begin{aligned} \Delta\bar{\varphi}^i_0 &= Z(u^i{}_{\mathrm{A}}a_{\mathrm{A}} + u^i{}_{\mathrm{A}}{}^{*}a_{\mathrm{A}}{}^{*}) + Q^i{}_{\alpha}[\varphi_0]\xi^{\alpha} , \\ \Delta\bar{\phi}^i_0 &= \hat{Z}(u^i{}_{\mathrm{A}}a_{\mathrm{A}} + u^i{}_{\mathrm{A}}{}^{*}a_{\mathrm{A}}{}^{*}) + Q^i{}_{\alpha}[\varphi_0]\hat{\xi}^{\alpha} . \end{aligned}\right\} \tag{9.32}$$

$Q^i{}_{\alpha}[\varphi_0]\xi^{\alpha}$ and $Q^i{}_{\alpha}[\varphi_0]\hat{\xi}^{\alpha}$ are (generally divergent) gauge terms, and Z and $\hat{Z}$ are renormalization constants that must be adjusted to unity by appropriate choice of counter-terms in the classical action. Evidently $\Delta\bar{\varphi}^i_0$ and $\Delta\bar{\phi}^i_0$ may both be identified (modulo a gauge transformation) with the wave packet $u^i{}_{\mathrm{A}}a_{\mathrm{A}} + u^i{}_{\mathrm{A}}{}^{*}a_{\mathrm{A}}{}^{*}$ itself.

Since the amplitude of this wave packet becomes infinitely small at infinity it follows, from the fall-off of the Green's functions $\Gamma^{ij}[\varphi_0]$, $(\hat{\Gamma}^{ij}[\varphi_0,0])_{\rightarrow}$ and the Feynman boundary conditions they impose on the second terms on the right of eqns (9.20) and (9.29), that the positive (negative) frequency parts of both $\Delta\bar{\varphi}^i$ and $\Delta\bar{\phi}^i$ reduce, modulo a gauge transformation, to $u^i{}_A a_A$ $(u^i{}_A{}^* a_A{}^*)$ in the remote past (future). Since $\Delta\bar{\varphi}$ and $\Delta\bar{\phi}$ have identical physical boundary conditions at infinity and satisfy identical equations (eqns (9.19) and (9.31)) they must be equal everywhere, modulo a gauge transformation (9.23). Since expression (9.22) is invariant under gauge transformations it follows that the tree functions $T_{i_1\ldots i_n}$, $\hat{T}_{i_1\ldots i_n}$, $(\hat{T}_{i_1\ldots i_n})_{\rightarrow}$ all yield the same renormalized S-matrix.

If $\bar{\varphi} \neq \varphi_0$ then we have a non-vacuum background, and the asymptotic reference states of our theory are relative vacua. But a relative vacuum state is a *coherent state* when viewed with reference to the absolute vacuum state. That is, it can be decomposed into a coherent superposition of particle states in the Fock space based on the absolute vacuum. The S-matrix with $\bar{\varphi} = \varphi_0$ is just as relevant for describing transitions between such superpositions as it is for describing transitions between individual particle states. Thus we do not really need to work with tree amplitudes evaluated at $\bar{\varphi} \neq \varphi_0$. It would be nice to know, and is very probably true, that $\Delta\bar{\varphi}_0^i$ and $\Delta\bar{\varphi}_0^i$ are physically equivalent even when $\bar{\varphi} \neq \varphi_0$, so that expressions (9.23) and (9.27) are equal in the general case, but this remains to be proved.

10. THE SIGNIFICANCE OF THE GAUGE-INVARIANT EFFECTIVE ACTION

The tree functions $T_{i_1\ldots i_n}$ constructed from the gauge invariant effective action Γ involve the derivatives $\Gamma_{,ijk\ldots}$ as vertex functions. When the indicated differentiations are carried out on the diagrams of which Γ is composed (Fig. 1 with shift operation $(\ )_{\rightarrow}$ understood) new bare vertex functions are produced, which do not occur in the original diagrams. These are: (i) $F_{ij,kl\ldots}[\bar{\varphi}]$, where one or more external lines meet a pair of internal solid lines; (ii) $\gamma^{\alpha}{}_{\beta,ij\ldots}[\bar{\varphi}]$ where one or more external lines meet a pair of internal dotted lines; (iii) $V^{\alpha}{}_{\beta i,jk\ldots}[\bar{\varphi}]$ where one or more external lines meet a pair of internal dotted lines and a single internal solid line. If one or more external lines meet three or more internal solid lines the associated vertex function is $S_{,ijkl\ldots}[\bar{\varphi}]$.

A line appearing in a diagram belonging to the perturbation expansion of the full propagator Γ^{ij}, or of a complete tree function (many-particle propagator)

$$\Gamma^{i_1\ldots i_n} \stackrel{\text{def}}{=} \Gamma^{i_1j_1}\ldots\Gamma^{i_nj_n}T_{j_1\ldots j_n} \quad \text{(arguments suppressed)}, \tag{10.1}$$

is 'external' in the above sense if, when cut, it leaves the diagram separated into two parts. Such lines are always recognizable and the values to be assigned to any vertices to which they may be attached are unambiguous. No special marks are needed on the diagrams themselves to distinguish these vertices from the internal vertices.

The addition of these new bare vertex functions to the theory is a small price to pay for the simplification that is thereby achieved in the renormalization programme. The simplification consists in the fact that the full radiatively corrected vertex functions $\Gamma_{,ijk\ldots}$ are now all related by simple Ward identities obtained by repeatedly differentiating eqn. (8.7) (cf. eqns (1.9):

$$\left.\begin{aligned} \Gamma_{,ijk}Q^k{}_\alpha &\equiv -\Gamma_{,kj}Q^k{}_{\alpha,i} - \Gamma_{,ik}Q^k{}_{\alpha,j}, \\ \Gamma_{,ijkl}Q^l{}_\alpha &\equiv -\Gamma_{,ljk}Q^l{}_{\alpha,i} - \Gamma_{,ilk}Q^l{}_{\alpha,j} - \Gamma_{,ijl}Q^l{}_{\alpha,k}, \text{ etc.} \end{aligned}\right\} \tag{10.2}$$

For renormalizable theories this means that only a very few adjustable constants are present, all of them physical, and that only a very few counter-terms are needed. Since Γ is gauge invariant these counter-terms must themselves be gauge invariant and must be found among the few integral invariants that have the right dimension. It is not necessary to renormalize non-gauge-invariant parts of the effective action separately.

The gauge invariant effective action also has a special significance for quantum gravity. Although this theory yields finite scattering processes in one-loop order when matter is absent, the effective action itself is not finite. Moreover, to renormalize the effective action in this order one must introduce an arbitrary scale factor, which is usually chosen to be the Planck mass and which reflects the essential non-renormalizability of the theory. On the other hand, with this scale factor included, the one-loop contribution to Γ appears, in several important cases, to be probably computable as an explicit functional of the geometry of space-time. In the most tractable of these cases the geometry is cosmological, with 'closed' topology. For such models there is no asymptotic region and one may question the relevance for them of the effective action, which, up to now, has been discussed only in the context of S-matrix theory. Consider, however, the following argument.

Suppose we choose, in an asymptotically Minkowskian space-time, a background geometry φ which represents a dispersed classical gravitational wave packet in the remote past. Then the relative 'in' vacuum is a coherent state $|\psi\rangle$ with $\langle\psi|\boldsymbol{\varphi}|\psi\rangle =$ φ in the remote past. Suppose this dispersed wave packet has the potentiality of developing, as time passes, into a strong concentrated packet containing regions of large curvature. Suppose, however, that it is smooth enough and of long enough wavelength that the incoherent quantum pairs produced by it are negligible. Then we may choose φ in the remote future in such a way that it not only satisfies the classical field equations but is also (approximately) equal to $\langle\psi|\boldsymbol{\varphi}|\psi\rangle$ there. With this choice of φ the 'in' and 'out' relative vacua are (approximately) equal (Heisenberg picture), and we have

$$\langle\psi|\boldsymbol{\varphi}|\psi\rangle \approx \langle\boldsymbol{\varphi}\rangle = \bar{\varphi} \ , \qquad (10.3)$$

which holds not only in the past and future but for all times. The field $\bar{\varphi}$, however, is governed not by the classical action but by the effective action. This means that if the curvature gets strong enough for virtual quantum processes to become important, the expectation value (10.2) will satisfy not the classical field equations (1.2) but the effective field equations (9.4).

The implications for cosmology are obvious. There the chief

role of the effective action is not to generate an S-matrix but to describe the quantum corrections to the dynamics of the universe as a whole, especially in its early stages.

At this point one must recall that the effective action incorporates Feynman boundary conditions. There will be a difficulty with these boundary conditions if the universe possesses a geometrical singularity in the past or future. One must begin by assuming that there is no singularity. (Remember, $\Gamma[\bar{\varphi}]$ must be computed for arbitrary $\bar{\varphi}$.) Only afterward are the effective field equations imposed. If these equations maintain freedom from singularities, all is well. If not then one must conclude that quantum gravity is as incomplete a theory as classical relativity.

In my opinion no believable potential cure for the ills of classical relativity besides quantization is presently visible. The problem of computing the effective action in cosmological settings is therefore of prime importance.

I do not believe that invoking 'topological fluctuations' is going to help us. In fact, I believe that to talk of 'space-time foam' and 'changing topology' is nonsense. Topology is built in at the beginning, in the very definition of the base manifold of any theory. It is not a dynamical object, and to this date no one has come within twenty-five light years of making it so, despite a quarter century of handwaving about it.

I do not insist that space-time necessarily be viewed as a manifold in the classical sense, only that whatever replaces the manifold idea be definable with mathematical precision and that its meaning in experimental physical terms be reasonably clear. The theory of supermanifolds and supergravity is a good example of such an extension of classical ideas. If quantum gravity does not save us perhaps supergravity will. The problem of constructing gauge invariant effective actions for locally supersymmetric theories therefore also deserves serious attention.

ACKNOWLEDGEMENT

This work was supported by grants from the U.S. National Science Foundation.

REFERENCES

DeWitt, B.S. (1964). *Phys. Rev. Lett.* **13**, 114.
—— (1967). *Phys. Rev.* **162**, 1195.
—— (1965). *Dynamical Theory of Groups and Fields*. Gordon and Breach, New York.
Fadde'ev, L.D. and Popov, V.N. (1967). *Phys. Lett.* **25B**, 29.
Gribov, V.N. (1977). Lecture at the Twelfth Winter School of the Leningrad Nuclear Physics Institute. (SLAC translation No. 176.)
't Hooft, G. (1976). 'The Background Field Method in Gauge Field Theories', in *Acta Univ. Wratislavensis No. 368, XIIth Winter Sch. Theor. Phys. Karpacz Feb.-March 1975. Functional and Probabilistic Methods in Quantum Field Theory*. **I**.

21
THE COSMOLOGICAL CONSTANT IN QUANTUM GRAVITY AND SUPERGRAVITY

M.J. Duff

Physics Department, Imperial College, London SW7

1. INTRODUCTION

I would like to describe some recent work on the quantization of gravity and supergravity in the presence of a cosmological constant. (Christensen and Duff 1980; Christensen, Duff, Gibbons and Roček 1980*a*,*b*).

Now when Einstein first introduced a cosmological constant into his gravitational field equations, he was guided by various prejudices about the static nature of the universe which he later abandoned in the face of growing evidence for an expanding universe. Since Einstein's time the cosmological constant has had its ups and downs (both empirically and metaphorically) and current estimates put its physical value as very small and possibly zero ($< 10^{-57}$ cm^{-2}).

As far as quantum gravity is concerned, almost all discussions to date have focused their attention on the case of vanishing cosmological constant $\Lambda = 0$. This is not without reason. Whatever one's attitude to the value of Λ at the classical level, attempts to build a consistent quantum theory for non-vanishing Λ present new difficulties over and above the already formidable problems present when $\Lambda = 0$. Unfortunately both grand unified and super unified theories of nature predict, for one reason or another, an enormous cosmological constant, and in the opinion of the author, there is to date no satisfactory means by which the cosmological constant may be 'argued away'. The one arising from spontaneous symmetry breakdown may be cancelled by an *ad hoc* addition to the Lagrangian, but this will only result in a non-vanishing Λ in the early stages of the universe when the symmetry is restored through high-temperature effects. In general, moreover, a vanishing Λ at the tree level will not prevent its reappearance through closed loop effects. For a review of the status of the cosmological constant and the problems involved see Christensen and Duff (1980). In the present context therefore, the words of Zeldovich and Novikov seem particularly appropriate: 'After a genie is let out of a bottle (i.e. now that the possibility is admitted that $\Lambda \neq 0$) legend has it that the genie can be chased back in only with the greatest difficulty.' For the moment, then, let us not attempt to chase the genie back into the bottle; let us admit a non-vanishing Λ and see whether we can cope with it.

There are many questions one might ask of a quantum field theory with a cosmological constant, and I shall not attempt to answer all of them, but a very important one concerns ultraviolet divergences and renormalizability. Here the analysis turns out to be reasonably straightforward and explicit results for one-loop counter-terms and anomalous scaling behaviour have been given both for pure gravity and for gravity plus matter fields of spin 0, ½ and 1 in the presence of a cosmological constant. (Christensen and Duff 1980). For related work see Gibbons and Perry (1978). Summarized below are the calculations of a forthcoming publication (Christensen, Duff, Gibbons, and Roček 1980*a*,*b*) in which these techniques are generalized to spin-3/2 fields and hence to supergravity, with dramatic results for the $O(N)$ extended models.

These developments may appear as something of a luxury: if ordinary quantum gravity is non-renormalizable without a cosmological constant, it is not likely to become so with the additional complication of a non-vanishing Λ. However, such arguments require drastic revision in extended supergravity where the gauging of the $O(N)$ symmetry (Freeman and Das 1977) requires a (huge) cosmological constant $\Lambda = -6e^2/\kappa^2$ and gravitino mass parameter m, with $\Lambda = -3\,m^2$, (e is the gauge coupling constant and $\kappa^2 = 8\pi \times$ Newton's constant). Thus it is plausible that the ultra-violet behaviour of these models at higher loops, by virtue of their extra local symmetry, may even be an improvement over theories without a cosmological constant. Moreover, the strong empirical evidence in favour of a vanishing cosmological constant may be only an apparent discrepancy between theory and experiment if, as suggested in Hawking's 'space-time foam' (Hawking 1978), one reinterprets Λ as a measure of the average small scale curvature of space-time.

2. ONE-LOOP COUNTER-TERMS

Let us first recall the pure gravity results (Christensen and Duff 1980). If, at the classical level, we take the Einstein action

$$S = -\frac{1}{2\kappa^2}\int d^4x\sqrt{g}\,(R-2\Lambda) \tag{2.1}$$

then, using the background field method, the one-loop counter-terms will be a linear combination of $R_{\mu\nu\rho\sigma}R^{\mu\nu\rho\sigma}$, $R_{\mu\nu}R^{\mu\nu}$, R^2, ΛR and Λ^2 but with gauge dependent coefficients. Gauge invariance is achieved by use of the field equations $R_{\mu\nu} = \Lambda g_{\mu\nu}$. Alternatively, terms which vanish with the field equations may be removed by gauge-dependent field redefinition. Either way, the resulting counter-term ΔS may then be written

$$\Delta S = -\frac{1}{\varepsilon}\gamma \tag{2.2}$$

where $\varepsilon = n - 4$ is the dimensional regularization parameter. γ is given by

$$\gamma = A\chi + B\delta \tag{2.3}$$

where A and B are numerical coefficients and where

$$\chi \equiv \frac{1}{32\pi^2}\int d^4x\sqrt{g}\,{}^*R_{\mu\nu\rho\sigma}{}^*R^{\mu\nu\rho\sigma} \tag{2.4}$$

$$\delta \equiv \frac{1}{12\pi^2}\int d^4x\sqrt{g}\,\Lambda^2 = -\frac{\kappa^2\Lambda}{12\pi^2}\,S \tag{2.5}$$

The star denotes the duality operation, and $\sqrt{g}\;{}^*R_{\mu\nu\rho\sigma}{}^*R^{\mu\nu\rho\sigma} = \sqrt{g}\,(R_{\mu\nu\rho\sigma}R^{\mu\nu\rho\sigma} - 4R_{\mu\nu}R^{\mu\nu} + R^2)$ is a total divergence which is sometimes discarded. Its integral over all space, however, yields the Euler number χ, a topological invariant which takes on integer values in spaces with non-trivial topology. The explicit calculations yield $A = 106/45$ and $B = -87/10$. Thus, in contrast to the case $\Lambda = 0$, pure gravity with a Λ term is no longer one-loop 'finite' (in the non-topological sense) because $B \neq 0$.

One may now repeat the exercise for simple supergravity with a gravitino mass term (Macdowell and Mansouri 1977; Deser and Zumino 1977; Townsend 1977; van Nieuwenhuizen 1979):

$$S = \int d^4x(\det e^a_\mu)\Big[-\frac{1}{2\kappa^2}R + \frac{1}{2}\varepsilon^{\mu\nu\rho\sigma}\psi_\mu\gamma_5\gamma_\nu D_\rho\psi_\sigma + m\psi_\mu\sigma^{\mu\nu}\psi_\nu + \frac{1}{3}(s^2 + p^2 - A^\mu A_\mu) + \frac{2m}{\kappa}s\Big]. \qquad (2.6)$$

Elimination of the auxiliary field s yields a cosmological constant $\Lambda = -3m^2$. The one-loop counter-terms will now be given by the appropriate supersymmetric completion of those encountered in pure gravity, i.e., ΔS is again given by $-\varepsilon^{-1}(A\chi + B\delta)$ with $\delta = -\kappa^2(12\pi^2)^{-1}S$ (on-shell) but where S is now given by eqn (2.6). The topological invariant χ, on the other hand, acquires no extra terms. (Townsend and van Nieuwenhuizen, 1979). The coefficients A and B will now receive contributions both from the graviton and the gravitino (with its appropriate mass parameter). Explicit calculations (Christensen, Duff, Gibbons, and Roček 1980*a*,*b*) yield A = 41/24 and B = -77/12 and, once again in contrast to the case $\Lambda = 0 = m$, simple supergravity is no longer one-loop finite.

3. EXTENDED SUPERGRAVITY

Having obtained the contribution to A and B for spins 2 and 3/2, it is now tempting to combine these results with those for spins 1, ½, and 0, and apply them to the extended $O(N)$ theories with gauged internal symmetry, especially because the renormalization of Λ takes on a new significance: by supersymmetry the coefficient B also determines the renormalization of the gauge coupling constant e.

The supersymmetric completion of δ now contains the spin-1 gauge field contribution $e^2 \operatorname{Tr} F_{\mu\nu}F^{\mu\nu}$. Note that this arises from two different sources: in addition to the usual charge renormalization effects, there will also be one-loop counter-terms of the form $\kappa^2 R \operatorname{Tr} F_{\mu\nu}F^{\mu\nu}$. On using the field equations $R = 4\Lambda + \ldots$ with $\kappa^2\Lambda = -6e^2$ this is converted into an extra $e^2 \operatorname{Tr} F_{\mu\nu}F^{\mu\nu}$ term.

Before displaying our results, some qualifications are

required. Although the construction of consistent $O(N)$ supergravity Lagrangians has been successfully achieved for all N up to $N = 8$, the corresponding Lagrangians with gauged internal symmetry have to date been written down explicitly only for $N = 2,3$ (Freedman and Das 1977) and $N = 4$ (Das, Fischler, and Rocek 1977; Freedman and Schwartz 1978). It is thus an assumption on our part that such Lagrangians exist for $N = 5$, 6,7, and 8. As far as we are aware, there are no theoretical reasons preventing such a construction since the appropriate supersymmetry algebras for $N > 4$ are perfectly respectable. (We refrain from going beyond $N = 8$ for the usual reason of requiring no spin higher than 2.) The crucial observation however, is that by restricting our attention to the gravitational part of the on-shell counter-terms at the one-loop level, the details of the interaction terms in such Lagrangians are not relevant: all that is required to determine the coefficients A and B is the pure spin-2 Lagrangian itself together with that part of the remaining Lagrangian quadratic in the lower-spin fields. Having calculated the gravitational contribution to ΔS on shell, the remainder is determined by the supersymmetry which guarantees that (with $\kappa^2\Lambda = -6e^2$)

$$\Delta S = \frac{-1}{\varepsilon}\left[A\chi + B\,\frac{e^2}{2\pi^2}\,S\right] \tag{3.1}$$

where S is the classical action. The signal for asymptotic freedom is $B > 0$.

Only the kinetic terms are needed to fix the contributions to A from fields of different spin. These have been calculated before (Christensen and Duff 1978*a*, 1979*a*). To calculate B we also require knowledge of the mass terms. All particles must be massless for all N if, as we are assuming, supersymmetry is not spontaneously broken. For $N > 4$ there is an 'apparent mass' parameter m for the gravitinos given by $\Lambda = -3\,m^2$ which we assume to remain the same for $N > 4$. Similarly we assign no such parameters to the spin-1 and spin-½ fields for $N > 4$ since they are absent for $N \leqslant 4$. The scalar fields, which first make their appearance at $N = 4$ require greater care. The spin-2, spin-0 coupling in the $N = 4$ model is known to be of the form

$$L = -\frac{1}{2\kappa^2}\sqrt{g}(R - 2\Lambda) + \frac{1}{2}\sqrt{g}\ \phi^i\left[-\Box + \frac{2\Lambda}{3}\right]\phi^i + O(\phi^3) \qquad (3.2)$$

i.e., minimal coupling with a mass term. However, one could equally well use

$$L = -\frac{1}{2\kappa^2}\sqrt{g}(R - 2\Lambda) + \frac{1}{2}\sqrt{g}\ \phi^i\left[-\Box + \frac{R}{6}\right]\phi^i + O(\phi^3) \qquad (3.3)$$

i.e., conformal coupling with no mass term. The equivalence is seen by making a Weyl rescaling in the Lagrangian (3.3) of the form

$$g_{\mu\nu} \to \Omega^2 g_{\mu\nu},\ \phi^i \to \Omega^{-1}\phi^i;\ \Omega^2 = 1 + \frac{\kappa^2}{6}\ \phi^i\phi^i \qquad (3.4)$$

which yields the Lagrangian (3.2). Both versions yield the same B coefficient on mass shell since the field equations imply $R = 4\Lambda + \ldots$. We therefore adopt a conformal coupling with no mass term for all $N \geqslant 4$. With the above assumptions, one finds the contributions to A and B shown in Table 3.1.

TABLE 3.1

S	$360A$	$60B$
0	4	-1
1/2	7	-3
1	-52	-12
3/2	-233	137
2	848	-522

The combined results for $O(N)$ supergravity then follow from the well known particle content shown in Table 3.2. The most remarkable feature is clearly the vanishing of the B coefficient for all $N > 4$, though the integral value of A for all $N > 2$ is not without interest. We now discuss the implications of these results.

TABLE 3.2

S	2	3/2	1	1/2	0	*A*	*B*
N							
1	1	1				41/24	-77/12
2	1	2	1			11/12	-13/3
3	1	3	3	1		0	-5/2
4	1	4	6	4	2	-1	-1
5	1	5	10	11	10	-2	0
6	1	6	16	26	30	-3	0
7	1	8	28	56	70	-5	0
8	1	8	28	56	70	-5	0

4. RENORMALIZABILITY

Apart from the topological χ counter-term, the $N > 4$ theories are seen to remain one-loop finite on shell even when the internal symmetry is gauged and $\Lambda \neq 0$. In particular, the one-loop contribution to the renormalization group $\beta(e)$ function vanishes! This is reminiscent of the $N = 4$ Yang–Mills multiplet in flat space, whose β function is known to vanish to two-loop order (Jones 1977; Pendleton and Poggio 1977). The vanishing β function in $N > 4$ gauged supergravity is no less mysterious than in $N = 4$ Yang–Mills and, at the time of writing, is understood only as a 'miraculous' cancellation of numerical coefficients. There have been earlier speculations (de Wit and Ferrara 1979) that $N > 4$ theories might show improved ultraviolet behaviour, but we do not know their connection, if any, with the concrete calculations presented here. These cancellations can hardly be accidental, however, and provide something of an *a posteriori* justification for our previous assumptions on $N > 4$ theories.

For $N \leqslant 4$, we do not have one-loop finiteness but rather one-loop renormalizability. Moreover, the negative value of B indicates that these theories are not asymptotically free (inasmuch as asymptotic freedom is meaningful for theories which may not be renormalizable at higher loops). One will find two-loop renormalizability for all N when $\Lambda \neq 0$ for the same reason one finds two-loop finiteness when $\Lambda = 0$ (Grisaru 1977) and it would be interesting to know the β-function. Three-loops and beyond is still a mystery (Deser, Kay, and Stelle 1977).

5. TOPOLOGY

Another remarkable feature peculiar to $N > 4$ models is that $\gamma = A\chi + B\delta$ = integer (since A is an integer, B is zero and χ takes on integer values). If previous experience is any guide this may be indicative of a new 'Super Index Theorem' (Christensen and Duff 1979; Hawking and Pope 1978) for $N > 4$. Let us recall the significance of γ. At the one-loop level γ counts the total number of eigenmodes (boson minus fermion) of the differential operators whose determinants govern the one-loop functional integral. (It is closely related in the anomalous trace of the energy-momentum tensor.) The number of zero eigenvalue modes will be finite and given by an integer; the number of non-zero modes is formally infinite. After regularization (e.g. by the zeta-function method), this number is rendered finite but not necessarily an integer. In certain circumstances, however, there may be a mutual cancellation of the non-zero modes between the bosons and fermions, in which case γ = integer. Such a cancellation does indeed take place in $\Lambda = 0$ supergravity (Hawking and Pope 1978) when the space is self-dual, i.e., $R_{\mu\nu\rho\sigma} = \pm {}^*R_{\mu\nu\rho\sigma}$ (which implies $R_{\mu\nu} = 0$). If, in addition, the space is compact with spin structure (i.e., fermions can be globally defined) then χ = integer $\times$ 24. Consistent with this is the result in Table 3.2 that A = integer/24 for all N. We do not know whether any similar mechanism can take place when $\Lambda \neq 0$ and χ is not so restricted, but our results indicate that $N > 4$ are the most likely candidates. One might also ask whether there is anything special, from a topological point of view, about gauged supergravity when $N > 4$. There is one remark, brought to our attention by C.J. Isham: on a four-dimensional manifold with non-trivial topology, SO(N) gauge theories are 'topologically stable' only for $N > 4$, in the sense that there is a natural one to one correspondence between the SO(N) bundles (or, equivalently, between the topological sectors of the gauge theory). Again, we do not yet know whether this is mere coincidence.

Finally, we know that the signs of A and B in simple and extended supergravity reinforce the conclusions concerning 'space-time foam' reached in the context of pure gravity (Christensen and Duff 1980). If these one-loop results are taken seriously the sign of γ would seem to imply that space-time becomes 'foamier and foamier' the shorter the length scale, in contrast to the picture of 'one unit of topology per Planck volume' expected if γ were positive definite (Hawking 1978).

6. ANTISYMMETRIC TENSORS

It has recently been discovered (Duff and van Nieuwenhuizen 1980; van Nieuwenhuizen in this volume) that different field representations for particles of given spin and number of degrees of freedom, although naively equivalent (Sezgin and van Nieuwenhuizen 1980), lead to different quantum effects in spaces with non-trivial topology. Thus the gauge theory of an antisymmetric rank-2 tensor $\phi_{\mu\nu}$ (with 1 degree of freedom) differs from that of a scalar ϕ; and that of an antisymmetric rank-3 tensor $\phi_{\mu\nu\rho}$ (with 0 degrees of freedom) differs from nothing. These effects show up in the A coefficient of the Euler number which appears in the counter-terms and trace anomalies. In fact,

$$A[\phi_{\mu\nu}] = A[\phi] + 1$$
$$A[\phi_{\mu\nu\rho}] = -2 \qquad (6.1)$$

Another feature of the rank-three potential $\phi_{\mu\nu\rho}$ coupled to gravity is that its field strength $F_{\mu\nu\rho\sigma}$ appears in the Einstein equation as

$$G_{\mu\nu} - g_{\mu\nu}F^2 = 0 \qquad (6.2)$$

But its equation of motion

$$D^{\mu}F_{\mu\nu\rho\sigma} = 0 \tag{6.3}$$

implies

$$F_{\mu\nu\rho\sigma} = \varepsilon_{\mu\nu\rho\sigma}\Lambda^{\frac{1}{2}} \tag{6.4}$$

where Λ is a constant. Substitution into (6.2) therefore yields Einstein's equation with a cosmological constant! Moreover, the one-loop counter-terms of the gravity—$\phi_{\nu\mu\rho}$ system calculated by Sezgin and van Nieuwenhuizen (1980) yield, upon using (6.2) and (6.4), the same result (except for the above mentioned topological terms) as those for pure gravity with a cosmological constant calculated by Christensen and Duff (1980). Thus the cosmological constant may be reformulated as a gauge theory of $\phi_{\mu\nu\rho}$.

Both the topological and cosmological aspects of antisymmetric tensors have interesting application to supergravity:

1. As already pointed out by Ogievetsky and Sokatchev (1980), their version of the superspace formalism of $N = 1$ supergravity naturally yields a cosmological constant (without the usual trick of adding on a term linear in the auxiliary field S). This is because in their formulation the traditional S and P auxiliary fields (Ferrara and van Nieuwenhuizen 1978; Stelle and West 1978) are replaced by S_{μ} and P_{μ} (Stelle and West 1978). Setting $A_{\mu\nu\rho} = \varepsilon_{\mu\nu\rho\sigma}S^{\sigma}$ etc., then yields a kinetic $F_{\mu\nu\rho\sigma}F^{\mu\nu\rho\sigma}$ term of the kind described above, and hence a cosmological constant in the field equations.

2. Christensen and Duff (1979) had noted that 'the combination of fields appearing in the Super Index Theorems are just those one encounters in physical supermultiplets. The only exception seems to be the appearance of the antisymmetric tensors (1, 0) and (0, 1) which are not normally considered as elementary fields in a Lagrangian'. Indeed, in deriving the supergravity axial and conformal anomalies from the super-theorem, we found it necessary, at the time, to discard these representations in what seemed a rather unnatural way because in the supergravity theories under consideration, the spin-0 particles were realized as scalar fields ϕ. However, in the light of quantum inequivalences of van Nieuwenhuizen and myself described above,

Siegel (private communication) has pointed out to me that in the version of the $N = 8$ theory obtained by dimensional reduction (Cremmer and Julia 1977) the 70 spin-0 fields actually appear as $63\phi + 7\phi_{\mu\nu} + \phi_{\mu\nu\rho}$. (Previously it had been assumed that this was equivalent to the version with 70ϕ fields.) The importance of this, as noted by Siegel, is that the A coefficient of -5 in Table 3.2 becomes, on using (6.1),

$$A = -5 + 7 - 2 = 0$$

(From Table 3.2, the vanishing of the A coefficient would also occur for all $N > 2$, if

$$3 - N[\psi_{\mu}] + N[\phi_{\mu\nu}] - 2N[\phi_{\mu\nu\rho}] = 0$$

but I have been unable, so far, to truncate this version of the $N = 8$ theory to lower N.) Thus even the topological counter-term, and hence the trace anomaly, now vanish! With hindsight, we may now re-derive this result from the Super Index theorems *with* the (1,0) and (0,1) representations included. The hitherto unrealized importance of anti-symmetric tensors may also help to clear up a mystery in the axial and conformal anomalies of supergravity calculated by Christensen and Duff (1978*a*). By working with spin-0 fields described by scalars, we found that these anomalies did not appear to form a supermultiplet (along with that of the supercurrent) as one would naively expect (Ferrara and Zumino 1975). This is a point of which Marc Grisaru has constantly reminded me. However, just as the trace anomaly result (6.1) followed as a consequence of the Gauss–Bonnet theorem, so the Hirzebruch signature theorem should provide an additional antisymmetric tensor contribution to the axial anomaly (Duff and van Nieuwenhuizen 1980). This is presently being investigated.

I have also checked that there is a cancellation of zero-modes for ALE instantons in this $N = 8$ theory, using the results of Hawking and Pope (1978). This represents an additional cancellation because ALE spaces have boundaries, and means that even the boundary contributions to the one-loop counter-terms (Christensen and Duff 1978*b*), which are normally ignored, also cancel. In fact, there are four ways in which a quantum theory of gravity might not be finite: (i) conventional counter-terms like $R_{\mu\nu}R^{\mu\nu}, R^2$ and matter contributions (ii) topological Euler number counter-terms (iii) cosmological constant counter-terms of the kind discussed in this paper and (iv) boundary contributions of the kind mentioned above. In all calculations carried out to date, $N = 8$ supergravity has passed all four tests and, indeed, is the only theory known to do so!

ACKNOWLEDGEMENTS

I am very grateful for conversations with S.M. Christensen, G.W. Gibbons, M. Rocek, and P. van Nieuwenhuizen, in collaboration with whom this work was carried out, and for correspondence with W. Siegel.

REFERENCES

Christensen, S.M. and Duff, M.J. (1978*a*). *Phys. Lett.* 76B, 571.
—— —— (1978*b*). *Phys. Lett.* **79B**, 213.
—— —— (1979). *Nucl. Phys.* **B154**, 301.
—— —— (1980). *Nucl. Phys.* **B170**, 480.
—— ——, Gibbons, G.W., and Roček, M. (1980*a*). *Phys. Rev. Lett.* **45**, 161.
—— —— —— —— (1980*b*). One-loop effects in supergravity with a cosmological constant (to appear).
Cremmer, E. and Julia, B. (1979). *Nucl. Phys.* **B159**, 141.
Das, A., Fischler, M., and Roček, M. (1977). *Phys. Rev.* **D16**, 3427.
Deser, S. and Zumino, B. (1977). *Phys. Rev. Lett.* **38**, 1433.
——, Kay, J., and Stelle, K. (1977). *Phys. Rev. Lett.* **38**, 527.
de Wit, B. and Ferrara, S. (1979). *Phys. Lett.* **81B**, 317.
Duff, M.J. and van Nieuwenhuizen, P. (1980). *Phys. Lett.* **94B**, 179.
Ferrara, S. and van Nieuwenhuizen, P. (1978). *Phys. Lett.* **74B**, 303.
—— and Zumino, B. (1975). *Nucl. Phys.* **B87**, 207.
Freedman, D.Z. and Das, A. (1977). *Nucl. Phys.* **B120**, 221.
—— and Schwarz, J.H. (1978). *Nucl. Phys.* **B137**, 333.
Gibbons, G.W. and Perry, M.J. (1978). *Nucl. Phys.* **B146**, 90. (This paper contains numerical errors corrected by Christensen and Duff 1980.)
Grisaru, M. (1977). *Phys. Lett.* **66B**, 75.
Hawking, S.W. (1978). *Nucl. Phys.* **B144**, 349.
—— and Pope, C.N. (1978). *Nucl. Phys.* **B154**, 381.
Jones, D.R.T. (1977). *Phys. Lett.* **72B**, 199.
Macdowell, S.W. and Mansouri, F. (1977). *Phys. Rev. Lett.* **38**, 739.
Ogievetsky, V. and Sokotchev, E. (1980). Dubna Preprint.
Pendeleton, H. and Poggio, E. (1977). *Phys. Lett.* **72B**, 200.
Sezgin, E. and van Nieuwenhuizen, P. (1980). *Phys. Rev.* D22, 301.
Stelle, K. and West, P. (1978). *Phys. Lett.* **74B**, 409.

Townsend, P.K. and van Nieuwenhuizen, P. (1979). *Phys. Rev.* **D19**, 3592.
van Nieuwenhuizen, P. (1979). Recent developments in gravitation. (ed Deser and Levy) Plenum Press, New York.

22

QUANTUM (SUPER)GRAVITY

P. van Nieuwenhuizen

Institute for Theoretical Physics, State University of New York, at Stony Brook, Long Island, New York 11794

1. QUANTUM SUPERGRAVITY*

At the first conference, in early 1974, I reported on results concerning the finiteness of the S-matrix at the one-loop level when one couples gravity to matter. The situation seemed quite bleak at the time. My collaborator at that time wrote about that in the proceedings (Deser 1975).

Today there is a dramatic improvement: particular combinations of matter fields coupled to gravity by minimal coupling and to each other by particular couplings (such as Pauli couplings between photons and electrons) yield a finite S-matrix, not only at the one-loop level (Grisaru, van Nieuwenhuizen and Vermaseren 1976) but even at the two-loop level (Grisaru 1977).† These combinations are, of course, just such that they yield models of supergravity, namely the N-extended pure supergravities. It would have been difficult to discover supergravity by looking for such matter combinations, because one needs the particular couplings mentioned above. In fact, prior to the discovery of supergravity, particular matter combinations were investigated, in the hope that they would have a one-loop finite S-matrix, notably QED coupled to gravity, and Dirac electrons coupled to gravity plus torsion, but these cases led to a divergent S-matrix.

The reason that one has finiteness in the models of extended

---

* See Freedman and van Nieuwenhuizen (1978) for an elementary introduction to supergravity. Lectures by the author are found in *Physics Reports*, **68**, 4 (1981).

† At the three-loop level candidates for divergences in the S-matrix have been proposed, in the N = 1 and N = 2 model up to terms quadratic in the gravitino fields (Deser, Kay, and Stelle 1977), and to all orders in the gravitino fields (Ferrara and van Nieuwenhuizen 1974), and recently in the N = 8 model to all orders (Howe, Lindström and Götenberg in a preprint; Brink and Howe 1979). I will be discussing, however, why I believe that the N = 8 model is finite to all loops.

pure supergravity is due to the fact that in these models all particles are connected by symmetries to each other. Now processes with only external gravitons (but any internal fields) must have divergences of the form

$$\Delta\mathscr{L}(1 - \text{loop}) = \frac{1}{n-4}\left[\alpha R^2_{\mu\nu} + \beta R^2 + \gamma R^2_{\mu\nu\rho\sigma}\right]. \quad (1.1)$$

In topologically trivial spaces the last term can be dropped due to the Gausz–Bonnet theorem

$$R^2_{\mu\nu\rho\sigma} - 4R^2_{\mu\nu} + R^2 = \text{total derivative}. \quad (1.2)$$

For the S-matrix one has $R_{\mu\nu} = R = 0$. This means that the external gravitons have physical momenta and polarizations, and this follows from the background field method. (For a discussion of the relation between the background field method and normal field theory see Grisaru, van Nieuwenhuizen, and Wu (1975*a*) where also various examples are worked out.) Since one can use the symmetries of supergravity to rotate away gravitons and replace them by any other matter particles, processes with other particles than gravitons as external lines are also finite.

It is clear that if one has a second set of particles (not containing the graviton) which all rotate into each other under the symmetries of the theory, but which cannot reach the first set of particles (the set containing the graviton), then processes with second-set external lines need not be finite. A moment of thought will convince the reader that one needs explicit computations to know for sure that such S-matrices are really divergent, but these computations have been performed and have confirmed this anticipation (Vermaseren and van Nieuwenhuizen 1976; Fischler 1979).

Two-loop finiteness of supergravity was first proven by Grisaru (1976). He combined two observations. First of all, the only possible two-loop invariant for processes with only external gravitons is given by

$$\Delta\mathscr{L} = \alpha\left[\frac{\omega^2}{(n-1)^2} \text{ or } \frac{\omega^2}{n-1}\right]\left[R_{\mu\nu\rho\sigma}R^{\rho\sigma\alpha\beta}R_{\alpha\beta}{}^{\mu\nu}\right] \quad (1.3)$$

if $R_{\mu\nu} = R = 0$. This is easily proved by writing down all possible invariants of the generic form $\kappa^2 R^3$ or $\kappa^2 DDR^2$ etc., and then by using all kinds of identities. If one constructs Borngraphs for graviton–graviton scattering with one vertex coming from eqn (1.3) and all other vertices given by the Hilbert action $\sqrt{g}R$, then one finds that only processes where helicity of the gravitons is flipped, are non-zero (van Nieuwenhuizen and Wu 1977). For example, $(2) + (2) \rightarrow (-2) + (\pm 2)$ is non-zero, but $(2) + (2) \rightarrow (2) + (2)$ is zero. The second observation is that in supergravity helicity is conserved. One can now connect these two arguments as follows. The one-loop divergences of the S-matrix can be thought of as the above-mentioned Borngraphs with the one vertex given by eqn (1.3) due to contracting the divergent loops to a point. Since both the Lagrangian and the counter-Lagrangian (eqn 1.3)) are separately invariant under supersymmetry, and supergravity preserves helicity, these Borngraphs must preserve helicity. However, the only non-vanishing Borngraphs flip helicity, so that α eqn (1.3) must be equal to zero. For the proof that supergravity conserves helicity see Grisaru *et al* (1976). Any theory that preserves helicity (not only supergravity) is thus two-loop finite. For pure gravity this is not known, but it could very well be the case that pure gravity is also two-loop finite.

At the three-loop level nothing is definitely known about finiteness. There are many conjectures, but no solid results. However, in the $N = 4$ globally supersymmetric Yang–Mills model, the β-function vanishes at the one, two and even at the three loop level.* In a superspace formulation of this model there is only one type of divergence (only one independent Z-factor) and vanishing of the β-function means vanishing of this divergence. Hence the superspace model has finite *Green's* functions up to and including 3-loops! This model has an important property in common with the $N = 8$ supergravity model: whereas the $N = 4$ globally supersymmetric model is the largest with spins $j \leqslant 1$, the $N = 8$ locally supersymmetric model is the largest with spins $j \leqslant 2$. Some people believe that also

---

* Grisaru, M.T., Rocek, M., and Siegel, W. *Phys. Rev. Lett.* to appear. Caswell, C. in Maryland preprint. Tarasov.

the $N = 8$ model will be (at least) three-loop finite. This model has also another peculiar property. It can be obtained from the $N = 1$ model in 11 dimensions by dimensional reduction, and one can then either make duality transformations which transform all axial vectors into vectors, or not make these transformations. Only in the latter case do all trace anomalies cancel. Thus there are two versions of this model, and in only one of the two do the counter-terms in topologically non-trivial spaces also cancel (see below). This cancellation is again unexpected and seems to herald good news for the $N = 8$ model.

A puzzling feature is that one needs the duality transformations in order to extend the naive SO(7) local x SL(7,R) global symmetry (which is due to the space-time symmetries in 11 dimensions) to the maximal SU(8) local x E_7 global.*

One would think: the more symmetries, the better, but at least for the trace anomaly this seems not to be true.

2. QUANTUM GRAVITY

Pure Einstein gravity is one-loop finite. This follows from eqn (1.1) in the same way as we proved one-loop finiteness for supergravity. If helicity is conserved at the quantum level, pure Einstein gravity will also be two-loop finite; this is unknown at present. In fact, in a study of the higher-loop renormalizability of Einstein gravity based on Reggeization and what is technically called Mandelsterm counting, it was concluded that if helicity is conserved in ordinary gravity, then it is finite (at least) up to and including four loops (Grisaru *et al.* 1975*b*). This Mandelsterm counting is based on an analysis of kinematical zeros due to Lorentz invariance in the S-matrix and dynamical poles, generated by exchange of gravitons. The usefulness of such an approach was demonstrated when it was shown that at the Born-level all helicity amplitudes for graviton–graviton scattering are uniquely determined from the knowledge of such kinematical singularities

---

* For details, see the six lectures by the author on supergravity in the Proceedings of the 1980 Nuffield workshop: *Superspace and supergravity* (Cambridge University Press; S. Hawking and M. Rocek, ed.)

alone, without detailed Feynman graph calculations, reflecting thus the uniqueness of Einstein theory (Grisaru *et al.* 1975*c*).

As soon as one couples matter fields to gravity even one-loop finiteness is lost (except when one takes precisely the matter combinations *and* the couplings of supergravity). This was shown for scalars by 't Hooft and Veltman (1974) for photons or electrons by Deser and van Nieuwenhuizen (1974), and for Yang–Mills bosons by Deser, van Nieuwenhuizen, and Tsao (1974). A natural question often asked at that time was: is there some combination of fields such that all infinities cancel 'miraculously'? This word 'miraculously' has become a codeword for: 'there are extra symmetries which are not yet known but which will explain these cancellations'. The most interesting study in this direction was an investigation of the coupling of the two most elegant field theories we have: QED and general relativity. Again, as I already said, the S-matrix turned out to be divergent at the one-loop level (Grisaru *et al.* 1975*b*).

Another study was suggested by a remark of Veltman that torsion might play a role in a future quantum theory of gravity. To study this possibility, spin 1/2 electrons were coupled to gravity and the well-known four-fermion couplings were added to the action, which result from torsion. (By this is is meant that one takes in both the Hilbert action and the Dirac action the spin connection as an independent field, and then one proceeds to solve its algebraic field equation. This leads to torsion

$$\omega_{\mu}{}^{mn} = \omega_{\mu}{}^{mn}(e) + \bar{\lambda}\gamma_5\gamma_{\mu}\sigma^{mn}\lambda \ . \qquad (2.1)$$

Substituting this result back into the action leads to the mentioned four-fermion seagull couplings). However, again it was found that this system also was no longer one-loop finite (van Nieuwenhuizen 1975). It is interesting to note that supergravity is a theory with torsion: one can write it either in first order formalism without four-fermion couplings (the fermion is in supergravity a real spin 3/2 electron called gravitino, rather than a complex spin 1/2 electron) or in second order formalism with $\omega(e)$ as spin connection, but then with four-fermion couplings.

Antisymmetric tensor fields

Although certain spins were studied using particular field representations for these spins, the question arises whether different field representations for a given spin might give different renormalizability results. Two cases were studied: spin (0) represented by $A_{\mu\nu} = A_{\nu\mu}$ and no dynamical modes represented by $A_{\mu\nu\rho}$ (totally antisymmetric). Let us first show that $A_{\mu\nu}$ describes really spin 0. This can be deduced from the field equations

$$\mathscr{L} = F^2_{\mu\nu\rho}, \quad F_{\mu\nu\rho} = \partial_\mu A_{\nu\rho} + \text{cyclic terms} \tag{2.2}$$

as follows. The curl $F_{\mu\nu\rho}$ equals in four dimensions $\varepsilon_{\mu\nu\rho\sigma}\phi^\sigma$ so that the Maxwell equations imply $\phi_\sigma = \partial_\sigma\phi$. Reinserting this into the action yields the Klein–Gordon action for ϕ. If one does not like to insert field equations into the action, one can use a path-integral approach

$$Z = \int \mathrm{d}A_{\mu\nu}\mathrm{d}F_{\rho\sigma\tau}\exp \mathrm{i}\left[F^2_{\mu\nu\rho} + F^{\mu\nu\rho}\partial_\mu A_{\nu\rho}\right] \tag{2.3}$$

Integration first over F yields (2.2) but integration over $A_{\nu\rho}$ yields $\delta(\partial^\mu F_{\mu\nu\rho})$ and leads to the Klein–Gordon action. In a similar way one may show that in four dimensions $A_{\mu\nu\rho}$ contains no dynamical modes.

One can calculate the one-loop $\Delta\mathscr{L}$ using the background field formalism and then *one finds different results* for ϕ coupled to gravity and for $A_{\mu\nu}$ coupled to gravity, even if one uses the ϕ and $A_{\mu\nu}$ field equations as well as the Einstein equations. In both cases one has

$$\Delta\mathscr{L} = \alpha R^2 + \beta\chi, \quad \chi = \text{Euler invariant} \tag{2.4}$$

and one finds that α is the same in both cases (Sezgin and van Nieuwenhuizen 1980*b*) but β is not. In fact (Duff and van Nieuwenhuizen 1980)

$$\Delta\mathscr{L}(A_{\mu\nu} + \text{gravity}) - \Delta\mathscr{L}(\phi + \text{gravity}) = \chi \tag{2.5}$$

Similarly one finds for the three index photon

$$\Delta\ (A_{\mu\nu\rho} + \text{gravity}) = -\ 2\chi\ . \tag{2.6}$$

Since $\Delta\mathscr{L}$ yields the trace anomaly, we conclude: different field representations of a given spin lead to different trace anomalies, and to different counter-terms in spaces with non-trivial topologies.

For details of quantizing the coupling of $A_{\mu\nu}$ and $A_{\mu\nu\rho}$ to gravity see the next section. For a more detailed discussion and topological aspects, see the contribution of M. Duff in this volume.

3. TOPICS OF COVARIANT QUANTIZATION

Gauge theories are quantized by means of covariant quantization techniques. A number of new aspects have been discovered in the past few years which we now enumerate:

(i) *Open gauge algebras*. Let us explain what an open gauge algebra is. The commutator of two symmetries of a given action is, of course, again an invariance. However, for supergravity, one finds for the commutator of two local supersymmetry transformations on the gravitino field (Freedman and van Nieuwenhuizen 1976)

$$[\delta_{\text{sup}}(\varepsilon_1(x)),\ \delta_{\text{sup}}(\varepsilon_2(x))]\psi_\mu = [\delta_{\text{g.c.}}(\xi^\nu(x)) +$$

$$\delta_{\text{Lorentz}}(\xi^\nu\ (x)\omega_\nu^{mn}(e,\psi)) + \delta_{\text{sup}}(-\ \xi^\nu(x)\psi_\nu)]\psi_\mu$$

+ extra terms with ξ^μ and ξ^{mn} proportional to the gravitino field equation. (3.1)

In here, $\xi^\mu = \frac{1}{2}\bar{\varepsilon}_2\gamma^\mu\varepsilon_1$ and $\xi^{mn} = \frac{1}{2}\bar{\varepsilon}_2\sigma^{mn}\varepsilon_1$.

For the vierbein, no such extra terms are present. These extra terms are separately an invariance of the action, but they are not of the form of the three other symmetries, namely g (general) c (co-ordinate) transformations, local Lorentz transformations and local supersymmetry, even if one allows the local parameters to be field dependent. One can, in principle, define these extra terms to be a fourth local symmetry (depending on, say, $\varepsilon_1(x)$ for fixed constant ε_2). If one considers commutators of these four local symmetries one then finds again extra terms, etc. So one ends up with an infinite dimensional Lie algebra and, worse, the elements of this algebra contain external constants (for example, the constant ε_1 above).

(ii) *Quantization of gauge theories with open gauge algebras.* If one makes a Hamiltonian analysis and starts with a path-integral in phase-space which is such that it takes the standard (naively expected) form, and if one reduces the variables pq to only physical modes p^*q^*, then Fradkin and Vassiliev (1977) showed that upon integrating out the ps, one is left with four-ghost couplings of the Faddeev–Popov ghosts. Thus, in theories with open gauge algebras, unitarity requires that the quantum action contains in addition to gauge fixing and terms quadratic in ghosts, extra four-ghost seagulls. The same thing was found by Kallosh (1978), who tried to construct a quantum action which is invariant under so called Becchi–Rouet–Stora–Tyutin rules. In the proof of invariance one needs at a given stage the commutator of two local symmetries, and the non-closure of the gauge algebra again led directly to the four-ghost couplings. This is thus not a Hamiltonian approach. Sterman, Townsend, and van Nieuwenhuizen (1978) also used BRST formalism, but analysed Feynman diagrams to directly verify what extra couplings were needed to restore unitarity; they also found the four-ghost couplings. There is now a very simple way to understand the origin of the four-ghost couplings, as we show.

(iii) *Auxiliary fields.* One can close the open gauge algebra for $N = 1$ (Ferrara and van Nieuwenhuizen 1978; Stelle and West 1978) and (recently) for $N = 2$ pure extended supergravity, (Fradkin and Vasileev 1979; De Witt and van Holten 1979; Breitenlohner and Sohnius 1980) by adding extra fields. In the action these fields are auxiliary fields. For example, in simple ($N = 1$) supergravity, one finds

$$\mathscr{L}^{\text{class}}(\text{aux}) = \frac{\det e}{3}(S^2 + P^2 - A_m{}^2) \qquad (3.2)$$

In the gauge algebra, they are just three new fields. Nevertheless, one can also read off from the gauge algebra that they are auxiliary fields in the action, because S, P and A_m rotate under local supersymmetry into the gravitino field equation. Indeed, $S = P = A_m = 0$ is itself a field equation, and field equations rotate into field equations. One can deduce, reading this argument backwards, that S,P,A_m will be auxiliary, because they rotate into a field equation. The precise new transformation rules are not very difficult. In particular

$$\delta\psi_\mu = \frac{1}{\kappa}(D_\mu + \frac{i\kappa}{2}A_\mu\gamma_5)\varepsilon + \frac{1}{6}\gamma_\mu(S - i\gamma_5 P - i\not{A}\gamma_5)\varepsilon \qquad (3.3)$$

If one now covariantly quantizes $N = 1$ supergravity in the gauge $\gamma^\mu\psi_\mu = 0$, then one finds the usual Faddeev–Popov gauge fixing terms by varying $\gamma^\mu\psi_\mu$ and sandwiching the result with an anti-ghost and a ghost field. In this way one finds extra terms containing S,P,A_m in the action

$$\mathscr{L}^{\text{ghost}}(\text{aux}) = \bar{C}\left(\frac{i}{2}A_\mu\gamma_5 + \frac{1}{6}\gamma^\mu(S - i\gamma_5 P - i\not{A}\gamma_5)\right) \qquad (3.4)$$

and eliminating S,P,A_m from $\mathscr{L}^{\text{class}}(\text{aux}) + \mathscr{L}^{\text{ghost}}(\text{aux})$ by means of their *still non-propagating* field equation, one finds the four-ghost couplings.

(iv) *Lorentz ghosts cannot be neglected*. One often hears statements that since one can fix the local Lorentz invariance by an *algebraic* fixing term

$$\mathscr{L}^{\text{fix}}(\text{Lorentz}) = (e_{a\mu}\delta^\mu_b - e_{b\mu}\delta^\mu_a)^2 \qquad (3.5)$$

also the Lorentz ghosts are non-propagating and hence, to all intents and purposes, they can be dropped from consideration. Actually, in supergravity this argument is incorrect, while in gravity it is correct (Jones, Fung, and van Nieuwenhuizen 1980). The reason is that in supergravity as well as in gravity the variation of the Lorentz gauge fixing term yields the following term in the Lorentz ghost section

$$\mathscr{L}^{\text{ghost}}(\text{Lorentz}) = C^{*mn}(C_{mn} + \partial_n C_m) \tag{3.6}$$

where C_m is the co-ordinate ghost and C_{mn} the Lorentz ghost, so that elimination of C^{*mn} yields $C_{mn} - \frac{1}{2}(\partial_m C_n - \partial_n C_m) = 0$. In gravity C^{*mn} and C^{mn} only appear in this way, since the terms containing C^{*m} are obtained by varying the metric. But in supergravity the field C_{mn} appears somewhere else, namely in the supersymmetry antighost sector, obtained by varying the supersymmetry gauge $\gamma^\mu \psi_\mu = 0$. One finds in fact

$$\mathscr{L}^{\text{ghost}}_{(\text{sups})} = \overline{C}(\gamma^\mu D_\mu C - \frac{1}{2}\sigma^{mn} C_{mn} \gamma^\mu \psi_\mu + \text{more}) \tag{3.7}$$

and if one substitutes the C^{*mn} field equation, one replaces C_{mn} by $-\frac{1}{2}(\partial_m C_n - \partial_n C_m)$ and thus one generates a non-vanishing coupling of the form $(\overline{C}\partial_m C_n \psi)$. Thus, even though the Lorentz ghosts themselves can be eliminated (just as the antisymmetric part of the vierbein field), they lead to extra couplings in the effective quantum action.

Historically, these extra couplings were found because a certain Ward identity (transversality of the gravitino self-energy) was broken (due to not including the extra couplings).

(v) *Quantization of antisymmetric tensor fields.* Coupling $A_{\mu\nu} = -A_{\nu\mu}$ to gravity, one quantizes by adding a gauge-fixing term which fixes the gauge $\delta A_{\mu\nu} = D_\mu \Lambda_\nu - D_\nu \Lambda_\mu$

$$\Delta_F^{(1)} \delta(D^\mu A_{\mu\nu} - b_\nu) \tag{3.8}$$

The normalization $\Delta_F^{(1)} = \det(D^\lambda D_\lambda \delta^\mu{}_\nu - D^\mu D_\nu)$ yields upon exponentiation an anticommuting complex vector *gauge* ghost. In fact, both $C^{*\mu}$ and C_μ are gauge fields. Thus we need ghosts for ghosts (Namazie and Storey 1979; Townsend 1980): we once more fix the gauge of this vector gauge ghost C_μ and C^*_μ by adding as gauge fixing term

$$\Delta_F^{(2)} \delta(D^\mu C_\mu - a)\Delta_F^{(2)} \delta(D^\mu C^*_\mu - a^*) \tag{3.9}$$

where a is anticommuting. The normalization $\Delta_F^{(2)} = (D^\mu D_\mu)^{-1}$ yields upon exponentiation two commuting complex scalar ghosts.

Now we must average the gauge fixing delta-functions with 't Hooft averaging functionals. First of all, the last delta function: we multiply as follows by unity

$$\int da^*da \exp(-a^*a) \tag{3.10}$$

Performing the da^*da integration, one finds the standard term $C^*_\mu D^\lambda D_\lambda C^\mu$ for the complex vector ghost in the complete quantum action.

Finally, we average the first delta-function. At this point something subtle happens (Siegel 1980): b_ν is constrained

$$D^\nu b_\nu = D^\nu D^\mu A_{\mu\nu} = R^{\mu\nu} A_{\mu\nu} = 0 \tag{3.11}$$

because $R^{\mu\nu}$ is symmetric. Thus we use as averaging functional not simply a square of b_ν but use only its transverse part: $(b_\nu{}^T)^2$. In covariant form this can be written as

$$\int db_\sigma \exp[b_\mu(g^{\mu\nu} - D^\mu(D^\lambda D_\lambda)^{-1}D^\nu)b_\nu] \;. \tag{3.12}$$

(Note that, indeed, $b_\nu = D_\nu f$ yields zero in the exponential.)

Since this last functional is gauge invariant under $\delta b_\nu = D_\nu f$, we once more fix the gauge by

$$\Delta_F^{(3)}\delta(D^\mu b_\mu - g),\ \Delta_F^{(3)} = D^\mu D_\mu \tag{3.13}$$

One finds that $\Delta_\mathrm{F}^{(3)}$ yields an anticommuting complex scalar Faddeev–Popov ghost. We average this last delta-function by a particular averaging functional

$$M\int dg \exp[-g(D^\lambda D_\lambda)^{-1}g],\ M = \det(D^\lambda D_\lambda)^{-\frac{1}{2}} \tag{3.14}$$

because then in the product of the two last functionals one finds only the naively expected result b_μ^2.

The one unexpected ghost is the one following from M. It is real (because of the factor 1/2, see below) commuting Nielsen–Kallosh ghost. This extra ghost arose 'from normalizing the functional which averages the delta-function which fixes the gauge of the functional which averages $\delta(D^\mu A_{\mu\nu} - b_\nu)$'. It was first found by Siegel (1980) and used by Sezgin and van Nieuwenhuizen (1980) to show the before-discussed equivalence (up to trace anomalies) of $A_{\mu\nu}$ and ϕ at the quantum level.

(vi) *Actions for real ghost fields.* A problem, well known among quantum supergravity specialists, has been that a real commuting spinor ghost appears if one exponentiates the normalization in the averaging functional

$$N \int da \exp(\bar{a}\not{D}a),\ N = \det(\not{D})^{-\frac{1}{2}} \tag{3.15}$$

which is used to exponentiate the gauge fixing term

$$(\det \not{D})^{-1}\delta(\gamma^{\mu}\psi_{\mu} - a)$$

The factor $(\det \not{D})^{-1}$ yields a complex commuting ghost action upon exponentiation

$$\mathcal{L} = \sqrt{g}\bar{E}\ \not{D}E,\ \bar{E} = E^{\dagger}\gamma_0 \tag{3.16}$$

but the factor $(\det \not{D})^{-\frac{1}{2}}$ would seem to yield as action the same result but with a real (Majorana) spinor in order to get the square root. The problem is, however, that such an action cannot be used, since it is a total derivative

$$\mathcal{L} = \sqrt{g}\bar{F}\ \not{D}F = \frac{1}{2}\partial_{\mu}(\sqrt{g}\bar{F}\gamma^{\mu}F) = 0,\ \bar{F} = F^{T}C \tag{3.17}$$

On the other hand, Ward identities come out only correctly if one treats F as having opposite statistics from an anticommuting real spinor.

The solution was found by various people independently. One simply writes

$$(\not{D}^{-\frac{1}{2}}) = (\not{D}^{\frac{1}{2}})(\not{D}^{-1}) \tag{3.18}$$

and now one exponentiates to find: one real anticommuting spinor ghost and one complex commuting spinor ghost. The sum acts in all applications just as if one had one real commuting spinor ghost.

Similar problems and solutions exist for integer spin ghosts.

See the factor $M = (D^{\lambda}D_{\lambda})^{-\frac{1}{2}}$ under (v).

4. HIGHER SPIN FIELD THEORY

The success of supergravity which for the first time led to a constant coupling of spin 3/2 has rekindled the interest in higher spin field equations. At the free field level one can write down an equation for any spin actions with positive energy (Fang and Fronsdal 1978) (more technically: actions whose propagators have only first order poles with positive residues). One and a half years ago, a study was made of massless and massive spin 5/2 theory. Requiring that the free field action be free from ghosts led to a unique solution (representing the spin 5/2 by a symmetric tensorial spinor $\psi_{\mu\nu}$) which has a local gauge invariance (Berends, De Witt, van Holten, and van Nieuwenhuizen 1979*a*,*b*, 1980).

$$\mathscr{L} = -\frac{1}{2}\bar{\psi}_{\mu\nu}\not{\partial}\psi_{\mu\nu} + (\bar{\psi}\cdot\gamma)^{\mu}[2\partial\cdot\psi_{\mu} - \not{\partial}(\gamma\cdot\psi)_{\mu} - \partial_{\mu}\psi]$$

$$+ \frac{1}{4}\bar{\psi}\not{\partial}\psi, \quad \psi = \psi^{\lambda}{}_{\lambda} \text{ and } \gamma\cdot\psi_{\mu} = \gamma^{\nu}\psi_{\nu\mu} \tag{4.1}$$

The local gauge invariance is in terms of a spin 3/2 parameter

$$\delta\psi_{\mu\nu} = \partial_{\mu}\varepsilon_{\nu} + \partial_{\nu}\varepsilon_{\mu}, \quad \gamma^{\mu}\varepsilon_{\mu} = 0 \tag{4.2}$$

This action was derived by Schwinger from source theory, but not requiring that the external sources be conserved still leads to only this action.

One would like to be able to couple this system to gravity. If one simply replaces ordinary derivatives by covariant derivatives, the spin 5/2 action is no longer invariant, but its variation is proportional to curvature terms. It is trivial to substitute

$$\delta\psi_{\mu\nu} = D_{\mu}\varepsilon_{\nu} + D_{\nu}\varepsilon_{\mu}, \quad \gamma^{\mu}\varepsilon_{\mu} = 0 \tag{4.3}$$

into (4.1), and one finds terms with $R_{\mu\nu}$ and R (hence proportional to $G_{\mu\nu}$ and terms proportional to the full Rieman curvature $R_{\mu\nu\rho\sigma}$ (Berends *et al.* 1979*a*,*b*, 1980). The $G_{\mu\nu}$ terms can be cancelled by adding the Hilbert action and choosing the vierbein variation appropriately, since the variation of the Hilbert action is equal to $G_{\mu\nu}$ times δe^m_μ. However, the $R_{\mu\nu\rho\sigma}$ terms cannot be cancelled. Thus it seems not feasible to make the coupling gauge invariant.

Also couplings to matter systems have been investigated by Berends *et al.* (1979*a*,*b*, 1980) and, again, no invariant action seems to exist.

Thus it seems that nature stops at spin 2. However, one aspect which has not been fully investigated is the possibility of off-diagonal couplings. In the massive spin 5/2 system one finds in addition to (4.1) terms involving an auxiliary spin 1/2 field λ which couples like $\bar{\lambda}\not{\partial}\psi$ etc., and has mass terms $\bar{\lambda}\lambda$, $\bar{\lambda}\psi$, $\psi_{\mu\nu}\psi^{\mu\nu}$ and $\bar{\psi}\psi$. Perhaps one can construct consistent theories with one or more such auxiliary off-diagonal fields. It might also be that one needs spin 3 together with spin 5/2.

5. HIGHER DERIVATIVES AND R^2 THEORIES

Actions starting with squares of curvatures can lead to genuinely renormalizable theories. For example,

$$\mathscr{L} = \alpha R^2_{\mu\nu} + \beta R^2 + \gamma R \tag{5.1}$$

leads to a propagator in which all terms are proportional to $\kappa^{-2}k^{-4}$ instead of k^{-2}, provided one excludes three cases (Stelle 1971):

$$\mathscr{L} = R^2_{\mu\nu} - \frac{1}{3}R^2, \quad \mathscr{L} = \beta R^2 + \gamma R, \quad \mathscr{L} = \gamma R \; . \tag{5.2}$$

However, these three cases are the only cases where unitarity holds. Thus, for higher derivative actions involving only $g_{\mu\nu}$, one has either unitarity or renormalizability, but never both.

An attractive alternative is to consider theories with propagating torsion but which are not of higher derivative type. In this approach one considers $g_{\mu\nu}$ and $\omega_\mu{}^{mn}$ as independent fields, In the Einstein action one can solve $\omega_\mu{}^{mn}$ using the Palatini formalism, but in R^2 actions, the spin connection has a propagating field equation, and one has two dynamical fields in the theory.

If one writes down the most general action containing only second (or first, or none) derivatives of $g_{\mu\nu}$ and $\omega_\mu{}^{mn}$, one can analyse whether these theories can be unitary. That is to say, one writes down the propagator, coupled to sources which satisfy all source constraints which correspond to gauge invariances in the action (there are about 30 gauge invariances possible!), and then one investigates for which parameters there are only first order poles with non-negative residues.

The result of this detailed analysis, based on spin projection operators which diagonalize the problem into separate spin blocks, is that there are several many-parameter solutions which are unitary, but none is renormalizable. The reason can be understood as follows. Any propagator of the form $A(k^2)k^{-2}(k^2 + \kappa^{-2})^{-1}$ can be decomposed into $(-\kappa^2)A(-\kappa^{-2})k^{-2} + \kappa^{-2}A(0)(k^2 + \kappa^{-2})^{-1}$. Contrary to what many people think, this in itself constitutes no problem. Negative residue poles in lower spin sectors can combine with similar terms in higher spin sectors in such a way that the sum has again positive residues. An example is the action

$$\mathscr{L} = \alpha\kappa^{-2}R(g) + \beta R^2(g) \tag{5.3}$$

which can be decomposed into spin projection operators for spin 2 and spin 0 as follows

$$= \alpha\Box\kappa^{-2}(P^2 - 2P^0) + \beta\Box^2 P^0$$

$$P^2 = \frac{1}{2}(\theta_{\mu\rho}\theta_{\nu\sigma} + \theta_{\mu\sigma}\theta_{\nu\rho}) - \frac{1}{3}\theta_{\mu\nu}\theta_{\rho\sigma}, \quad \theta_{\mu\nu} = \delta_{\mu\nu} - \partial_\mu\partial_\nu\Box^{-1}$$

$$P^0 = \frac{1}{3}\theta_{\mu\nu}\theta_{\rho\sigma} \tag{5.4}$$

The sources still satisfy $\partial_\mu T^{\mu\nu} = 0$ as a result of general co-ordinate invariance, and the propagators have positive residues

$$\text{Propagator} = \frac{P^2}{\alpha\Box} + \frac{P^0}{(-2\alpha + \beta\Box)\Box}\ . \tag{5.5}$$

The negative residue in $P^0(-2\alpha\Box)^{-1}$ combines with $P^2(\alpha\Box)^{-1}$ to form a positive residue, as one easily checks. However, in the highest spin sector, such negative residues can evidently not be combined with positive residues of yet higher spin sectors, and this explains qualitatively that in the spin 2 sector propagators of unitary theories go like k^{-2} for large momenta. It follows that unitary theories are non-renormalizable, since one needs k^{-4} in *all* spin sectors for renormalizability (Neville 1978, 1980; Sezgin and van Nieuwenhuizen 1980*a*).

There exist many R^2 theories in the literature, and with the complete set of spin projection operators we have analyzed all of those we are aware of. None of these theories turned out to be unitary; all had ghosts (Sezgin and van Nieuwenhuizen 1980*a*).

If one insists on renormalizability and unitarity, R^2 theories seem out. But perhaps renormalizability is not such an important criterion.

6. REGULARIZATION BY DIMENSIONAL REDUCTION

In the usual dimensional regularization scheme one lets the indices of $g_{\mu\nu}$ become n-dimensional. This leads to an n-dependent propagator for the graviton

$$P^{(2)}_{\mu\nu\rho\sigma} = \left(\frac{-i}{2p^2}\right)\left[\delta_{\mu\rho}\delta_{\nu\sigma} + \delta_{\mu\sigma}\delta_{\nu\rho} - \frac{2}{n-2}\delta_{\mu\nu}\delta_{\rho\sigma}\right] .$$

Similarly, the propagator for the gravitino ψ^a_μ in supergravity becomes n-dependent if one uses the standard action in n-dimensions

$$\mathcal{L}^{\text{class}} = \frac{1}{2}\det \bar{\psi}_\mu\gamma^{[\mu}\gamma^\rho\gamma^{\sigma]}D_\rho\psi_\sigma$$

$$\mathcal{L}^{\text{fix}} = \frac{1}{4}\bar{\psi}_\mu\gamma^\mu\not{\partial}\gamma^\sigma\psi_\sigma$$

and the anticommutators $\{\gamma^\mu,\gamma^\nu\} = 2\eta^{\mu\nu}$. Indeed, the field equation in n-dimensions is still n-independent, namely

$$\mathcal{L}^{\text{class}} + \mathcal{L}^{\text{fix}} = \frac{1}{4}\bar{\psi}_\mu\gamma^\sigma\not{\partial}\gamma^\mu\psi_\sigma + \text{vertices} \qquad (6.1)$$

but inverting this, using $\gamma^\mu\gamma_\mu = n$, yields as propagator

$$P^{3/2}_{\mu\nu} = \frac{1}{(n-1)p^2}\left[\gamma_\nu \not{p} \gamma_\mu + (4-n)\left(\delta_{\mu\nu}\not{p} - 2\frac{p_\mu p_\nu \not{p}}{p^2}\right)\right]$$

It is possible to use a modified version of dimensional regularization where indices of $g_{\mu\nu}$ and ψ_μ are not n-dimensional but 4-dimensional. This is the regularization scheme by dimensional reduction (Siegel 1979). One formally descends from 4 to $n < 4$ dimensions with momenta p_μ but keeps indices of fields 4-dimensional. It has been shown in one- and two-loop calculations that this does not lead to a violation of gauge Ward identities (Capper, Jones, and van Nieuwenhuizen 1980). This justifies the use of typical 4-dimensional identities in quantum gravity, such as the Gauss–Bonnet theorem.

For supersymmetry, this scheme is very useful because actions which are supersymmetric (either globally or locally) in 4 dimensions are not supersymmetric in n dimensions. For example, one needs in some models the identity for real spinors λ

$$(\bar\lambda\lambda)(\bar\lambda\lambda) = -(\bar\lambda\gamma_5\lambda)(\bar\lambda\gamma_5\lambda)$$

which is a rearrangement of the components of λ^a, and remains true as long as $a = 1,4$, even when λ^a depends on x^μ which are n-dimensional. It has been checked explicitly that Ward identities in global and local supersymmetry remain satisfied when one uses this new scheme, but are violated in the old version of dimensional regularization where μ in ψ_μ becomes n-dimensional (Capper *et al.* 1980; Jones *et al.* 1980).

Since μ in x^μ and p_μ is continued to $n < 4$, one might wonder how one still finds anomalies. In the approach of 't Hooft and Veltman, for example, $n > 4$ and they use the identity $[\gamma_5, \gamma_\mu] = 0$ for $\mu > 4$ while $\{\gamma_\mu, \gamma_\mu\} = 0$ for $\mu \leqslant 4$ so that

$$(q^1_\mu + q^2_\mu)\gamma_\mu\gamma_5 = (\not{q}^1 + \not{p})\gamma_5 - \gamma_5(\not{q}_2 - \not{p}) - 2\hat{\not{p}}\gamma_5$$

where $\hat{p}^\mu$ is non-zero only for $\mu > 4$, and it is $\hat{p}$ which yields the axial anomaly.

As shown by Nicolai and Townsend (1980) one can get the correct anomalies for Yang–Mills theories in this modified version of dimensional regularization, if one no longer assumes the cyclicity of the trace: tr $AB \neq$ tr BA. Thus now

$$(q^1_\mu + q^2_\mu)\gamma_\mu\gamma_5 = (\not{q}^1 + \not{p})\gamma_5 - \gamma_5(\not{q}^2 - \not{p})$$

but $\mathrm{tr}(\not{q}^1 + \not{p})\gamma_5 A$ is not equal to tr $\gamma_5 A(\not{q}^1 + \not{p})$. Rather, one must commute $\not{q}^1 + \not{p}$ to the right-hand end. Doing this, and evaluating the usual momentum integrals, one finds the standard axial anomaly.

For the supersymmetry anomaly one finds also the correct result (Nicolai and Townsend 1980; Majunder, Poggio, and Schnitzer 1980). The vertex in Yang–Mills theory is given by

$$S^\mu = G^a_{\alpha\beta}\sigma^{\alpha\beta}\gamma^\mu\lambda^a$$

In the Yang–Mills tensor only $\partial_\alpha W^a_\beta$ contributes to the triangle graph from which one finds the anomaly and $0 \leqslant \alpha \leqslant n$ but $0 \leqslant \beta \leqslant 4$. Taking the index μ of S^μ n-dimensional, one finds that $\gamma_\mu S^\mu$ contains a factor $\gamma_\mu\sigma^{\alpha\beta}\gamma^\mu = -(2 - n)\sigma^{\alpha\beta}$. The fields W^a_β with $n \leqslant \beta \leqslant 4$ yield extra contributions (these represent so-called ε-scalars) and it would seem that one now obtains a contribution at the one-loop level from the ε-scalars. Actually, their contribution cancels and one finds the same supersymmetry anomaly in $\gamma^\mu S_\mu$ while $\partial_\mu S^\mu = 0$. This is important because it says that the proofs of finiteness of supergravity, which assumed that supergravity is still locally supersymmetric at the one, two, etc., loop level, are correct. No anomalies are present in $\partial^\mu S_\mu = 0$. Finally, the trace anomaly is the same as in ordinary dimensional regularization since (supersymmetric) Yang–Mills theory is conformally invariant, so that $T^\alpha{}_\alpha = (n-4)$ while ε-scalars yield a second factor $(n - 4)$, meaning that they do not contribute at the one-loop level.

REFERENCES

Berends, F.A., De Witt, B., van Holten, J.W., and van Nieuwenhuizen, P. (1979*a*). *Phys. Lett.* **B83**, 188.
—— —— —— —— (1979*b*). *Nucl. Phys.* **B154**, 261.
—— —— —— —— (1980). *J. Phys.* **A13**, 1643.
Breitenlohner, P. and Sohnius, M. (1980). *Nucl. Phys.* **B165**, 483.
Brink, L. and Howe, P. *Phys. Lett.* **B88**, 268.
Capper, D., Jones, D.R.T., and van Nieuwenhuizen, P. (1980). *Nucl. Phys.* **B167**, 479.
Deser, S. (1975). In *Quantum gravity: an Oxford symposium.* (ed. C.J. Isham, R. Penrose, and D.W. Sciama). Oxford University Press.
—— and van Nieuwenhuizen, P. (1974). *Phys. Rev.* **D10**, 401, 411.
——, Kay, J., and Stelle, K. (1977). *Phys. Rev. Lett.* **38**, 527.
——, van Nieuwenhuizen, P., and Tsao, H.S. (1974). *Phys. Rev.* **D10**, 3337.
Duff, M. and van Nieuwenhuizen, P. (1980). *Phys. Lett.* **94B**, 179.
Fang, J. and Fronsdal, C. (1978). *Phys. Rev.* **D18**, 3630.
Ferrara, S. and van Nieuwenhuizen, P. (1978). *Phys. Lett.* **B74**, 333.
Fischer, M. (1979). *Phys. Rev.* **D20**, 396.
Fradkin, E. and Vasileev, M. (1977). *Phys. Lett.* **B72**, 70.
—— —— (1979). *Phys. Lett.* **B85**, 47.
Freedman, D.Z. and van Nieuwenhuizen, P. (1976). *Phys. Rev.* **D14**, 912.
—— —— (1978). *Scient. Am.* Feb. 1978.
Grisaru, M.T. (1977). *Phys. Lett.* **B66**, 75.
——, van Nieuwenhuizen, P., and Vermaseren, J.A.M. (1976). *Phys. Rev. Lett.* **37**. 1662.
—— ——, and Wu, C.C. (1975*a*). *Phys. Rev.* **D12**, 3203.
—— —— —— (1975*b*). *Phys. Rev.* **D12**, 1563.
—— —— —— (1975*c*). *Phys. Rev.* **D12**, 397.
't Hooft, G. and Veltman, M. (1974). *Annls Inst. Henri Poincaré.* XX, **1**, 69.
Jones, D.R.T., Fung, M.K., and van Nieuwenhuizen, P. (1980). *Phys. Rev.* **D22**, 2364.
Kallosh, R. (1978). *Nucl. Phys. B.* **141**, 141.
Majunder, P., Poggio, E., and Schnitzer, H.J. (1980). Brandeis preprint.
Namazie, M.A. and Storey, D. (1979). *Nucl. Phys.* **B157**, 170.
Neville, D.E. (1978). *Phys. Rev.* **D18**, 3535.
—— (1980). *Phys. Rev.* **D21**, 867 and 2075.
Nicolai, H. and Townsend, P.K. (1980). *Unification of the fundamental interactions.* Proc. Europhys. Study Conf. (ed. S. Ferrara, J. Ellis, and P. van Nieuwenhuizen). Plenum Press.
van Nieuwenhuizen, P. (1975). In *Spin and torsion in quantum gravity.* Proc. Caracas Conf. 1975.
—— and Wu, C.C. (1977). *J. math. Phys.* **18**, 182.
Sezgin, E. and van Nieuwenhuizen, P. (1980*a*). *Phys. Rev.* **D21**, 3269.
—— —— (1980*b*). *Phys. Rev.* **D22**, 301.
Siegel, W. (1979). *Phys. Lett.* **B84**, 193.
—— (1980). *Phys. Lett.* **93B**, 170.
Stelle, K.S. (1977). *Phys. Rev.* **D16**, 953.
—— and West, P.C. (1978). *Phys. Lett.* **B74**, 330.
Sternman, G., Townsend, P.K., and van Nieuwenhuizen, P. (1978). *Phys. Rev.* **D17**, 1501.
Townsend, P.K. (1980). In *Supergravity: Proceedings of the Stony Brook Conference.* North-Holland, Amsterdam.
Vermaseren, J.A.M. and van Nieuwenhuizen, P. (1976). *Phys. Lett.* **B65**, 263.

23

A REVIEW OF BROKEN SUPERGRAVITY MODELS

S. Ferrara

CERN, Geneva, and Laboratori Nazionali di Frascati, INFN, Frascati

1. WHY SUPERGRAVITY?

The present theoretical approach to high-energy particle physics is based on some general principles which have received considerable experimental support in recent years. These guidelines can be summarized as follows: the dynamical framework of our theoretical understanding of the basic forces of nature relies on Lagrangian quantum field theory. This framework is supplemented by symmetry principles, i.e., the dynamics should not change when we perform some symmetry operations on the field variables of a model under consideration. These symmetry principles, when the symmetry operations are continuous, are further constrained by the requirement that symmetries should be made local. More interestingly this last requirement often gives some additional information on the underlying dynamics. In fact in recent time there has been increasing evidence that gauge quantum field theory may have some unconventional properties which may show up beyond perturbation theory such as confinement, dynamical Higgs mechanism and topological effects. If one accepts such a framework for modern particle physics it seems natural to further use some economy principle, that is, different low energy symmetries are eventually unified at energies higher than the energy scales which correspond to the strength of the interactions needed today in order to understand 'low energy' phenomena (Iliopoulos 1979).

In the above mentioned framework a possible scenario for the unification of particle interactions emerges:

(a) Unification of electromagnetic and weak forces at 100 GeV;

(b) Grand Unification of electroweak and nuclear (strong) forces at 10^{14}-10^{15} GeV;

(c) Superunification of electroweak, strong and gravitational forces at the Planck energy $\sim 10^{19}$ GeV.

Although this framework is highly questionable, especially in (b) and (c), it is important that the requirements of Grand and Superunifications can already give constraints and predictions on physical phenomena at present energies. Examples in Grand Unified Theories are the proton decay rate, neutrino oscillations and cosmological implications (Ellis 1979; Ellis, Gaillard, Maiani, and Zumino 1980*a*; Barbieri 1980; Nanopoulos 1980). In Superunified Theories these constraints are more speculative but in some models one can get a modification of the Newton law at short distances which may give detectable effects in laboratory experiments (Scherk 1979, 1980). Another prediction arises in a pre-constituent model for a superunified theory based on spontaneously broken supergravity with a local SU(8) invariance (Ellis *et al.* 1980*a*). This model restricts to three the number of quark-lepton families in a GUT based on SU(5). SU(5) is also the maximal allowed subgroup of this theory which may remain unbroken in a first step of sequential breakings. We recall that SU(5) is the minimal simple group which contains $SU(3)_{COLOUR} \otimes (SU(2) \otimes U(1))_{ELECTROWEAK}$.

In the attempts at superunifications, in the framework of gauge quantum fields theory, it is inevitable that the unifying gauge group must contain the space-time symmetry group and the internal symmetry group all together. The former is related to gravitational interactions, the latter to the non-gravitational sector which is described by a Yang–Mills theory. Once this gauge principle is accepted the only known way these groups can be related is at present in the framework of supergravity (Freedman, van Nieuwenhuizen, and Ferrara 1976; Deser and Zumino 1976).

Supergravity is the gauge theory of a graded (super) algebra. Graded (super) algebras whose even (bosonic) part contains the space-time symmetry algebra (Poincaré, Conformal or De Sitter) are usually called supersymmetries (Fermi-Bose symmetries). Supergravity is therefore the gauge theory of supersymmetry. What is supersymmetry? Supersymmetry (Gol'fand and Likhtman 1971; Wess and Zumino 1974; Corwin, Ne'eman, and Sternberg 1975; Fayet and Ferrara 1977; Salam and Strathdee 1978) is a set of symmetry operations which transform particles of different statistics into each others. More specifically it is a continuous symmetry of local quantum field theory, based on a GLA (Graded Lie Algebra), which enables us to put bosons and fermions in irreducible multiplets of particle states. In supersymmetry the odd part of the GLA contains N-spinorial charges of Majorana type Q^i_α $\alpha = 1\ldots4$, $i = 1\ldots N$. These charges form the so called grading representation of the GLA. They are a representation of the even part of GLA. The Lie algebra of GLA contains the Poincaré algebra and one has

$$[Q^i_\alpha, P_\mu] = 0 \;, \tag{1.1}$$

$$[Q^i_\alpha, M_{\mu\nu}] = i(\sigma_{\mu\nu})_\alpha^{\ \beta} Q^i_\beta \quad , \tag{1.2}$$

$$\left\{Q^i_\alpha, \bar{Q}^j_\beta\right\} = -\,2\gamma^\mu_{\alpha\beta} P_\mu \delta^{ij} + Z^{ij}\delta_{\alpha\beta} + Z'^{ij}\gamma^5_{\alpha\beta} \;, \tag{1.3}$$

the generators Z^{ij} (Z'^{ij}), both antisymmetric in the i,j indices, belong to the centre of GLA and are called central charges. $M_{\mu\nu}$, P_μ are the usual generators of the Poincaré group. In local quantum field theory the central charges correspond to additional quantum numbers. They may correspond to global or local symmetries depending on the particular theory under consideration. These operators have non-vanishing eigenvalues only when they act on massive supermultiplets. Moreover when supersymmetry is gauged, i.e. in supergravity, these massive multiplets are minimally coupled to some spin 1 partners (graviphotons) of the graviton with a dimensionless coupling constant g_z proportional to their mass M and the gravitational constant K (Ferrara, Scherk, and Zumino 1977; Zachos 1978; Scherk 1979) $g_z \propto MK$. This peculiar relation, as pointed out by Scherk (1979), can give a substantial modification of the Newton law at short distances for identical particles because the gravitational force is attractive, the vectorial force is repulsive and being of comparable strength, they may compensate at short distances.

We now come back to the supersymmetry algebra whose relevant commutation rules are given by (1.1), (1.2) and (1.3). In local quantum field theory the fermionic generators Q^i_α are obtained as space-integrals of local current densities

$$Q^{i}_{\alpha} = \int d^3\underline{x} J^{i}_{o\alpha}(\underline{x}, t) \tag{1.4}$$

and current conservation

$$\partial^{\mu} J^{i}_{\mu\alpha}(x) = 0 = \frac{d}{dt} Q^{i}_{\alpha} = 0$$

ensures that the corresponding action

$$I = \int d^4x \, \mathscr{L}(x) \tag{1.5}$$

is invariant under supersymmetry transformations

$$\delta_{\varepsilon} I = \int d^4x \delta_{\varepsilon} \mathscr{L} = 0, \qquad \delta_{\varepsilon} \mathscr{L} = [\bar{\varepsilon}^{i} Q^{i}, \mathscr{L}] \ . \tag{1.6}$$

We note that in order to define $\delta_{\varepsilon} \mathscr{L} = (\delta\mathscr{L}/\delta\Phi)\delta_{\varepsilon}\Phi$ anticommuting spinorial parameters ε^{i}_{α} having the same transformation properties of Q^{i}_{α} must be introduced. These parameters are odd elements of a Grassman algebra. They anticommute with any fermionic quantity while they commute with ordinary c-numbers as well as with bosonic operators.

We now give some motivations for the introduction of supersymmetry in local quantum field theory. There are indeed several reasons and the one alluded to at the beginning of this section, namely the unification programme, is just one of them.

1. Supersymmetry is the only known symmetry consistent with relativistic QFT which relates particles with different spin. This is a major breakthrough because it gives a way to overcome previous no-go theorems on the possibility of non-trivial mixing between space-time and internal symmetries.

2. When supersymmetry is realized as a global (rigid) symmetry, supersymmetric theories are the less divergent QFTs known today. For instance the super Yukawa theory, which is a particular combination of renormalizable couplings among scalars and spin-1/2 fermions, needs only one (logarithmically divergent) renormalization constant. Other remarkable examples are the supersymmetric Yang–Mills theories: the maximally extended N = 4 theory has a vanishing $\beta(g)$ function at the first three loops and it may lead to a completely finite (calculable) theory.

3. At the local level gauge supersymmetry requires gravity. It gives a new type of extension of the Einstein theory in which a subtle interplay between the quantum mechanical concept of spin and of space-time geometry occurs. General co-ordinate invariance is not the primary (local) symmetry principle: co-ordinate transformations are obtained by merely repeating twice a local supersymmetry transformation.

4. Supergravity theories have better quantum properties than Einstein theory (van Nieuwenhuizen and Grisaru 1977). They offer examples of model field theories in which interaction of matter receives finite gravitational radiative corrections in the first two orders of perturbation theory. This is a major advance with respect to all previous unsuccessful attempts in ordinary Einstein theory.

5. Supergravity and especially extended supergravities offer possible schemes for superunification (Gell-Mann 1977). Yang–Mills and Einstein Lagrangians are linked under the invariance principle of local supersymmetry.

6. Supergravity theories provide the unification of interactions of gauge particle of different spin like the graviton, Rarita–Schwinger spin-3/2 fields (gravitinos) and Yang–Mills vector bosons. Gauge particles of half-integral spin therefore emerge for the first time. As a by-product supergravity provides the first example of a consistent interaction for Rarita–Schwinger fields. In extended supergravity spin 0 and 1/2 matter fields are unified with gauge fields in a single irreducible multiplet. Matter fields and geometrical fields are therefore unified through the local supersymmetry principle.

2. MULTIPLET STRUCTURE

Supersymmetric theories and in particular supergravity theories are described in terms of field supermultiplets. These are collections of ordinary fields with different spins, statistics, and internal symmetry properties. In this section we will consider the multiplet structure of one particle states which are supposed to be described by the asymptotic fields of supersymmetric quantum field theories.

We consider the representation of the supersymmetry algebra

given by (1.1), (1.2), (1.3) on one-particle states. The construction of particle supermultiplets uses the Wigner method of induced representations (Salam and Strathdee 1974*a*, 1975; Nahm 1978; Freedman 1978). We will temporarily consider the supersymmetry algebra without the central charge operators Z^{ij}, Z'^{ij}. Because these operators belong to the centre of the algebra it is consistent to set them equal to zero.

We consider first the stability subalgebra of a time-like momentum $P^\mu = (M,\vec{O})$. M is the common mass of the different spin states of a massive multiplet. If we use two component Weyl spinors the stability subalgebra is

$$\left\{Q^i_\alpha, \bar{Q}_{\beta j}\right\} = \delta^i_j \delta_{\alpha\beta} \qquad \alpha,\beta = 1,2 \qquad i,j = 1 \ldots N \tag{2.1}$$

$$\left\{Q^i_\alpha, Q^j_\beta\right\} = 0 \tag{2.2}$$

It is nothing but the Clifford algebra for $2N$ creation and destruction fermionic operators.

If $M = 0$ we may choose $P^\mu = (1,0,0,-1)$ and we have

$$\left\{Q^i_2, \bar{Q}_{2j}\right\} = \delta^i_j \, , \tag{2.3}$$

$$\left\{Q^i_1, \bar{Q}^j_1\right\} = 0 \, , \tag{2.4}$$

$$\left\{Q^i_\alpha, Q^j_\beta\right\} = 0 \, . \tag{2.5}$$

From (2.4), (2.5) we can put $Q^i_1 = 0$, $Q^i_2 = Q^i$ and then we obtain the Clifford algebra for N creation and destruction operators. If we define the Clifford vacuum Ω as the one-particle state having the property

$$Q^i_\alpha \Omega = 0 \tag{2.6}$$

then we obtain irreducible representations of the supersymmetry algebra of dimension $2^{2N} \times d_\Omega$ for $M \neq 0$ and $2^N \times d_\Omega$ for $M = 0$. d_Ω is the dimensionality of the representation of the Clifford vacuum. This is a representation of the even part of the stability supersymmetry algebra. The smallest representation (fundamental representation) is obtained when Ω is a singlet

($d_\Omega = 1$). The generic state of the fundamental representation is obtained by repeatedly applying the creation operators Q^i_α to the vacuum state Ω.

$$\Omega,\ \bar{Q}^i_\alpha \Omega,\ \bar{Q}^i_\alpha \bar{Q}^j_\beta \Omega,\ \ldots \tag{2.7}$$

For M $\neq$ 0 we get states

$$\bar{Q}^{i_1}_{\alpha_{i_1}} \ldots\ldots \bar{Q}^{i_{N-n}}_{\alpha_{i_{N-n}}} \Omega \tag{2.8}$$

The state of given spin $s = \frac{N-n}{2}$ $(0 \leqslant n \leqslant N)$ belongs to the n-fold (traceless) antisymmetric representation of Sp($2N$) (Ferrara and Zumino 1979a): $[2N \times \ldots 2N]_n$, $2N$ is the defining vector representation of Sp($2N$). The highest spin state is a singlet with spin $S = N/2$. If Ω is not a singlet but has spin J and belongs to a given representation R then a generic state will belong to a reducible representation of SU(N) given by $[2N \times \ldots 2N]_{N-n}\, x\, R$ and it will contain several spin states from $\frac{N-n}{2} + J$ down to $\frac{N-n}{2} - J$. Irreducible representations of the supersymmetry algebra which are not the fundamental one have therefore the property that a state of given spin S lies in a reducible representation of Sp($2N$). For $M = 0$ the generic state

$$\bar{Q}^{i_1} \ldots\ldots \bar{Q}^{i_n}\, \Omega \tag{2.9}$$

($\bar{Q}^{i_1}$ is a lowering helicity operator) belongs to the N-fold antisymmetric representation of SO(N) if Ω is a singlet. SO(N) can be enlarged to SU(N) if we allow the internal symmetry part not to commute with parity. In this case $\bar{Q}^i$ belongs to the $\bar{N}$ representation of SU(N). If the Clifford vacuum Ω_λ has helicity λ the state (2.9) has helicity $\lambda - n/2$. If we require a given massless representation to be TCP self-conjugate we must add to the states given by (2.9) the states

$$\Omega_{-\lambda},\ Q_{i_1}\Omega_{-\lambda},\ Q_{i_1}Q_{i_2}\Omega_{-\lambda},\ \ldots \tag{2.10}$$

or equivalently the states

$$\Omega_{\frac{N}{2}-\lambda}\,,\quad \bar{Q}^{i_1}\Omega_{\frac{N}{2}-\lambda}\,,\quad \bar{Q}^{i_1}\bar{Q}^{i_1}\Omega_{\frac{N}{2}-\lambda}\quad \ldots . \tag{2.11}$$

Note that Q^i belongs to the N representation of SU(N) and $[N\times\ldots N]_n = [N\times\ldots\bar{N}]_{N-n}$. We have then that a TCP conjugate massless supermultiplet has dimension 2^{N+1} with helicity content $|\lambda|\ldots.|\lambda - N/2|$. There are however special cases in which the representation given by (2.9) is automatically PCT self-conjugate. This happens when $\lambda = \frac{N}{2} - \lambda$ i.e. $\lambda = \frac{N}{4}$. In this case the representation has dimension 2^N. Particular cases of self-conjugate multiplets are the $\lambda = 1$ gauge supermultiplet of $N = 4$ maximally extended Yang–Mills theory and the $\lambda = 2$ gauge supermultiplet of $N = 8$ maximally extended supergravity theory. We note, as far as internal symmetry properties are concerned, that SO(N) is the maximal internal symmetry of a massless supermultiplet which gives states which are parity preserving. Any SO(N) representation in a given CPT self-conjugate supermultiplet is automatically parity preserving. However if we enlarge SO(N) to SU(N), the SU(N) representations are generally not invariant under parity. A given SU(N) representation acts on a state of a given chirality. For instance if the Clifford vacuum has helicity (chirality) λ and belongs to the representation R of SU(N), the generic state given by (2.9) will have helicity (chirality) $\lambda - \frac{n}{2}$ and will belong to the representation $[\bar{N}\times\bar{N}\times\ldots\bar{N}]_n\times R$ of SU(N). It follows that a given SU(N) representation (irreducible or not) can be parity preserving if and only if it is self-conjugate. This implies that in a given supermultiplet one can get states of both chiralities only if the corresponding SU(N) representation is self-conjugate. An irreducible CTP self-conjugate multiplet is never SU(N) self-conjugate so it cannot be invariant under parity because it will contain some states of fixed chirality. However one can add several massless supermultiplets in order to obtain only SU(N) self-conjugate representations and therefore states with both chiralities. This is for instance what happens if we decompose a massive representation of N-extended supersymmetry with respect to states of given helicities. We obtain in this way a set of massless (CPT self-conjugate) multiplets which are classified according to SU(N)representations coming from the reduction $Sp(2N) \to SU(N)$.

We can give an explicit representation of the $Sp(2N)$ and of $SU(N)$ generators for massive and massless supermultiplets if we consider the enveloping algebras of the stability subalgebras given by (2.1)–(2.5).

Consider first $M \neq 0$, then the following matrices give a realization of the $N(2N+1)$ dimensional compact Lie algebra

$$S^{ij} = Q^{i}_{\alpha}\varepsilon^{\alpha\beta}Q^{j}_{\beta} \tag{2.12}$$

$$A^{ij} = \frac{1}{2}\left(Q^{i}_{\alpha}\delta^{\alpha\dot{\alpha}}\bar{Q}^{j}_{\dot{\alpha}} - \bar{Q}^{j}_{\dot{\alpha}}\delta^{\dot{\alpha}\alpha}Q^{i}_{\alpha}\right) . \tag{2.13}$$

Any $Sp(2N)$ generator can be written in the form

$$\Lambda = \begin{pmatrix} A^{ij} + A^{ji} & S^{ij} + S^{*ij} \\ S^{ij} + S^{ij} & A^{ij} - A^{ji} \end{pmatrix} + i \begin{pmatrix} A^{ij} - A^{ji} & S^{ij} - S^{*ij} \\ S^{*ij} - S^{ij} & A^{ij} - A^{ji} \end{pmatrix} \tag{2.14}$$

Equation (2.14) is in fact the most general form of an element which belongs to the $SU(2N)$ algebra and which satisfies the symplectic condition $\Lambda^{T}\Omega + \Omega\Lambda = 0$, Λ $SU(2N)$, $\Omega = \begin{pmatrix} 0 & -I \\ I & 0 \end{pmatrix}$.

For massless representations $S^{ij} = 0$, because $Q^{i}_{1} = 0$ and we have only the $U(N)$ subalgebra of $Sp(2N)$. The $SO(N)$ subalgebra of $U(N)$ is given by

$$\frac{i}{2}\left(Q^{i}_{2}\bar{Q}^{j}_{2} - Q^{j}_{2}\bar{Q}^{i}_{2}\right) . \tag{2.15}$$

The matrices S^{ij}, A^{ij} satisfy the following set of commutation relations

$$\begin{aligned} [S^{ij}, S^{*lm}] &= A^{im}\delta^{jl} + A^{jm}\delta^{il} + A^{jl}\delta^{im} + A^{il}\delta^{jm} \\ [S^{ij}, S^{lm}] &= 0 \\ [A^{lm}, S^{ij}] &= S^{lj}\delta^{im} + S^{li}\delta^{jm} \\ [A^{ij}, A^{lm}] &= \delta^{jl}A^{im} - \delta^{im}A^{lj} . \end{aligned} \tag{2.16}$$

From the previous considerations we obtain the following general result: N-extended supersymmetry massless representations contain a set of particle states whose helicity must reach at least $\frac{N}{4}$ $(\frac{N+1}{4})$. It follows that scalar–spinor multiplets can exist up to $N = 2$ and scalar–spinor–vector multiplets up to $N = 4$.

Perturbatively renormalizable QFT limits the spin content of fields to 0, 1/2, and 1. We conclude that these theories can tolerate at most $N = 4$ extended supersymmetry. However if we allow helicity-2 states, suitable to describe the graviton, as it happens in supergravity theory, we have much larger supermultiplets. Supergravities can tolerate at most $N = 8$ extended supersymmetry. We also note that the maximally extended theories, which occur at $N = 4$ for Yang–Mills interactions and at $N = 8$ for gravitational interactions correspond to PCT self-conjugate particle supermultiplets.

In Tables 2.1 and 2.2 the list of all massless multiplets which correspond to maximal helicities $\lambda = \pm 1$ and $\lambda = \pm 2$ are reported.

TABLE 1.1

Particle spectrum of extended Yang–Mills theories

The gauge group G is arbitrary and commutes with the supersymmetry generators.

N \ λ	1	$\frac{1}{2}$	0
1	1	1	
2	1	2	$2 = 1 \oplus 1$
3	1	$3 \oplus 1$	$3 \oplus 3$
4	1	4	$3 \otimes 1 \oplus 1 \otimes 3$

From Table 2.2 we see the remarkable fact that the (± 1) helicity states embedded in the gravitational multiplet occur in the adjoint representation of SO(N) as a consequence of the fact that the maximal helicity state (± 2) is a singlet. This circumstance suggests that, using the supermultiplet of the graviton, one can unify gravity with an SO(N) Yang–Mills theory whose gauge fields just correspond to the ± 1 helicity partners of the graviton. This possibility has been shown to occur (Freedman and Das 1977; Freedman and Schwartz 1978;

Fradkin and Vasiliev 1977) in extended supergravities up to $N = 4$ by explicit construction of the Lagrangian. For $N > 4$ the Lagrangians with an SO(N) gauge coupling have not been constructed so far and it remains to be seen whether this is possible. As far as the representation content is concerned exceptional cases are $N = 6$ and $N = 7$ extended supergravities. In these cases the ±1 helicity states belong to reducible representations of SO(N). For $N = 6$ the adjoint + a singlet (15+1), for $N = 7$ to the adjoint + the fundamental (21+7) representations respectively. In these special cases one expects the symmetries to be enlarged to SO(6) × SO(2) and SO(8) respectively.

TABLE 2.2

Particle spectrum of extended supergravity

N \ λ	2	$\frac{3}{2}$	1	$\frac{1}{2}$	0
1	1	1			
2	1	2	1		
3	1	3	3		
4	1	4	$3 \otimes 1 \oplus 1 \otimes 3$	4	$2 = 1 \oplus 1$
5	1	5	10	$10 \oplus 1$	$5 \oplus 5$
6	1	6	$15 \oplus 1$	$20 \oplus 6$	$15 \oplus 15$
7	1	$7 \oplus 1$	$21 \oplus 7$	$35 \oplus 21$	$35 \oplus 35$
8	1	8	28	56	$35 \oplus 35$

The unification picture which comes out by gauging SO(N) is not satisfactory with the present understanding of 'low energy' particle phenomenology (Gell-Mann 1977). If we want to identify the (±1), (±1/2) and 0 helicity partners of the gravitons with the fundamental particles of a unified theory of electromagnetic, weak and strong interactions we must recover at least $SU(3)_{COLOUR} \times (SU(2) \times U(1))_{ELECTROWEAK}$ but this is not a subgroup of SO(8), the orthogonal symmetry group of the maximally extended supergravity theory. Also if we confine ourselves to the unbroken gauge group $SU(3)_{COLOUR} \otimes U(1)_{em} \subset SO(8)$ many light particles (like the muon) are missing in the fundamental gauge multiplet (Gell-Mann 1977).

We now consider again massive representations. We have seen that the fundamental representation, in absence of central charges, has dimension 2^{2N}. The spin content runs from 0 up to $N/2$, therefore in N-extended supersymmetry with massive multiplets without central charges, one must have at least particle states with spin $N/2$ and lower.

Massive representations can be smaller in presence of central charges (Haag, Lopuszanski, and Sohnius 1975; Fayet 1979). We give a few examples. In $N = 2n$ extended supersymmetry, if one central charge does vanish, one can get massive multiplets of dimension 2^{2n+1} (instead of 2^{4n}), whose maximum spin is $n/2$ (instead of n). These 2^{2n+1} states are a doublet of massive representations of n-extended supersymmetry without central charges. The internal symmetry breaks from $Sp(4n)$ down to $Z \times Sp(2n)$. Z is the central charge operator which rotates the two real multiplets of maximum spin $S = n/2$. In Tables (2.3) and (2.4) we report some massive representations of extended supersymmetry without and with central charge respectively.

Massive representations of extended supersymmetry with non-vanishing central charges occur in the $N = 8$ broken supergravity model through dimensional reduction (Scherk and Schwartz 1979; Cremmer, Scherk, and Schwartz 1979*a*). From this model one gets massive multiplets with central charge for $N = 6, 4, 2$ extended supersymmetry (Ferrara and Zumino 1979*a*). Other examples for $N = 4,2$ are given by Yang–Mills theories with a spontaneous breakdown of the Yang–Mills group (Fayet 1979).

TABLE 2.3

Some massive representations of N-extended supersymmetry (without central charges)

$N \backslash J$	$\frac{5}{2}$	2	$\frac{3}{2}$	1	$\frac{1}{2}$	0
					1	2
				1	2	1
1			1	2	1	
		1	2	1		
				1	4	5
2			1	4	$5 \oplus 1$	4
		1	4	$5 \oplus 1$	4	1
			1	6	14	14
3		1	6	$14 \oplus 1$	$14 \oplus 6$	14
4		1	8	27	48	42
5	1	10	44	110	165	132

TABLE 2.4

Some massive representations of N-extended supersymmetry (with central charges). Complex representations

The central charge acts as a phase transformation of these complex representations.

$N \backslash J$	$\frac{3}{2}$	1	$\frac{1}{2}$	0
			1	2
2		1	2	1
	1	2	1	
		1	4	5
4	1	4	$5 \oplus 1$	4
6	1	6	14	14

3. SPONTANEOUS SUPERSYMMETRY BREAKING: THE SUPER-HIGGS EFFECT

The basic feature of supersymmetric field theories lies in the possibility of having interacting multiplets unifying particles of different spin and internal quantum numbers. This remarkable property has its counterpart in the fact that all particles in a given multiplet are degenerate in mass. This is a bad property because such a degeneracy has not been observed in nature. If supersymmetry is relevant for the physical world it must have two properties:

(a) It must be a broken symmetry;

(b) It must be realized as a gauge symmetry.

The symmetry breaking can be either spontaneous or explicit but renormalizability and simplicity arguments favour a spontaneous symmetry breaking. In fact non-supersymmetric interactions would probably spoil good renormalizability properties and lose all predictive power of the symmetry. The requirement (b) is due to the particular role which gravity plays as gauge theory of the Poincaré group. The supersymmetry algebra given by (1.1), (1.2), (1.3) implies that if we gauge the Poincaré group then also supersymmetry must be gauged. In fact gravity + global supersymmetry imply local supersymmetry and vice versa local supersymmetry implies gravity. From the previous arguments we can conclude that only spontaneously broken realization of supergravity may have the chance of being candidates for the superunification of the fundamental interactions.

When supersymmetry is spontaneously broken, particles arise in the spectrum of the theory which are massless and carry the same quantum numbers of the broken generators Q^i_α. The signal for spontaneous breakdown of supersymmetry is a non-linear term in the transformation law of some spin 1/2 fields (Volkov and Akulov 1973*a*; Iliopoulos and Zumino 1974; Salam and Strathdee 1974*b*; Fayet and Iliopoulos 1974; O'Raifeartaigh 1975).

$$\delta\lambda^i_\alpha(x) = a\varepsilon^i_\alpha + q\text{-number terms.} \qquad (3.1)$$

The parameter a which has the dimension of (mass)² is related to the size of spontaneous symmetry breaking. In model field theories the parameter a is proportional to the vacuum expectation value of some auxiliary field H of the basic field representation of the supersymmetry algebra

$$\langle 0|H|0\rangle \sim a\,. \tag{3.2}$$

The spin -1/2 particle fields which transform as in (3.1) are the Goldstone fermions (Goldstinos) of the theory. These massless particles can be identified with new types of neutrinos or are eventually eaten up by the spin 3/2 gauge particles (gravitinos) of local supersymmetry. It is evident from (3.1) that if $\varepsilon^{i}_{\alpha}(x)$ is space-time dependent we can perform a supersymmetry transformation such that $\lambda^{i}_{\alpha}(x) = 0$. The lower ±1/2 helicity states of the massive gravitinos are nothing but the would be Goldstone fermions of spontaneously broken supersymmetry. This mass generation for spin 3/2 particles is called the super-Higgs mechanism (Volkov and Sorakova 1973; Deser and Zumino 1977) and is entirely analogous to the more conventional Higgs–Kibble mechanism for vector particles in spontaneously broken Yang–Mills theories. Because Rarita–Schwinger fields can have consistent interactions only through gravitons we encounter the remarkable situation that the Higgs mechanism for spin 3/2 particles can only occur in curved space-time. Due to this fact it is important to have the possibility that the spontaneous breakdown does not induce a cosmological constant which would affect the particle interpretation of the physical states. It will turn out that the constraint of zero cosmological term in broken supergravity is a quite strong one and it will have additional consequences on the properties of the broken theory.

If the gravitino is a physical particle one may wonder what is the magnitude of its mass. This is controversial. If spontaneous supersymmetry breaking takes place in the energy range of hadron interactions then the mass of the gravitino is very small (Deser and Zumino 1977)

$$m_{\psi} \sim Km^{2}_{\text{proton}} \sim 10^{-19}\ \text{GeV} \tag{3.3}$$

This is also the case in the supersymmetric version of electroweak interactions where Fayet (1977, 1980) found

$$m_\psi \sim \frac{KM_w^2}{e} \sim \frac{Ke}{G_F} \sim 10^{-5} - 10^{-6} \text{ eV} . \tag{3.4}$$

A supersymmetry breakdown at scales of electroweak interactions would imply the existence of new families of hadrons (R-hadrons) associated to a new (R) quantum number in the GeV-mass range. More importantly gravitational amplitudes involving massive gravitinos would be of strength comparable to electroweak amplitudes involving the would be Goldstone fermions and they could give detectable effects at moderate energies.

The alternative picture is that supersymmetry breaking occurs at the Planck scale. This would in particular explain why there is no trace of a supersymmetry pattern at present energies. In this case exact supersymmetry would manifest at the superunification scale and in the gauge hierarchy it would be prior to the grand unification scale which is believed to be in the 10^{14}-10^{15} GeV range. In such a situation the gravitino, if not confined, would conceivably have a mass $m_\psi \gtrsim 10^{19}$ GeV. An intermediate possibility, which may occur in extended supergravity, is that the different supersymmetry generators Q^i_α are broken at different scales. In particular some of them at superhigh energies ($10^{15} \sim 10^{19}$ GeV range) and some at 'low energies' ($\sim 10^2$ GeV).

Recently it has been suggested (Cremmer and Julia 1978; Ellis *et al.* 1980) that the elementary fields of $N = 8$ extended supergravity, with the exception of the graviton, may be preconstituent fields of a superunified theory. In particular the gravitino would be superconfined and would not correspond to any physical state. We will comment on this possibility in the last section of this review.

We now return to the question of spontaneous breaking and super-Higgs effect in supergravity. Using a non-linear realization of supersymmetry (Volkov and Akulov 1973), which is suitable to describe Goldstone fermions (Volkov and Akulov 1973; Iliopoulos and Zumino 1974; Salam and Strathdee 1974; Fayet and Iliopoulos 1974; O'Raifeartaigh 1975), it has been shown, to lowest order in the gravitational constant K, that the spin 3/2 gravitino can become massive with vanishing cosmological term. Moreover the gravitino mass m_ψ is related to the symmetry breaking parameter a (see eqn. (3.1) by the following universal relation (Deser and Zumino 1977)

$$m_\psi = \frac{1}{\sqrt{6}} Ka \tag{3.5}$$

In the class of models considered in (Fayet 1977, 1980) one has

$$a = \frac{\Delta m^2}{e_G} \tag{3.6}$$

where Δm^2 is the fermion-boson (mass)^2 splitting of a given multiplet and e_G is the Yukawa or vector coupling of the Goldstone spinor λ.

The super-Higgs effect and the general validity of eqn. (3.5) has been explicitly exhibited (Cremmer, Julia, Scherk, Ferrara, Girardello, and van Nieuwenhuizen 1978, 1979) to all order in K in the minimal $N = 1$ supergravity model which describes the interaction of the gauge multiplet of helicity content (± 2, $\pm 3/2$) with the (super) Higgs multiplet of spin content ($1/2$, 0^+, 0^-). The most general interaction is described by a real function of two variables ($A(0^+)$,$B(0^-)$). If a canonical energy term for the scalar fields is demanded, then the degree of arbitrariness is reduced to a function of the complex variable $A + iB$. After spontaneous breakdown of supersymmetry and super-Higgs effect (without induced cosmological constant) the spectrum consists of a massless graviton (± 2 helicity states), a massive gravitino and two massive super-Higgs particles. Moreover the following general mass formula holds

$$4m_\psi^2 = m_A^2 + m_B^2 . \tag{3.7}$$

It is interesting to note that eqn. (3.7) ensures that the one-loop induced cosmological term by radiative corrections is at most logarithmically divergent, in contrast with non-supersymmetric field theories where it is quartically divergent. We recall that the one-loop induced cosmological term for a particle of spin J and mass M_J is given, according to De Witt (1965)

$$G = -\frac{(-)^{2J}(2J+1)}{8\pi^2}\left[\frac{1}{\sigma^2} - \frac{M_J^2}{2\sigma} - \frac{M_J^4}{4}\left(\frac{1}{2}\lg|2M_J^2\sigma| + \gamma - \lg 2 - \frac{5}{4}\right)\right] \quad (3.8)$$

where $\Lambda = 1/\sigma$ is an ultraviolet cutoff. The previous mass formula (eqn. (3.7)) is a particular mass of a more general mass formula which holds in a large class of supersymmetric theories (Ferrara, Girardello, and Palumbo 1979)

$$\sum_J (-)^{2J}(2J + 1)M_J^2 d_J = 0 \quad (3.9)$$

d_J is the degeneracy of the state of mass M_J and spin J. Equation (3.9) is the statement that the graded trace of the square mass matrix vanishes

$$\mathrm{GrTr}M^2 = 0 \quad (3.10)$$

It is interesting to note that in spontaneously broken (renormalizable) globally supersymmetric models eqn. (3.9) has been shown to be preserved by radiative corrections up to finite terms of second order in the symmetry breaking parameter (Girardello and Iliopoulos 1979).

The question of an induced cosmological constant is a standard problem in ordinary gauge theories and especially in Grand Unified Theories. These theories would predict a huge cosmological constant in contrast with experimental evidence. One could cure this by adding to the Lagrangian a compensating cosmological term which however must be tuned with a fantastically high accuracy. To our knowledge supersymmetry is the only principle which implies (Zumino 1975)

$$\langle 0|T_{\mu\nu}(x)|0\rangle = 0 \quad (3.11)$$

provided supersymmetry is unbroken. The very simple explanation of (3.11) is that $T_{\mu\nu}(x)$ belongs to a supermultiplet which contains the supersymmetry current $J^i_{\mu\alpha}(x)$ and under a supersymmetry transformation

$$\delta J^i_{\mu\alpha}(x) \sim \delta^{ij}(\bar{\varepsilon}\gamma^\nu)_\alpha T_{\mu\nu}(x) + q\text{-number terms} \quad (3.12)$$

by taking the V.E.V. of (3.12) and using the fact that $Q^i_\alpha |0\rangle = 0$ one gets (3.11). It is more important that eqn. (3.11) is preserved by the spontaneous breakdown of supersymmetry. This is precisely what happens in the 'minimal' model previously considered with one super-Higgs multiplet. It must be pointed out that the vanishing of the cosmological term in a spontaneously broken theory is not automatic but one has to select some potential which has this property. The mass relations (3.7) and (3.9) are automatically fulfilled once this choice has been made.

The super-Higgs mechanism considered in this section is the simplest case of spontaneously broken supergravity model. This mechanism can probably be enlarged to more complicated (and interesting) situations like $N = 1$ supergravity with Yang–Mills multiplets or $N = 2$, $N = 4$ supergravity coupled to matter multiplets. Work along these lines has not been done yet but is strongly recommended.

4. ALTERNATIVE WAYS OF SPONTANEOUS BREAKDOWN IN SUPERGRAVITY

The spontaneous supersymmetry breaking and super-Higgs effect discussed in the previous section is the natural generalization of the standard Higgs mechanism of spontaneously broken Yang–Mills theories. It takes place because the scalar potential has minima which correspond to non-supersymmetric solutions of the equations of motion. The field which takes a non-vanishing expectation value is one of the auxiliary fields of the theory and it provides the symmetry breaking parameter which is in turn related to the gravitino mass.

Two other types of spontaneous supersymmetry breaking can occur in extended supergravity models. The first one uses generalized dimensional reduction to get masses in lower dimensions (Scherk and Schwartz 1979). The second one (Ellis 1980) advocates a dynamical mechanism responsible for the supersymmetry breakdown and it is based on the assumption that (extended) supergravity theories make sense as quantum field theories. The latter is a bold hypothesis because supergravity theories fall in the class of (power-counting) non-renormalizabl field theories.

Let us first consider the case of supersymmetry breaking through dimensional reduction (Scherk and Schwartz 1979). This method uses the fact that if a Lagrangian has a global compact symmetry G in $D+E$ dimensions, this symmetry can be used to obtain masses in D dimensions by requiring that the $D+E$ dimensional fields depend on the extra E coordinates through an element of the group G. If the global symmetry is non-compact then G is a maximal compact subgroup. If the theory has general covariance in $D+E$ dimensions, the dimensionally reduced theory (in D dimensions) has a mass matrix which depends at least on a number of parameters which equal the rank of $G \times SO(E)$, $SO(E)$ being the maximal compact subgroup of $SL(E,R)$. Dimensional reduction from $D+E$ to D dimensions in a theory invariant under general coordinate transformations induces in addition a minimal (broken) gauge symmetry whose global Lie algebra has dimension E. The E vector fields gauging this internal symmetry are the vector fields coming from the gravitational field in $D+E$ dimensions. In ordinary dimensional reduction (without dependence of fields on the extra E coordinates) this gauge group is unbroken and reduces to $U(1)^E$. In the case of generalized dimensional reduction this group is non-trivial and if one demands that no cosmological term is induced in the 4-dimensional theory, one obtains a 'flat group' according to Scherk and Schwarz (1979). We observe that this group always contains an unbroken U(1) subgroup whose gauge field has been called by Scherk (1979) graviphoton. In extended supergravity the flat group has dimensions higher than E because the theory has additional (abelian) gauge symmetries in $D+E$ dimensions. For instance in $D = 4$, $N = 8$ spontaneously broken supergravity the flat group has dimensions 28 and in lower N-extended models it has dimension $N(N-1)/2$. Spontaneously broken $N = 8$ supergravity in four-dimensions can be obtained through generalized dimensional reduction of $N = 8$ extended supergravity in $D = 5$ dimensions (Cremmer *et al.* 1979*a*).

In five dimensions the supersymmetry algebra has a global $Sp(N)$ symmetry (Cremmer *et al.* 1979*a*).

$$\left\{\bar{Q}^{a}_{\alpha}, Q^{b}_{\beta}\right\} = \Omega^{ab}\gamma^{\mu}_{\beta\alpha}P_{\mu} \qquad \begin{array}{r} a,b = 1 \ldots N \\ \alpha,\beta = 1 \ldots 4 \\ \mu = 0 \ldots 4 \,. \end{array} \tag{4.1}$$

N has to be even because the charges satisfy a generalized Majorana condition

$$Q^{a}_{\alpha} = C\bar{Q}^{Ta}_{\alpha}, \quad C = \gamma_0\gamma_5, \quad \bar{Q}^{a}_{\alpha} = (Q^{+}_{a}\gamma_0)_{\alpha} \,,$$
$$Q_{a} = \Omega_{ab}Q^{b}, \quad \Omega_{ab} = -\,\Omega_{ba} \,. \tag{4.2}$$

Massless five-dimensional multiplets of $N = 8$ supersymmetry have the same spin content as massive four-dimensional multiplets of $N = 4$ supersymmetry (without central charges). This is because the little group (SO(3)) of a massless particle of given momentum in $D = 5$ dimensions is the same as the little group of a massive particle in $D = 4$ dimensions.

The particle fields of the supergravity multiplet occur in antisymmetric traceless representations of Sp(N). For $N = 8$ we have a singlet graviton, an octet of spin 3/2, a 27-plet of spin 1, a 48-plet of spin 1/2 and a 42-plet of spin 0.

$$e_{\mu a}, \quad \psi^{a}_{\mu\alpha}, \quad A^{ab}_{\mu}, \quad \chi^{abc}_{\alpha}, \quad \Phi^{abcd}. \tag{4.3}$$

In the full non-linear theory this Sp(8) is the diagonal subgroup of $\mathrm{Sp}(8)_{\mathrm{global}} \otimes \mathrm{Sp}(8)_{\mathrm{local}} \subset \mathrm{E}_{6\,\mathrm{global}} \otimes \mathrm{Sp}(8)_{\mathrm{local}}$, the last being the full group of invariance of the five-dimensional Lagrangian.

The mass matrix of the spontaneously broken four-dimensional theory is obtained by introducing x_5-coordinate dependence on the five-dimensional fields in the form (Scherk and Schwarz 1979; Cremmer *et al.* 1979*a*)

$$\Phi_{J}(x,x_5) = \exp(\mathrm{i}\,\mathcal{M}_{J}x_5)\Phi_{J}(x) \tag{4.4}$$

$\mathcal{M}_J$ being a representative of the Cartan subalgebra of Sp(8) in the representation appropriate to the particle field of spin J. The generic element can be written as follows

$$\mathcal{M} = \sum_{k=1}^{4} m_{k}\lambda_{m} \tag{4.5}$$

where λ_k is a basis in the (4-dimensional) Sp(8) Cartan subalgebra. The mass spectrum of the broken theory depends therefore on four arbitrary (real) parameters. They equal the rank of the global symmetry group G = Sp(8). The mass spectrum is given in Table 4.1. We list some properties of the $N = 8$ broken theory:

(a) The mass spectrum satisfies three mass relations (for all values of m_k)

$$\sum_J (-)^{2J}(2J+1)M_J^{2r} = \mathrm{GrTr}\mathscr{M}^{2r} = 0 \qquad r = 0,1,2,3 \tag{4.6}$$

In general the number of these mass relations depends on the rank of G. These mass relations are an extension, for $r > 1$, of the mass formula (3.9).

(b) The theory has an unbroken U(1) gauge group whose gauge field is the (4μ) component of the five dimensional graviton. This field (graviphoton) is minimally coupled to all massive particles with charge coupling

$$Q_i = \pm\ 2M_iK \tag{4.7}$$

(c) If n parameters out of the m_ks vanish the theory has an unbroken $N = 2n$ local supersymmetry and massive multiplets have non-vanishing central charge (Ferrara and Zumino 1979). These multiplets contain charged particles which lie in representations of Sp($2n$).

(d) The introduction of the parameter m_k in the broken theory breaks the SO(8) global symmetry of the $N = 8$ unbroken theory down to SO(2) ~ U(1). However if h of the m_ks ($h \leqslant 4$) are equal, the theory has an additional SU(h) global symmetry and the 28-dimensional flat group contains at least an unbroken $U(1)^{h^2}$ factor.

We now make some comments on the properties listed above. Property (a) ensures that, according to (3.8) the one-loop induced cosmological term is finite for $N = 8$ and $N = 6$ broken supergravity. The vector field minimally coupled to massive particles is the gauge field of the central charge operator which appears in the supersymmetry algebra for N = 2, 4 and 6. This explains eqn. (4.7). The property alluded to in (c) and the symplectic structure of $\mathscr{M}$ are responsible for the mass relations given in (a). The quantity $\mathrm{GrTr}\mathscr{M}^{2r}$ is in fact an homogeneous symmetric polynomial of order $2r$ in the m_k variables. $\mathrm{GrTr}\,\mathscr{M}^{2r} = 0$ for all r if at least one of the m_ks vanishes. Then it follows that $\mathrm{GrTr}\,\mathscr{M}^{2r} = 0$ identically for $r = 0 \ldots n - 1$ for an $\mathrm{Sp}(2n)$ group. As a by-product any truncation of $N - 8$ supergravity with $2n < 8$ supersymmetries will fulfill $n - 1$ mass relations.

TABLE 4.1

Mass spectrum of the broken $N \frac{3}{8} 8$ theory

Spin	Mass	Degeneracy	Number of states
2	0	1	1
$\frac{3}{2}$	$\lvert m_i \rvert$	2	32
	0	4	8
1	$\lvert m_i \pm m_j \rvert \quad i < j$	2	72
	$\lvert m_i \rvert$	4	32
$\frac{1}{2}$	$\lvert m_i \pm m_j \pm m_k \rvert \quad i < j < k$	2	64
	0	6	6
0	$\lvert m_i \pm m_j \rvert \quad i < j$	2	24
	$\lvert m_1 \pm m_2 \pm m_3 \pm m_4 \rvert$	2	16

In the final part of this report we would like to mention the possibility of a dynamical spontaneous symmetry breaking in extended supergravity. This situation has been recently envisaged by Ellis, Gaillard, Maiani, and Zumino (1980) and Ellis, Gaillard, and Zumino (1980) following a previous result obtained by Cremmer and Julia (1978) for the unbroken N-extended supergravity models.

Cremmer and Julia have pointed out that in N-extended supergravities there are hidden internal symmetries which do not commute with the supersymmetry charges. These symmetries are always of the form $G_{\text{global}} \otimes G_{\text{local}}$ where G_{global} is non-compact and G_{local} is compact. Moreover $G_{\text{global}} \supset G = G_{\text{local}}$ as its maximal compact subgroup. For example in $N = 8$ supergravity in $D = 5$ dimensions $G_{\text{global}} = E_6$, $G_{\text{local}} = \text{Sp}(8)$. In $N = 8$ supergravity in $D = 4$ dimensions $G_{\text{global}} = E_7$, $G_{\text{local}} = \text{SU}(8)$. In particular the scalar fields always belong to the coset space G_{global}/G. Note that for $N < 4$ $G_{\text{global}} = G$. In $N = 8$, $D = 4$ supergravity the scalar fields can be represented by a 56 x 56 matrix, a group element of E_7, of the form

$$V = \begin{pmatrix} U_{[AB]}{}^{[MN]} & V_{[AB][MN]} \\ \bar{V}^{[AB][MN]} & \bar{U}^{[AB]}{}_{[MN]} \end{pmatrix} \tag{4.8}$$

where square brackets mean antisymmetrization of indices and $(A,B) = 1...8$, $(M,N) = 1...8$ transform under SU(8) and E_7 respectively. From (4.8) one can construct an E_7 invariant composite vector field

$$\partial_\mu V V^{-1} = \begin{pmatrix} Q_{\mu[A}{}^{[\mathbb{C}}\delta_{B]}{}^{D]} & P_{\mu[ABCD]} \\ \bar{P}_\mu{}^{[ABCD]} & \bar{Q}_{\mu[A}{}^{[C}\delta_{B]}{}^{D]} \end{pmatrix} \tag{4.9}$$

where $Q_{\mu A}{}^B$ and $P_{\mu[ABCD]}$ transform according to the self-conjugate (adjoint) 63 and 70 dimensional representations of SU(8). $Q_{\mu A}{}^B$ is a connection for SU(8) so that the Lagrangian for the scalar fields can be written

$$(D_\mu V V^{-1})^2 \sim P_{\mu[ABCD]}\bar{P}^{\mu[ABCD]} \tag{4.10}$$

and contains only 70 degrees of freedom. From (4.10) it is evident that the scalar sector of the theory has a σ-model structure and the theory has an SU(8) local invariance. This is analogous to the CP^{N-1} models in $D = 2$ dimensions when $G_{\text{global}} = \text{SU}(N)$ and $G_{\text{local}} = \text{SU}(N-1) \otimes \text{U}(1)$.

Cremmer and Julia conjectured in analogy to CP^{N-1} models that the composite operator $Q_{\mu A}{}^B$ may become dynamical because of quantum effects. In CP^{N-1} models this indeed happens

(D'Adda, Di Vecchia, and Lüscher 1978, 1979; Witten 1979) as a non-perturbative phenomenon and it can be studied using the $1/N$ expansion thanks to the arbitrariness of the SU(N) symmetry of these models. Moreover the renormalizability of CP^{N-1} models is crucial. In extended supergravity none of these two points are fulfilled but one may take the optimistic point of view that $N = 8$ supergravity makes sense as a quantum theory and then one might explore the consequences of such an hypothesis.

In a supersymmetric field theory the SU(8) gauge connection $Q_{\mu A}{}^{B}$ must belong to some $N = 8$ supermultiplet. In particular if $Q_{\mu A}{}^{B}$ becomes a propagating field and its propagator develops a pole at zero mass one should get quantum excitations which fall in a $N = 8$ massless supermultiplet which contains helicity ± 1 states in the adjoint representation of SU(8). This is under the assumption that the quantum effect responsible for these new particle states is prior to supersymmetry breaking. Ellis *et al.* (1980) conjecture that this composite supermultiplet contains the standard gauge fields of low energy physics, the photon, coloured gluons, the $W^{\pm}, Z$ vector bosons of weak interactions as well as the gauge fields of the GUT and additional gauge fields associated with the generation group. They also conjecture that the same supermultiplet must contain all leptons and quarks of electroweak and strong interactions.

The fundamental multiplet of the $N = 8$ supergravity Lagrangian is assumed to be a multiplet of preconstituent fields (preons) with the exception of the ± 2 helicity states which correspond to the graviton. Therefore the interpretation of the basic multiplet of extended supergravity is entirely different from the models considered in the previous section and the spin 3/2 gravitinos as well as the other elementary fields are supposed to be superconfined at least for energies up to the Planck mass 10^{19} GeV.

The composite CPT self-conjugate multiplet which contains the adjoint representation of SU(8) is assumed to be that of (4.11) (Ellis *et al.* 1980). Ellis, Gaillard, and Zumino (1980) show that extended supergravities with $N \leqslant 6$ have not sufficient states in an analogous supermultiplet to accommodate the observed spin-1/2 spectrum of fermions, namely three $(10 + \overline{5})$

Helicity	$\mp\frac{5}{2}$	∓ 2	$\mp\frac{3}{2}$	∓ 1	$\mp\frac{1}{2}$	∓ 0
SU(8)	8	28	$\bar{8}$	63	8	28
		36	56	1	216	420
content			168	70	$\overline{56}$	$\overline{28}$
				378	$\overline{504}$	$\overline{420}$

(4.11)

SU(5) representations of left-handed fermions. They focus the attention on the spin-1/2 and SU(5) content of (4.11) and neglect the SU(5) representations which are not contained in the SU(8) representations of the type $\bar{8} \times [8\times\ldots 8]_N - [8\times\ldots 8]_{N-1}$. Under these circumstances these authors conclude that the maximal unbroken subgroup of SU(8) below 10^{19} GeV is SU(5) and that the maximal anomaly free renormalizable gauge theory which can be constructed out of the multiplet given by (4.11) contains exactly three families of $(10 + \bar{5})$ SU(5) representations for left-handed spin-1/2 particles plus a set of self-conjugate representations

$$(45 + \overline{45})_L + 4(24)_L + 9(10 + \overline{10})_L + 3(5 + \bar{5})_L + 9(1)_L \quad (4.12)$$

which may acquire a big SU(5) invariant mass. The other unwanted states in the supermultiplet given by (4.11) are supposed to have become massive with masses $\gtrsim 10^{19}$ GeV through an as yet unspecified dynamical symmetry breaking and to be decoupled from the low energy (10^{15} GeV) renormalizable theory (GUT). This last argument (Veltman theorem) is required because the full SU(8) supermultiplet would have anomalies both at the composite and at the preon level. Moreover many unwanted high spin states ($J \geqslant 1$) could not become massive through conventional symmetry breaking due to the absence in (4.10) of the required helicity partners with the same properties under the exact gauge group $SU(3)_{colour} \otimes U(1)_{e.m.}$.

It is interesting to further remark that the requirement of chirality (left-handed spin-1/2 states without their right-handed partners) and a vector-like subgroup $SU(3)_{colour} \otimes U(1)_{e.m.}$

of the bigger group SU(8) almost uniquely favour the anomaly free subset of spin-1/2 particles given by $3(\overline{5} + 10)$ plus the self-conjugate states given by (4.12). If one considers for instance a subset of (4.11) without SU(6) (instead of SU(5)) anomalies and a vector like subgroup $SU(3)_{colour} \otimes U(1)_{e.m.}$ then the only solutions are completely vector like. Another possibility would be to go to the low energy theory through a sequential breaking which does not contain SU(5), for example: $SU(8) \supset Sp(8) \supset SU(3) \otimes SU(2) \otimes U(1)$. The Sp(8) group is of some interest in supergravity because it plays a role in attempts to formulate this theory in a consistent way off mass-shell (Cremmer, Ferrara, Stelle, and West 1980). Unfortunately this decomposition is completely vector-like and it would also imply a wrong charge assignment for quarks and leptons.

Some possible alternatives can be envisaged in the framework considered by Ellis and his co-workers. One possibility is to look for anomaly free subset of (4.11) without disregarding the SU(8) trace-representations contained in $\overline{8} \times [8 \times \ldots 8]_N$. Another possibility is to try to restore conventional symmetry breaking by considering more supermultiplets other than (4.11). It is evident that if one adds sufficiently many massless supermultiplets to (4.11) one can give in fact a (supersymmetric invariant) mass to every state by enlarging SU(8) to Sp(16) (see Section 2). This is too naive but one could contemplate intermediate situations and it is to be seen if one can give masses to the unwanted (would be) massless states in a way which is still consistent with the constraints imposed by 'low energy' phenomenology.

REFERENCES

Barbieri, R. (1980). *Proc. Europhys. Study Conf. Unification of the Fundamental Interactions*. Erice (Eds. S. Ferrara, J. Ellis, P. van Nieuwenhuizen) Plenum Press, p. 17.

Corwin, L., Ne'eman, Y., and Sternberg, S. (1975). *Rev. Mod. Phys.* **47**, 573.

Cremmer, E. (1980). *Proc. Europhys. Study Conf. Unification of the Fundamental Interactions*. Erice (Eds. S. Ferrara, J. Ellis, P. van Nieuwenhuizen) Plenum Press, p. 137.

—— and Julia, B. (1978). *Phys. Lett.* **80B**, 48.

—— —— (1979*b*). *Nucl. Phys.* **B159**, 141.

——, Scherk, J., and Schwartz, J.H. (1979*a*). *Phys. Lett.* **48B**, 83.

——, Ferrara, S., Stelle, K.S., and West, P.C. (1980). *Phys. Lett.* **94B**, 349.

Cremmer, E., Julia, B., Scherk, J., Ferrara, S., Giradello, L., and van Nieuwenhuizen, P. (1978*b*). *Phys. Lett.* **79B**, 231.
——— ——— ——— ——— ——— (1979*c*). *Nucl. Phys.* **B147**, 105.
D'Adda, A., Di Vecchia, P., and Lüscher, M. (1978). *Nucl. Phys.* **B146**, 63.
——— ——— ——— (1979). *Nucl. Phys.* **B149**, 285.
Deser, S. and Zumino, B. (1976). *Phys. Lett.* **62B**, 335.
——— ——— (1977). *Phys. Rev. Lett.* **38**, 1433.
De Witt, B.S. (1965). *Dynamical theory of groups and fields*. 231. Gordon and Breach, New York.
Ellis, J. (1979). *Proc. EPS Intern. Conf. High Energy Phys.* Vol. II, 940. Geneva.
——— (1980). *Proc. Europhys. Study Conf. Unification of the Fundamental Interactions*. Erice (Eds. S. Ferrara, S. Ellis, P. van Nieuwenhuizen) Plenum Press, p. 461.
———, Gaillard, M.K., and Zumino, B. (1980*b*). *Phys. Lett.* **94B**, 343.
——— ———, Maiani, L., and Zumino, B. (1980). *Proc. Europhys. Study Conf. Unification of the Fundamental Interactions*. Erice (Eds. S. Ferrara, J. Ellis, P. van Nieuwenhuizen) Plenum Press, p. 69.
Fayet, P. (1977). *Phys. Lett.* **70B**, 461.
——— (1979). *Nucl. Phys.* **B149**, 137.
——— (1980). *Proc. Europhys. Study Conf. Unification of the Fundamental Interactions*. Erice (Eds. S. Ferrara, J. Ellis, P. van Nieuwenhuizen) Plenum Press, p. 587.
——— and Ferrara, S. (1977). *Phys. Rep.* **32C**, 251.
——— and Iliopoulos, J. (1974). *Phys. Lett.* **51B**, 461.
Ferrara, S. and Zumino, B. (1979). *Phys. Lett.* **86B**, 279.
———, Girardello, L., and Palumbo, F. (1979). *Phys. Rev.* **D20**, 403.
———, Scherk, J., and Zumino, B. (1977). *Nucl. Phys.* **B121**, 393.
Fradkin, E. and Vasiliev, M. (1977). Lebedev Institute Preprint.
Freedman, D.Z. (1978). *Recent Development in Gravitation*. (ed. M. Levy and S. Deser). Plenum Press, New York.
——— and Das, A. (1977). *Nucl. Phys.* **B120**, 221.
——— and Schwartz, J. (1978). *Nucl. Phys.* **B137**, 333.
———, van Nieuwenhuizen, P., and Ferrara, S. (1976). *Phys. Rev.* **D13**, 3214.
Gell-Mann, M. (1977). Talk given at the Washington Meeting of the American Physical Society, Washington. (April).
Girardello, L. and Iliopoulos, J. (1979). *Phys. Lett.* **88B**, 88.
Gol'fand, Yu.A. and Likhtman, E.P. (1971). *JETP Lett.* **13**, 323.
Haag, R., Lopuszanski, J.T., and Sohnius, M. (1975). *Nucl. Phys.* **B88**, 257.
Iliopoulos, J. (1979). *Proc. EPS Intern. Conf. High Energy Phys.* Vol. I, 371. Geneva.
——— and Zumino, B. (1974). *Nucl. Phys.* **B76**, 310.
Nahm, W. (1978). *Nucl. Phys.* **B135**, 149.
Nanopoulos, D.V. (1980). Preprint CERN TH 2866. To appear in *Proc. Europhys. Study Conf. Unification of the Fundamental Interactions*. Erice.
van Nieuwenhuizen, P. and Grisaru, M.T. (1977). In *Deeper Pathways in High-Energy Physics*. (ed. A. Perlmutter and L.F. Scott).
O'Raifeartaigh, L. (1975). *Nucl. Phys.* **B96**, 331.
Salam, A. and Strathdee, J. (1974*a*). *Nucl. Phys.* **B80**, 499.
——— ——— (1974*b*). *Phys. Lett.* **49B**, 465.
——— ——— (1975). *Nucl. Phys.* **B84**, 127.
——— ——— (1978). *Fortsch. Phys.* **26**, 56.
Scherk, J. (1979). In *Proceedings of the Supergravity Workshop*. (ed. P. van Nieuwenhuizen, and D.Z. Freedman). North-Holland, Amsterdam, p. 43. *Proc. Europhys. Study Conf. Unification of the Fundamental Interactions*. Erice (Eds. S. Ferrara, J. Ellis, P. van Nieuwenhuizen) Plenum Press, p. 381.
——— and Schwartz, J.H. (1979). *Nucl. Phys.* **B153**, 61.

Volkov, D.V. and Akulov, V.P. (1973*a*). *Phys. Lett.* **46**B, 109.
—— and Soroka, V.A. (1973*b*). *Pisma Zh. exps. teor. Fiz.* **18**, 529. (*JEPT Lett.* **18**, 213.)
Wess, J. and Zumino, B. (1974). *Nucl. Phys.* **B70**, 39.
Witten, E. (1979). *Nucl. Phys.* **B149**, 285.
Zachos, K. (1978). *Phys. Lett.* **76B**, 329.
Zumino, B. (1975). *Nucl. Phys.* B89, 535.

24

REALIZING THE SUPERSYMMETRY ALGEBRA

K.S. Stelle

The Blackett Laboratory, Imperial College, London SW7

and

P.C. West

Department of Mathematics, King's College, London WC2

1. INTRODUCTION

In this lecture we wish to show how supersymmetry is realized in concrete terms. That is, we will find sets of fields and their transformations which form a representation of the supersymmetry group. In particular, we will show how the algebra associated with the supersymmetry group is realized.

Although supersymmetry is now six years old, it is still at the stage where it is the construction of supersymmetric theories which presents the central difficulties to the development of the subject. It is only once the construction of these theories is completed that one will be able to extract their physical properties. The problem stems from the difficulty in finding the representations of supersymmetry out of which theories can be constructed. Indeed, it is only after the discovery of these representations that one is able to formulate supersymmetric theories in a way which makes the supersymmetry manifest. To stress the importance of such formulations, we shall first consider their role in answering some of the interesting questions concerning supersymmetry at the present. These questions also indicate how it is hoped that supersymmetry could contribute to our understanding of the world and to the consistency of our current theories.

1. Does supersymmetry lead to the unification of gravity and quantum mechanics? It is well known that there is a considerable reduction in the number of divergencies of supersymmetric theories. Part of this reduction is due to the fact that renormalization respects the supersymmetry and so only supersymmetric counter-terms are allowed. However, there is also the

phenomenon of 'miraculous' cancellations. These are situations where, although there exist supersymmetric counter-terms, their coefficients are zero. An example of this restriction of counter-terms by supersymmetry is the case of two-loop simple supergravity where there exists no supersymmetric counter-term. However, at three-loops there does exist a supersymmetric counter-term and it would require a 'miraculous' cancellation for its coefficient to be zero. The analogous situation for the larger extended theories is unknown and it is hoped that the additional symmetry could lead to more promising results. What is required in these theories is a systematic method to construct the possible supersymmetric counter-terms as well as a technique for calculating the Feynman graphs necessary for evaluation of the coefficients of these counter-terms.

2. Does supersymmetry lead to a finite four-dimensional field theory? The maximally extended supersymmetric Yang–Mills theory has a β function which vanishes up to the two-loop order. This is an example of one of the 'miraculous' cancellations referred to above. It will be extremely interesting to know whether the β function is zero to all orders.

3. Does supersymmetry lead to a unification of all the forces of nature? The maximally extended supergravity theory contains particles of all spins from spin 0 to spin 2.

It is unknown how the fundamental particles that occur in this theory could correspond to those in nature, although it is clear that there is not a sufficient number of them to correspond to all the quarks, leptons and gauge bosons now seen. It has been suggested that the particles now observed could be accounted for by the bound states of this theory. This hope has been encouraged by the discovery that the theory possesses an SU(8) symmetry and by the further speculation that the possible bound states fall into representations of this SU(8) symmetry.

None of these three questions was voiced in the early investigations of supersymmetry and it is possible that the aspirations for supersymmetry will change further as the subject develops. However, in order to answer these questions we need to be able to calculate effectively, particularly at the quantum level.

To perform all but the simplest of such calculations requires a formulation of these theories in which the supersymmetry is manifest. There are two reasons for this. First, from a practical viewpoint, calculations performed in a non-supersymmetric way are very lengthy because one must take into account every particle of the theory separately. Calculating in a manifestly supersymmetric fashion, on the other hand, treats many particles as one entity. For example, the maximally extended supergravity theory contains particles of many different spins but a manifestly supercovariant formulation would treat them as just one entity. Second, if one does not work in a manifestly supersymmetric formalism one cannot guarantee that the quantum theory possesses the same symmetries as the classical theory. This is particularly true in the cases where the fields of the theory carry a representation of the supersymmetry group only if the classical field equations are imposed. That is, the supersymmetry is only realized on the on-shell states of the theory.

It has often been the case that theories of supersymmetry are initially constructed in such a way that the supersymmetry is only realized on the on-shell states and not on the fields out of which the Lagrangian is constructed. Unfortunately, the maximally extended supergravity and all the other supersymmetry theories with more than two supersymmetries have been formulated in just such a way. We will discuss later the one recently found exception, the off-shell formulation of the maximally extended Yang–Mills theory (Sohnius, Stelle, and West 1980*a*,*b*).

To make progress it is necessary to find formulations of these theories in which the Lagrangians are constructed out of sets of fields which form representations of the supersymmetry group. That is, the theory should carry a representation of supersymmetry regardless of whether we impose the classical field equations or not. Only then can the supersymmetry be made manifest.

The above considerations are somewhat similar to the need for manifestly Lorentz covariant theories. One can imagine the difficulties encountered in Yang–Mills theories if we regarded each component of the vector potential as distinct rather than as a four-vector. At the quantum level, the development of Lorentz-covariant Feynman rules or path integrals was necessary to be able to calculate with speed and confidence.

There are two ways in which supersymmetric theories can be formulated in a manifestly supersymmetric fashion. The first method relies upon finding sets of component fields whose transformations form a representation of the supersymmetry group. Given such sets of fields (supermultiplets) one can find, as with any other symmetry, rules for combining these supersymmetric multiplets together to form new supermultiplets, and one can find rules for constructing invariants. This method, called tensor calculus, does not make use of extra coordinates beyond those of ordinary space-time. The difficulty in this construction is the initial step, that of finding the supermultiplets. This difficulty stems from a phenomenon unique to supersymmetry. Given a particular theory in which the fields form a representation of the supersymmetry group on-shell (that is when the classical field equations of the theory hold) it is necessary to add additional fields to the on-shell fields in order to find a representation of the supersymmetry group without the use of any field equations. These additional fields are called auxiliary fields and will not propagate in the particular theory being considered, although they may propagate in other theories constructed from the same supermultiplet. The auxiliary fields are often very difficult to find and their absence represents the main stumbling block to the further development of supersymmetric theories.

The second method is to work in a space having additional fermionic dimensions, called superspace. This construction will be explained later. Although in a superspace formulation one always has a representation of the algebra, there is a difficulty corresponding to that of finding the auxiliary fields, namely the necessity of finding superspace constraints.

These two formulations are equivalent and we wish to stress that there is now a well known path from one formulation to the other. Given a theory formulated in superspace we can find the auxiliary fields in ordinary space and then the associated tensor calculus. Conversely, given a theory with its auxiliary fields and hence a tensor calculus formulation, we can, in a straightforward way, use the technique called gauge completion to find its superspace formulation. It is worth noting that although almost all theories were first found in their auxiliary field formulation, it is the superspace formulation which allows the easy calculation of quantum processes.

In this article, we will first discuss the representations of supersymmetry which are relatively simple to find, namely those for the on-shell states of the theory. We will then discuss in detail the two types of manifestly supersymmetric formalism in the context of the simplest model of supersymmetry (the Wess—Zumino model) and also in the context of simple supergravity. Finally, we will discuss the methods of extending the on-shell representations on states to the off-shell representations on fields. We begin by describing the supersymmetry group on which all discussions are based.

2. THE SUPERSYMMETRY GROUP

This group occupies a unique place in theoretical physics: it is the only known group that combines space-time and internal symmetries in a Lorentz-covariant way. That is, it is the only known non-trivial extension of the Poincaré group to include internal symmetries. It has been shown that no Lie group could achieve this result (Coleman and Mandula 1967). The algebra associated with the supersymmetry group has anti-commutators as well as commutators (Gol'fand and Likhtman 1971; Volkov and Akulov 1973; Wess and Zumino 1974*a*). It is a graded Lie algebra.

The generators of this algebra are the translations P_μ and the Lorentz rotations $J_{\mu\nu}$ of the Poincaré group, and in addition N spinorial charges $Q^i_\alpha, Q_{\dot\beta j}$ ($i,j = 1 \to N$) plus the central charges Z^e and the internal symmetry generators T_a. Here we use two-component spinor notation. The graded Lie algebra is, in addition to the Poincare algebra,

$$\{Q_\alpha{}^i, Q_{\dot\beta j}\} = 2\delta^i_j (\sigma^\mu)_{\alpha\dot\beta} P_\mu \tag{2.1}$$

$$\{Q_\alpha{}^i, Q_\beta{}^j\} = 2\varepsilon_{\alpha\beta}(a^e)^{ij} Z_e$$

$$\{Q_{\dot\alpha i}, Q_{\dot\beta j}\} = 2\varepsilon_{\dot\alpha\dot\beta}(a^{e}*)_{ij} Z_e\ . \tag{2.2}$$

$$[Q_\alpha{}^i, P_\mu] = 0 = [Q_{\dot\beta j}, P_\mu] \tag{2.3}$$

$$[Q_{\alpha i}, J_{\mu\nu}] = \frac{i}{2}(\sigma_{\mu\nu})_\alpha{}^\beta Q_{\beta i} \tag{2.4}$$

$$[T_a, Q_\alpha{}^i] = f_a{}^i{}_j Q_\alpha{}^j \tag{2.5}$$

$$[T_a, T_b] = f_{ab}{}^c T_c \tag{2.6}$$

$$[Z^e, Q_\alpha{}^i] = 0 = [Z^e, Q_{\dot\beta j}] = [Z^e, P_\mu] = [Z^e, J_{\mu\nu}] \tag{2.7}$$

where $(a^e)^{ij} = -(a^e)^{ji}$, $(a^{e*})_{\dot{i}\dot{j}}$ is the complex conjugate of $(a^e)^{ij}$ and $(\sigma_\mu)_{\alpha\dot\beta} = (1, \sigma)$ where σ are the Pauli matrices. $(\sigma^{\mu\nu})_\alpha{}^\beta = -\frac{1}{2}((\sigma_a)_{\alpha\dot\delta}(\sigma_b)^{\beta\delta} - (a \leftrightarrow b))$.

Equation (2.1) tells us that P_μ can be written in terms of the two spinorial generators $Q_\alpha{}^i$ and $Q_{\dot\beta j}$. This is what is meant by the often quoted statement that 'the supersymmetry group is the square root of the Poincaré group'. In particular, it is a simple consequence of this equation that

$$P^0 = \frac{1}{4N}(\sigma^0)^{\alpha\dot\beta}\{Q_\alpha{}^i Q_{\dot\beta i} + Q_{\dot\beta i} Q_\alpha{}^i\} \tag{2.8}$$

Since P^0 is the Hamiltonian of the theory, it follows that the Hamiltonian in all supersymmetric theories is given by eqn. (2.8) and so is positive definite. The fact that the spinorial generators and the four-translations commute (eqn. (2.3)) tells us that the on-shell states which belong to any irreducible representation of supersymmetry must have the same mass. It is this property that makes it essential for supersymmetry to be broken in some way if it is to apply to nature. Clearly, it is only through the mechanism of spontaneous symmetry breaking that the strongly predictive power of supersymmetric theories can remain. As with other symmetries, the spontaneous breaking of supersymmetry leads to a Goldstone mode, in this case a Goldstone spinor.

Equation (2.4) tells us that the spinorial generators form

a spin ½ representation of the Lorentz group. This implies that they generate supersymmetry transformations which alter the spin of the state or field on which they act by ½ unit. Consequently, supersymmetry is a symmetry which mixes bosons and fermions. In fact, to specify a given supersymmetric theory, it is sufficient to give only its fermionic structure, after which the bosonic structure is determined. An extremely useful guide in constructing supersymmetric theories is the following rule which follows from the algebra: there are, in any supersymmetric representation, the same number of bosons as fermions. This applies to the on-shell degrees of freedom of states and to the off-shell description in terms of fields.

The central charges Z^e commute with all the generators of the graded Lie algebra and only occur in the anti-commutator of two undotted (or dotted) spinorial charges (Haag, Lopuszanski, and Sohnius 1975). It is quite consistent to set them to zero without affecting the rest of the algebra, (a Wigner—Inönoü contraction). This result, together with the historical accident that most experience with supersymmetry is based on the $N = 1$ case, where there are no central charges, led to the feeling that central charges were bizarre objects not encountered in the usual run of events. In fact, this is not the case and setting them to zero in the algebra at the outset will, more often than not, make vital constructions impossible.

It is interesting to note that if one wishes to combine space-time and internal symmetries, one has to accept not only the spinorial generators of supersymmetry but also a bosonic central charge generator. We will later show that the non-vanishing of this generator is equivalent to a dependance of the fields on a fifth dimension. The central charge generates translations in this fifth dimension. This is in contrast to the mechanism of dimensional reduction which uses extra dimensions to generate internal symmetries.

The internal symmetry generators act on the spinorial generators and not on those of the Poincaré group. What the internal group is depends upon the particular system under study and whether or not that system has a central charge.

3. ON-SHELL IRREDUCIBLE REPRESENTATIONS OF SUPERSYMMETRY

These representations may be calculated from the algebra using an extension of Wigner's method of induced representations. We refer the reader to Salam and Strathdee (1974) for the details of this simple construction and record here only the results. The nature of the on-shell representation, of course, depends on the on-shell characteristics being considered, namely, on whether the multiplet is massive or massless, has zero or non-zero central charge, and on the number N of spinorial symmetry generators.

If the multiplet is massless and has no central charge on-shell, then the irreducible representation contains states with helicities from a maximum helicity λ to a minimum helicity $\lambda - {}^{N}/2$. The demand of a CPT invariant theory requires us to add to this the irreducible representation with helicities from $-\lambda$ to $-(\lambda - {}^{N}/2)$ if these helicities are not already contained in the first representation.

If the multiplet is massive but has no central charge on-shell, then the irreducible representation with the lowest maximal spin contains a maximal spin of ${}^{N}/2$.

If the multiplet is massive but does possess an on-shell central charge, then the irreducible representation with the lowest maximal spin has a maximal spin of $\frac{N+1}{4}$ if N is odd and $\frac{N}{4}$ if N is even. It is also possible to calculate the number of on-shell states with each helicity.

The table below gives the results for a massless on-shell representation without central charge which has maximal helicity 1.

spin \ N	1	2	4
1	1	1	1
½	1	2	4
0		2	6

We see that as N increases, the multiplicities of each spin and the number of different types of spin increase. The simplest theory is that with $N = 1$. It contains one spin 1 and one spin ½, consistent with the formula for the lowest helicity $\lambda - {}^{N}/_{2}$, which in this case gives $1 - \frac{1}{2} = \frac{1}{2}$. The $N = 4$ multiplet is CPT self conjugate, since in this case we have $\lambda - {}^{N}/_{2} = 1 - {}^{4}/_{2} = -1$. The table has stopped at N equal to 4 since when N is greater than 4 we must have particles of spin greater than 1. (Clearly, $N > 4$ implies that $\lambda - {}^{N}/_{2} = 1 - {}^{N}/_{2} < -1$.) This leads us to the well-known statement that the $N = 4$ supersymmetric theory is the maximally extended Yang–Mills theory. It is the supersymmetric Yang–Mills theory which has the greatest number of supersymmetries and which is hoped to be finite.

The content for a massless on-shell representation without central charge which has maximal helicity 2 is given below

spin \ N	1	2	8
2	1	1	1
$^{3}/2$	1	2	8
1		1	28
½			56
0			70

The $N = 1$ supergravity theory contains only one spin 2 graviton and one spin $^{3}/2$ graviton. It is often referred to as the simple supergravity theory. For the $N = 8$ supergravity theory, $\lambda - \frac{N}{2} = 2 - \frac{8}{2} = -2$. Consequently it is CPT self conjugate and so contains particles from spin 2 to spin 0. Clearly, for theories in which N is greater than 8, particles of higher than spin 2 will occur. Thus, the $N = 8$ theory is the maximally extended supergravity theory.

It is claimed that this theory is in fact the largest possible supergravity theory. This contention rests on the widely held belief that it is impossible to consistently couple particles of spin $\frac{5}{2}$ to other particles. If true, this leaves the burden of realizing supersymmetry in nature on the shoulders of the maximally extended supergravity theory.

4. THE WESS–ZUMINO MODEL

In order to focus our thoughts let us consider the simplest possible supersymmetric theory. This is the theory for which the on-shell states are massless, form a representation of $N = 1$ supersymmetry (and so automatically have no central charge) and have the smallest possible maximal helicity ($\lambda = \frac{1}{2}$). Hence it contains helicities from $\lambda = \frac{1}{2}$ to $\lambda - \frac{N}{2} = 0$. The *on-shell* content can easily be deduced from the supersymmetry algebra and is one spin ½ and two spin 0s.

Unfortunately, it is not easy to find an off-shell formulation of the theory, that is, to deduce the set of fields which carry a representation of the supergroup and from which we can construct a Lagrangian that has the above on-shell states. We now present the solution which was found by Wess and Zumino (1974*a*). It involves the fields A, B, χ_α, F, and G where A, B, F, and G are spin zero fields and χ_α is a Majorana spinor.

Their supersymmetry transformations, which form a representation of the $N = 1$ supersymmetry algebra, are given by

$$\delta A = i\bar{\zeta}\chi, \qquad \delta B = i\bar{\zeta}\gamma_5\chi$$

$$\delta\chi = \{F + \gamma_5 G + \not\partial(A + \gamma_5 B)\}\zeta$$

$$\delta F = i\bar{\zeta}\not\partial\chi, \qquad \delta G = i\bar{\zeta}\gamma_5\not\partial\chi \tag{4.1}$$

The action, which is invariant under these transformations, is

$$A = \int d^4x\left\{-\frac{1}{2}(\partial_\mu A)^2 - \frac{1}{2}(\partial_\mu B)^2 - \frac{i}{2}\bar{\chi}\not\partial\chi + \frac{1}{2}F^2 + \frac{1}{2}G^2\right\}. \tag{4.2}$$

This formulation may come as something of a surprise, for although the fields A, B, and χ_α correspond to the two spin-0s and one spin-½ on-shell states, the fields F and G are of dimension two and the equations of motion set them to zero. Since they vanish on shell, it follows that their variations must also vanish on-shell and consequently they must vary into equations of motion. Such fields are called auxiliary fields. Their inclusion is necessary for the fields of the theory to carry a representation of the supersymmetry algebra irrespective of any particular dynamics. Their presence is essential if the theory is to have a representation off-shell. Although the auxiliary fields F and G do not propagate in the Wess–Zumino model, it is perfectly possible to build other models out of the supermultiplet (A,B,χ,F,G) in which F and G do propagate. The need for auxiliary fields to provide a representation of the group algebra on fields is a phenomenon unique to supersymmetry and, except in very rare cases, necessary in all supersymmetric theories. Unfortunately, although the on-shell representations of supersymmetry are easy to find, the off-shell representations, i.e. the auxiliary fields, can be very difficult to find.

There is a simple argument which demonstrates the need for auxiliary fields in supersymmetric theories. It follows from the supersymmetry algebra that, in a supersymmetry theory where the fields do form a representation, the number of fermionic fields equals the number of bosonic fields. Since imposing the equations of motion is a supersymmetric procedure, there are the same number of fermionic and bosonic on-shell states. For example, in the Wess–Zumino model we have on-shell two bosonic degrees of freedom for A and B and two fermionic degrees of freedom for χ_α. Off-shell, we have 4 bosonic fields, A, B, F, and G and 4 fermionic fields, χ_α. The auxiliary fields F and G are necessary to make up the four bosonic fields required to balance the four fermionic ones. In general, we see that the auxiliary fields are necessary to account for the fact that bose and fermi fields yield different numbers of degrees of freedom when one goes on-shell.

If we set F and G to zero, the transformations on the remaining fields are given by

$$\delta A = i\bar{\zeta}\chi, \qquad \delta B = i\bar{\zeta}\gamma_5\chi$$
$$\delta\chi = \{\not{\partial}(A + \gamma_5 B)\}\zeta \tag{4.3}$$

Although these transformations do not form a representation of the supergroup, they are a symmetry of the equations of motion. This is because the fields A, B, and χ lead to an on-shell supersymmetric representation. Consequently, the commutator of two supersymmetry transformations δ_{ζ_2} and δ_{ζ_1} of eqn. (4.3) is given by

$$[\delta_{\zeta_1},\delta_{\zeta_2}] = + 2i\bar{\zeta}_2\partial\!\!\!/\zeta_1 + \text{equation of motion terms} \qquad (4.4)$$

However, the set of transformations of eqn. (4.3) are an infinitesimal invariance of the Lagrangian of eqn. (4.2) with F and G set to zero. As already mentioned, with the exception of the very simplest theories, all supersymmetry theories have first been constructed with the auxiliary fields absent. In fact, the auxiliary fields for the maximally extended supergravity theory are still unknown. Clearly, only if we have a theory written in terms of a set of fields which do form a realization of the algebra can the supersymmetry be made manifest. The importance of these formulations has already been explained and we next discuss the two types of manifestly supersymmetric formalism, the tensor calculus and the superspace formulation.

5. THE TENSOR CALCULUS

Given sets of fields (supermultiplets) which form a representation of the supersymmetry group we can proceed as with any other symmetry and find rules to

(a) Combine two supermultiplets to give another supermultiplet.

(b) Differentiate a supermultiplet to obtain another supermultiplet.

(c) Construct invariants from supermultiplets. Such sets of operations are what is normally referred to as a tensor calculus. Once we have established the validity of these rules we need only work with the supermultiplets and never with the individual component fields which make up the supermultiplets. All expressions are then written in terms of supermultiplets and the formalism is manifestly supersymmetric. We must stress that even to begin such a procedure, that is to obtain the supermultiplets, we must know the auxiliary fields.

We illustrate the above rules with the supermultiplet of the preceding section (Wess and Zumino 1974*a*,*b*) which contains the fields (A, B, χ, F, G) denoted collectively by S, with the transformation rules of eqn. (4.1).

(a) Given two such multiplets, $(A_1, B_1, \chi_1, F_1, G_1)$ and $(A_2, B_2, \chi_2, F_2, G_2)$ denoted by S_1 and S_2 respectively, we can form a third supermultiplet $S_1.S_2$, which also has the transformations of eqn. (4.1), by the rule

$$S_1.S_2 = (A_1A_2 - B_1B_2, A_1B_2 + A_2B_1, (A_1 - \gamma_5B_1)\chi_2$$
$$+ (A_2 - \gamma_5B_2)\chi_1, A_1F_2 + A_2F_1 + B_1G_2 + B_2G_1 - i\bar{\chi}_1\chi_2,$$
$$A_1G_2 + A_2G_1 - B_1F_2 - B_2F_1 + i\bar{\chi}_1\gamma_5\chi_2) \quad (5.1)$$

(b) Given the supermultiplet S we can differentiate to form a third supermultiplet of the same type by the rule

$$\hat{\mathrm{D}}S = (F, G, \not{\partial}\chi, \partial^2 A, \partial^2 B) \quad (5.2)$$

(c) Lastly, given the supermultiplet S we can construct a supersymmetric invariant by inserting its F or G component, denoted $[S]_F$ or $[S]_G$ into a spatial integral, viz

$$\int \mathrm{d}^4x[S]_F \quad \text{or} \quad \int \mathrm{d}^4x[S]_G \quad (5.3)$$

This works because the F and G components vary into total divergences.

As an example of these rules we now can construct the Wess–Zumino action. Take a supermultiplet S and apply the differential operator of rule (b) to S, $\hat{\mathrm{D}}S$, then combine this supermultiplet with S itself according to rule (a), $S\cdot\hat{\mathrm{D}}S$, finally construct an invariant according to rule (c)

$$\int \mathrm{d}^4x[S.\hat{\mathrm{D}}S]_F \,. \quad (5.4)$$

The reader may easily verify that this is the Wess–Zumino action. The mass and interaction terms are given by the obviously supersymmetric invariant terms

$$\int \mathrm{d}^4x m[S.S]_F \text{ and } \int \mathrm{d}^4x\lambda[S.(S.S)]_F \quad (5.5)$$

Clearly, we could also find rules which involve more than one type of representation and construct other supersymmetric theories.

Finally, we record some facts about Majorana spinors which have been used to derive some of the above formulae. A Majorana spinor is defined by the condition that $\bar{\chi}^{\alpha} = ((\chi)^{+}\gamma^{0})^{\alpha} = C^{\alpha\beta}\chi_{\beta}$ where $C^{\alpha\beta} = -C^{\beta\alpha}$ is the charge conjugation matrix. A consequence of this is that two Majorana spinors satisfy

$$\bar{\chi}\Gamma_R\lambda = \bar{\lambda}\Gamma_R\chi \text{ if } \Gamma_R = \mathbf{1},\gamma_5 \text{ or } \gamma_\mu\gamma_5$$

and

$$\bar{\chi}\Gamma_R\lambda = -\bar{\lambda}\Gamma_R\chi \text{ if } \Gamma_R = \gamma_\mu,\sigma_{\mu\nu} \quad .$$

6. SUPERSPACE

For simplicity we consider only the case of $N = 1$ superspace (Salam and Strathdee 1974, 1975). This eight dimensional superspace is labelled by coordinates $(x^\mu, \theta^\alpha, \theta^{\dot\beta})$. The four x^μs are the coordinates of the usual Minkowski space. The $\theta^\alpha, \theta^{\dot\beta}$ are four anti-commuting coordinates, namely

$$\theta^\alpha\theta^\beta + \theta^\beta\theta^\alpha = 0 = \theta^{\dot\alpha}\theta^{\dot\beta} + \theta^{\dot\beta}\theta^{\dot\alpha} = \theta^\alpha\theta^{\dot\beta} + \theta^{\dot\beta}\theta^\alpha$$
$$x^\mu\theta^\alpha - x^\mu\theta^\alpha = 0 = x^\mu\theta^{\dot\beta} - x^\mu\theta^{\dot\beta} \quad . \tag{6.1}$$

Supersymmetry is realized on superspace by the transformations

$$x^\mu \to x^\mu + i\theta^\alpha(\sigma^\mu)_{\alpha\dot\beta}\zeta^{\dot\beta} - i\zeta^\alpha(\sigma^\mu)_{\alpha\dot\beta}\theta^{\dot\beta}$$
$$\theta^\alpha \to \theta^\alpha + \zeta^\alpha, \qquad \theta^{\dot\beta} \to \theta^{\dot\beta} + \zeta^{\dot\beta} \tag{6.2}$$

which form a representation of the $N = 1$ supersymmetry algebra. In fact, superspace is none other than the coset space of the $N = 1$ supersymmetry group divided by the Lorentz group. The transformations of eqn. (6.2) are the natural action of the group on its own coset. Superspace is an extention of the coset space formed from the Poincaré group divided by the Lorentz group, which is just the usual Minkowski space.

We can consider functions on superspace. A simple example

is a general scalar function, $\phi(x^\mu,\theta^\alpha,\theta^{\dot\beta})$. We can expand ϕ in a Taylor series in terms of its fermionic coordinates;

$$\begin{aligned}\phi(x^\mu,\theta^\alpha,\theta^{\dot\beta}) &= C(x) + \theta^\alpha\chi_\alpha(x) + \chi_{\dot\beta}(x)\theta^{\dot\beta}\\ &+ \tfrac{1}{2}\theta^\alpha\theta_\alpha M(x) + \tfrac{1}{2}\theta_{\dot\beta}\theta^{\dot\beta}N(x) + \theta^\alpha(\sigma^\mu)_{\alpha\dot\beta}\theta^{\dot\beta}A_\mu(x)\\ &+ \tfrac{1}{2}\theta_\alpha\theta^\alpha\lambda_{\dot\beta}(x)\theta^{\dot\beta} + \tfrac{1}{2}\theta^{\dot\beta}\theta_{\dot\beta}\theta^\alpha\lambda_\alpha(x) + \tfrac{1}{4}\theta^\alpha\theta_\alpha\theta^{\dot\beta}\theta_{\dot\beta}D(x)\ . \end{aligned} \tag{6.3}$$

This expansion terminates at terms quartic in $\theta^\alpha,\theta^{\dot\beta}$, for a term quintic in $\theta^\alpha,\theta^{\dot\beta}$ vanishes identically since the θs anti-commute (i.e. $\theta_\alpha\theta_\beta\theta_\gamma = 0$). The fields C,χ,M,N,A_μ,λ, and D will eventually be identified with the usual component fields in Minkowski space. It is important to observe that this superfield contains a spin 1 component field. Superfields which carry Lorentz labels contain component fields of spin even higher than spin 1. The supersymmetry transformations of the component fields is given by the specification that ϕ be a scalar under the supersymmetry transformations of eqn. (6.2)

$$\phi(x^\mu,\theta^\alpha,\theta^{\dot\beta}) = \phi'(x^\mu+i\theta^\alpha(\sigma^\mu)_{\alpha\dot\beta}\zeta^{\dot\beta}-i\zeta^\alpha(\sigma^\mu)_{\alpha\dot\beta}\theta^{\dot\beta},\theta^\alpha+\zeta^\alpha,\theta^{\dot\beta}+\zeta^{\dot\beta}) \tag{6.4}$$

It turns out that the ordinary derivatives in superspace are not supercovariant. A set of supercovariant derivatives is given by

$$D_\mu = \partial_\mu,\ D_\alpha = \frac{\partial}{\partial\theta^\alpha} + i(\not\partial)_{\alpha\dot\beta}\theta^{\dot\beta},\ D_{\dot\beta} = -\frac{\partial}{\partial\theta^{\dot\beta}} - i\theta^\alpha(\not\partial)_{\alpha\dot\beta}\ . \tag{6.5}$$

These non-trivial structures reflect the fact that superspace, which is entirely specified by the supersymmetry group, is a space with zero curvature, but non-zero torsion.

The appearance of a spin 1 component field in $\phi(x^\mu,\theta^\alpha,\theta^{\dot\beta})$ is at first sight not consistent with the fact that the simplest supersymmetric theory, the Wess–Zumino model, does not contain a spin 1 field. The resolution of this apparent dilemma is the realization that superspace carries a reducible representation of the Lorentz group. Therefore, we may impose a supersymmetric constraint on ϕ, namely

$$D_\alpha \phi = 0 \quad . \tag{6.6}$$

The above equation has as solution a non-trivial superfield which is denoted ϕ_+ and which is called a chiral superfield. This, of course, implies that the superfield ϕ forms a reducible representation of supersymmetry. The superfield ϕ_+ is, in fact, an irreducible representation of supersymmetry and has the form

$$\phi_+(x^\mu, \theta^\alpha, \theta^{\dot\beta}) = C(x) + \chi_{\dot\alpha}(x)\theta^{\dot\alpha} + \tfrac{1}{2}\theta_{\dot\beta}\theta^{\dot\beta}N(x) - i\theta^\alpha(\not\partial)_{\alpha\dot\beta}\theta^{\dot\beta}C(x)$$

$$- \tfrac{i}{2}\theta^{\dot\beta}\theta_{\dot\beta}\theta^\alpha(\not\partial)_{\alpha\dot\gamma}\chi^{\dot\gamma}(x) - \tfrac{1}{4}\theta^\alpha\theta_\alpha\theta^{\dot\beta}\theta_{\dot\beta}\partial^2 C(x)$$

where we can write $C = A + iB$ and $N = F + iG$. (6.7)

This superfield has exactly the correct type and number of component fields for the Wess–Zumino theory, including the auxiliary fields. The supersymmetry transformations of eqn. (4.1) are a consequence of the fact that ϕ_+ is a scalar under supersymmetry.

To construct models we utilize the following facts

(a) The product of two superfields is clearly also a superfield.

(b) The application of a covariant derivative to a superfield is by construction a superfield.

(c) Supersymmetric invariants are given by

$$\int d^4x \int d\theta^\alpha d\theta_\alpha d\theta^{\dot\beta} d\theta_{\dot\beta}\phi \text{ and } \int d^4x \int d\theta^{\dot\beta} d\theta_{\dot\beta}\phi_+ \tag{6.8}$$

for a general superfield ϕ and for a chiral superfield ϕ_+, respectively.

The Wess–Zumino action is given by the expression

$$A = \int d^4x d\theta^{\dot\beta} d\theta_{\dot\beta}\phi_+ D^\alpha D_\alpha(\phi_+)^* + \text{hermitian conjugate} \tag{6.9}$$

(or equivalently $\int d^4x d^4\theta \phi_+(\phi_+)^*$).

The invariant mass and interaction terms are given by

$$A = \int d^4x d\theta^{\dot\beta} d\theta_{\dot\beta}\{\phi_+{}^2 + \lambda\phi_+{}^3\} + \text{hermitian conjugate.} \tag{6.10}$$

The occurrence of the high spin fields and the resulting necessity for a constraint is not peculiar to this example. It is a phenomenon which occurs in all other supersymmetric theories. The superspace formalism automatically carries a representation of the supersymmetry group on fields and so must encode the auxiliary fields. Whereas in order to construct the tensor calculus we require a knowledge of the auxiliary fields, the superspace formalism requires a knowledge of the necessary superspace constraints. The discovery of these constraints is a problem of comparable difficulty to that of finding the component auxiliary fields required for the tensor calculus approach.

The superspace and tensor calculus formalisms lead to the same results and are of course completely equivalent. Given both formalisms, the manipulations in one formalism can easily be translated into the other. The tensor calculus rules correspond to the rules given in this section.

We would like to stress that there is now a well known path called 'gauge completion' (Arrowitt and Nath 1975; Ferrara and van Nieuwenhuizen 1980; van Nieuwenhuizen and West 1980) which permits one to go from one formalism to the other. Given a theory with its auxiliary fields, we can easily find the superspace constraints and so the superspace formalism. Conversely, given the superspace formalism including the superspace constraints of a particular theory, we can find the auxiliary fields and so the tensor calculus.

Generally speaking, it has been the auxiliary field and tensor calculus formalism which has been discovered before the superspace formalism, although it is superspace which is most convenient in the end for quantum calculations.

7. SIMPLE SUPERGRAVITY

Supergravity theories are theories of supersymmetry that involve a graviton, that is, have a maximum helicity of 2. So according to our previous discussion, the on-shell states of the $N = 1$ supergravity is the theory of a graviton and a gravitino, or Rarita–Schwinger particle. Our considerations with the Wess–Zumino model lead us to suspect that the fields (the vierbien, $e_\mu{}^a$ and gravitino field $\psi_\mu{}^\alpha$) which correspond to these on-shell states will not form a representation of supersymmetry; we will be required to find additional auxiliary fields.

A set of fields which does form a representation of supersymmetry is

$$e_\mu{}^a,\ \psi_\mu{}^\alpha,\ M,\ N \text{ and } b_\mu \quad . \tag{7.1}$$

The fields M, N and b_μ must be the auxiliary fields. One may check, in this supermultiplet, the rule for representations of supersymmetry that the number of boson and fermion fields are equal. In doing this one must subtract the gauge degrees of freedom, which are not to be considered as part of the off-shell multiplet. We have 16 -4 -6 = 6 for $e_\mu{}^a$, 1 for M, 1 for N and 4 for b_μ, making 12 bosonic fields. Taking into account the Rarita–Schwinger gauge invariance, $\psi_\mu \to \psi_\mu + \partial_\mu \varepsilon$, we have 16 -4 = 12 fermionic fields. Clearly, the vierbein and the Rarita–Schwinger field alone could not have formed a representation of supersymmetry. We refer the reader to Stelle and West 1978*a* and van Nieuwenhuizen 1978*a* for the transformations of these fields, the corresponding graded Lie algebra and the invariant Lagrangian, which was found to be

$$\mathscr{L}_{SG} = \frac{1}{2\kappa^2} eR(e,\omega(e,\psi)) - \frac{1}{2} i\bar{\psi}_\mu R^\mu(\omega(e,\psi)) - \frac{1}{3}e^2M^2 - \frac{1}{3}e^2N^2 + \frac{1}{3}eb_\mu^2$$

$$\text{where } \omega_{\mu ab}(e,\psi) = \omega_{\mu ab}(e) + \frac{1}{4} i\kappa^2(\bar{\psi}_\mu\gamma_a\psi_b + \bar{\psi}_a\gamma_\mu\psi_b - \bar{\psi}_\mu\gamma_b\psi_a)$$

$$D_\mu = \partial_\mu + \frac{1}{2}\omega_{\mu ab}\sigma^{ab},\quad R_{\mu\nu}{}^{ab}\frac{1}{2}\sigma_{ab} = [D_\mu, D_\nu] \ ,$$

$$R = e_a{}^\mu e_b{}^\nu R_{\mu\nu}{}^{ab},\quad R^\mu = \varepsilon^{\mu\nu\rho\kappa}\gamma_5\gamma_\nu D_\rho\psi_\kappa \ . \tag{7.2}$$

In one important aspect, supergravity theories differ from other supersymmetric theories. To obtain the linearized $N = 1$ supergravity theory we consider the fields $\kappa h_{\mu a} = e_{(\mu a)} - \eta_{\mu a}$, $\psi_\mu{}^\alpha$, M, N and b_μ and neglect in their transformations all non-linear terms and also regard all transformation parameters as constants. That is, the linearized theory possesses only a rigid supersymmetry. These transformations, when supplemented by the gauge invariance

$$\delta h_{\mu\nu} = \partial_\mu \xi_\nu + \partial_\nu \xi_\mu \tag{7.3}$$

$$\delta \psi_\mu{}^\alpha = \partial_\mu \eta^\alpha \tag{7.4}$$

then form a representation of the algebra of eqns. (2.1) to (2.7) and are an invariance of the Lagrangian of eqn. (7.2) when we retain only terms bilinear in these fields. However, we wish to consider a theory in which the rigid translations become Einstein's general coordinate transformations

$$e_{\mu a} \to e_{\mu a} + e_{\nu a} \partial_\mu \xi^\nu + \xi^\nu \partial_\nu e_{\mu a} \; . \tag{7.5}$$

In the supersymmetric theory we now expect the commutator of two supersymmetry transformations to give a general coordinate transformation. Consequently in order to produce a local symmetry, the supersymmetry transformations must themselves be local. Just as the rigid translations and the linearized gauge transformations of eqn. (7.3) combine to produce general coordinate transformations, the rigid supersymmetry transformations and the linearized gauge transformations combine to produce a local supersymmetry transformation. The corresponding algebra is a local version of the algebra given in eqns. (2.1) to (2.7).

An interesting historical note is that it was two years after the discovery of supersymmetry before a formulation of $N = 1$ supergravity without auxiliary fields was found (Ferrara, Freedman, and van Nieuwenhuizen 1976; Deser and Zumino 1976). It required a further two years to find a formulation with a simple set of auxiliary fields.

Once the auxiliary fields were known, a tensor calculus for $N = 1$ supergravity was constructed (Ferrara and van Nieuwenhuizen 1978*b*,*c*; Stelle and West 1978*b*,*c*,*d*). This tensor calculus enables, in $N = 1$ supergravity, the simple discussion of Feynman rules, the easy construction of models, a discussion of Yang–Mills invariance in supergravity, the construction of explicit models in which spontaneous breaking of local supersymmetry occurs (Cremmer, Ferrara, Girardello, Julia, Scherk, and van Nieuwenhuizen 1979) and the construction to all orders of counter-terms for quantum processes with more than two loops.

We conclude this discussion with a very brief account of the superspace formulation of $N = 1$ supergravity (Wess and Zumino 1977, 1978; Siegel and Gates 1979; Brink, Gell-Mann, Ramond, and Schwartz 1978; Bedding, Downes-Martin, and Taylor 1979; Ogievetsky and Sokatchev 1977*a,b*). We introduce, in analogy with general relativity, a vierbein $E_M{}^A(x^\mu, \theta^\alpha, \theta^{\dot\beta})$ and a connection $\Omega_{MA}{}^B(x^\mu, \theta^\alpha, \theta^{\dot\beta})$. Here M represents the eight base indices and A, B the eight tangent space indices.

The transformations under which the theory is to be invariant are general coordinate transformations (reparameterization of superspace) and local Lorentz transformations. These latter transformations are restricted so that the tangent space sectors labelled by $\underset{\sim}{a}$, by α and by $\dot\beta$ are always rotated by the same amount.

Just as with the Wess–Zumino model, we find that the field $E_M{}^A$ contains component fields of spin higher than 2. This difficulty is overcome by the introduction of constraints on the only covariant objects of the theory; the torsion tensor $T_{AB}{}^C$ and the curvature tensor $R_{AB}{}^{CD}$. Subject to setting some of these tensors to zero, it is found that the action (Wess and Zumino 1978)

$$I = \int d^4x d^4\theta (\det E_M{}^A) \tag{7.6}$$

yields the $N = 1$ supergravity action of eqn. (7.2).

The superspace formalism is relatively straightforward, apart from the difficulty in finding the relevant superspace constraints. A systematic procedure for finding the superspace constraints for both Yang–Mills and supergravity theories that has so far been successfully applied to the $N = 1$ and 2 theories has been found (Gates 1979; Stelle and West 1979; Gates, Stelle, and West 1980; Stelle and West 1980). One of the essential ingredients in this procedure is the requirement that the generic representations of rigid supersymmetry be preserved in the case of local invariances. The connection between the preservation of the chiral superfield and certain superspace constraints was noticed by Gates and Siegel (1979). We refer the reader to the above references for an exposition of this method.

As already explained, given either the auxiliary field or superspace formalism it is a simple matter to find the other. Unfortunately, the auxiliary fields are still unknown for any of the supergravity theories with $N > 2$. We now present a new method of finding auxiliary fields.

8. DIMENSIONAL REDUCTION BY LEGENDRE TRANSFORMATION

Several supersymmetric theories have been found by the technique of dimensional reduction. This technique is a method of constructing actions starting from actions in higher dimensions. Although it can be used to go from an action in any dimension to an action in any lower dimension, we take the case of a supersymmetric five dimensional action

$$I^{(5)} = \int \mathscr{L}(\phi, \partial_a \phi) d^5x, \quad a = 1 \to 5 , \tag{8.1}$$

where ϕ denotes collectively the bosonic and fermionic fields of the five-dimensional theory. We now demand that the fields not be dependent on x_5, i.e. $\partial_5 \phi = 0$. The derived four dimensional action is given by

$$I^{(4)} = \int \mathscr{L}(\phi, \partial_\mu \phi, \partial_5 \phi = 0) d^4x; \quad \mu = 1 \to 4 . \tag{8.2}$$

The action, $I^{(4)}$, apart from a $\int dx_5$ factor, equals the action $I^{(5)}$ when $\partial_5 \phi = 0$. Consequently, $I^{(4)}$ is invariant under the five dimensional supersymmetry transformations with $\partial_5 \phi = 0$. The advantage of this technique is that the symmetries of the higher dimensional space-time group which operate on the abandoned coordinates become internal symmetries in the lower dimensional theory. Hence, it is a 'natural' way of generating internal symmetries. For example, if one starts with an $N = 1$ supersymmetric Yang–Mills theory in 6 or 10 dimensions we obtain in four dimensions the $N = 2$ or $N = 4$ supersymmetric Yang–Mills theories, respectively (Gliozzi, Scherk, and Olive 1978; Brink, Schwarz, and Scherk 1977). Clearly, if we start with a theory without (with) auxiliary fields we will end up with a theory without (with) auxiliary fields. Unfortunately, the auxiliary fields are as hard to find in the higher dimensional theory as they are in the resulting lower dimensional theory.

The aesthetic character of the beautiful dimensional reduction technique is impaired by the requirement that one ignore all dependence on the higher coordinates. We could ask the question, 'Is there an alternative dimensional reduction technique which maintains a field dependence on the higher coordinates?' Or, in our particular case, can we keep the dependence on x_5 and still find a four-dimensional supersymmetric action. This possibility poses a difficulty, however.

The transformations in the four-dimensional theory are obtained from the five-dimensional transformations with the enforcement of whatever restrictions on the fields is demanded by the dimensional reduction procedure. Hence, the four-dimensional group will include shifts in the x_5 coordinate. Although this poses no problem for the action of eqn. (8.2) since x_5 does not occur, if we wish to retain a dependence of the fields on x_5, we must then demand that our new four-dimensional action be invariant under shifts in x_5 and so be independent of x_5. The difficulty is that the action is a *four*-dimensional integral of an expression involving fields that are dependent on x_5.

In fact, every physicist knows the resolution of this dilemma in the context of going from a four-dimensional theory to a three-dimensional theory. It is just the canonical transition from the Lagrangian to the Hamiltonian formulation of classical field theory. Consider a four-dimensional theory, with $I^{(4)}$ invariant under some symmetry group, G

$$I^{(4)} = \int d^4x \, \mathscr{L}(\phi, \partial_\mu \phi) \tag{8.3}$$

The canonical momenta are defined to be

$$\pi = \frac{\partial \mathscr{L}}{\partial(\partial_0 \varphi)} \tag{8.4}$$

and the Hamiltonian is given by

$$H(\pi,\varphi,\partial_i\varphi) = \int \pi\partial_0\varphi d^3x - \int \mathscr{L} d^3x \equiv \int \mathscr{H} d^3x; \quad (i = 1 \to 3) \tag{8.5}$$

The Hamiltonian H is taken to be a function of the fields π, φ, $\partial_i\varphi$ and is an integral of the Hamiltonian density, $\mathscr{H}$ over only three-space. Despite this, provided we impose Hamilton's equations of motion

$$\frac{d\varphi}{dt} = \frac{\delta H}{\delta\pi}, \qquad \frac{d\pi}{dt} = -\frac{\delta H}{\delta\varphi}, \tag{8.6}$$

the Hamiltonian is time independent,

$$\frac{dH}{dt} = 0, \tag{8.7}$$

and is invariant under the symmetry group G. (For this to be true, G is not allowed to have an explicit dependence on t, such as occurs in the Galilean transformation.)

We now apply this procedure to obtain an invariant four-dimensional action from a five-dimensional action. In this case, G is replaced by the supersymmetry group, the time t is replaced by x_5 and the Hamiltonian becomes the negative of the invariant four-dimensional action that we seek.

The detailed method is as follow. We start with a supersymmetric action $I^{(5)}$ in five dimensions

$$I^{(5)} = \int d^5x\, \mathscr{L}(\phi,\partial_a\phi) \tag{8.8}$$

The 'canonical momenta' are defined by

$$\pi = \frac{\partial\mathscr{L}}{\partial(\partial_5\phi)} \tag{8.9}$$

and the 'Hamiltonian' is given by

$$H^{(4)}(\phi,\pi,\partial_\mu\phi) = \int (\pi\partial_5\phi - \mathscr{L})d^4x .$$

This 'Hamiltonian' is a function of ϕ, π and $\partial_\mu\phi$ and is an integral over four-space. We then impose the five-dimensional equations of motion,

$$\frac{d\phi}{dx_5} = \frac{\delta H^{(4)}}{\delta \pi}, \qquad \frac{d\pi}{dx_5} = \frac{\delta H^{(4)}}{\delta \varphi} . \tag{8.10}$$

These equations in general just restrict the x_5 dependence of the fields π and ϕ in the same way that Hamilton's equations, in their usual context, determine the time evolution of the fields. We wish to stress that they do not in any way impose the four-dimensional equations of motion. We then identify the required four-dimensional action with the negative of this Hamiltonian

$$I^{(4)} = - H^{(4)} . \tag{8.11}$$

Having imposed the equations of motion as constraints, we find that $I^{(4)}$ is invariant under the supersymmetry transformations obtained by imposing the equations of motion (8.10) upon the five-dimensional supersymmetry transformations, and in particular it is invariant under central charge transformations,

$$\frac{dI^{(4)}}{dx_5} = 0 . \tag{8.12}$$

We note that this new dimensional reduction procedure, which we call 'dimensional reduction by Legendre Transformation' (Sohnius, Stelle, and West 1980*b*) gives us more fields than the old dimensional reduction technique does, namely the π fields.

We may wonder what is the significance of the translations in x_5? The supersymmetry algebra in five dimensions is

$$\{Q_{\alpha i}, \bar{Q}^{\beta j}\} = 2i\delta_i^{\,j}\gamma^a P_a = 2i\delta_i^{\,j}(\gamma^\mu P_\mu + \gamma^5 P_5)$$

where $a = 1 \to 5$, $\mu = 1 \to 4$.

The last term generates the x_5 translations, but from the point of view of the four-dimensional algebra it will be interpreted as the central charge generator. Thus we find that central charge transformations are just translations in an additional bosonic dimension.

This new technique has one distinct advantage over the

original dimensional reduction procedure. Suppose the auxiliary fields for $I^{(5)}$ are not known. That is, suppose the supersymmetry transformations of the five-dimensional fields only form a representation of the supersymmetry transformations once we impose the five-dimensional equations of motion. However, it is these equations that are imposed when obtaining the four-dimensional action by Legendre transformation. Hence, the resulting four-dimensional action must have a set of fields which form a true realization of the supersymmetry algebra. As such, we must have introduced into the theory the necessary auxiliary fields. The only fields which are introduced in addition to those obtained by the usual dimensional reduction technique are the canonical momenta π. Clearly, these fields must be the auxiliary fields. This suspicion is confirmed by the fact that the Hamiltonian is generically of the form

$$H^{(4)} = \int \left\{\frac{1}{2}\pi^2 + \frac{1}{2}(\partial_\mu\phi)^2 + \ldots\right\}d^4x \quad . \tag{8.14}$$

Consequently, we see that the auxiliary fields are none other than the canonical momenta for the physical fields.

We give an application of this new technique for dimensional reduction in the next section.

9. OFF-SHELL FORMULATION OF $N = 4$ SUPERSYMMETRIC YANG–MILLS THEORY

As explained in the introduction, the $N = 4$ Yang–Mills theory is of importance since it is the maximal supersymmetric Yang–Mills theory and there is some evidence that it could be the first finite four-dimensional quantum field theory. To prove this conjecture, it will be necessary to find the supersymmetric Feynman rules and use them to analyse the ultraviolet behaviour of the theory. These Feynman rules, however, can only be found from an off-shell formulation of the theory. This formulation was found for the Abelian theory by a new method of discovering auxiliary fields which enables one in certain cases to find off-shell representations of supersymmetry from a knowledge of the easily found on-shell representations of supersymmetry. We refer the reader to Sohnius,

Stelle, and West (1980*a*) for an exposition of this method and its application to the $N = 4$ supersymmetric Abelian theory.

This theory and its Yang–Mills counterpart can also be derived by the technique of dimensional reduction by Legendre transformation (Sohnius, Stelle, and West 1980*b*). We start from the known supersymmetric Yang–Mills theory in ten dimensions (Gliozzi, Scherk, and Olive 1978; Brink, Schwartz, and Scherk 1977)

$$I^{(10)} = \int d^{10}x \operatorname{Tr}\left\{-\frac{1}{4}F_{mn}F^{mn} - \frac{i}{4}\bar{\lambda}\Gamma^m \mathscr{D}_m\lambda\right\}; \quad n,m = 1 \to 10 \;. \tag{9.1}$$

The auxiliary fields for this ten-dimensional theory are unknown. We perform the usual dimensional reduction from ten to five dimensions to obtain the Lagrangian $L^{(5)}$,

$$A^{(5)} = \int d^5x L^{(5)} = \int d^5x \operatorname{Tr}\left\{-\frac{1}{4}F_{ab}F^{ab} - \frac{1}{2}\mathscr{D}_a\phi_s\mathscr{D}^a\phi_s - \frac{i}{4}\bar{\lambda}^i\gamma_a\mathscr{D}^a\lambda_i + \frac{1}{2}\bar{\lambda}^i(t_s)_i{}^j[\lambda_j,\phi^s] + \frac{1}{4}[\phi_r,\phi_s][\phi^r,\phi^s]\right\},$$

where $a,b = 1 \to 5$; $r,s = 1 \to 5$; $i,j = 1 \to 4$; $(t_s)_i{}^j$ are the 'gamma matrices' of SP(4) and $\phi_s = A_{5+s}$, (9.2)

and then a dimensional reduction by Legendre transform from five to four dimensions. We obtain, in the Abelian case, an action which has an SP(4) internal symmetry (by definition, the subgroup of SU(4) which preserves the antisymmetric metric $\Omega_{ij} = -\Omega_{ji}$). The component fields are one vector V_μ, four spin ½ Majorana fields $\lambda_\alpha{}^i$, five spin 0s represented by scalar fields ϕ_{ij} ($\phi_{ij} = -\phi_{ij}$, $\phi_{ij}\Omega^{ij} = 0$), one spin 0 represented by an antisymmetric tensor, $A_{\mu\nu}$, and finally five scalar auxiliary fields, H_{ij} ($H_{ij} = -H_{ji}$, $H_{ij}\Omega^{ij} = 0$). Under the symmetry group SP(4) the vector V_μ and antisymmetric tensor $A_{\mu\nu}$ are singlets, the spinors $\lambda_\alpha{}^i$ are in the fundamental representation and the five scalars ϕ_{ij} and the auxiliary fields H_{ij} are in the second rank antisymmetric traceless representation.

The action is given by

$$A^{(4)} = \int d^4x\left\{-\frac{1}{4}(F_{\mu\nu})^2 - \frac{1}{2}(\partial_\mu\phi_{ij})^2 + \frac{1}{2}(\partial_\nu{}^*A^{\nu\mu})^2 - \frac{i}{2}\bar{\lambda}^i\partial\!\!\!/\lambda_i + \frac{1}{2}(H_{ij})^2\right\}. \tag{9.3}$$

and is invariant under SP(4) as well as the following

supersymmetry, central charge and gauge transformations

$$\delta V_\mu = i\bar{\zeta}^i\gamma_\mu\lambda_i + \omega\partial_\nu{}^*A^\nu{}_\mu + \partial_\mu\Lambda$$

$$\delta\lambda_i = \{-\sigma_{\mu\nu}F^{\mu\nu}\zeta_i - 2i\not{\partial}\phi_i{}^j\zeta_j + i\gamma_5\gamma^\mu(\partial^\nu{}^*A_{\nu\mu})\zeta_i + 2\gamma_5 H_i{}^j\zeta_j\} + i\omega\gamma_5\not{\partial}\lambda_i$$

$$\delta\phi_{ij} = \bar{\zeta}_i\lambda_j - \bar{\zeta}_j\lambda_i - \frac{1}{2}\Omega_{ij}\bar{\zeta}^k\lambda_k - \omega H_{ij}$$

$$\delta H_{ij} = -i\bar{\zeta}_i\gamma_5\not{\partial}\lambda_j + i\bar{\zeta}_j\gamma_5\not{\partial}\lambda_i + \frac{1}{2}i\Omega_{ij}\bar{\zeta}^k\gamma_5\not{\partial}\lambda_k + \omega\partial^2\phi_{ij}$$

$$\delta A_{\mu\nu} = -\bar{\zeta}^i\sigma_{\mu\nu}\lambda_i + \omega{}^*F_{\mu\nu} + \partial_\mu\Lambda_\nu - \partial_\nu\Lambda_\mu\,. \qquad (9.4)$$

where $\zeta^i, \omega, \Lambda, \Lambda_\mu$ are the parameters of supersymmetry, central charge, vector gauge and antisymmetric gauge transformations. These fields indeed form a representation of the supersymmetry group with the transformations given in eqn. (9.4). An alternative derivation of these results has been given (Taylor 1980) using a projection operator method (Sokatchev 1976; Ogievetsky and Sokatchev 1977*b*). This method constructs the projection operators for the irreducible representations of supersymmetry from a superspace realization of the Casimir operators of the supersymmetry algebra. The Lagrangian is then built using these projection operators.

The SP(4) symmetry arises because the ten-dimensional group SO(9,1) has been reduced to the five-dimensional Lorentz group SO(4,1) by the usual dimensional reduction, producing an internal symmetry SP(4), which is locally isomorphic to SO(5). The dimensional reduction by Legendre transformation introduces the central charge transformation but no extra internal symmetry group.

The details of the method of dimensional reduction by Legendre transformation and its application to the above $N = 4$ supersymmetric Maxwell and Yang–Mills theory and other theories is contained in Sohnius, Stelle, and West (1980*b*). The application to obtain an off-shell formulation of $N = 8$ supergravity is contained in Cremmer, Ferrara, Stelle, and West (1980). Given the auxiliary fields of these theories we can then easily find their superspace formulation using the method of gauge completion.

10. CONCLUSION

We have emphasized the importance of manifestly supersymmetric formulations of supersymmetric theories in attempting to extract the physics from these theories. There are two such formulations, the auxiliary field formulation with its associated tensor calculus, and the superspace formulation. The discovery of these formulations requires a knowledge of the representations of the supersymmetry group. A new technique of dimensional reduction which retains the dependence of the fields on the higher dimensional coordinate has been described. When applied to supersymmetric theories it leads to the discovery of new representations of supersymmetry. This has been applied to find the off-shell formulation of the maximally extended supersymmetric Yang–Mills theory.

A more detailed account of many of the topics considered here can be found in the Erice Lectures by M.F. Sohnius, K.S. Stelle, and P.C. West (1980*c*).

REFERENCES

Arnowitt, R. and Nath, P. (1975). *Phys. Lett.* **56B**, 117.
Bedding, S., Downes-Martin, S., and Taylor, J.G. (1979). *Ann. Phys.* **1**, 175.
Brink, L., Gell-Mann, M., Ramond, P., and Schwarz, J. (1978). *Phys. Lett.* **74B**, 336.
——, Schwartz, J. and Scherk, J. (1977). *Nucl. Phys.* **B121**, 77.
Coleman, S. and Mandula, J. (1967). *Phys. Rev.* **159**, 1251.
Cremmer, E., Ferrara, S., Girardello, L., Julia, B., Scherk, J., and van Nieuwenhuizen, P. (1979). *Nucl. Phys.* **B147**, 105.
—— ——, Stelle, K.S., and West, P.C. (1980). *Phys. Lett.* **94B**, 349.
Deser, S. and Zumino, B. (1976). *Phys. Lett.* **62B**, 335.
Ferrara, S. and van Nieuwenhuizen, P. (1978*a*). *Phys. Lett.* **74B**, 333.
—— —— (1978*b*). *Phys. Lett.* **76B**, 404.
—— —— (1978*c*). *Phys. Lett.* **78B**, 573.
—— —— (1980). In *Supergravity*. (ed. P. van Nieuwenhuizen and S. Ferrara). North-Holland, Amsterdam.
——, Freedman, D.Z., and van Nieuwenhuizen, P. (1976). *Phys. Rev.* **D13**, 3214.
Gates, S.J., Jr, (1979). In *Supergravity*. (ed. P. van Nieuwenhuizen and D.Z. Freedman). North-Holland, Amsterdam.
—— and Siegel, W. (1979). *Nucl. Phys.* **B147**, 77.
——, Stelle, K.S., and West, P.C. (1980). *Nucl. Phys.* **B169**, 347.
Gliozzi, F., Scherk, J., and Olive, D. (1978). *Nucl. Phys.* **B133**, 253.
Gol'fand, Y.A. and Likhtman, E.P. (1971). *JETP Letts.* **13**, 323.
Haag, R., Lopuszanski, J.T. and Sohnius, M.F. (1975). *Nucl. Phys.* **B88**, 257.

Ogievetsky, V. and Sokatchev, E. (1977*a*). *Nucl. Phys.* **B87**, 207.
—— —— (1977*b*). *Nucl. Phys.* **B124**, 309.
van Nieuwenhuizen, P. and West, P.C. (1980). *Nucl. Phys.* **B169**, 501.
Salam, A. and Strathdee, J. (1974). *Nucl. Phys.* **B80**, 499.
—— —— (1975). *Phys. Rev.* **D11**, 1521.
Siegel, W. and Gates, S.J., Jr. (1979). *Nucl. Phys.* **B87**, 207.
Sohnius, M.F., Stelle, K.S., and West, P.C. (1980*a*). *Phys. Lett.* **92B**, 123.
—— —— —— (1980*b*). *Nucl. Phys.* **B173**, 127.
—— —— —— (1980*c*). In *Unification of the Fundamental Particle Interactions* (ed. S. Ferrara, J. Ellis, and P. van Nieuwenhuizen). Plenum, New York.
Sokatchev, E. (1976). *Nucl. Phys.* **B138**, 109.
Stelle, K.S. and West, P.C. (1978*a*). *Phys. Lett.* **B74**, 330.
—— —— (1978*b*). *Phys. Lett.* **77B**, 376.
—— —— (1978*c*). *Nucl. Phys.* **B140**, 285.
—— —— (1978*d*). *Nucl. Phys.* **B145**, 179.
—— —— (1979). In *Supergravity*. (ed. P. van Nieuwenhuizen and D.Z. Freedman). North-Holland, Amsterdam.
—— —— (1980). *Phys. Lett.* **90B**, 393.
Taylor, J.G. (1980). 'A Superfield Formulation of Extended Supersymmetric Gauge Theories'. King's College preprint.
Volkov, D.V. and Akulov, V.P. (1973). *Phys. Lett.* **46B**, 109.
Wess, J. and Zumino, B. (1974*a*). *Nucl. Phys.* **B70**, 39.
—— —— (1974*b*). *Nucl. Phys.* **B78**, 1.
—— —— (1977). *Phys. Lett.* **B66**, 361.
—— —— (1978). *Phys. Lett.* **B74**, 51.

25

SOME REMARKS ON TWISTORS AND CURVED-SPACE QUANTIZATION

R. Penrose

Mathematical Institute, Oxford University

1. INTRODUCTORY REMARKS

Let me start by making some general comments concerning twistor theory (Penrose 1967, 1975; Penrose and MacCallum 1972; Penrose and Ward 1980; Hughston and Ward 1979). Some people at this conference might perhaps be puzzled that the theory seems to be trying to play two diametrically opposite roles. An impression that may have been gained from Chris Isham's clear and comprehensive opening survey, for example (Isham in this volume), is that twistor theory is something possibly to be resorted to if all else fails: if our accepted physical pictures of space-time and of quantization procedures finally land us with irresolvable contradictions, then perhaps (before giving up altogether!) we might (briefly?) turn to twistor theory, as we seem driven to seek radical alternative viewpoints as regards the ultimate nature of space and time. On the other hand, in Richard Ward's elegant lecture (Ward in this volume), twistor theory was seen in quite a different light, namely as something that has a mathematical role to play as an aid in the solution of certain systems of partial differential equations of some possible interest to physics, twistors being, in effect, regarded as a source of 'mathematical tricks', but without necessarily having any deep significance in terms of physics.

In attempting to explain this apparent conflict of purpose, I should begin by stressing that, while twistor theory is undoubtedly striving to do something 'new' as regards one's descriptions of space-time, it does *not* throw overboard existing physical theory. Twistor concepts are firmly based in established physics – notably those parts of physics (special relativity, Maxwell theory and the basic formalism of quantum mechanics) that have become most forcefully established in experiments performed over the past century. Twistors present, at least in the first instance, merely a mathematical reformulation of physics, so, in a clear sense, the excellent experimental support that this physics enjoys is support also for twistor theory! What is new in the twistor approach (and when I say 'new', I mean merely 16 years old) is a shift in emphasis in what are to be regarded as the primary elements of physical theory. Thus, while in the normal view space-time points are the primary ingredients in terms of which things are to be described, according to the twistor view, space-time points are *derived* rather than primary objects. So, whereas the twistor picture is initially equivalent to the normal space-time view (at least as regards special relativity), it offers scope for generalizations in which this ceases to be the case, where space-time points have become 'fuzzy' objects or have perhaps disappeared altogether.

The dual nature of the role of twistor theory should be clear from this. Inasmuch as it is equivalent to conventional physical theory, twistor theory provides us with a mathematical reformulation of the equations of physics – at least those which have so far been successfully transcribed – often yielding novel insights that are hard to come by in a more direct approach. In so far as the theory suggests new developments for physics, it does indeed offer scope for a radical change, the nature of which would differ in principle from any change that could plausibly be anticipated from the standpoint of conventional space-time theory. It should be emphasized, however, that while such a change is very much a part of the motivation behind the twistor approach, almost all of the work that has so far actually been carried out in the subject has been concerned with mathematical reformulations. The idea is that once an entirely twistorial description of some part of physics has been achieved, then the natural road to its most promising generalizations should become apparent!

I ought to make one comment by way of qualification of some of the above remarks. This concerns the twistor transcription of *general* relativity. It may be said that twistor theory owes much to Einstein's view of gravity. The idea of covariance, of the geometrization of physics, and of the importance of allowing the basic manifolds of the theory to become curved; all these have had essential roles to play in twistor theory. And in more than one place in the development of the theory, key ideas have arisen out of certain partially successful attempts to transcribe general relativity. Yet Einstein's theory has so far remained stubbornly resistant to a complete transcription into twistor terms. At present, it is not clear that a satisfactory transcription is possible at all, although in my view (not shared by all my colleagues!) there are some remarkably promising indications. My earlier comments about physical laws being in principle re-expressible in twistor terms do not, strictly speaking, apply to general relativity as things stand.

2. STANDARD TWISTOR DESCRIPTIONS

Let us recall, first, the standard flat-space twistor correspondence. The geometry is clearest in terms of the *projective* twistor space $\mathbb{PT}$, whose points are labelled by homogeneous complex coordinates $Z^0:Z^1:Z^2:Z^3$, where the Zs are the components of a twistor Z^α. A point of Minkowski space, with standard Minkowski coordinates t,x,y,z (metric diag (+1,-1, -1,-1)) is said to be *incident* with the twistor Z^α whenever the equation

$$\begin{pmatrix} Z^0 \\ Z^1 \end{pmatrix} = \frac{i}{\sqrt{2}} \begin{pmatrix} t+z & x+iy \\ x-iy & t-z \end{pmatrix} \begin{pmatrix} Z^2 \\ Z^3 \end{pmatrix} \tag{2.1}$$

holds. This equation may also be written in the 2-spinor form

$$\omega^A = ix^{AB'}\pi_{B'}, \tag{2.2}$$

where $(\omega^0,\omega^1,\pi_{0'},\pi_{1'}) = (Z^0,Z^1,Z^2,Z^3)$ and where $(x^{00'},x^{01'},x^{10'},x^{11'}) = 2^{-\frac{1}{2}}(t+z,\ x+iy,\ x-iy,\ t-z)$. The condition for (2.1) or (2.2) to have solutions in real Minkowski space $\mathbb{M}^I$ (i.e. t,x,y,z real, i.e. $x^{AB'}$ Hermitian) for a given twistor Z^α (with $\pi_{A'} \neq 0$), is

$$Z^\alpha \bar{Z}_\alpha = 0$$

where $(\overline{Z}_0, \overline{Z}_1, \overline{Z}_2, \overline{Z}_3) = (\overline{Z^2}, \overline{Z^3}, \overline{Z^0}, \overline{Z^1})$. The twistor norm $Z^\alpha \overline{Z}_\alpha$ can also be written

$$Z^\alpha \overline{Z}_\alpha = Z^0 \overline{Z^2} + Z^1 \overline{Z^3} + Z^2 \overline{Z^0} + Z^3 \overline{Z^2}$$

$$= \omega^A \overline{\pi}_A + \pi_{A'} \overline{\omega}^{A'}$$

which is a Hermitian form of signature (++--). The solutions for $x^{AB'}$ in M^I of (2.1) or (2.2) constitute a null straight line whenever $Z^\alpha \overline{Z}_\alpha = 0$, $\pi_{A'} \neq 0$. If $\pi_{A'} = 0$, we also say that $Z^{\underline{\alpha}}$ determines a null straight line, but this time a generator of $\mathscr{I}$, the light cone at infinity which completes $\mathbb{M}^I$ to a compact manifold $\mathbb{M}$. The real 5-dimensional hypersurface in $\mathbb{PT}$ that is given when $Z^\alpha \overline{Z}_\alpha = 0$ is called $\mathbb{PN}$.

We have just seen that the points of $\mathbb{PN}$ represent null geodesics in $\mathbb{M}$. In the reverse correspondence, any point of $\mathbb{M}$ represents a projective line in $\mathbb{PT}$ which lies entirely within $\mathbb{PN}$. This may be seen from the incidence relations (2.1) and (2.2) by holding the Minkowski point fixed and allowing Z^α to vary. The space of null geodesics through a fixed Minkowski point (i.e. the generators of its light cone) correspond in $\mathbb{PN}$ to this projective line (Fig. 2.1). The topology of each

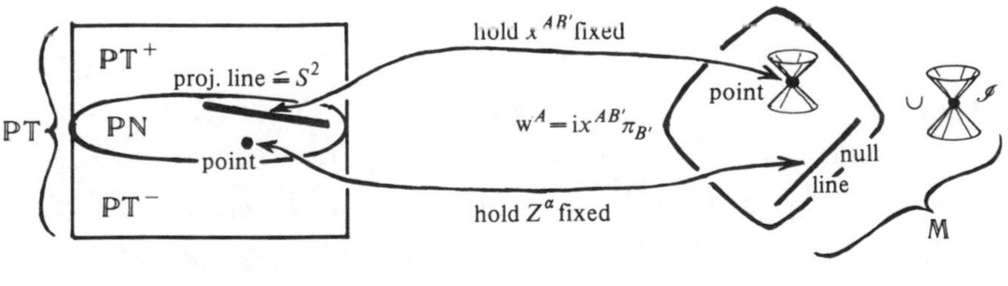

FIG. 2.1. The standard correspondence between $\mathbb{PN}$ and $\mathbb{M}$

such a projective line is a sphere S^2.

We also use the correspondence (2.2) or (2.1) in the case of *complexified* points ($x^{AB'}$) not necessarily Hermitian). Holding $x^{AB'}$ fixed we still get a projective line in $\mathbb{PT}$ to represent it, but where the line does not necessarily lie in $\mathbb{PN}$. In fact it lies entirely in $\mathbb{PT}^+$ ($Z^\alpha \overline{Z}_\alpha > 0$) or $\mathbb{PT}^-$ ($Z^\alpha \overline{Z}_\alpha < 0$) if the complexified Minkowski point lies in the *forward* or *backward tube*, respectively. (Holding Z^α fixed, we now find that the complexified Minkowski points which satisfy (2.1) or (2.2) constitute a type of totally null plane referred to as an α-plane.)

Note that the correspondence is a non-local one. A small, say convex, region in (complexified) space-time is represented by a non-local region in $\mathbb{PT}$ (swept out by projective lines) whose topology has the homotopy type of a sphere S^2. If we were to choose a 'smaller' region of $\mathbb{PT}$ not containing such a suitable topological sphere, or else some region of $\mathbb{PT}$ that is, in some sense, too 'bent' (so that none of the spheres that it contains has the holomorphic structure of a projective line) then the space-time 'points' that it attempts to represent would have somehow 'disappeared'. If we were to deform the twistor space itself, we might find that it then contains *no* global region which is sufficiently extensive to enable us to define a space-time point — even though the *local* holomorphic structure of the (projective) twistor space may be identical with that of $\mathbb{PT}$. Such 'deformed' twistor spaces in fact, play an essential role in the twistor description of general relativity (Penrose 1976), but where, in the normal situation, the deformations are not so great that the corresponding space-time points cease to exist.

3. MASSLESS FIELDS AND SHEAF COHOMOLOGY

The standard correspondence, whereby (massless) fields in space-time are defined in terms of holomorphic 'twistor functions', is also an essentially non-local one. We envisage that the relevant portion of twistor space (say $\mathbb{PT}^+$, for positive frequency fields) is covered by a locally finite system of open sets $\mathcal{U}_i$, the 'twistor function' being a system of holomorphic functions f_{ij} defined on the intersections $\mathcal{U}_i \cap \mathcal{U}_j$ of *pairs* of these open sets. The fs satisfy

$$f_{ij} = - f_{ji} \tag{3.1}$$

and are subject to the *cocycle condition*

$$f_{ij} + f_{jk} + f_{ki} = 0 \tag{3.2}$$

on each triple overlap $\mathcal{U}_i \cap \mathcal{U}_j \cap \mathcal{U}_k$. This system of fs is taken to be reduced *modulo* those f_{ij}s of the form of differences

$$h_i - h_j \tag{3.3}$$

where h_i is defined on $\mathcal{U}_i$ and h_j on $\mathcal{U}_j$. This means that the fs are really defining an element of the *first holomorphic sheaf cohomology group* of the space, with respect to the covering $\{\mathcal{U}_j\}$, (Morrow and Kodaira 1971; Wells 1979). To get rid of the dependence on the specific covering $\{\mathcal{U}_j\}$ we should take the direct limit with respect to finer and finer coverings. However, in practice, it is usually not necessary to take such a limit, where the sets $\mathcal{U}_i$ are taken to be appropriately 'convex' (Stein manifolds).

The passage from the twistor f_{ij} description to a space-time description can now be achieved by means of a *contour integral*. A point in $\mathbb{M}$, or else in its complexification $\mathbb{CM}$, is represented by a complex projective line in $\mathbb{PT}$, whose topology is S^2. This S^2 intersects the cover $\{\mathcal{U}_j\}$ in some system of open sets which cover S^2. In the simplest case there are just two (connected) sets in the cover $\{S^2 \cap \mathcal{U}_i\}$, which intersect in an annular region, and the contour integration is achieved using a contour running round this annulus (Fig. 3.1). In more complicated cases, this simple closed contour must be replaced by a suitable branched contour.

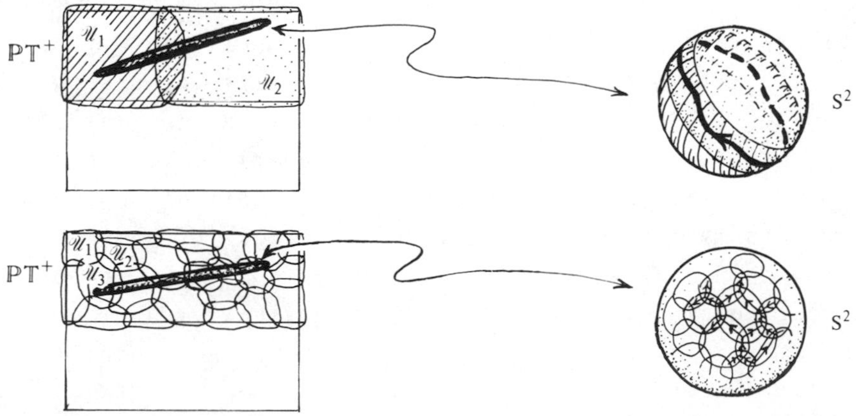

FIG. 3.1. The field at a space-time point is evaluated by contour integration.

The field defined in this way (Penrose 1969; 1979*a,b*; Penrose and Ward 1980; Ward this volume; Eastwood, Penrose, and Wells 1981) automatically satisfies the massless free field equations for helicity $\frac{1}{2}n\hbar$, where the twistor function f_{ij} is taken to be homogeneous of degree $-n-2$ in the twistor Z^{α}.

4. DEFORMED TWISTOR SPACES

A description in terms of first sheaf cohomology groups is very suggestive. For such cohomology groups may also play a role as *linearized deformations*. There is, indeed, a close similarity between functions defined on overlaps of pairs of open sets, subject to the relations (3.1) and (3.2), and *transition functions* used to build up a manifold out of coordinate patches. The key difference is that the relations (3.1) and (3.2) are linear, whereas the corresponding functional relations for transition functions are almost always non-linear. We may, indeed, obtain certain sheaf cohomology group elements as descriptions of *linearized* first-order deformations of manifolds. Then the 'gauge freedom' (3.3) arises from those linearized deformations which are trivial because they correspond simply to reparameterizations of the patches, defined by the h_is. In fact, finite (as opposed to infinitesimal) deformations can be thought of as being described by 'non-linear sheaf cohomology'. Then, in a sense, it is (3.3) which corresponds to the coordinate freedom in building a manifold, and (3.1) and (3.2) which correspond to the various compatibility relations for the transition functions.

This suggests that the sheaf cohomology group elements on twistor space that describe linear or non-interacting massless fields may perhaps be thought of as the linearized versions of some more general (non-linear) manifold construction. Indeed, this is precisely the case for twistor functions of homogeneity +2, describing linear massless particles of helicity $-2\hbar$ ($n = -4$): a non-linear version of such a twistor function gives rise to the 'non-linear graviton' construction whereby the general (if necessary, positive frequency) *anti-self-dual* solution of Einstein's vacuum equations is obtained from a deformed twistor space construction (Penrose 1976; Ward in this volume). Thus, in a certain sense, we have 'one half' of (vacuum) general relativity precisely coded into twistor terms. That is, we can produce the general Ricci-flat 'space-time' with anti-self-dual Weyl curvature. To complete the picture we would have also to produce the *self-dual* part of the curvature by a twistor space construction. (Of course, the above term 'one-half' must be interpreted suitably liberally. One may note that the only space-times which are both anti-self-dual and Lorentzian are flat!)

The nature of the non-linear-graviton construction is such as to suggest, initially, that this programme is totally unpromising. At first sight it would seem that one ought to pass to the *dual* twistor space in order to have a non-trivial self-dual curvature, and to some kind of deformed *product* of the twistor space with its dual in order that *both* self-dual *and* anti-self-dual parts to the (Weyl) curvature may co-exist. However, this could not be viewed as a 'non-linear' version of the twistor function cohomology construction. For in the linearized theory, both the self-dual ($n = 4$) and anti-self-dual ($n = -4$) fields are described on the same *twistor* space, the dual space being not required. Of course the entire construction could be replaced by one in which the dual space $\mathbb{T}^*$ is used *throughout*, instead of $\mathbb{T}$. This makes no difference except to interchange the concepts of right- and left-handedness. But for a consistent quantum theory, one should not have to *change* the quantum description that is being used, merely because the helicity changes sign (which would be analogous to having to pass from a position to a momentum description whenever certain quantum numbers lie outside some range). Thus, whether we choose $\mathbb{T}$ or $\mathbb{T}^*$ we ought to be able to choose just one of them *consistently*, and to describe the solutions of the *full* Einstein vacuum equations, without having to resort to a hybrid description in which $\mathbb{T}$ and $\mathbb{T}^*$ are each used for part of the construction.

We recall now, that for the anti-self-dual case, a suitable non-linear version of the +2 homogeneity twistor function exists ($-n-2 = +2$, when $n = -4$). What we require next is a corresponding non-linear version of the -6 homogeneity twistor function ($-n-2 = -6$, when $n = 4$). To understand what is presumably entailed, we think in terms of the (Poincaré invariant) exact sequence

$$\begin{array}{l} 0 \to \mathbb{S} \to \mathbb{T} \to \tilde{\mathbb{S}}^* \to 0 \\ \omega^A \mapsto (\omega^A, 0) \\ \qquad (\omega^A, \pi_{A'}) \mapsto \pi_{A'} \end{array}$$

in the first instance, where $\mathbb{S}$ stands for the 'unprimed' spin space, and $\tilde{\mathbb{S}}^*$ for the dual 'primed' spin space. The standard non-linear graviton construction is a 'deformed' version of the interpretation of a space-time point as a cross-section of the fibration $\mathbb{T} \to \tilde{\mathbb{S}}^*$, i.e. a deformed version of the flat-space map

$$\begin{array}{c} \mathbb{T} \underset{x}{\leftarrow\text{-}\text{-}} \tilde{\mathbb{S}}^* \\ (\mathrm{i}x^{AB'}\pi_{B'}, \pi_{A'}) \leftarrow\!\shortmid \pi_{A'} \end{array} \qquad (4.1)$$

What we now require is a suitably 'deformed' version (referred to as the 'googly' graviton construction, Penrose 1979*a,b*) of the conjugate dual map

$$\begin{array}{c} \mathbb{S} \underset{\tilde{x}}{\leftarrow\text{-}} \mathbb{T} \\ \omega^A - \mathrm{i}\tilde{x}^{AB'}\pi_{B'} \leftarrow\!\shortmid (\omega^A, \pi_{A'}) \end{array} \qquad (4.2)$$

The idea seems sound on very general grounds, but the detailed structure of the type of deformation that is required (which should depend on a non-linear version of a -6 homogeneity twistor function) seems remarkably elusive. At the present stage of understanding, there is no guarantee that such an idea will succeed. This is very much 'work in progress'. And it should be further remarked that even if it does succeed the problem of *combining* the deformed versions of (4.1) and (4.2) into a single scheme — wherein the most troublesome of the non-linearities of Einstein's equations would appear to reside (including those responsible for the violation of Huygens' principle, Ward in this volume) — would still remain to be solved.

In view of these most formidable difficulties, it may very reasonably be asked whether this programme has any chance at all of success. The non-linear graviton construction was found after all, in about 1975. If a solution to the remaining 'half' (or 'two-thirds') of the problem is to have any reasonable solution, then surely some clear indication of it should have become apparent in the five intervening years? To this I can only reply by pointing out some history. Twistor theory, as such, dates back essentially to 1964, while the 'twistor function' description of positive helicity (or self-dual) massless fields emerged in 1966. The description of negative helicity (or anti-self-dual) fields had to wait seven further years to appear! A simple suggestion made by Lane Hughston in about 1973 led to a new viewpoint in which the two helicities could be treated together within the same twistor scheme. Once having seen the suggestion, then, of course, all seemed obvious. But for seven years a somewhat misleading viewpoint had prevented us from noticing it! It should be pointed out also that it took us until 1976 to realize that we had been doing sheaf cohomology since 1966. (This was partly a matter of simple ignorance — though only partly. Michael Atiyah was very instrumental in educating us.) So it is, to me, not so surprising that we can as yet not see around whatever subtleties may be preventing us from understanding how general relativity is to be properly transcribed into twistor terms. In my own view, it is *necessary* for the theory that a solution can eventually be found. The theory ultimately stands or falls by it.

5. HYPERSURFACE TWISTORS

There are, however, various interesting things that twistor theory has to say about curved space-times, and about quantization in such backgrounds, even in the absence of a complete solution. One line of development concerns the construction of *hypersurface twistor space* (Penrose 1975). If we consider a normal Lorentzian space-time $\mathcal{M}$ (preferably globally hyperbolic) with a hypersurface $\mathcal{S}$ lying in it (preferably a spacelike Cauchy hypersurface — or else $\mathcal{I}^+$ or $\mathcal{I}^-$) then we can perform an analogous construction to that which produces $\mathbb{PN}$, from $\mathbb{M}$, to obtain a manifold $\mathbb{P}\mathscr{N}$ as the space whose points represent null geodesics in $\mathscr{M}$. The local structure that $\mathbb{P}\mathscr{N}$ acquires – essentially the structure that arises when a 5-real-dimensional hypersurface resides in a 3-complex-dimensional ambient space (although here the structure can be defined intrinsically) – is called a CR-structure (Folland and Kohn 1972; Nirenberg 1973). In space-time geometrical terms, it describes the *shear* that systems of null geodesics possess, just as they encounter the hypersurface $\mathscr{S}$ (see Fig. 5.1).

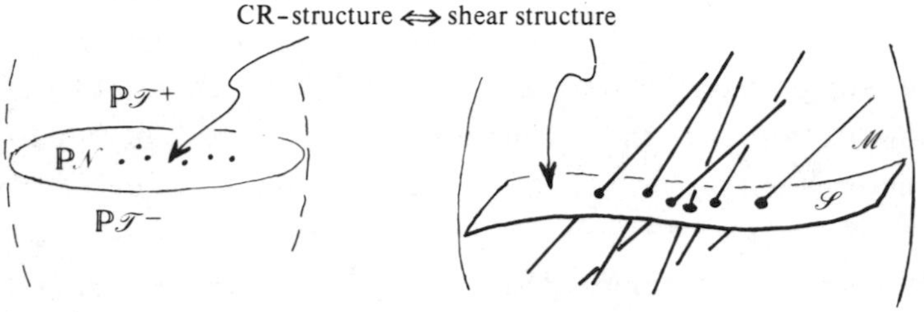

FIG. 5.1. The hypersurface twistor space for a hypersurface $\mathscr{S}$.

Assuming that $\mathscr{M}$ and $\mathscr{S}$ are analytic, it turns out that $\mathbb{P}\mathscr{N}$ is, indeed, a hypersurface in some ambient (projective) curved twistor space $\mathbb{P}\mathscr{T}$, which $\mathbb{P}\mathscr{N}$ divides into two portions $\mathbb{P}\mathscr{T}^+$ and $\mathbb{P}\mathscr{T}^-$. (In fact the space $\mathscr{T}$ can be interpreted as the space of solutions of the spinor ordinary differential equation $\pi^{C'}\pi_{D'}n^{BD'}\nabla_{BC'}\pi_{A'} = 0$, in the complexified space $\mathbb{C}\mathscr{S}$, n^a being the normal to $\mathbb{C}\mathscr{S}$ in $\mathbb{C}\mathscr{M}$. Factoring out by the equivalence $\lambda\pi_{A'} \equiv \pi_{A'}$ $(\lambda\pi_{A'} \neq 0)$ we obtain $\mathbb{P}\mathscr{T}$ (Penrose 1975).)

An essential difference, however, between the general construction and that which arises when $\mathscr{M}$ is (conformally) flat is that generally the CR-structure that $\mathbb{P}\mathscr{N}$ possesses (and therefore the very identities of points of $\mathbb{P}\mathscr{T}^{\pm}$) depends essentially on the location of the hypersurface $\mathscr{S}$. This is because, with conformal curvature, the Sachs equations (Sachs 1961; Penrose 1968) will entail a change in the shear as we pass along the various null geodesics. Thus, the CR-structure of $\mathbb{P}\mathscr{N}$ 'evolves with time' whenever $\mathscr{M}$ is conformally curved. (This is, strictly speaking, true even when $\mathscr{M}$ is static. In such cases, the CR-structure evolves *pointwise* on $\mathbb{P}\mathscr{N}$, as $\mathscr{S}$ is translated in time, even though as an entire abstract CR-manifold, the structure of $\mathbb{P}\mathcal{N}$ remains unaltered with time.)

6. AN APPROACH TO CURVED-SPACE QUANTIZATION

For the moment, let us set aside the fact that the CR-structure of $\mathbb{P}\mathcal{N}$ evolves and consider, instead, what the CR-structure of $\mathbb{P}\mathcal{N}$ can do for us. It is known (Ward 1977) that initial data for a (suitable) massless field at $\mathcal{S}$ can be given in terms of the first holomorphic sheaf cohomology of (portions of) $\mathbb{P}\mathcal{T}$. By analogy with the situation with $\mathbb{PT}$, we may expect to have some sort of positive-/negative-frequency splitting, depending upon whether the cohomology refers to $\mathbb{P}\mathcal{T}^+$ or $\mathbb{P}\mathcal{T}^-$. To be able to define such a splitting is a key requirement for building up a quantum field theory.

Now there is a deep theorem (Andreotti and Hill 1972; Hill and MacKichan 1977) which can be used to relate the cohomology of (any portion of) $\mathbb{P}\mathcal{T}$ to a kind of cohomology (referred to as hyperfunction-$\bar{\partial}_B$-cohomology; see Folland and Kohn 1972; Nirenberg 1973) which resides entirely 'within' the hypersurface $\mathbb{P}\mathcal{N}$. Referring to this cohomology as $\mathcal{H}^p(\mathcal{Q})$ $(p = 0,1,\ldots)$ and the corresponding holomorphic sheaf cohomology in $\mathbb{P}\mathcal{T}$ by $H^p(\mathcal{R})$, where $\mathcal{Q}$ and $\mathcal{R}$ are open subsets of $\mathbb{P}\mathcal{N}$ and of $\mathbb{P}\mathcal{T}$, respectively, the theorem implies

$$\mathcal{H}^1(\mathcal{U}\cap\mathbb{P}\mathcal{N}) = \frac{H^1(\mathcal{U}\cap\mathbb{P}\mathcal{T}^+)\oplus H^1(\mathcal{U}\cap\mathbb{P}\mathcal{T}^-)}{H^1(\mathcal{U}\cap\mathbb{P}\mathcal{T})} \tag{6.1}$$

where $\mathcal{U}$ is any open set in $\mathbb{P}\mathcal{T}$. Now consider first the special case $\mathbb{P}\mathcal{T} = \mathbb{PT}$. Provided that the various regions in $\mathbb{P}\mathcal{T}$ occurring on the right-hand side of this equation are of the appropriate kind, namely (in essence) that they can be described as the union of a set of projective lines, then the H^1 cohomology groups on the right all refer to massless fields in the corresponding regions of $\mathbb{CM}$. (I am supposing that the homogeneity degree is fixed throughout, so we have massless fields of a specific fixed helicity.) If, in fact, $\mathcal{U}$ is the whole of $\mathbb{PT}$, then we have $H^1(\mathcal{U}\cap\mathbb{PT}) = H^1(\mathbb{PT}) = 0$, so (9) becomes

$$\mathcal{H}^1(\mathbb{PN}) = H^1(\mathbb{PT}^+)\oplus H^1(\mathbb{PT}^-)$$

the two terms on the right describing, respectively, the positive- and negative-frequency fields. A similar splitting will occur in (6.1) in any case for which $H^1(\mathscr{U} \cap \mathbb{P}\mathscr{T}) = 0$. George Sparling has suggested, in fact, that (6.1) be used as a basis for an approach to quantum field theory in curved space-time. He shows first, that the left-hand-side of (6.1) can be interpreted (for suitable regions $\mathscr{U}$, and in particular for a region $\mathscr{U}$ which covers the whole of $\mathbb{P}\mathscr{N}$) as the space of initial data on the corresponding region of $\mathscr{S}$ for the appropriate massless fields (say Maxwell fields, so as to be safe from consistency conditions: Buchdahl 1958; Plebanski 1965). Thus, apart from the awkwardness of the 'denominator' in (6.1), this expression provides us with a *definition* of the required positive-/negative-frequency splitting.

We have two problems. First, we may have $H^1(\mathscr{U} \cap \mathbb{P}\mathscr{T}) \neq 0$, so that the positive-/negative-frequency splitting is not quite uniquely defined. Second, the CR-structure of $\mathbb{P}\mathscr{N}$ 'evolves with time' in the sense of being dependent on the choice of $\mathscr{S}$, so the positive-/negative-frequency splitting may be expected also to evolve with time. This is the phenomenon which is normally taken to be responsible for particle creation effects in curved space-time.

But, the first problem of the 'denominator' term in (6.1) still remains. One way to ensure the condition $H^1(\mathscr{U} \cap \mathbb{P}\mathscr{T}) = 0$ is to require that $\mathscr{U}$ is a Stein manifold. But, Stein manifolds contain no projective lines, so now the interpretation in terms of fields in space-time necessarily breaks down. But, it must still be the case that $\mathcal{H}^1(\mathscr{U} \cap \mathbb{P}\mathscr{N})$ has some space-time interpretation. We can choose arbitrarily small neighbourhoods of any point $\mathbf{p} \in \mathbb{P}\mathscr{N}$ which are Stein, and the points of such a neighbourhood represent null geodesics in $\mathscr{M}$ lying close to the particular null geodesic p in $\mathscr{M}$ which corresponds to $\mathbf{p}$. Thus, whatever interpretation $\mathcal{H}^1(\mathscr{U} \cap \mathbb{P}\mathscr{T})$ has in space-time, it must have to do with the propagation of fields in the neighbourhood of p (Penrose 1979*b*). In fact, $\mathcal{H}^1(\mathscr{U} \cap \mathbb{P}\mathscr{N})$ is infinite-dimensional, even in the limit when $\mathscr{U}$ shrinks down to the point $\mathbf{p} \in \mathbb{P}\mathscr{N}$ itself. This limiting infinite-dimensional space seems to correspond to propagation non-analytic information along p. There is a unique positive-/negative-frequency splitting in this limit – as, indeed, there is whenever $\mathscr{U}$ is Stein. But the significance of this for curved-space quantum field theory is not at all clear. It is not even understood, at present, how the evolution in CR-structure that occurs when $\mathscr{S}$ is moved up in time affects this splitting. However, it should be a much easier question to examine than the general global question of the evolution of positive-/negative-frequency splitting since here we do not need to examine the structure of the entire hypersurface $\mathscr{S}$ and its embedding in $\mathscr{M}$, but merely at its intersection with the null geodesic p.

It would be interesting to examine the relation between Sparling's type of approach and the more conventional attacks on the problem of quantization (of massless fields) in a curved-space background. Clearly much more work needs to be done here.

One aspect of all this which I find intriguing is its possible relation to something that has always struck me as an irony inherent in the subject of quantum field theory. Classically, the concept of 'particle' arose out of attempts to localize the structure of matter into its smallest indivisible units. 'Locality' is, after all, central to the notion of a classical particle. Even in quantum field theory, a 'particle' ought intuitively to represent the most highly localizable of individual physical objects. Yet as theory stands, the concept of particle requires, strictly speaking, a knowledge of the structure of space-time even in its most distant regions. The detailed effect of the shape of distant regions can presumably be generally ignored. But it is an oddity that the concept 'particle' cannot really be *defined* in quantum field theory, without reference to them. Perhaps there may be new insights to be gained by use of an approach in which the concepts 'local' and non-local' have shifted their meanings!

REFERENCES

Andreotti, A. and Hill, C.D. (1972). *Ann. Scu. norm. sup. Pisa* CI. *Sci.* **26**, 325, 747.
Buchdahl, H.A. (1958). *Nuovo Cim.* **10**, 96.
Eastwood, M.G., Penrose, R., and Wells, R.O., Jr. (1981). *Commun. Math. Phys.* **78**, 305.

Folland, G.B. and Kohn, J.J. (1972). *The Neumann Problem for the Cauchy-Riemann Complex*. Princeton University Press.
Hill, C.D. and MacKichan, B. (1977). *Ann. Scu. norm. sup. Pisa* CI. *Sci.* **4**, 577.
Hughston, L.P. and Ward, R.S. (Eds) (1979). *Advances in Twistor Theory*. Pitman, London.
Morrow, J. and Kodaira, K. (1971). *Complex Manifolds*. Holt, Rienhard, and Winston, New York.
Nirenberg, L. (1973). *Lectures on Partial Differential Equations*. Amer. Math. Soc., Providence, Rhode Island.
Penrose, R. (1967). *J. Math. Phys*. **8**, 345.
—— (1968). In *Battelle Rencontres*. (ed. C.M. DeWitt and J.A. Wheeler). Benjamin, New York.
—— (1969). *J. Math. Phys*. **10**, 38.
—— (1975). In *Quantum Gravity, an Oxford Synposium*. (ed. C.J. Isham, R. Penrose, and D.W. Sciama). Oxford University Press.
—— (1976). *Gen. rel. Grav*. **7**, 31.
—— (1979*a*). In *Complex Manifold Techniques in Theoretical Physics*. (ed. D.E. Lerner and P.D. Sommers). Pitman, London.
—— (1979*b*). In *Advances in Twistor Theory*. (ed. L.P. Hughston and R.S. Ward). Pitman, London.
—— and MacCallum, M.A.H. (1972). *Phys. Reports*. 6, 241.
—— and Ward, R.S. (1980). In *General Relativity and Gravitation, One Hundred Years after the Birth of Albert Einstein*. Vol. 2. (ed. A. Held). Plenum, New York.
Plebanski, J. (1965). *Acta Phys. Polon*. **27**, 361.
Sachs, R. (1961). *Proc. R. Soc. Lond*. **A264**, 309.
Wells, R.O., Jr. (1979). *Differential Analysis on Complex Manifolds*. Springer-Verlag, New York.
Ward, R.S. (1977). *Curved Twistor Spaces*. D. Phil. thesis, Oxford.

26

THE TWISTOR APPROACH TO DIFFERENTIAL EQUATIONS

R.S. Ward

Dept. of Mathematics, Trinity College, Dublin

1. INTRODUCTION

The two most important beliefs that motivate twistor theory are, first, that the space-time continuum is an inappropriate arena for discussing quantum field theory and quantum gravity; and, secondly, that complex numbers and holomorphic (i.e. complex-analytic) functions ought to play a more important role than is customary in relativity (Penrose 1975). The idea is that physics should be formulated not in space-time, but in a complex manifold called twistor space. Perhaps the most remarkable achievement of the twistor programme is that it has provided a link between certain classical field equations, and modern algebraic geometry and complex analysis; a link which has led to a greater understanding of these equations, and techniques for solving them. The object of my lecture is to review this aspect of the programme, and to summarize the results that have been obtained by several people over a number of years.

This part of twistor theory is the best known, but it is nevertheless only a small part; and in addition to being interesting and useful in its own right, it may be viewed as providing support for the far more ambitious (though as yet unfulfilled) aims of the twistor programme, which include the achievement of a better understanding of quantum gravity.

2. THE WAVE EQUATION AND TWISTORS IN FLAT SPACE

In relativistic field theories one deals with hyperbolic equations in space-time: for example, the wave equation, the Dirac equation, Maxwell's equations, the Yang–Mills equations, Einstein's equations, and so forth. For purposes of quantization, one is also interested in the elliptic equations obtained by analytically continuing these equations from Lorentzian space-time to positive-definite 4-space. What all these equations have in common is that the general solution of the equation depends on one or more *arbitrary* functions of *three* variables. For example, in the hyperbolic case, the specification of initial data on a (three-dimensional) space-like hypersurface uniquely determines a solution throughout space-time.

Another example of this is the following solution formula, due to Bateman (1904): the general real-analytic solution of the wave equation $\Box\phi = 0$ in flat space-time is

$$\phi(x,y,z,t) = \int_{-\pi}^{\pi} F(x\cos\theta + y\sin\theta + iz,\ y + iz\sin\theta + t\cos\theta, \theta)\,d\theta\ ,$$

where $F(.,.,.)$ is an arbitrary function of three variables, complex-analytic in the first two. The function F is, in effect, a function on (projective) *twistor space*, and twistor theory provides a way of understanding and generalizing this solution formula.

The essential features of the twistor-space/space-time correspondence are as follows. More detailed descriptions may be found in Penrose (1975) and Penrose and Ward (1980). The letters $a,b,c,\ldots$ are used for space-time 4-vector indices, while $A,B,C,\ldots$ and $A',B',C',\ldots$ are 2-component spinor indices. In Minkowski space-time, with metric $ds^2 = (dx^0)^2 - (dx^1)^2 - (dx^2)^2 - (dx^3)^2$, a 4-vector v^a has as its spinor equivalent the spinor $v^{AA'}$, where

$$\begin{bmatrix} v^{00'} & v^{01'} \\ v^{10'} & v^{11'} \end{bmatrix} = 2^{-\frac{1}{2}} \begin{bmatrix} v^0 + v^1 & v^2 + iv^3 \\ v^2 - iv^3 & v^0 - v^1 \end{bmatrix} .$$

A point x in Minkowski space-time is represented by its position vector x^a (or $x^{AA'}$) with respect to some origin. Note that the condition that x^a be *real* is equivalent to the condition that the 2 x 2 matrix $x^{AA'}$ be *Hermitian*. To obtain *complexified* Minkowski space-time, we just allow the x^a to become complex, or (equivalently) allow the matrix $x^{AA'}$ to be *non*-Hermitian.

Projective twistor space PT is the three-dimensional complex projective space $\mathbb{C}P^3$. [$\mathbb{C}P^n$ is obtained by taking the space $\mathbb{C}^{n+1}$ of $(n+1)$-tuples of complex numbers $(z^1, z^2, \ldots, z^{n+1})$ and factoring out by the proportionality relation $(\lambda z^1, \ldots, \lambda z^{n+1}) \sim (z^1, \ldots, z^{n+1})$, with $\lambda \in C$, $\lambda \neq 0$. The numbers $z^1, \ldots, z^{n+1}$ serve as *homogeneous co-ordinates* on $\mathbb{C}P^n$.] In the case of $PT \cong \mathbb{C}P^3$, let us call the homogeneous co-ordinates $(\omega^0, \omega^1, \pi_{0'}, \pi_{1'}) \equiv (\omega^A, \pi_{A'})$.

The basic correspondence between space-time and twistor space is given by the equation

$$\omega^A = ix^{AA'}\pi_{A'} \, . \tag{2.1}$$

This equation can be read in two different ways. First, consider $(\omega^A, \pi_{A'})$ to be fixed, assuming for the moment that $\pi_{A'} \neq 0$ (meaning that at least one of $\pi_{0'}$ and $\pi_{1'}$ is non-zero). Then (2.1), considered as an equation for $x^{AA'}$, has as its solution a complex 2-plane in complexified Minkowski space, whose tangent vectors are of the form $v^{AA'} = \xi^A\pi^{A'}$ for some ξ^A. Such a 2-plane is called an *α-plane*. We could also refer to α-planes as *self-dual totally null* planes, because if v and u are any two vectors tangent to an α-plane, then v and u are orthogonal and $v \otimes u - u \otimes v$ is a self-dual bivector. [A skew-symmetric tensor $F^{ab} = -F^{ba}$ is self-dual if

$$^*F^{ab} := \frac{1}{2}\varepsilon^{ab}{}_{cd}F^{cd} = iF^{ab} \, ,$$

where ε_{abcd} is a totally skew tensor with $\varepsilon_{0123} = 1$. The spinor version of the self-duality condition is that $F^{AA'BB'}$ should have the form $F^{AA'BB'} = \varepsilon^{AB}\phi^{A'B'}$, where $\phi^{A'B'}$ is symmetric and ε^{AB} is skew, $\varepsilon^{01} = 1$. See Penrose and MacCallum (1973). The bivector $v \otimes u - u \otimes v$ has this form, where $\phi^{A'B'}$ is proportional to $\pi^{A'}\pi^{B'}$.] So $(\omega^A, \omega_{A'})$ determines an α-plane in complexified space-time; because (2.1) is homogeneous in $(\omega^A, \pi_{A'})$, this α-plane is unchanged if we replace $(\omega^A, \pi_{A'})$ by $(\lambda\omega^A, \lambda\pi_{A'})$, where $\lambda \neq 0$. Finally, *every* α-plane arises in this way, namely as a solution of (2.1).

If we look for *real* solutions x^a of eqn (2.1), it turns out that a solution exists only if

$$\omega^A\bar{\pi}_A + \bar{\omega}^{A'}\pi_{A'} = 0 \, . \tag{2.2}$$

Here $\bar{\pi}_A$ denotes the complex conjugate of $\pi_{A'}$, i.e. $\bar{\pi}_0 = \overline{\pi_{0'}}$ and $\bar{\pi}_1 = \overline{\pi_{1'}}$; and similarly for $\bar{\omega}^{A'}$. If (2.2) is satisfied, and as before $\pi_{A'} \neq 0$, then the solution space of (2.1) in *real* Minkowski space-time is a *null geodesic*; and every null geodesic arises in this way.

If $\pi_{A'} = 0$, the point $(\omega^A, \pi_{A'}) = (\omega^A, 0)$ in twistor space can be interpreted in space-time as an α-plane (or null geodesic) at infinity. To understand these one has to go to *compactified* space-time. We shall not enter into the details of this here.

The other way of interpreting equation (2.1) is that of regarding $x^{AA'}$ as fixed and solving for $(\omega^A, \pi_{A'})$. Clearly the solutions are obtained by letting $\pi_{A'}$ take on any value and taking ω^A to be $ix^{AA'}\pi_{A'}$. So the solution space is a complex 2-plane; after factoring out by the proportionality relation $(\lambda\omega^A, \lambda\omega_{A'}) \sim (\omega^A, \pi_{A'})$ this becomes a complex projective 1-space $\mathbb{C}P^1$ (which is topologically a 2-sphere S^2). So the space-time point x determines a Riemann sphere $L_x \cong \mathbb{C}P^1$ in projective twistor space. If x is a *real* point, then L_x lies entirely within the space PN of *null twistors*, i.e. those twistors whose homogeneous co-ordinates $(\omega^A, \pi_{A'})$ satisfy (2.2).

To sum up, the correspondence between space-time and twistor space is expressed by:

(complex) α-plane ↔ point in PT
(real) null geodesic ↔ point in PN
complex space-time point x ↔ sphere L_x in PT
real space-time point x ↔ sphere L_x in PN.

See Fig.1. A crucial feature of the correspondence is the way in which it encapsulates the conformal (i.e. the light-cone) structure of space-time. We say that two points x and y are *null-separated* if the vector $y^a - x^a$ has zero length with respect to the Minkowski metric, i.e. if y lies on the null cone of x. And it turns out that

$$x \text{ and } y \text{ are null-separated} \Leftrightarrow L_x \text{ and } L_y \text{ intersect.} \quad (2.3)$$

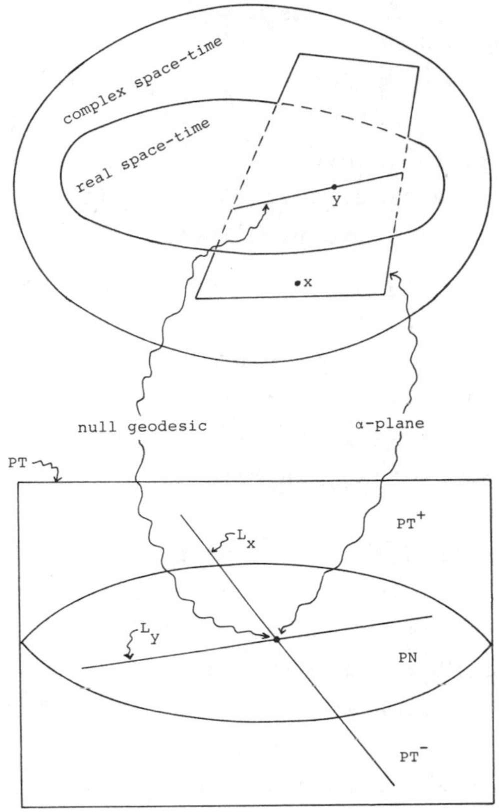

FIG.1.

Two special versions of the correspondence are worth mentioning. First, the points of positive-definite (Euclidean) 4-space correspond to a certain subfamily of the spheres L_x in PT, and this positive-definite version has been extensively developed; see, for example, Atiyah, Hitchin, and Singer (1978). Secondly, the spheres L_x which lie entirely in the 'top half' PT^+ of twistor space (consisting of points satisfying $\omega^A\bar{\pi}_A + \bar{\omega}^{A'}\pi_{A'} > 0$) correspond to points x lying in the 'forward tube' $\mathbb{C}M^+$ of complexified Minkowski space-time. The forward tube is the natural domain of definition of positive frequency fields (Streater and Wightman 1964).

We can now re-write Bateman's solution formula for the wave equation in a more natural-looking form. Let $f(\omega^A,\pi_{A'})$ be a complex-analytic function, homogeneous of degree -2 [i.e. $f(\lambda\omega^A,\lambda\pi_{A'}) = \lambda^{-2}f(\omega^A,\pi_{A'})$], and possibly having singularities.

Define a field $\phi(x^a)$ by:

$$\phi(x^a) = \frac{1}{2\pi i} \oint f(ix^{AA'}\pi_{A'}, \pi_{B'})\pi_{C'}d\pi^{C'} ,$$

the integral being taken over any closed one-dimensional contour that avoids the singularities of f. It is easy to check that $\Box\phi = 0$, and it is also true (but more difficult to prove) that *every* solution of $\Box\phi = 0$ arises in this way. The reason for assuming f to have homogeneity -2 is that then the 1-form $f\ \pi_{C'}d\pi^{C'}$ will have homogeneity 0 and hence be a 1-form on the *projective* space PT (or rather on some subregion of PT, because in general it will have singularities somewhere). The homogeneity means that, in effect, f is a free function of *three* variables.

The fact that f is not defined on the whole of PT, and also that ϕ does not determine f uniquely (we can replace f by $f+h$, where h is any function whose integral is zero), seem to make matters rather complicated. It turns out to be correct to regard f not as a function in the usual sense, but as an element of a *sheaf cohomology group* (see, for example, Penrose 1979*a*). One then obtains results such as the following, which relates to positive frequency fields.

Theorem. There is a natural, conformally invariant isomorphism between the following two complex vector spaces:

(a) the space of holomorhpic solutions of $\Box\phi(x^a) = 0$ on the forward tube $\mathbb{C}M^+$; and
(b) the sheaf cohomology group $H^1(PT^+, \mathcal{O}(-2))$

In effect, elements of $H^1(PT^+, \mathcal{O}(-2))$ are just arbitrary* functions of three variables, not subject to any differential

---

* By 'arbitrary function' is meant 'arbitrary holomorphic (i.e. complex-analytic) function'. The role played by the complex-analyticity is crucial.

equation; the differential equation $\Box\phi = 0$ has disappeared in passing from the space-time picture to the twistor space picture. In the next section we shall consider to what extent this result can be generalized.

3. HUYGENS' PRINCIPLE

In the previous section we saw that solutions of the wave equation in flat space-time correspond to free functions on three-dimensional twistor space. It seems likely that the reason why this works is closely tied up with the way in which the light-cone structure of space-time is coded into the space-time/twistor-space correspondence (cf. 2.3), and with the fact that the wave equation in flat space-time satisfies *Huygens' Principle* (HP). Let us begin by recalling what HP is.

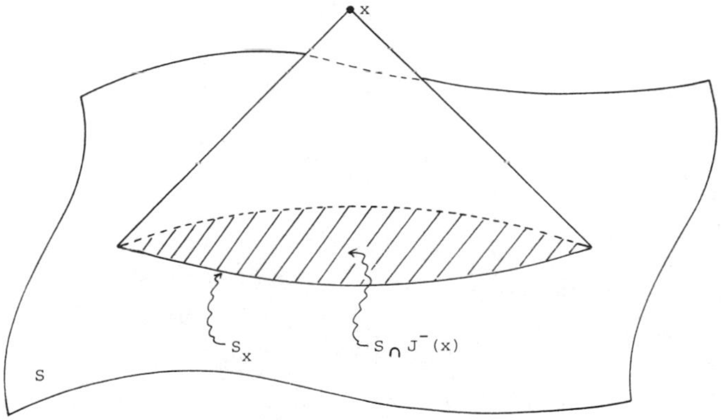

FIG. 2.

Suppose we have a hyperbolic equation in a (fixed) Lorentzian space-time. Then initial data on a Cauchy surface S determines

the field everywhere else in space-time. More precisely, the field at the space-time point x is determined by the initial data on $S \cap J^{-}(x)$. [$J^{-}(x)$ denotes the causal past of x (Hawking and Ellis 1973). For the sake of simplicity we are assuming that the space-time is globally hyperbolic.] See Fig. 2. The physical interpretation is that signals may travel at, or more slowly than, the speed of light, but no faster. This is a property of all hyperbolic systems.

In some hyperbolic systems, however, the propagation of signals is particularly neat and simple: signals travel at *exactly* the speed of light, and no slower. In other words:

HP1 The field at x is determined by the initial data on the boundary S_x of $S \cap J^{-}(x)$.

See Fig. 2. This is one possible formulation of Huygens' principle. To study some other possible definitions, let us consider as an example the conformally-invariant wave equation in curved space-time:

$$\Box\phi + \frac{1}{6}R\phi = 0 \ . \tag{3.1}$$

HP2 For each x_0, (3.1) has a solution of the form $\phi(x) = f(x)/\Gamma(x,x_0)$, where f is smooth and $\Gamma(x,x_0)$ is the square of the geodesic distance from x to x_0.

HP3 For each x_0, (3.1) has a (distributional) solution supported on the null cone of x_0.

HP4 For each x_0, there exist smooth functions $g(x), h(x)$ such that the hypersurface $g(x) = 0$ is the future null cone of x_0, and such that $\phi(x) = h(x)F(g(x))$ is a solution of (3.1) for every function $F:\mathbb{R} \to \mathbb{R}$.

HP2 says that there exists an elementary solution in the sense of Hadamard; HP3 refers to elementary solutions in the more modern sense; and HP4 refers to simple progressing waves (Friedlander 1975). HP4 implies that if a source is abruptly switched off, then the signal observed at a field point will also end abruptly: it will not 'ring on'. In the case we are considering, namely that of the wave equation, the definitions HP1,2,3,4 are all equivalent (cf. McLenaghan 1969; Friedlander

1975), and there are theorems which tell us when HP holds. For example, there is the following.

Theorem. (McLenaghan 1969). Among vacuum space-times ($R_{ab} = 0$), the wave equation satisfies HP if and only if the space-time is either flat or a plane wave.

Remarks

(a) This theorem emphasizes the fact that 'HP equations' are rather rare: almost every hyperbolic equation does *not* satisfy HP. That they crop up so often is because they describe the simplest situations, in which the field propagates 'cleanly', without back-scattering.

(b) The above definitions of HP are all essentially *local* in space-time — one may talk of HP holding in some neighbourhood. But HP has global consequences: essentially it implies that no global scattering occurs, that the in-data at past infinity equals the out-data at future infinity. This can be made more precise, for example as follows. Suppose we are dealing with a conformally invariant equation in flat space-time. Regard $\mathscr{I}^-$ and $\mathscr{I}^+$, the null cones at past and future infinity (Hawking and Ellis 1973), as the initial- and final-data surfaces. There is a natural way of matching up points on $\mathscr{I}^+$ with points on $\mathscr{I}^-$ (Penrose 1974), so we can compare the initial data with the final data. For the wave equation in flat space-time, it turns out that

$$\{\text{data on } \mathscr{I}^+\} = -\{\text{data on } \mathscr{I}^-\}$$

(Grgin 1966; Lerner and Clarke 1977). Similar results hold for other HP equations.

Let us move on now to consider some of these other equations, and their description in the twistor picture.

1. Massless fields of helicity $s\hbar$, for $s = \pm\frac{1}{2}, \pm 1, \ldots$:

$$\nabla^{AA'}\phi_{AB\ldots D} = 0\,,$$

$$\nabla^{AA'}\phi_{A'B'\ldots D'} = 0\,. \qquad (3.2)$$

Here, as before, one can write down definitions of Huygens' Principle, analogous to HP1–4. These need no longer be equivalent; for example Maxwell's equations ($s = \pm 1$), in a gravitational plane-wave background, satisfy HP1 (Künzle 1968) but not HP3 or HP4 (Penrose 1972). However, in flat space-time the eqns. (3.2) satisfy all of HP1–4; and there is a twistor space description of these massless free fields which is very similar to that of the scalar field in Section 2. It simply involves working with twistor functions $f(\omega^A, \pi_{A'})$ homogeneous of degree $-2s-2$ rather than -2. For details, see Penrose (1979*a*). The 'no global scattering' statement becomes

$$\{\text{data on } \mathscr{I}^+\} = i^{2s-2}\,\{\text{data on } \mathscr{I}^-\}\ .$$

2. The self-dual Yang–Mills equations:

$$^*F_{ab} = iF_{ab}\ . \tag{3.3}$$

Here F_{ab} is the Yang–Mills field strength, a 2-form taking values in the Lie algebra of the gauge group G (which need not be SU(2)). The gauge fields satisfying (3.3) form a proper subset of the space of solutions of the Yang–Mills equations

$$D^a F_{ab} = 0\ . \tag{3.4}$$

Self-dual Yang–Mills fields have cropped up in several different areas: for example in connection with instantons or pseudoparticles (Belavin, Polyakov, Schwartz and Tyupkin 1975), Yang–Mills–Higgs monopoles (Manton 1978), and coherent helicity eigenstates (Duff and Isham 1980).

The eqns. (3.3) are a set of coupled first-order non-linear partial differential equations for the gauge potential A_a, since F_{ab} is defined in terms of A_a by $F_{ab} = \nabla_a A_b - \nabla_b A_a + [A_a, A_b]$. I claim that (3.3) satisfies Huygens' Principle, although I shall neither prove this claim here nor define exactly what I mean by it. Essentially, (3.3) satisfies both HP1 (if one takes gauge freedom into account) and HP4 (cf. Kovaks and Lo 1979). In addition, there is, as before, no global scattering:

$$\{\text{field on } \mathscr{I}^+\} = \{\text{field on } \mathscr{I}^-\} .$$

So even though the field, being non-linear, interacts with itself, this interaction has completely unravelled by the time the field has arrived at future infinity.

The twistor translation of self-dual Yang–Mills theory in flat space-time involves vector bundles over (subregions of) twistor space PT (Ward 1977). As an example, for positive frequency fields there is the following result.

Theorem. There is a one-to-one correspondence between
(a) anti-self-dual Yang–Mills fields on the forward tube $\mathbb{C}M^+$, with gauge group $GL(n,\mathbb{C})$; and
(b) holomorphic n-dimensional vector bundles E over PT^+, such that

$$\text{E } \textit{restricted to any } L_x \textit{ in } PT^+ \textit{ is analytically trivial.} \tag{3.5}$$

(Here L_x means the sphere in PT corresponding to the space-time point x: cf. Section 2). So again one finds that the differential equation in space-time disappears when one passes to the twistor picture. A version of the theorem adapted to compactified, positive-definite 4-space S^4 and to the gauge group SU(2) has proved to be very useful in the study of instanton solutions (Atiyah and Ward 1977; Atiyah, Hitchin, Drinfield and Manin 1978; Christ, Weinberg, and Stanton 1978; Corrigan, Fairlie, Templeton, and Goddard 1978).

3. Several other examples are known of systems of equations satisfying HP and possessing a twistor solution procedure. First, there are combinations of the cases discussed above, such as a scalar field minimally coupled to a self-dual Yang–Mills field:

$$D_a D^a \phi = 0 ,$$

where $D_a = \nabla_a + A_a$ (Hitchin 1980). Secondly, there are systems of equations analogous to the self-dual Yang–Mills equations, but involving a mixture of fields of different helicities.

These were first noticed by Sparling (1977). An example is the following system: three fields ψ_A, $\phi_{AA'}$ and $\xi_{AA'B'}$ with helicities $\frac{1}{2}\hbar$, $\hbar$, and $\frac{3}{2}\hbar$ respectively, satisfying

$$\nabla^A{}_{A'}\psi_A + \phi^A{}_{A'}\psi_A = 0 \ ,$$

$$\nabla^A{}_{(A'}\phi_{B')A} + 2\psi^A\xi_{AA'B'} = 0 \ ,$$

$$\nabla^A{}_{(A'}\xi_{B'C')A} - \phi^A{}_{(A'}\xi_{B'C')A} = 0 \ , \tag{3.6}$$

the round brackets denoting symmetrization over the enclosed indices. These equations possess the gauge freedom

$$\psi_A \mapsto e^{-\theta}\psi_A \ ,$$

$$\phi_{AA'} \mapsto \phi_{AA'} + \nabla_{AA'}\theta - \psi_A\mu_{A'} \ ,$$

$$\xi_{AA'B'} \mapsto e^{\theta}\{\xi_{AA'B'} - \nabla_{A(A'}\mu_{B')} + \phi_{A(A'}\mu_{B')} - \psi_A\mu_{A'}\mu_{B'}\} \ ,$$

where θ and $\mu_{A'}$ are arbitrary functions of x^a. Solutions of the system (3.6), modulo the gauge freedom, correspond to holomorphic 2-dimensional vector bundles E over twistor space, satisfying instead of (3.5) the conditions

> E *restricted to* L_x *is isomorphic to* $1 \oplus H$, *and*
> $\det E = H$.

Here 1 denotes the trivial line bundle, H the Hopf bundle over $\mathbb{C}P^1$ (i.e. the line bundle with Chern class +1), and det E the determinant line bundle of E.

But what about the equations that do *not* satisfy Huygens' Principle? The implication of earlier comments would seem to be that any twistor space version of such equations would be qualitatively different from that of systems satisfying HP. Not enough is known to make any definite statement about this. Let me just mention two examples of equations that do not satisfy HP. First, linear equations describing particles with mass, such as the Klein–Gordon and Dirac equations, are usually dealt with in twistor theory by using a product of two or more twistor spaces, rather than just one twistor space. See, for example, Hughston (1979). In this case, the differential equations are still present in the twistor picture, and do not disappear. Secondly, the Yang–Mills equations (3.4) (*without* the self-duality condition) have a twistor description which involves vector bundles over a *five*-dimensional complex manifold, and satisfying a rather awkward condition (Isenberg, Yasskin, and Green 1978; Witten 1978). This description is not yet fully understood, but certainly it seems to be much more complicated than all the previously-mentioned cases.

4. TWISTORS FOR CURVED SPACE-TIME

The previous sections dealt with the twistor approach to differential equations in *flat* space-time. Two related questions now present themselves: is there a useful twistor translation of differential equations in a (fixed) *curved* space-time background, and is there a twistor description of Einstein's equations for the space-time itself? To begin with, let us consider some features of curved space-time that have a bearing on these questions.

Recall that a congruence (i.e. a three-parameter family) of null geodesics (light rays) is said to be *shear-free* if a small circular image is not distorted into an ellipse as it propagates along the congruence (Penrose 1968). In Minkowski space-time, a congruence which 'starts off' shear-free, remains shear-free as one moves along the null geodesics. There is a neat twistor description of shear-free congruences in flat space, due to Kerr (Penrose 1967). A congruence, being a three-parameter family of null geodesics, corresponds to a three-real-dimensional surface C in PN, the space of null twistors. Kerr's theorem states (essentially) that the congruence is shear-free if and only if C is the intersection with PN of a *complex-analytic-surface* in PT. So the complex structure of PT, the fact that we are able to talk about *holomorphic* twistor functions, is closely connected with the existence and behaviour of shear-free congruences in flat space-time.

In curved space-time, however, the situation is very different. A congruence of null geodesics which starts off shear-free and then enters a region of conformal curvature, will generally pick up shear. To put this another way, a generic space-time possesses no shear-free congruences at all. This would seem to indicate that a twistor space with a natural complex structure can no longer be defined. This conclusion is supported by the observation that in general Huygens' Principle does *not* hold, either for (say) the wave equation on a general curved background, or for Einstein's equations. So how is one to proceed?

Several approaches to the problem were reported at the first Oxford Quantum Gravity Symposium: local twistors, global null twistors, hypersurface twistors, and asymptotic twistors (Penrose 1975). Of these, the last two appear to be the most promising. They involve choosing a hypersurface S in the space-time and defining a twistor space PT(S) with respect to this hypersurface. This twistor space has a natural complex structure, which however depends on S: if one changes S, then the complex structure of PT(S) is 'shifted'. See Penrose (1975) and Penrose and Ward (1980) for more details.

Returning to the question of existence or non-existence of shear-free congruences, recall that the shear σ of a congruence is a *complex-valued* function on space-time (Penrose 1968). This means that if we pass from real to complex space-times, then the shear σ and its complex conjugate $\bar{\sigma}$ are replaced by two *independent* complex-valued functions σ and $\tilde{\sigma}$, which we may call respectively the left- and right-handed shear of a congruence of complex null geodesics. In particular, we may now look for right-shear-free congruences that are not necessarily left-shear-free (i.e. $\tilde{\sigma} = 0$, $\sigma \neq 0$).

As was remarked above, shear is created by conformal curvature. In a real, Lorentzian space-time, the Weyl conformal curvature tensor C_{abcd} may be decomposed into its self-dual and anti-self-dual parts as follows:

$$C_{abcd} = C^{+}{}_{abcd} + C^{-}{}_{abcd}$$

$$^{*}C^{\pm}{}_{abcd} \equiv \tfrac{1}{2}\varepsilon_{ab}{}^{ef}C^{\pm}{}_{efcd} = \pm\, iC^{\pm}{}_{abcd}\ ;$$

and C^{+} and C^{-} are complex conjugates of each other. Either they both vanish (in which case the space-time is conformally flat) or neither of them vanishes (in which case the space-time is conformally curved). But when we consider complex space-times, then C^+ and C^- become *independent* tensors and one of them may vanish without the other necessarily vanishing as well. It turns out that right-handed shear is created by the self-dual part C^+ of the Weyl tensor, and left-handed shear by the anti-self-dual part. So in a complex space-time in which $C^+ = 0$, there are plenty of right-shear-free congruences, although (if $C^- \neq 0$) there may not be any left-shear-free congruences.

All of this suggests that there might be a very natural generalization of flat-space twistor theory to complex space-times where the Weyl tensor is either self-dual or anti-self-dual. That this is indeed the case was discovered by Penrose (1976), who showed that such space-times correspond to *complex-analytic deformations* of suitable subsets of the 'flat' twistor space PT. Furthermore, he showed that the twistor translation of Einstein's vacuum equations $R_{ab} = 0$ involves no differential equations at all; again we find that complicated differential equations 'disappear' on passing to the twistor picture. Of course, this procedure does not give the general solution of Einstein's equations: it applies only to space-times with self-dual or anti-self-dual Weyl tensor. Such space-times cannot be real with Lorentzian signature (because, as remarked above, C^+ and C^- are complex conjugates of each other in a Lorentzian space), but they may be complex, or real with signature ++++ or ++--. Self-dual spaces have cropped up in several areas, notably in the $\mathcal{H}$-space theory of Newman (1976) and in the 'positive definite' approach to quantum gravity (Hawking 1979).

The basic idea of Penrose's construction is that of generalizing the concept of an α-plane to that of an α-*surface* which is totally null and self-dual, in the same sense as these terms were used in Section 2. The condition for the existence of α-surfaces is the same as that for the existence of right-shear-free congruences, namely the vanishing of C^+ (Penrose 1976). So given a space-time M with $C^+ = 0$, we simply define the corresponding twistor space PT(M) to be the space of α-surfaces in M. Then, as long as we are careful to avoid global problems, PT(M) will be a three-dimensional complex manifold which is a complex-analytic deformation of the 'flat' twistor space PT. Furthermore, it contains in its complex structure all the information about the *conformal* structure of M; this involves a generalization of the statement (2.3). So one has the following one-to-one correspondence:

space-times with conformal metric satisying $C^+ = 0$	$\leftrightarrow$	deformed twistor spaces.

If we want to characterize metrics (and not just conformal metrics), then some additional structure has to be added to the deformed twistor space. The remarkable fact that Penrose discovered is that this can be done in such a way that the metric necessarily satisfies Einstein's equations $R_{ab} = 0$. It is possible to generalize his result so as to allow for a 'cosmological constant', and one then obtains a correspondence which may be summarized as follows (Ward 1980):

space-times with metric satisfying $C^+ = 0$ and $R_{ab} = \lambda g_{ab}$	$\leftrightarrow$	deformed twistor spaces possessing a 1-form τ and a 3-form ρ, with $\tau\wedge d\tau = \lambda\rho$

The forms τ and ρ are essentially arbitrary ($\tau\wedge d\tau$ is necessarily a constant multiple of ρ for global reasons, so the relation $\tau\wedge d\tau = \lambda\rho$ merely identifies this constant as λ). Thus once more we find that the differential equations in the space-time picture disappear on passing to the twistor picture.

The twistor method is therefore useful for studying anti-self-dual solutions of Einstein's equations; for some applications, see Ward (1978); Tod and Ward (1979); Hitchin (1979). The existence of this neat space-time/twistor-space correspondence is again due, in some sense, to the fact that the anti-self-dual Einstein equations satisfy a Huygens' Principle. In addition, equations such as the conformally-invariant wave equation, and Maxwell's equations, on an anti-self-dual space-time background, also satisfy HP. (This is not inconsistent with McLenaghen's theorem quoted in Section 2, since his theorem applies only to real, Lorentzian space-times). Consequently, one would expect there to be twistor solution procedures for these equations, and this indeed turns out to be the case: see, for example, Hitchin (1980); Atiyah (1980).

5. SUMMARY AND OUTLOOK

We have seen that the twistor programme (of formulating physics in twistor space rather than space-time) gives special emphasis to, and works very well for, the 'simplest' field theories – those where the propagation of fields satisfies Huygens' Principle. Much work remains to be done in understanding and applying these techniques. But clearly the major challenge is to achieve a greater understanding of more general situations, involving massive particles, scattering, general curved space-time, and of course quantization. Some approaches to these problems have been mentioned earlier: the multi-twistor approach to massive particles, a description of general Yang–Mills solutions, and the hypersurface-twistor approach to general curved space-times. Others include the 'twistor diagram' method of computing scattering amplitudes (Penrose 1975), and some recent thoughts on the curved-space problem (Penrose 1979*b*). These ideas have opened up new and exciting possibilities towards understanding gravity and quantum theory.

REFERENCES

Atiyah, M.F. (1980). Green's functions for self-dual 4-manifolds (preprint, Oxford).
—— and Ward, R.S. (1977). *Commun. Math. Phys.* **55**, 117.
——, Hitchin, N.J., and Singer, I.M. (1978). *Proc. Roy. Soc.* **A362**, 425.
—— ——, Drinfield, V.G., and Manin, Yu.I. (1978). *Phys. Lett.* **65A**, 185.
Bateman, H. (1904). *Proc. Lond. Math. Soc.* **2**, 451.
Belavin, A.A., Polyakov, A.M., Schwartz, A.S., and Tyupkin, Yu.S. (1975). *Phys. Lett.* **59B**, 85.
Christ, N.H., Weinberg, E.J., and Stanton, N.K. (1978). *Phys. Rev.* **D18**, 2013.
Corrigan, E.F., Fairlie, D.B., Templeton, S., and Goddard, P. (1978). *Nucl. Phys.* **B140**, 31.
Duff, M.J. and Isham, C.J. (1980). *Nucl. Phys.* **B162**, 271.
Friedlander, F.G. (1975). The wave equation on a curved space-time. Cambridge University Press.
Grgin, E. (1966). Ph.D. Thesis, Syracuse University.
Hawking, S.W. (1979). In *General Relativity*. (ed. S.W. Hawking and W. Israel). Cambridge University Press.
—— and Ellis, G.F.R. (1973). *The large scale structure of space-time.* Cambridge University Press.

Hitchin, N.J. (1979). *Math. Proc. Camb. Phil. Soc.* **85**, 465.
—— (1980). *Proc. Roy. Soc.* **A370**, 173.
Hughston, L.P. (1979). *Twistors and particles*. Springer lecture notes in physics. No. 97.
Isenberg, J., Yasskin, P.B., and Green, P.S. (1978). *Phys. Lett.* **78B**, 462.
Kovaks, E. and Lo, S.-Y. (1979). *Phys. Rev.* **D19**, 3649.
Künzle, H.P. (1968). *Proc. Camb. Phil. Soc.* **64**, 779.
Lerner, D.E. and Clarke, C.J.S. (1977). *Commun. Math. Phys.* **55**, 179.
Manton, N.S. (1978). *Nucl. Phys.* **B135**, 319.
McLenaghan, R.G. (1969). *Proc. Camb. Phil. Soc.* **65**, 139.
Newman, E.T. (1976). *Gen. Rel. Grav.* **7**, 107.
Penrose, R. (1967). *J. Math. Phys.* **8**, 345.
—— (1968). The Structure of space-time. In *Battelle Rencontres*. (ed. C.M. De Witt and J.A. Wheeler). Benjamin, New York.
—— (1972). The geometry of impulsive gravitational waves. In *General Relativity*. (ed. L. O'Raifeartaigh). Clarendon Press, Oxford.
—— (1974). Relativistic symmetry groups. In *Group theory in non-linear problems*. (ed. A.O. Barut). D. Reidel, Dordrecht.
—— (1975). Twistor theory, its aims and achievements. In *Quantum gravity*. (ed. C.J. Isham, R. Penrose, and D.W. Sciama). Clarendon Press, Oxford.
—— (1976). *Gen. Rel. Grav.* 7, 31.
—— (1979*a*). Twistor functions and sheaf cohomology; (1979*b*). A googly graviton? In *Advances in twistor theory*. (ed. L.P. Hughston and R.S. Ward). Pitman, London.
—— and MacCallum, M.A.H. (1973). *Phys. Reports*. **6**, 241.
—— and Ward, R.S. (1980). Twistors for flat and curved space-time. In *General relativity and gravitation*. (ed. A. Held). Plenum, New York.
Sparling, G.A.J. (1977). Dynamically broken symmetry and global Yang–Mills theory in Minkowski space (preprint, Pittsburgh).
Streater, R.F. and Wightman, A.S. (1964). *PCT, spin and statistics, and all that*. Benjamin, New York.
Tod, K.P. and Ward, R.S. (1979). *Proc. Roy. Soc.* **A368**, 411.
Ward, R.S. (1977). *Phys. Lett.* **61A**, 81.
—— (1978). *Proc. Roy. Soc.* **A363**, 289.
—— (1980). Self-dual space-times with cosmological constant (to appear in *Commun. Math. Phys.*).
Witten, E. (1978). *Phys. Lett.* **77B**, 394.

27

QUANTUM MECHANICS FOR COSMOLOGISTS

J.S. Bell

CERN — Geneva

1. INTRODUCTION

Cosmologists, even more than laboratory physicists, must find the usual interpretive rules of quantum mechanics (Dirac 1947) a bit frustrating:

'...any result of a measurement of a real dynamical variable is one of its eigenvalues...'

'...if the measurement of the observable ... is made a large number of times the average of all the results obtained will be...'

'...a measurement always causes the system to jump into an eigenstate of the dynamical variable that is being measured...'

It would seem that the theory is exclusively concerned with 'results of measurement' and has nothing to say about anything else. When the 'system' in question is the whole world where is the 'measurer' to be found? Inside, rather than outside, presumably. What exactly qualifies some subsystems to play this role? Was the world wave function waiting to jump for thousands of millions of years until a single-celled living creature appeared? Or did it have to wait a little longer for some more highly qualified measurer — with a Ph.D.? If the theory is to apply to anything but idealized laboratory operations, are we not obliged to admit that more or less 'measurement-like' processes are going on more or less all the time more or less everywhere? Is there ever then a moment when there is no jumping and the Schrödinger equation applies?

The concept of 'measurement' becomes so fuzzy on reflection that it is quite surprising to have it appearing in physical theory *at the most fundamental level*. Less surprising perhaps is that mathematicians, who need only simple axioms about otherwise undefined objects, have been able to write extensive works on quantum measurement theory — which experimental physicists do not find it necessary to read. Mathematics has been well called (Russell 1953) 'the subject in which we never know what we are talking about'. Physicists, confronted with such questions, are soon making measurement a matter of degree, talking of 'good' measurements and 'bad' ones. But the postulates quoted above know nothing of 'good' and 'bad'. And does not any *analysis* of measurement require concepts more *fundamental* than measurement? And should not the fundamental theory be about these more fundamental concepts?

One line of development towards greater physical precision, would be to have the 'jump' in the equations and not just in the talk — so that it would come about as a dynamical process in dynamically defined conditions. The jump violates the linearity of the Schrödinger equation, so that the new equation (or equations) would be non-linear. It has been conjectured (Wigner 1962*) that such non-linearity might be especially important precisely in connection with the functioning of conscious organisms - i.e. 'observers'.

It might also be that the non-linearity has nothing in particular to do with consciousness, but becomes important (Ludwig 1961) for any large object, in such a way as to suppress superposition of macroscopically different states. This would be a mathematical realization of at least one version of the 'Copenhagen interpretation', in which large objects, and especially 'apparatus', must behave 'classically'. Cosmologists should note, by the way, that the suppression of such macroscopic superpositions is vital to Rosenfeld's notion (Rosenfeld 1963) of an unquantized gravitational field — whose source (roughly speaking) would be the quantum expectation value of the energy density. If this were attempted with wave functions grossly ambiguous about, say, the relative positions of sun and planets, serious problems would quickly appear.

There have been several studies (de Broglie 1960; Laurent and Roos 1965; Shapiro 1973; Marinov 1974; Kupczynski 1974; Mielnik 1974; Pearle 1976; Bialnicki-Birula and Mycielski 1976;

* For a still more central role for the observer, see, C.M. Patton and J.A. Wheeler, in *Quantum Gravity* (ed. C. Isham, R. Penrose, and D. Sciama, Oxford 1975).

Shimony 1979; Kibble 1979; Kibble and Randjbar-Daemi 1980) of non-linear modifications of the Schrödinger equation. But none of these modifications (as far as I know) has the property required here, of having little impact for small systems but nevertheless suppressing macroscopic superpositions. It would be good to know how this could be done.

No more will be said in this paper about such hypothetical non-linearities. I will consider rather theories in which a linear Schrödinger equation is held to be exactly and universally correct. There is then no 'jumping', no 'reducing', no 'collapsing', of the wave function. Two such theories will be analyzed, one due to de Broglie (1956) and Bohm (1952) and the other to Everett (1957). It seems to me that the close relationship of the Everett theory to the de Broglie–Bohm theory has not been appreciated, and that as a result the really novel element in the Everett theory has not been identified. This really novel element, in my opinion, is a repudiation of the concept of the 'past', which could be considered in the same liberating tradition as Einstein's repudiation of absolute simultaneity.

It must be said that the versions presented here might not be accepted by the authors cited. This is to be feared particularly in the case of Everett. His theory was for long completely obscure to me. The obscurity was lightened by the expositions of De Witt (1970, 1971). But I am not sure that my present understanding coincides with that of De Witt, or with that of Everett, or that a simultaneous coincidence with both would be possible.*

---

* In particular it is not clear to me that Everett and De Witt conceive in the same way the division of the wave function into 'branches'. For De Witt this division seems to be rather definite, involving a specific (although not very clearly specified) choice of variables (instrument readings) to have definite values in each branch. This choice is in no way dictated by the wave function itself (and it is only after it is made that the wave function becomes a complete description of De Witt's physical reality). Everett on the other hand (at least in some passages) seems to insist on the significance of assigning an arbitrarily chosen state to an arbitrarily chosen subsystem and evaluating the 'relative state' of the remainder. It is when arbitrary mathematical possibilities are given equal status in this way that it becomes obscure to me that any physical interpretation has either emerged from, or been imposed on, the mathematics.

The following starts with a review of some relevant aspects of conventional quantum mechanics, in terms of a simple particular application. The problems to which the unconventional versions are addressed are then stated in more detail, and finally the de Broglie—Bohm and Everett theories are formulated and compared.

2. COMMON GROUND

To illustrate some points which are not in question, before coming to some which are, let us look at a particular example of quantum mechanics in actual use. A nice example for our purpose is the theory of formation of an α particle track in a set of photographic plates. The essential ideas of the analysis have been around at least since 1929 when Mott (1929) and Heisenberg (1930) discussed the theory of Wilson cloud chamber tracks.* Yet somehow many students are left to rediscover for themselves ideas of this kind. When they do so it is often with a sense of revelation; this seems to be the origin of several published papers.

Let the α particle be incident normally on the stack of plates and excite various atoms or molecules in a way permitting development of blackened spots. In a first approach (Heisenberg (1930)to the problem only the α particle is considered as a quantum mechanical system, and the plates are thought of as external measuring equipment permitting a sequence of measurements of transverse position of the α particle. Associated with each such measurement there is a 'reduction of the wave packet'

---

* The particularly instructive nature of this example has been stressed by E.P. Wigner.

* cont'd from p. 613

The five papers (Everett 1957; Wheeler 1957; De Witt 1970, 1971; Cooper and van Vechten 1969), a longer exposition by Everett, and a related paper by N. Graham (1973), are collected in *The many-worlds interpretation of quantum mechanics* (1973) (ed. B.S. De Witt and N. Graham). Princeton, N.J.

See also: De Witt, B.S. and others, *Physics Today* (1971). **24**, 36. Bell, J.S. (1976). In *Quantum mechanics, determinism, causality and particles*. (ed. M. Flato *et al.*). Reidel, Dordrecht.

in which all of the incident de Broglie wave except that near the point of excitation is eliminated. If the 'position measurement' were of perfect precision the reduced wave would emerge in fact from a point source and, by ordinary diffraction theory, then spread over a large angle. However, the precision is presumably limited by something like the atomic diameter a Then the angular spread can be as little as

$$\Delta\theta \approx (ka)^{-1}$$

With an α particle of about one MeV, for example, $k \approx 10^{13}\ \text{cm}^{-1}$, and with $a \approx 10^{-8}$ cm

$$\Delta\theta \approx 10^{-5} \text{ radians.}$$

In this way, one can understand that the sequence of excitations in the different plates approximate very well to a straight line pointing to the source.

This first approach may seem very crude. Yet in an important sense it is an accurate model of all applications of quantum mechanics.

In a second approach we can regard the photographic plates also as part of the quantum mechanical system. As Heisenberg remarks 'this procedure is more complicated than the preceding method, but has the advantage that the discontinuous change in the probability function recedes one step and seems less in conflict with intuitive ideas'. To minimize the increased complication we will consider only highly simplified 'photographic plates'. They will be envisaged as zero temperature mono-atomic layers of atoms each with only one possible excited state, the latter supposed to be rather long-lived. Moreover, we will continue to neglect the possibility of scattering without excitation, (i.e., elastic scattering), which is not very realistic.

Suppose that the α particle originates in a long-lived radioactive source at position $\vec{r}_0$ and can be represented initially by the steady state wave function

$$\psi(\vec{r}) = \frac{\exp(ik_0|\vec{r} - \vec{r}_0|)}{|\vec{r} - \vec{r}_0|}$$

Let ϕ_o denote the ground state of the stack of plates. Let n (= 1,2,3,...) enumerate the atoms of the stack and let

$$\phi(n_1, n_2, n_3, \ldots)$$

denote a state of the stack in which atoms $n_1, n_2, n_3, \ldots$ are excited. In the absence of α particle stack interaction the combined state would be simply

$$\phi_o \frac{\exp(ik_o|\vec{r} - \vec{r}_o|)}{|\vec{r} - \vec{r}_o|}$$

To this must be added, because of the interaction, scattered waves determined by solution of the many-body Schrödinger equation. In a conventional multiple scattering approximation the scattered waves are

$$\sum_N \sum_{n_1, n_2, \ldots n_N} \phi(n_1, n_2, \ldots n_N) \frac{\exp(ik_N|\vec{r} - \vec{r}_N|)}{|\vec{r} - \vec{r}_N|} f_N(\theta_N) \times \tag{2.1}$$
$$\frac{\exp(ik_{N-1}|\vec{r}_N - \vec{r}_{N-1}|)}{|\vec{r}_N - \vec{r}_{N-1}|} f_{N-1}(\theta_{N-1}) \times \ldots \frac{\exp(ik_o|\vec{r}_1 - \vec{r}_o|)}{|\vec{r}_1 - \vec{r}_o|} .$$

The general term here is a sum over all possible sequences of N atoms, with $\vec{r}_1$ denoting the position of atom n_1, $\vec{r}_2$ of atom n_2, and so on; $k_n = (k_{n-1} - \varepsilon)^{\frac{1}{2}}$ where ε is a measure of atomic excitation energy; θ_n is the angle between $\vec{r}_n - \vec{r}_{n-1}$, and $\vec{r}_{n+1} - \vec{r}_n$ (or $\vec{r} - \vec{r}_N$ for $n = N$). Finally $f_n(\theta)$ is the inelastic scattering amplitude for an α particle of momentum k_{n-1} incident on a single atom; in the Born approximation for example we could give an explicit formula for $f(\theta)$ in terms of atomic wave functions, and would indeed find for it an angular spread

$$\Delta\theta \approx (ka)^{-1}$$

The relative probabilities for observing that various sequences of atoms $n_1, n_2, \ldots$ have been excited are given by the squares of the moduli of the coefficients of

$$\phi(n_1, n_2, \ldots)$$

It is again clear that because of the forward peaking of $f(\theta)$ excited sequences will form essentially straight lines pointing towards the source.

We considered here only the location, and not the timing, of excitations. If timing also had been observed then in the first kind of treatment the reduced wave after each excitation would have been an appropriate solution of the time-dependent Schrödinger equation, limited in extent in time as well as space. In the second kind of treatment some physical device for registering and recording times would have been included in the system. We will not go further into this here. The comparison between the first and second kinds of treatment would still be essentially along the following lines. But before coming to this comparison it will be useful later to have pointed out two of the several general features of quantum mechanics which are illustrated in the example just discussed.

The first concerns the mutual consistency of different records of the same phenomenon. In the stack of plates of the above example we have a sequence of 'photographs' of the α particle, and because the particle is not *too* greatly disturbed by the photographing, the sequence of records is fairly continuous. In this way, there is no difficulty for quantum mechanics in the continuity between successive frames of a movie film nor in the consistency between two movie films of the same phenomenon. Moreover, if instead of recording such information on a film, it is fed into the memory of a computer (which can incidently be thought of as a model for the brain) there is no difficulty for quantum mechanics in the internal coherence of such a record – e.g., in the 'memory' that the α particle (or instrument pointer, or whatever) has passed through a sequence of adjacent positions. These are all just 'classical' aspects of the world which emerge from quantum mechanics at the appropriate level. They are called to attention here because later on we come to a theory which is fundamentally precisely about the contents of 'memories'.

The second point is the following. When the whole stack of plates is treated as a single quantum mechanical system, each α particle track is a single experimental result. To test the quantum mechanical probabilities requires then many such tracks.

At the same time a *single* track, if sufficiently long, can be regarded as a collection of many independent *single* scattering events, which can be used to test the quantum mechanics of the single scattering process. That this is so is seen to emerge from the more complete treatment whenever interactions between plates are negligible (and when the energy loss ε is negligible). Of course, there could be statistical freaks, tracks with all scatterings up, or all down, etc., but the *typical* track, if long enough, will serve to test predictions for $|f(\theta)|^2$. The relevance of this remark is that later we are concerned with theories of the universe as a whole. Then there is no opportunity to repeat the experiment; history is given to us once only. We are in the position of having a single track, and it is important that the theory has still something to say – provided that this single track is not a freak, but a typical member of the hypothetical ensemble of universes that would exhibit the complete quantum distribution of tracks. (For elaboration of this point see Everett 1957; De Witt 1970, 1971; Hartle 1968; Graham 1973.)

We return now to the comparison of the two kinds of treatment. The second treatment is clearly more serious than the first. But it is by no means final. Just as at first we supposed without analysis that the photographic plates could effect position measurements on the α particle, so we have now supposed without analysis the existence of equipment allowing the observation of atomic excitation. We can therefore contemplate a third treatment, and a fourth, and so on. Any natural end to this sequence is excluded by the very language of contemporary quantum theory, which never speaks of events in the system but only of the outcome of observations upon the system, implying always the existence of external equipment adapted to the observable in question. Thus the logical situation does not change in going from the first treatment to the second. Nor would it change on going further, although many people have been intimidated simply by increasing complexity into imagining that this might be so. In spite of its manifest crudity, therefore, we have to take quite seriously the first treatment above, as a faithful model of what we have to do in the end anyway.

It is therefore important to consider to what extent the first treatment is actually consistent with the second, and not simply superseded by the latter. The consistency is in fact quite high, especially if we incorporate into the rather vaguely 'reduced' wave function of the first treatment the correct angular factor $f(\theta)$ from the second. Then the first method will give exactly the same distribution of excitations, and the same correlations between those in different plates. However, it must be stressed that this perfect agreement is only a result of idealizations that we have made, for example, the neglect of interactions between atoms (especially in different plates). To take accurate account of these we are simply obliged to adopt the second procedure, of regarding α particle and stack together as a single quantum mechanical system. The first kind of treatment would be manifestly absurd if we were concerned with an α particle incident on two atoms forming a single molecule. It is perhaps not absurd, but it is not exact, when we have 10^{23} atoms with somewhat larger spaces between. Therefore, the placing of the inevitable split, between quantum system and observing world, is not a matter of indifference.

So we go on displacing this Heisenberg split to include more and more of the world in the quantum system. Eventually we come to a level where the required observations are simply of macroscopic aspects of macroscopic bodies. For example, we have to observe instrument readings, or a camera may do the observing, then we may observe the photographs of the instrument readings, and so on. At this stage, we know very well from everyday experience that it does not matter whether we think of the camera as being in the system or in the observer — the transformation between the two points of view being trivial, because the relevant aspects of the camera are 'classical' and its reaction on the relevant aspects of the instrument negligible. Then at this level it becomes of no *practical* importance just where we put the Heisenberg split — provided of course that these 'classical' features of the macroscopic world emerge also from the quantum mechanical treatment. There is no reason to doubt that this is the case.

This is already illustrated in the example that we analysed above. Thus the α particle is already largely 'classical' in its behaviour — preserving its identity, in a sense, as it is seen to move along a practically continuous and smooth path. Moreover, the different parts of the complete wave function (2.1) associated with different tracks can be to a considerable extent regarded as incoherent, as indicated by the success of the first kind of treatment. These 'classical' features can be expected to be still more pronounced for macroscopic bodies. The possibilities of seeing quantum interference phenomena are reduced not only by the shortness of the de Broglie wave length, which would make any such pattern extremely fine grained, but also by the tendency of such bodies to record their passage in the environment. With macroscopic bodies it is not necessary to ionize atoms; we have the steady radiation of heat for example, which would leave a 'track' even in the vacuum, and we have the excitation of the close packed low lying collective levels of both the body in question and neighbouring ones. (The high probability of exciting collective levels is emphasized by Zeh 1970.)

So there is no reason to doubt that the quantum mechanics of macroscopic objects yields an image of the familiar everyday world. Then the following rule for placing the Heisenberg split, although ambiguous in principle, is sufficiently unambiguous for practical purposes:

> *put sufficiently much into the quantum system that the inclusion of more would not significantly alter practical predictions.*

To ask whether such a recipe, however adequate in practice, is also a satisfactory formulation of fundamental physical theory, is to leave the common ground.

3. THE PROBLEM

The problem is this: quantum mechanics is fundamentally about 'observations'. It necessarily divides the world into two parts, a part which is observed and a part which does the observing. The results depend in detail on just how this division is made, but no definite prescription for it is given. All that we have is a recipe which, because of practical human limitations, is sufficiently unambiguous for practical purposes.

So we may ask with Stapp (1970, 1979): 'How can a theory which is *fundamentally* a procedure by which gross macroscopic creatures, such as human beings, calculate predicted probabilities of what they will observe under macroscopically specified circumstances ever be claimed to be a complete description of physical reality?'. Rosenfeld (1965) makes the point with equal eloquence: '... the human observer, whom we have been at pains to keep out of the picture, seems irresistibly to intrude into it, since after all the macroscopic character of the measuring apparatus is imposed by the macroscopic structure of the sense organs and the brain. It thus looks as if the mode of description of quantum theory would indeed fall short of ideal perfection to the extent that it is cut to the measure of man'.

Actually these authors feel that the situation is acceptable. As indicated by the quotations, they are among the more thoughtful of those who do so. Stapp finds reconciliation in the pragmatic philosophy of William James. On this view, the situation in quantum mechanics is not peculiar. But rather the concepts of 'real' or 'complete' truth are quite generally mirages. The only legitimate notion of truth is 'what works'. And quantum mechanics certainly 'works'. Rosenfeld seems to take much the same position, preferring however to keep academic philosophy out of it: 'we are not facing any deep philosophical issue, but the plain common sense fact that it takes a complicated brain to do theoretical physics'. That is to say, that theoretical physics is quite necessarily cut to the measure of theoretical physicists.

In my opinion, these views are too complacent. The pragmatic approach which they exemplify has undoubtedly played an indispensable role in the evolution of contemporary physical theory. However, the notion of the 'real' truth, as distinct from a truth that is presently good enough for us, has also played a positive role in the history of science. Thus Copernicus found a more intelligible pattern by placing the sun rather than the earth at the centre of the solar system. I can well imagine a future phase in which this happens again, in which the world becomes more intelligible to human beings, even to theoretical physicists, when they do not imagine themselves to be at the centre of it.

Less thoughtful physicists sometimes dismiss the problem by remarking that it was just the same in classical mechanics. Now if this were so it would diminish classical mechanics rather than justify quantum mechanics. But actually, it is not so. Of course, it is true that also in classical mechanics any isolation of a particular system from the world as a whole involves approximation. But at least one can *envisage* an accurate theory, of the universe, to which the restricted account *is* an approximation. This is not possible in quantum mechanics, which refers always to an outside observer, and for which therefore the universe as a whole is an embarrassing concept. It could also be said (by one unduly influenced by positivistic philsophy) that even in classical mechanics the human observer is implicit, for what is interesting if not experienced? But even a human observer is no trouble (in principle) in classical theory – he can be included in the system (in a schematic way) by postulating a 'psycho-physical parallelism' – i.e., supposing his experience to be correlated with some functions of the coordinates. This is not possible in quantum mechanics, where some kind of observer is not only essential, but essentially outside. In classical mechanics we have a model of a theory which is not *intrinsically* inexact, for it neither needs nor is embarrassed by an observer.

Classical mechanics does, however, have the grave defect, as applied on the atomic scale, of not accounting for the data. For this good reason it has been abandoned on that scale. However, classical concepts have not thereby been expelled from physics. On the contrary, they remain essential on the 'macroscopic' scale, for (Bohr 1949) "... it is decisive to recognize that, however far the phenomena transcend the scope of classical physical explanation, the account of all evidence must be expressed in classical terms'. Thus contemporary theory employs both quantum wave functions ψ and classical variables x, and a description of any sufficiently large part of the world involves both:

$$(\psi, x_1, x_2, \ldots)$$

In our discussion of the α particle track, for example, implicit classical variables specified the position of the various plates, and the degrees of excitation of the atoms were also considered as classical variables for which probability distributions could be extracted from the calculations. In a more thorough treatment the degrees of excitation of atoms would be replaced as classical variables by the degrees of blackening of the developed plates. And so on. It seems natural to speculate that such a description might survive in a hypothetical accurate theory to which the contemporary recipe would be a working approximation. The ψs and xs would then presumably interact according to some definite equations. These would replace the rather vague contemporary 'reduction of the wave packet' – intervening at some ill-defined point in time, or at some ill-defined point in the analysis, with a lack of precision which, as has been said, is tolerable only because of human grossness.

Before coming to examples of such theories I would like to suggest two general principles which should, it seems to me, be respected in their construction. The first is that it should be possible to formulate them for small systems. If the concepts have no clear meaning for small systems it is likely that 'laws of large numbers' are being invoked at a fundamental level, so that the theory is fundamentally approximate. The second, related, point is that the concepts of 'measurement', or 'observation', or 'experiment', should not appear at a fundamental level. The theory should of course allow for particular physical set-ups, not very well defined as a class, having a special relationship to certain not very well-defined subsystems – experimenters. But these concepts appear to me to be too vague to appear at the base of a potentially exact theory. Thus the xs would then not be 'macroscopic' 'observables' as in the traditional theory, but some more fundamental and less ambiguous quantities – 'beables' (Bell 1973).

The classical variables x were written just now as a discrete set. In relativistic theory continuous fields are likely to be more appropriate, in particular perhaps an energy density $T_{oo}(t,\vec{x})$. In the following we consider only the non-relativistic theory, with the particulate approximation

$$T_{oo}(t,x) = \sum m_n c^2 \delta(\vec{x} - \vec{x}_n(t))$$

This is parametrized by the finite set of all particle co-ordinates $\vec{x}_n$.

4. THE PILOT WAVE

The duality indicated by the symbol

$$(\psi,\ x)$$

is a generalization of the original wave-particle duality of wave mechanics. The mathematics had to be done with waves ψ extending in space, and then had to be interpreted in terms of probabilities for localized events. At an early stage de Broglie (1956) proposed a scheme in which particle and wave aspects were more closely integrated. This was reinvented by Bohm (1952). Despite some curious features it remains, in my opinion, well worth attention as a model of what might be the logical structure of a quantum mechanics which is not intrinsically inexact.

To avoid arbitrary division of the world into systems and apparatus, we must work straight away with some model of the world as a whole. Let this 'world' be simply a large number N of particles, with Hamiltonian

$$H = \sum_n \frac{\vec{p}_n^{\,2}}{2M_n} + \sum_{m>n} V_{mn}(\vec{r}_m - \vec{r}_n) \tag{4.1}$$

The world wave function $\psi(r,t)$, where r stands for all the $\vec{r}$s, evolves according to

$$\frac{\partial}{\partial t}\psi(r,t) = -\,\mathrm{i}H\psi \tag{4.2}$$

We will need the purely mathematical consequence of this that

$$\frac{\partial}{\partial t}\rho(r,t) + \sum_n \frac{\partial}{\partial \vec{r}_n} \cdot \vec{j}_n(r,t) = 0 \tag{4.3}$$

where

$$\rho(r,t) = |\psi(r,t)|^2 \tag{4.4}$$

$$\vec{j}_n(r,t) = M_n^{-1}\ \mathrm{Im}\left\{\psi^*(r,t)\frac{\partial}{\partial\vec{r}_n}\psi(r,t)\right\} . \tag{4.5}$$

We have to add classical variables. A democratic way to do this is to add variables $\vec{x}_1$, $\vec{x}_2$, ... $\vec{x}_N$ in one-to-one correspondence with the $\vec{r}$s. The $\vec{x}$s are supposed to have definite values at any time and to change according to

$$\frac{d}{dt}\vec{x}_n = \vec{j}_n(x,t)/\rho(x,t) = \frac{1}{M_n}\frac{\partial}{\partial\vec{x}_n}\mathrm{Im}\ \log\psi(x,t) . \tag{4.6}$$

We then have a deterministic system in which everything is fixed by the initial values of the wave ψ and the particle configuration x. Note that in this compound dynamical system the wave is supposed to be just as 'real' and 'objective' as say the fields of classical Maxwell theory — although its action on the particles, (4.6), is rather original. *No one can understand this theory until he is willing to think of* ψ *as a real objective field rather than just a 'probability amplitude'*, even though it propagates not in 3-space but in $3N$-space.*

From the 'microscopic' variables x can be constructed 'macroscopic' variables X

$$X_n = F_n(x_1, \ldots, x_N) \tag{4.7}$$

— including in particular instrument readings, image density on photographic plates, ink density on computer output, and so on. Of course, there is some ambiguity in defining such quantities — e.g., over precisely what volume should the discrete particle density be averaged to define the smooth macroscopic density? However, it is the merit of the theory that the ambiguity is not in the foundation, but only at the level of identifying objects of particular interest to macroscopic observers, and the ambiguity arises simply from the

* There is a problem with (4.6) where ρ vanishes. A cheap way of avoiding it is to replace ρ and j in (4.3), (4.6) and (4.9) by $\bar{\rho}$ and $\bar{j}$, obtained from ρ and j by folding with a narrow Gaussian distribution in the $(\vec{r}_1,\vec{r}_2, \ldots)$ space. Then $\bar{\rho}$ is always positive, while (4.3) remains valid. The deBB theory then gives $\bar{\rho}$ rather than ρ as probability distribution, but with sufficiently narrow Gaussian spread the difference is unimportant.

grossness of these creatures.

It is thus from the *x*s, rather than from ψ, that in this theory we suppose 'observables' to be constructed. It is in terms of the *x*s that we would define a 'psycho-physical parallelism — if we were pressed to go so far. Thus it would be appropriate to refer to the *x*s as 'exposed variables' and to ψ as a 'hidden variable'. It is ironic that the traditional terminology is the reverse of this.

It remains to compare the pilot-wave theory with orthodox quantum mechanics at the practical level, which is that of the *X*s. A convenient device for this purpose is to imagine, in the context of the orthodox approach, a sort of ultimate observer, outside the world and from time to time observing its macroscopic aspects. He will see in particular other, internal, observers at work, will see what their instruments read, what their computers print out, and so on. In so far as ordinary quantum mechanics yields at the appropriate level a classical world, in which the boundary between system and observer can be rather freely moved, it will be sufficient to account for what such an ultimate observer would see. If he were to observe at time t a whole ensemble of worlds corresponding to an initial state

$$\psi(\vec{r}_1, \ldots \vec{r}_N, 0)$$

he would see, according to the usual theory, a distribution of *X*s given closely by

$$\rho(X_1, X_2, \ldots) = \int d\vec{r}_1 d\vec{r}_2 \ldots d\vec{r}_N$$

$$\delta(X_1 - F_1(r))\delta(X_2 - F_2(r)) \ldots |\psi(r,t)|^2 \tag{4.8}$$

with $\psi(t)$ obtained by solving the world Schrödinger equation. It would not be exactly this, for his own activities cause wave-packet reduction and spoil the Schrödinger equation. But macroscopic observation is supposed to have not much effect on subsequent macroscopic statistics. Thus (4.8) is closely the distribution implied by the usual theory. Moreover, it is easy to construct in the pilot-wave theory an ensemble of worlds which gives the distribution (4.8) exactly. It is sufficient that the configuration x should be distributed according to

$$\rho(x,t)\,dx_1 dx_2 \ldots dx_N \tag{4.9}$$

It is a consequence of eqns. (4.3) and (4.6) that (4.9) will hold at all times if it holds at some initial time. Thus it suffices to specify in the pilot-wave theory that the initial configuration x is chosen at random from an ensemble of configurations in which the distribution is $\rho(x,o)$. It is only at this point, in defining a comparison class of possible initial worlds, that anything like the orthodox probability interpretation is invoked.

Then for instantaneous macroscopic configurations the pilot-wave theory gives the same distribution as the orthodox theory, insofar as the latter is unambiguous. However, this question arises: what is the good of *either* theory, giving distributions over a hypothetical ensemble (of worlds!) when we have only one world. The answer has been anticipated in the introductory discussion of the α particle track. A long track is on the one hand a single event, but is at the same time an ensemble of single scatterings. In the same way a single configuration of the world will show statistical distributions over its different parts. Suppose, for example, this world contains an actual ensemble of similar experimental set-ups. In the same way as for the α particle track it follows from the theory that the 'typical' world will approximately realize quantum mechanical distributions over such approximately independent components (Everett 1957; De Witt 1970, 1971; Hartle 1968; Graham 1973). The role of the hypothetical ensemble of worlds is precisely to permit definition of the word 'typical'.

So much for instantaneous configurations. Both theories give also trajectories, by which instantaneous configurations at different times are linked up. In the traditional theory these trajectories, like the configurations, emerge only at the macroscopic level, and are constructed by successive wave-packet reduction. In the pilot-wave theory macroscopic trajectories are a consequence of the microscopic trajectories determined by the guiding formula (4.6).

To exhibit some features of these trajectories, consider a standard example from quantum measurement theory — the measurement of a spin component of a spin ½ particle. A highly simplified model for this can be based on the interaction

$$H = g(t)\sigma \frac{1}{i}\frac{\partial}{\partial r} \tag{4.10}$$

where ρ is the Pauli matrix for the chosen component and r is the 'instrument reading' co-ordinate. For simplicity take the masses associated with both particle and instrument reading to be infinite. Then other terms in the Hamiltonian can be neglected, and the time-dependent coupling $g(t)$ can be supposed to arise from the passage of the particle along a definite classical orbit through the instrument. Let the initial state be

$$\psi_m(0) = \phi(r)a_m \tag{4.11}$$

where $\phi(r)$ is a narrow wave packet centred on $r = 0$ and m (= 1,2) is a spin index; we choose the representation in which σ is diagonal. The solution of the Schrödinger equation

$$\frac{\partial\psi}{\partial t} = -\,iH\psi$$

is

$$\psi_m(t) = \phi(r - (-1)^m h)a_m \tag{4.12}$$

where

$$h(t) = \int_{-\infty}^{t} dt'g(t') \tag{4.13}$$

After a short time the two components of (4.12) will separate in r space. Observation of the instrument reading will then, in the traditional view, yield the values $+h$ or $-h$ with relative probabilities $|a_1|^2$ and $|a_2|^2$, and with small uncertainties given by the width of the initial wave packet. Because of wave-packet reduction, subsequent observation will reveal the instrument continuing along whichever of the two trajectories, $\pm h(t)$, was in fact selected.

Consider now the pilot-wave version. Nothing new has to be said about the orbital motion of the particle, which was already taken to be classical and fixed. We do now have a classical variable x for the instrument reading. We could consider introducing classical variables for the spin motion, but in the simplest version (Bell 1966) this is not done; instead the spin indices of the wave function are just summed over in constructing densities and currents

$$\rho(r,t) = \psi^*(r,t)\psi(r,t) \tag{4.14}$$

$$j(r,t) = \psi^*(r,t)g\sigma\psi(r,t) \tag{4.15}$$

with the summation implied; the slightly surprising form of j follows from the gradient form of the coupling (4.10), and from the absence of the normal term (4.5) in the case of infinite mass. The motion of x is then determined by

$$\frac{dx}{dt} = (x,t)/\rho(x,t)$$

or explicitly

$$\frac{dx}{dt} = g\,\frac{\sum_m |a_m|^2|\phi(x - (-1)^m h)|^2(-1)^m}{\sum_m |a_m|^2|\phi(x - (-1)^m h)|^2}\,. \tag{4.16}$$

As soon as the wave packets have separated $\dot{x} = \pm g$, according to $x \approx \pm h$. Thus we have essentially the same two trajectories as the wave-packet reduction picture, and they will be realized with the same relative probabilities if x is supposed to have an initial probability distribution $|\phi(x)|^2$ – this is the familiar general consequence, for instantaneous configurations, of the method of construction. In any individual case which trajectory is selected is actually determined by the initial x value. But when that value is not known (when it is known only to lie in the initial wave packet) whether the particle is deflected up or down is indeterminate for practical purposes.

Consider now a slightly more complicated example, in which measurements of the above kind are made simultaneously on two spin ½ particles. Denote by r_1 and r_2 the co-ordinates of the two instruments. If the initial state is

$$\psi_{mn}(0) = \phi(r_1)\phi(r_2)a_{mn}$$

solution to the Schrödinger equation yields

$$\psi_{mn}(t) = \phi(r_1 - (-1)^m h_1)\phi(r_2 - (-1)^n h_2)a_{mn} \tag{4.17}$$

with

$$h_1(t) = \int_{-\infty}^{t} \mathrm{d}t' g_1(t'), \qquad h_2(t) = \int_{-\infty}^{t} \mathrm{d}t' g_2(t') .$$

In the wave-packet reduction picture one of four possible trajectories, $(\pm h_1, \pm h_2)$, will be realized, the relative probabilities being given by $|a_{mn}|^2$. The pilot-wave picture will give again an account identical for practical purposes, although the outcome is in principle determined by initial values of variables x_1 and x_2.

But when examined in detail the microscopic trajectories are quite peculiar during the brief initial period in which the different terms in (4.17) still overlap in (r_1,r_2) space. The detailed time development of the xs is given by

$$\dot{x}_1 = g_1 \frac{\sum_{m,n} (-1)^m |a_{mn}|^2 |\phi(x_1 - (-1)^m h_1)|^2 |\phi(x_2 - (-1)^n h_2)|^2}{\sum_{m,n} |a_{mn}|^2 |\phi(x_1 - (-1)^m h_1)|^2 |\phi(x_2 - (-1)^n h_2)|^2}$$

$$\dot{x}_2 = g_2 \frac{\sum_{m,n} (-1)^n |a_{mn}|^2 |\phi(x_1 - (-1)^m h_1)|^2 |\phi(x_2 - (-1)^n h_2)|^2}{\sum_{m,n} |a_{mn}|^2 |\phi(x_1 - (-1)^m h_1)|^2 |\phi(x_2 - (-1)^n h_2)|^2} \tag{4.18}$$

These expressions simplify greatly when the two spin states are uncorrelated, i.e. when a_{mn} factorizes

$$a_{mn} = b_m c_n .$$

The factors referring to the second particle then cancel out in the expression for $\dot{x}_1$, and those referring to the first particle cancel in the expression for $\dot{x}_2$, so that we have just two independent motions of the instrument pointers of the type already discussed. However, in general the spin state does not factorize. One can even envisage situations in which the two particles interact at short range and strong spin correlations are induced which persist when the particles subsequently move far apart. Then it follows from (4.18) that the detailed behaviour of x_1 and x_2 depends not only on the programmes h_1 and h_2 respectively of the local instruments, but also on those of the remote instruments h_2 and h_1. The detailed dynamics is quite non-local in character.

Could it be that this strange non-locality is a peculiarity of the very particular de Broglie–Bohm construction of the classical sector, and could be removed by a more clever construction? I think not. It now seems* that the non-locality is deeply rooted in quantum mechanics itself and will persist in any completion. Could it be that in the context of relativistic quantum theory c would be a limiting velocity and the strange long-range effects would propagate only subliminally? Not so. The aspects of quantum mechanics demanding non-locality remain in relativistic quantum mechanics. It may well be that a relativistic version of the theory, while Lorentz invariant and local at the observational level, may be necessarily non-local and with a preferred frame (or aether) at the fundamental level (Eberhard 1978). Could we not then just omit this fundamental level and restrict the classical variables to some 'observable' 'macroscopic' level? The problem then would be to do this with clean mathematics, and not just talk.

It can be maintained that the de Broglie–Bohm orbits, so troublesome in this matter of locality, are not an essential part of the theory. Indeed it can be maintained that there is no need whatever to link successive configurations of the world into a continuous trajectory. Keeping the instantaneous configurations, but discarding the trajectory, is the essential

---

* This question has been much discussed, and there has been an experimental programme to test the relevant aspects of quantum mechanics. Some papers, with many references, are: Clauser and Shimony 1978; Pipkin 1978; d'Espagnat 1978; Stapp 1979; Bell 1980.

feature (in my opinion) of the theory of Everett.

5. EVERETT (?)

The Everett (?) theory of this section will simply be the pilot-wave theory without trajectories. Thus instantaneous classical configurations x are supposed to exist, and to be distributed in the comparison class of possible worlds with probability $|\psi|^2$. But no pairing of configurations at different times, as would be effected by the existence of trajectories, is supposed. And it is pointed out that no such continuity between present and past configurations is required by experience.

I would really prefer to leave the formulation at that, and proceed to elucidate the last sentence. But some additional remarks must be made for readers of Everett and De Witt, who may not immediately recognize the formulation just made.

1. First there is the 'many-universe' concept given prominence by Everett and De Witt. In the usual theory it is supposed that only one of the possible results of a measurement is actually realized on a given occasion, and the wave function is 'reduced' accordingly. But Everett introduced the idea that *all* possible outcomes are realized every time, each in a different edition of the universe, which is therefore continually multiplying to accommodate all possible outcomes of every measurement. The psycho-physical parallelism is supposed such that our representatives in a given 'branch' universe are aware only of what is going on in that branch. Now it seems to me that this multiplication of universes is extravagant, and serves no real purpose in the theory, and can simply be dropped without repercussions. So I see no reason to insist on this particular difference between the Everett theory and the pilot-wave theory — where, although the *wave* is never reduced, only *one* set of values of the variables x is realized at any instant. Except that the wave is in configuration space, rather than ordinary three-space, the situation is the same as in Maxwell–Lorentz electron theory*.

---

* But the following difference of detail is notable. In the Maxwell–Lorentz electron theory particles and field interacted in a reciprocal

Nobody ever felt any discomfort because the field was supposed to exist and propagate even at points where there was no particle. To have multiplied universes, to realize all possible configurations of particles, would have seemed grotesque.

2. Then it could be said that the classical variables x do not appear in Everett and De Witt. However, it is taken for granted there that meaningful reference can be made to experiments having yielded one result rather than another. So instrument readings, or the numbers on computer output, and things like that, are the classical variables of the theory. We have argued already against the appearance of such vague quantities at a fundamental level. There is always some ambiguity about an instrument reading; the pointer has some thickness and is subject to Brownian motion. The ink can smudge in computer output, and it is a matter of practical human judgement that one figure has been printed rather than another. These distinctions are unimportant in practice, but surely the theory should be more precise. It was for that reason that the hypothesis was made of fundamental variables x, from which instrument readings and so on can be constructed, so that only at the stage of this construction, of identifying what is of direct interest to gross creatures, does an inevitable and unimportant vagueness intrude. I suspect that Everett and De Witt wrote as if instrument readings were fundamental only in order to be intelligible to specialists in quantum measurement theory.

3. Then there is the surprising contention of Everett and De Witt that the theory 'yields its own interpretation'. The hard core of this seems to be the assertion that the probability interpretation emerges without being assumed. In so far as this is true it is true also in the pilot-wave theory. In that theory our unique world is supposed to evolve in

---

* cont'd from p. 632.

way. In the pilot-wave theory the wave influences the particles but is not influenced by them. Finding this peculiar, de Broglie (1956) always regarded the pilot-wave theory as just a stepping-stone on the way towards a more serious theory which would be in appropriate circumstances experimentally distinct from ordinary quantum mechanics.

deterministic fashion from some definite initial state. However, to identify which features are details critically dependent on the initial conditions (like whether the first scattering is up or down in an α particle track) and which features are more general (like the distribution of scattering angles over the track as a whole) it seems necessary to envisage a comparison class. This class we took to be a hypothetical ensemble of initial configurations with distribution $|\psi|^2$. In the same way Everett has to attach weights to the different branches of his multiple universe, and in the same way does so in proportion to the norms of the relevant parts of the wave function. Everett and De Witt seem to regard this choice as inevitable. I am unable to see why, although of course it is a perfectly reasonable choice with several nice properties.

4. Finally there is the question of trajectories, or of the association of a particular present with a particular past. Both Everett and De Witt do indeed refer to the structure of the wave function as a 'tree', and a given branch of a tree can be traced down in a unique way to the trunk. In such a picture the future of a given branch would be uncertain, or multiple, but the past would not. But, if I understand correctly, this tree-like structure is only meant to refer to a temporary and rough way of looking at things, during the period that the initially unfilled locations in a memory are progressively filled, labelling the different branches of the tree only by the macroscopic-type variables describing the contents of the locations. When a more fundamental description is adopted there is no reason to believe that the theory is more asymmetric in time than classical statistical mechanics. There also apparent irreversibility can arise (e.g., the increase of entropy) when coarse-grained variables are used. Moreover, De Witt says '...every quantum transition taking place on every star, in every galaxy, in every remote corner of the universe is splitting our local world in myriads of copies of itself'. Thus De Witt seems to share our idea that the fundamental concepts of the theory should be meaningful on a microscopic level, and not only on some ill-defined macroscopic level. But at the microscopic level there is no

such asymmetry in time as would be indicated by the existence of branching and non-existence of debranching. Thus the structure of the wave function is not fundamentally tree-like. It does not associate a particular branch at the present time with any particular branch in the past any more than with any particular branch in the future. Moreover, it even seems reasonable to regard the coalescence of previously different branches, and the resulting interference phenomena, as *the* characteristic feature of quantum mechanics. In this respect an accurate picture, which does not have any tree-like character, is the 'sum over all possible paths' of Feynman.

Thus in our interpretation of the Everett theory there is no association of the particular present with any particular past. And the essential claim is that this does not matter at all. For we have no access to the past. We have only our 'memories' and 'records'. But these memories and records are in fact *present* phenomena. The instantaneous configuration of the xs can include clusters which are markings in notebooks, or in computer memories, or in human memories. These memories can be of the initial conditions in experiments, among other things, and of the results of those experiments. The theory should account for the present correlations between these present phenomena. And in this respect we have seen it to agree with ordinary quantum mechanics, in so far as the latter is unambiguous.

The question of making a Lorentz invariant theory on these lines raises intriguing questions. For reality has been identified only at a single time. This seems to be as much the case in the many universe version, as in the one universe version. In a Lorentz invariant theory would there be different realities corresponding to different ways of defining the time direction in the four-dimensional space?* Or if these various realities are to be seen as different aspects of one, and therefore correlated somehow, is this not falling back towards the notion of trajectory?

---

* Or would it be necessary to restrict memories to the here as well as the now? Point-sized reminiscers? See, H.D. Zeh, Heidelberg preprint, to be published in Foundations of Physics.

Everett's replacement of the past by memories is a radical solipsism – extending to the temporal dimension the replacement of everything outside my head by my impressions, of ordinary solipsism or positivism. Solipsism cannot be refuted. But if such a theory were taken seriously it would hardly be possible to take anything else seriously. So much for the social implications*. It is always interesting to find that solipsists and positivists, when they have children, have life insurance.

In conclusion it is perhaps interesting to recall another occasion when the presumed accuracy of a theory required that the existence of present historical records should not be taken to imply that any past had indeed occurred. The theory was that of the creation of the world in 4004 B.C. (at 6 o'clock in the evening on October 22nd. Usher 1660). During the 18th century growing knowledge of the structure of the earth seemed to indicate a more lengthy evolution. But it was pointed out that God in 4004 B.C. would quite naturally have created a going concern. The trees would be created with annular rings, although the corresponding number of years had not elapsed. Adam and Eve would be fully grown, with fully grown teeth and hair (they would have navels, although they had not been born. Gosse 1857). The rocks would be typical rocks, some occurring in strata and bearing fossils - of creatures that had never lived. Anything else would not have been reasonable (Chateaubriand 1802):

> Si le monde n'eut été à la fois jeune et vieux, le grand, le sérieux, le moral, disparaissaient de la nature, car ces sentiments tiennent par essence aux choses antiques. ... . L'homme-roi naquit lui-même à trente années, afin de s'accorder par sa majesté avec les antiques grandeurs de son nouvel empire, de même que sa compagne compta sans doute seize printemps, qu'elle n'avait pourtant point vécu, pour être en harmonie avec les fleurs, les oiseaux, l'innocence, les amours, et tout la jeune partie de l'univers.

---

* The present paper has much in common with an unpublished paper (CERN TH. 1424) presented at the International Colloquium on Issues in Contemporary Physics and Philosophy of Science, and their Relevance for our Society, Penn. State University, September 1971.

REFERENCES

Bell, J.S. (1966). *Rev. Mod. Phys.* **38**, 447.
—— (1973). In *The physicists' concept of nature.* (ed. J. Mehra). Reidel, Dordrecht.
—— CERN preprints TH. 2053 and TH. 2252. Also in Comments on atomic and molecular physics (1980), **9**, 121.
Bialnicki-Birula, I. and Mycielski, K. (1976). *Ann. Phys.* **100**, 62.
Bohm, D. (1952). *Phys. Rev.* **85**, 180.
Bohr, N. (1949). Discussion with Einstein in *Albert Einstein* (ed. P.A. Schlipp). Tudor, New York.
de Broglie, L. (1956). *Tentative d'interpretation causale et non-linéaire de la mecanique ondulatoire.* Gauthier-Villars, Paris.
—— (1960). *Non-linear wavemechanics.* Elsevier, Amsterdam.
de Chateaubriand, F. (1802). *Genie du Christianisme.*
Clauser, J.F. and Shimony, A. (1978). *Rep. prog. phys.* **41**, 1881.
Cooper, L.N. and van Vechtem, D. (1969). *Am. J. Phys.* **37**, 1212.
De Witt, B.S. (1970). *Physics Today.* **23**, 30.
—— (1971). In *Proc. Int. Sch. Phys. 'Enrico Fermi' Course IL; Foundations of Quantum Mechanics.* (ed. D. d'Espagnat). Benjamin, New York.
Dirac, P.A.M. (1947). *The principles of quantum mechanics.* (3rd edn.), Oxford University Press.
Eberhard, P.H. (1978). *Nuovo Cim.* **46B**, 392.
d'Espagnat, B. (1978). *Sci. Am.* Nov.
Everett, H. (1957). *Rev. mod. Phys.* **29**, 454.
Gosse, P.H. (1857). *Omphalos.*
Hartle, J.B. (1968). *Am. J. Phys.* **36**, 704.
Heisenberg, W. (1930). *Physical principles of the quantum theory.* Chicago.
Kibble, T.W.B. (1978). *Comm. Math. Phys.* **64**, 73.
—— (1979). *Comm. Math. Phys.* **65**, 189.
—— and Randjbar-Daemi, S. (1980). *J. Phys.* **A13**, 141.
Kupczynski, M. (1974). *Lett. Nuovo Cim.* **9**, 134.
Laurent, B. and Roos, M. (1965). *Nuovo Cim.* **40**, 788.
Ludwig, G. (1961). In *Werner Heisenberg und die physik underer zeit.* Vierweg, Braunschweig.
Marinov, M.S. (1974). *Sov. J. Nucl. Phys.* **19**, 173.
Mielnik, B. (1974). *Comm. Math. Phys.* **37**, 221.
Mott, N.F. (1929). *Proc. R. Soc.* **A126**, 79.
Pearle, B. (1976). *Phys. Rev.* D13, 857.
Pipkin, F.M. (1978). *Annual reviews of nuclear science.*
Rosenfeld, L. (1963). *Nucl. Phys.* 40, 353.
—— (1965). Suppl. *Prog. theor. Phys.* **222**.
Russell, B. (1953). *Mysticism and Logic.* Penguin, London.
Shapiro, I.R. (1973). *Soviet J. nucl. Phys.* **16**, 727.
Shimony, A. (1979). *Phys. Rev.* **A20**, 394.
Stapp, H. (1970). UCRL-20294.
—— (1979). *Found. Phys.* 9, 1, and (1980). *Found. Phys.* **10**, 767.
Usher, J. (1660). *Chronologia Sacra.* Oxford.
Wheeler, J.A. (1957). *Rev. mod. Phys.* **29**, 463.
Wigner, E.P. (1962). In *The scientist speculates.* (ed. R. Good). Heinemann, London.
Zeh, H.D. (1970). *Found. phys.* 1, 69.

28

QUANTUM THEORY WITHOUT AXIOMS

B. Mielnik

Institute of Theoretical Physics, Warszawa, Hoza 69, Poland

> It is known that dragons do not exist.
> However, each one does it in a different way.
>
> (Stanislaw Lem)

1. INTRODUCTION

Quantum gravity has been divided into a number of domains, as if it actually existed. In reality, the situation is more complicated. What certainly exists is an abyss. On one side of it there are the quantum theories which do not describe gravity. On the opposite side is GRT with some other theories trying to, but still not quite crossing the abyss to the quantum side. These include the new hypothesis about the non-linear graviton (Penrose 1975), the new algebraic development in the exact solutions (Plebanski 1975; Plebanski and Robinson 1976, 1977), superspaces (Misner, Thorne, and Wheeler 1973), etc. The situation on opposite sides of the abyss is very different.

The formalism of general relativity is flexible, its pseudo-Riemannian manifolds depend upon the matter distribution. An analogous flexibility is not seen on the quantum side (Finkelstein 1962). Here, all the models of the theory follow a certain rigid structural design. The variety of all physical situations (states) is represented by the 'density matrices' in a separable complex Hilbert space. The *observables* are linear self-adjoint operators. Though the details of the scheme vary and its domain is constantly extended, the basic design has never changed. It has neither been changed (though it was mathematically refined) by approaches dealing with positive functionals nor by *-algebras. In spite of names like 'strangeness', 'colour', 'charm' etc, the resulting scheme has some features of intrinsic monotony. After all, every infinite dimensional separable complex Hilbert space is isomorphic with any other such space. A disquieting question thus arises: why must nature repeat itself so much? This question is not irrelevant to attempts at gravity quantization. In most cases these attempts tend to reformulate general relativity in the spirit of quantum theory. However, one might wonder, whether the shape of quantum theories should be taken for granted. Should not the inverse programme be carried out first, by modifying the quantum theory in the spirit of general relativity? Should not quantum theories depart from their old linear formalism?

The shape of the present day quantum theory is secured by a sequence of requirements (axioms) stated by axiomatic quantum mechanics and axiomatic quantum field theory. They seem to show that the quantum structure cannot be changed without violating some deep phenomenological demands. However, past experiences show that the axioms can be very deceptive, even if they look obvious. (Imagine that the techniques of the axiomatic foundations were invented in physics before special relativity: would they prove anything but the necessity of Galilean space-time?) What indeed seems important in the axiomatic approaches are the concepts and not the axioms (Piron 1979). Significantly, the main concepts of quantum theory remain meaningful, even if the axioms are not assumed. They add up to a generalized scheme of non-linear statistical theory, which shares the spirit but not the form of quantum mechanics. This scheme is a solution of an old geometrization problem.

2. CONVEX GEOMETRY

Geometrization programmes may lead to various structures depending on the questions which they ask. Below I would like to ask a simple question. What picture does one see, looking at a physical theory from a distance, so that the details disappear? In the case of quantum theories various pictures have been proposed from the orthocomplemented lattices (Birkhoff, von Neumann, Piron) up to fibre bundles ('geometric quantization'). However, since quantum mechanics is a statistical theory, the most universal picture which remains after the details are forgotten is that of a convex set (Ludwig 1967; Mielnik 1969; Alfsen and Schultz 1976, 1978; Davies and Lewis 1970; Gudder 1973). I shall denote this set by S. The points $x, y, \ldots \in S$ represent the statistical ensembles (states) which are possible within the theory. The extreme points of S are the *pure states*. The other points of S are *state mixtures*. For any $x_1 \ldots, x_n \in S$ the symbol $p_1 x_1 + \ldots + p_n x_n$ means the mixture of $x_1 \ldots, x_n$ with the weight coefficients $p_1 \ldots, p_n$.

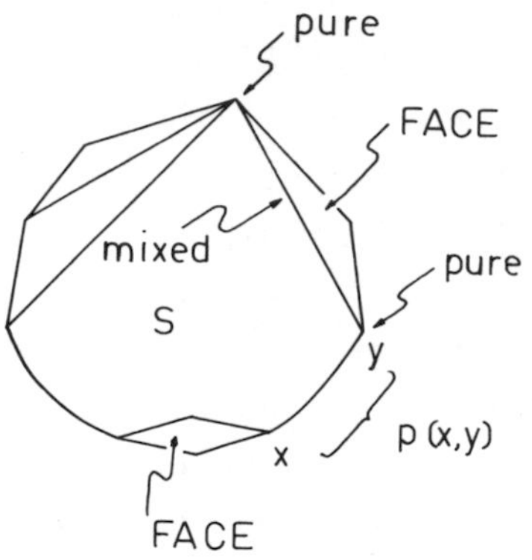

FIG. 2.1.

For the sake of physical interpretation S might be considered a convex set 'in itself' (Gudder 1973). However, for illustrative reasons, it is convenient to represent S as a closed subset of an affine topological space E. In the present day quantum theories S is the set of density matrices in a complex Hilbert space $\mathcal{H}$, $S = S_{\mathcal{H}} = \{x \in \mathcal{L}(\mathcal{H}) \colon x = x^* \geqslant 0, \ \mathrm{Tr}x = 1\}$, whereas in the classical theories S is a generalized simplex. Below, no particular geometry of S will be assumed. To say that statistical ensembles form a convex set does not yet imply any definite physical theory. The statistics of *anything* leads to a certain convex set (Alfsen). The information coded in S may seem weak: it tells only which states are mixtures of which other states. Recently, however, it was found that this information is rich enough to determine the analogues of the main geometric concepts of quantum theory. (Ludwig 1967, Mielnik, 1969, 1974, Alfsen and Schultz 1976, 1978.)

Transition Probabilities.

It turns out that S contains implicit information about the measuring devices. This is due to the following concept of a *normal functional.*

Definition 1. Given a convex set S in an affine topological space E, a *normal functional* is any linear, continuous real functional $\phi : E \to \mathbb{R}$ such that $0 \leqslant \phi(x) \leqslant 1$ for every $x \in S$. Geometrically, any non-trivial normal functional is represented by an ordered pair of closed hyperplanes in E (on which it takes the values 0 and 1) enclosing the convex set S. Physically, any normal functional may be interpreted as a counter endowed with the ability (either limited or perfect) to detect the ensemble individuals: for any state $x \in S$ the number $\phi(x)$ represents the average fraction of the individuals of the ensemble x which are 'noticed' by the counter ϕ. As is easily seen, the geometry of S, in general, imposes some limitations on the structure of normal functionals. Thus, for the convex set S in Fig. 2.1 no normal functional can exist which takes the value 1 at y and vanishes at x. For two arbitrary extremal points $x,y \in S$ define:

$$p(x,y) = \inf_{\phi(y)=1} \phi(x) \tag{2.1}$$

Obviously, $0 \leqslant p(x,y) \leqslant 1$ and $p(x,x) = 1$. The numbers $p(x,y)$ have been called *detection ratios* (Mielnik 1969). They determine an absolute selectivity limit for the statistical experiments. Given two pure states $x,y \in S$ with $p(x,y) > 0$. no filter can exist which would accept all systems in the state y but reject all systems in the state x. Every physical process which tolerates all y-systems must unavoidably tolerate at least the fraction $p(x,y)$ of x-systems. Although defined via normal functionals, the detection ratios depend only upon the geometry of S (Mielnik 1969, 1974). A special case of these quantities appears in orthodox quantum mechanics. Here, S is the set of the density matrices of $S = S_{\mathcal{H}}$, its extremal points being of the form $x = |\psi\rangle\langle\psi|$. Given any two extremal points $x = |\psi\rangle\langle\psi|$, $y = |\phi\rangle\langle\phi|$ (where ψ and ϕ are two unit vectors) the detection ratio becomes:

$$p(x,y) = |(\psi,\phi)|^2 \tag{2.2}$$

Hence, the quantities (2.1) are the proper analogues of the quantum mechanical transition probabilities. What seems significant, however, is that they exist for an arbitrary convex set and not only for $S = S_{\mathcal{H}}$.

'Quantum Logic'

In the description of quantum theory a special place belongs to the associated 'quantum logic' (Birkhoff and von Neumann 1936; Piron 1964, 1976). This aspect is also determined by the geometry of S. In that geometry one of the natural concepts is that of a *face*.

Definition 2. Given a convex set S, a *face* of S is any subset $R \subset S$ with two properties:

1. R too is convex, i.e. $x_1, x_2 \in R$, $p_1, p_2 \geqslant 0$, $p_1 + p_2 = 1 \Rightarrow p_1x_1 + p_2x_2 \in R$, and
2. $p_1x_1 + p_2x_2 \in R$ with $p_1, p_2 > 0$, $p_1 + p_2 = 1 \Rightarrow x_1, x_2 \in R$.

If S is a convex set in a topological affine space, the *closed* faces of S form a *complete lattice*. This lattice admits a simple physical interpretation (Mielnik 1976).

Statistical theory most easily describes the properties of the statistical ensembles. These ensembles correspond to the subsets $P \subset S$. The properties of single objects appear as a result of an abstraction process. The property P of the ensembles can define a property of a single object provided that two conditions hold: 1. whenever two ensembles possess the property P their mixtures must have it too, and 2. whenever a mixture has the property P, each of the mixture components must have it too. Mathematically this means that P should be a *face* of S. For reasons of regularity it seems plausible to assume that P is closed. Hence, the lattice of closed faces of S represents the properties of the single systems ordered according to their generality.

Definition 3. Two faces (properties) $P, R \subset S$ are called *orthogonal* ($P \perp R$) or *excluding*, if there is at least one normal functional taking the value 0 on P and 1 on Q (i.e. if there is a filter completely seperating objects with these two properties). The set $\mathbb{P}$ of all closed faces of S (properties) with the relation of inclusion ($\subset$) and exclusion ($\perp$) is a natural analogue of *quantum logic* (Finkelstein 1963). In general, however, this logic might not admit negation.

Definition 4. Let P be a closed face of S. If among all faces orthogonal to P the largest one exists, it will be denoted P' and called the *negation* of P. The existence of P' depends on the structure of the orthogonality relation ($\perp$). A particular case of the lattice $\mathbb{P}$ is obtained for $S = S_{\mathcal{H}}$. Here, the closed faces of $S_{\mathcal{H}}$ happen to be in one-to-one correspondence with the closed vector subspace of $\mathcal{H}$; the orthogonality of faces corresponds with the usual orthogonality of the subspaces. This means that the lattice of faces $\mathbb{P}$ is the right generalization of 'quantum logic'. What seems significant, however, is that this structure exists for an arbitrary convex S and not only for $S_{\mathcal{H}}$.

Present day quantum mechanics is merely a special case of this scheme. The convex set $S = S_{\mathcal{H}}$ here is distinguished by some exceptional symmetry properties. However, it seems strange that the whole scheme finds its application just for one type of convex geometry. This is as if the whole GRT was formulated to describe one cosmological model. Are the other models of a micro-universe impossible? The arguments aimed at reducing S to $S_{\mathcal{H}}$ already form an extensive domain of quantum theory. A number of axioms specifying the convex geometry of S have been formulated by Ludwig (1967); the most fundamental of them being that about the 'sensitivity increase'. The axioms of Ludwig have an advantage of being not 'unfairly obvious': they give a good description of the existing scheme without intimidating thoughts about other possibilities. In turn the 'spectral' axioms of Alfsen and Schultz (1976) describe S in a spirit of aesthetical principles. These axioms are equivalent to assuming that certain regularities typical of quantum theories should be generally preserved. Should they, however, be preserved? A special place in limiting the geometry of S belongs to the quantum logic axioms of Birkhoff and von Neumann (1936) and Piron (1964, 1976), which postulate the lattice structure of $\mathbb{P}$ and the existence of orthocomplementation. When thinking in terms of logic, they possess a certain suggestive naturality (e.g., if the 'properties' are the 'propositions' of a logic, it is natural to expect the existence of negation). In terms of convex geometry they mean some 'crystalline' symmetries of the face structure of S (each face defining exactly one 'opposite face' etc.). With these axioms Piron shows (1964, 1976) that the only irreducible model of 'quantum logic' with more than 3 mutually orthogonal pure states is isomorphic with the lattice of all closed subspaces of a Hilbert space $\mathcal{H}$. The Gleason theorem then implies $S = S_{\mathcal{H}}$. These results are sometimes taken as a proof that $S_{\mathcal{H}}$ is the only acceptable convex geometry for quantum theories. However, the axiomatic approach is a stick with two ends. As stated by Piron himself (1979) there is more than one game to be played with Piron's theorem. In fact, this theorem can be read back. Since the Hilbert space formalism is linear, Piron's result can be taken as a proof that the existence of negation in $\mathbb{P}$ (orthocomplementation) is essentially a feature of linear theories. Indeed, there are some strong indications that the orthocomplementation law has linearity at the bottom (Mielnik 1976). Hence, Piron's theorem might mean that whenever a non-linear variant of the theory is considered (e.g. Penrose 1975; Haag and Bannier 1978; Bialynicki-Birula and Mycielski 1976; Davies 1979*a*; Kibble 1978, 1979), the orthocomplementation is questionable, and an unrestricted geometry of S should be expected.

The most recent arguments restricting the shape of S are based on the axiom of symmetry of the transition probabilities: $p(y,x) = p(x,y)$, motivated by the intuition of time-reversal invariance (Haag (1980); Araki (1979)). A theorem about Jordan algebras then leads to $S = S_{\mathcal{H}}$ as the principal convex model. However, invariance under time reversal is a philosophical demand and the question arises, whether it must be generally valid. Significantly, the same point is being questioned at this conference (Penrose). This illustrates a certain natural limitation of the axiomatic approaches: after all, they can give only conditional evidence. However, it so happens, that no axioms are necessary to determine the geometry of *S*. Indeed, after looking at this geometry more closely one can see that it is of dynamical origin.

3. MOBILITY

Suppose, one has a set Φ of elements $\psi,\phi,\ldots$ denoting the pure states of a hypothetical system. Assume, that Φ is a topological space with a physically meaningful topology. Moreover, suppose Φ possesses also the structure of a generalized differential manifold (in general, of infinite dimension). No other structure (like that of a simplectic manifold) will for a moment be assumed on Φ. The manifold Φ is a phase space of the system.

In most physical theories the dynamics is introduced by postulating the existence of a certain one-parameter group of transformations of the phase space. This description, though plausible, is rather too narrow. The one-parameter group can only describe the behaviour of the system in fixed external surroundings (e.g. in vacuum). However, this is a poor description which cannot explain how the physical theory of the system has been created. To arrive at any dynamical theory, the physicist must be able to experiment with the system by changing the external conditions. This leads to a picture of an *open* system submerged in a variable exterior. To represent it properly I shall assume that *D* has a counterpart: a hypothetical set Ξ whose elements $\xi,\eta,\ldots$ denote the acts of 'external influences' which can be exerted on the system. (This means that the universe in which the system exists should contain some parameters which are not affected by the system itself but are in the experimenters' control.) It seems, that the proper relativistic image of ξs is that of asymptotic conditions. For a moment, however, I shall interpret every $\xi \in \Xi$ as a physical reality defined over all the space and acting on the system for a certain definite span of time (a good model is the impulse of an external field). The 'impulses' $\xi \in \Xi$ will be assumed dissipation free, so as not to introduce any statistical uncertainty into the behaviour of the system. Consistently, it will be assumed that every $\xi \in \Xi$ generates a unique transformation $a_\xi : \Phi \to \Phi$ mapping the pure states onto the pure states and representing the dynamical evolution of the system caused by the external input ξ. It seems natural to suppose that a_ξs should be diffeomorphisms of Φ. Now, the two philosophies of dynamics part. The traditional approach describes a system in a given external world. The only freedom left in a_ξ then is that of the time interval and one ends up with a one- or two-parameter family of transformations $\Phi \to \Phi$. This is no longer so in the case of the 'global picture' (Haag and Kastler 1964; Mielnik 1980). Here, one considers not the actual dynamical transformations, which the system happens to perform, but a wider class of all possible evolution transformations which the system is basically able to achieve under the influence of all 'external influences':

$$G = \{a_\xi : \xi \in \Xi\} . \tag{4.1}$$

Below, I shall assume that the 'external parameters' ξ can be applied at will, and when two are applied in succession ($\xi,\eta \to \xi\circ\eta$) the resulting transformation of the system is the superposition (product) of the two consecutive operations: $a_{\xi\circ\eta} = a_\xi a_\eta$. If this holds, the class of all achievable operations (3.1) becomes a semigroup. This semigroup, in general, is much wider than the traditionally employed one-parameter semigroups of the dynamical theories: it describes not actual motion of the system but its entire dynamical flexibility. It has been called the semigroup of *mobility* (Mielnik 1980).

Note, that the assumption about the possibility of superposing any two operations a_ξ, a_η is generally valid only in non-relativistic theories. It requires the assumption, that the external impulse η can be removed once the operation a_η is ended, and, as instantaneously, the impulse ξ created. This is true without restrictions only in the Galileo space-time. The modifications of the semigroup picture of mobility in the relativistic case will be discussed in Section 5.

In spite of the popularity of the fixed Hamiltonian, the mobility aspect is present in many dynamical theories. It exists in Haag and Kastler's (1964) approach to quantum field theory under the form of 'operations'. It intervenes in Schwinger's (1953) formalism of S-matrices depending on the external fields, as well as in the general concepts of open dynamical systems (Plebanski, Havas). It is closely related to the description of physical theories in terms of 'impotence principles' (Bergmann, Sudarshan). (In fact, an alternative way to define G would be to describe the transformations $a : \Phi \to \Phi$ which basically *cannot* be achieved.) The problem about the mobility of non-relativistic quantum systems in external fields has been raised by Lamb (1969) and Lubkin (1974) and solved in (Lubkin 1974; Mielnik 1977; Waniewski 1980) by showing that G coincides with the unitary group. This throws some light on the role of the unitary group in quantum mechanics: this group appears in the formalism of the theory since it represents the dynamical mobility. As a consequence, one can explain the Hilbert space structure of quantum mechanics as something conditioned by the mobility. This suggests, that in a general dynamical theory too, the geometry of the phase space is not fundamental but is of dynamical origin (Mielnik 1979). However, the principal consequence of the mobility concerns the *observables*.

4. FUNCTIONAL OBSERVABLES, AND CONSTRUCTION OF S.

The main structural element of present day theory is the *algebra of observables*. However, after looking for the motivations of that structure, one sees some peculiarities. The multiplication in the algebras of observables has no physical sense. The introduction of Jordan's algebras does not change that. Further, it can be seen that the quantum mechanical observable is an inhomogeneous concept. In it two aspects of reality are identified. One, is the operation performed by the measuring device. The other is the purely numerical output of the measurement. The self-adjoint operators in quantum theories are designed to bring a synthesis of these two aspects. The necessity of such a synthesis has never been clearly motivated and is a vulnerable point in the theory. Paradoxically, it is precisely this point which has become one of the most unquestionable features of quantum formalism. (Several generations of physicists already have trained themselves to feel that it is right that the observables are simultaneously the operators.) However, if the theory is to be indeed open this is the main point to be revised. Consistently I propose that the concept of the 'operator—observable' should be split and the original logical components restored. Below, *operations* will mean the transformations of the statistical ensembles induced by the macroscopic surroundings, an important part of them being represented by the mobility semigroup G. In turn, the *observables* will no longer stand for the operations, but will represent the purely numerical outputs of the measurements. This leads to the idea of a *functional observable* (Mielnik 1969, 1974; Alfsen and Schultz 1976, 1978; Haag and Bannier 1978; Kibble 1978, 1979) which seems applicable in all situations where the states of the system are not perceived directly but via some unpredictable statistical effects.

Definition 4. A *statistical experiment* is any mechanism assigning real numbers to the individuals of statistical ensembles. The mechanism is not assumed to have qualities of an idealized measurement: it may yield numerical answers in either a partly or completely random way. However, it is assumed that it *separates* the ensemble objects: i.e. it reacts with each single ensemble individual indepently of the existence and properties of the other ensemble individuals. Moreover, it is assumed that for every statistical ensemble (state) the expectation values for the numerical results are well defined.

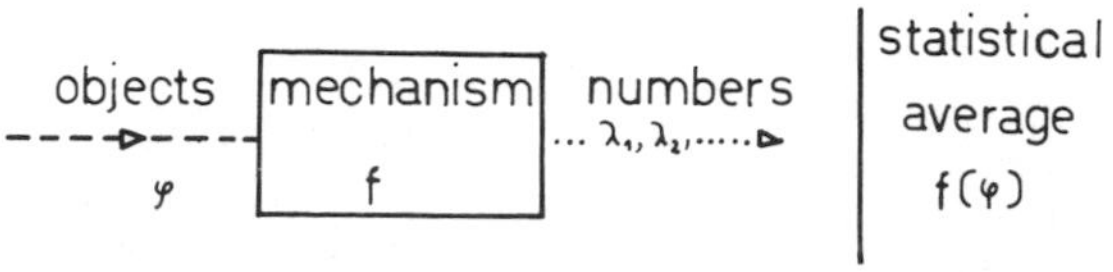

FIG. 4.1.

Definition 5. An *observable* is any continuous real function $f : \Phi \to \mathbb{R}$ whose values $f(\phi)$ are the statistical averages on the pure states $\phi \in \Phi$ of a certain statistical experiment.

In spite of a tolerant definition of the statistical

experiment in Fig. 4.1, it is not true that every real function on the phase space Φ is an observable. On the contrary, the observables, in general, form a narrow subclass of functions on Φ. This can be seen in orthodox quantum mechanics. Here, the manifold Φ is the unit sphere in a complex Hilbert space $\mathcal{H}$ and the only functions on Φ which are observables are the *quadratic forms*. The other functions of ψ, though they can be experimentally determined, cannot be directly measured as statistical averages. This impossibility is one of the most characteristic features of quantum mechanics. (It is also the true reason why observables here can be represented by the operators.) Since it is so, one can expect that an analogous distinguished class of functions should exist in a general theory. Given the manifold of pure states Φ of a hypothetical system, the nature of the system should be predicted by indicating which functions on Φ are the observables (Mielnik 1974, 1980). The class of these functions will be denoted by F. It might seem, that a class of functions is a poor structure compared to *-algebras. However, recent investigations show that the functional observables contain no less information than the orthodox algebras of observables (Alfsen and Schultz 1976, 1978).

The description of the theory in terms of functional observables happens to determine the convex set S. This is due to the following construction (Mielnik 1974). Given Φ, consider the set Π of all probability measures π on Φ. The elements $\pi \in \Pi$ may be interpreted as the *prescriptions* for producing the mixtures. The set Π has a natural convex structure (generalized simplex). The measures $\pi \in \Pi$, in general, are redundant in describing the mixed states, as two different prescriptions can define two equivalent mixtures. The observable properties of the mixed state prepared according to the prescription π are defined by: $f(\pi) = \int_\Phi f(\phi)d\pi(\phi)$. $(f \in F)$. Now, two measures $\pi, \pi' \in \Pi$ are equivalent $(\pi \equiv \pi')$ if $f(\pi) = f(\pi')$ for every $f \in F$. Having the equivalence $\equiv$ one can represent the *states* (pure or mixed) as the equivalence classes in Π. Thus, the convex set S emerges as the homomorphic image of the simplex Π as seen through the class of observables: $S = {}^{\Pi}/_{\equiv}$. In the case where Φ is the unit sphere and the observables f are the quadratic forms, the construction leads to $S = S_{\mathcal{H}}$. This seems to explain the origin of the particular convex geometry of quantum mechanics.

Now, the class of observables is conditioned by the mobility of the system. Given an operation a and an observable $f \in F$ corresponding to an arrangement in Fig. 4.1, one can arrange a new measurement by letting the system undergo the transformation a and then measuring f; the new statistical average being $(fa)(\phi) = f(a(\phi))$. Hence, F should be invariant under the mobility semigroup: $FG \subset F$. Knowing this, one can reconstruct F from more elementary statistical information. What one needs is a subclass $F_0 \subset F$ of 'elementary observables' which intervene at the end stages of arbitrary statistical experiments: the whole of F then being the linear span of GF_0. This is indeed so in orthodox quantum mechanics where the 'elementary observables' are Born's localization probabilities $f_\Omega(\phi) = \int_\Omega |\phi|^2 d_3x$, $\Omega \in \mathbb{R}^3$. Due to the quadratic character of these quantities, the linearity of the evolution processes and due to the mobility theorem (Mielnik 1977; Waniewski 1980) the resulting F is the class of all Hermitean quadratic forms on Φ and $S = S_{\mathcal{H}}$. In this way, the convex structure of orthodox quantum mechanics is fully explained. As it turns out, S is non-elementary and splits into three independent 'logical components':

$$\left.\begin{array}{l}\text{pure states}\\ \text{elementary observables}\\ \text{mobility}\end{array}\right\} \Rightarrow S\ .$$

The above construction can also be applied to the non-linear variants of wave mechanics which leads to certain convex geometries (Haag and Bannier 1978; Mielnik 1980). What one has here is a flexible mathematical mechanism interrelating the motion and the form of the theory. A variant of this mechanism exists also in a relativistic theory.

5. RELATIVISTIC MOBILITY

The idea of superposing operations, in its simplest form, is non-relativistic. Indeed, if the external conditions $\xi,\eta\ldots$ represent some physical fields in the space-time existing between space-like hyperplanes, the subsequent application of a_η and a_ξ would require, in general, a sudden change, namely that the remnants of the field ξ should be wiped out and some independent external field ξ established in the next strip of space-time. However, this is possible only in the Galilean space-time, where the changes to the external surrounding can be induced with infinite velocity and where the conditions on two sides of a hyperplane t = const are practically independent. This can no longer be accomplished in Minkowski space-time, where the external fields propagate with a finite velocity. However, there is here as well an element of freedom in the subsequent operations. It is best seen if the space-time is cut by a congruence of null and not space-like hyperplanes.

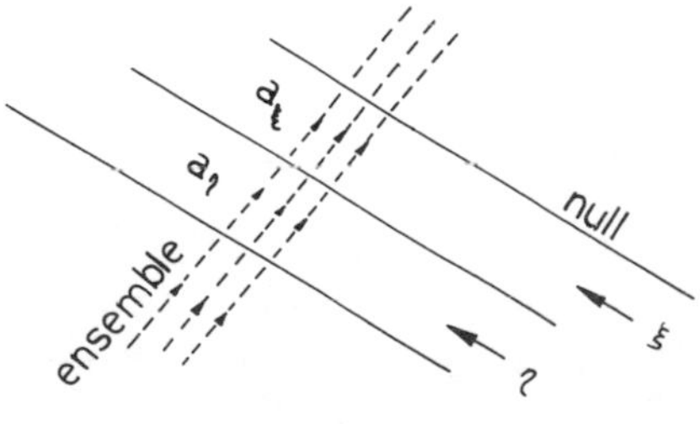

FIG. 5.1.

Assume, a particle beam crosses the null surfaces (Fig. 5.1) and the elements $\phi \in \Phi$ denote the states on the hypersurfaces of the congruence. Suppose the 'external influence' is due to certain physical fields; the parameters $\xi,\eta,\ldots$ symbolizing the experimenter's control are the asymptotic properties of the field in the past null infinity. If the experimenter originates the 'sandwich waves' propagating in between the characteristic surfaces in Fig. 5.1, the external conditions in the subsequent null strips are completely independent and the corresponding transformations a_ξ, a_η of particle states can be superposed. Thus, the aspect of mobility here reappears as a semigroup of transformations.

The situation is different if there is something in the space-time which spoils the independence of the subsequent operations. This might happen, for example, if the space-time is crossed by world-tubes of macroscopic bodies which scatter the external field and which cannot be removed by the experimenter (Fig. 5.2).

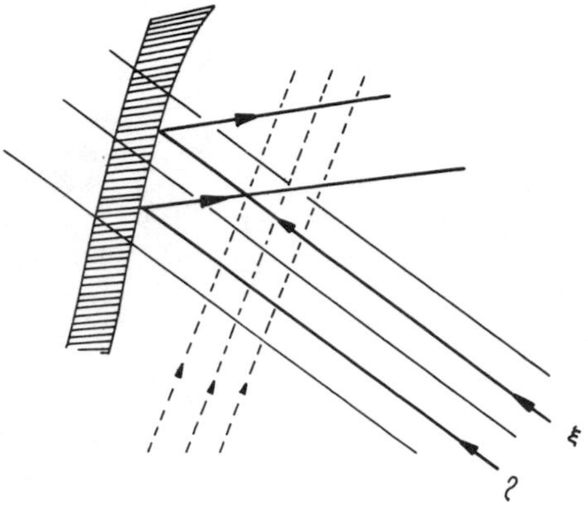

FIG. 5.2.

In this case, the effective external field in every null strip depends not only on its own asymptotic input, but also on the previous inputs. If a_ξ denotes the operation due to ξ alone, then the operation due to the composition $\xi \circ \eta$ of two subsequent inputs will no longer be the operator product: $a_{\xi \circ \eta} \neq a_\xi a_\eta$. The class of operations G is no longer the semigroup in the sense of operator multiplication and neither can be faithfully represented by the operator algebras. Still, G is a semigroup with respect to a modified composition law $a_\xi \circ a_\eta := a_{\xi \circ \eta}$ which no longer means the simple operator product. This makes G close to Randal and Foulis manuals with the asymptotic conditions $\xi, \eta, \ldots$ playing the role of the 'instructions' (Foulis and Randall 1979).

A still more general situation arises if the space-time is not flat and its geometry depends upon the external conditions $\xi, \eta, \ldots$ Given an 'initial state' defined on a certain null surface, the freedom of the subsequent dynamical operations means the branching of space-time into a variety of manifolds ('space-time leaves') endowed with different geometries and different contents of the external fields: the dynamical operations $a \in G$ bring the initial state ϕ into new spaces of states lying on different leaves (Fig. 5.3).

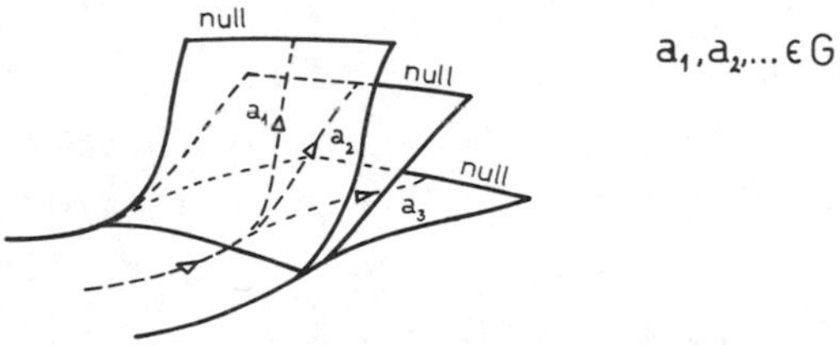

FIG. 5.3.

As the dynamical operations here cannot be meaningfully superposed, the mobility G is no longer a semigroup but just a class of operations. It reflects an aspect of freedom in the relativistic space-time, which cannot be adequately described by the formalism of the operator algebras.

The above picture of mobility suggests a revision of the quantum mechanical concept of a *state*. In present day quantum mechanics statistical interpretation tells of instantaneous checks on a particle's position at a certain given time. The resulting probability distribution is assigned to a certain space-like hyperplane, and so is the quantum mechanical concept of *pure state*. However, after looking at what indeed happens in quantum mechanical experiments, one may doubt whether this interpretation is realistic. How could a sudden check on the particle position be carried out? Apparently, the only possibility would be to have a macroscopic medium (vapour etc.) which could become suddenly 'aware' of the microparticle as a result of an internal phase transition. If the transition happened simultaneously in the whole of the medium, the result might lead to a sudden attempt at detecting the particle over all the space. This picture, though basically possible, is a gedanken—experiment and does not reflect the typical laboratory experience. What one has in the laboratory is usually a particle beam and a certain sensitive macroscopic medium (fluorescent screen, photographic emulsion etc). The medium does not make any sudden check at a given time. Instead, it is waiting. This is the micro-object which triggers the detection act. The moment in time when this happens is not in the experimenter's control.

The surface of the waiting screen in the space-time has one time and two space dimensions. An idea thus arises that the statistical interpretation of quantum mechanics should be reformulated: instead of defining the probability distribution for the space-like hypersurfaces it should do it for the 'vertical' surfaces marking the boundaries of the macroscopic bodies. In spite of interesting developments (Kijowski 1974; Piron 1978) this idea encounters some basic difficulties. If one assumes that the particle should be absorbed by the surface of the medium and not by its volume, the medium should be infinitely sensitive. Then, however, the effect of reflection dominates and the particle is rejected from the surface instead of being absorbed, making the whole problem about the surface probability distribution pointless (Lindbladt 1979; Davies 1979*b*). A similar conclusion follows from the 'Zeno paradox' for the repeated quantum measurement (Misra and Sudarshan 1977). It thus seems that there is no strictly surface-like probability distribution on the vertical surfaces. However, this counter-argument does not concern the null surfaces which therefore remain a tempting alternative. Significantly, a similar indication follows from the 'gedanken experiment' in which a hypothetical medium undergoes a transition from its 'particle blind' to its 'particle sensitive' phase. Here, in turn, it seems unlikely that the boundary between both phases in the space-time will be space-like. The internal processes preparing the phase transition in the medium have usually some external source (like heating a pot) and since the signals in the relativistic space-time propagate inside the light cones (except for special mechanisms, Plebanski 1970) it is unlikely that the progress of the internal state changes in the medium will be marked by space-like surfaces. They will be null surfaces at best. Hence, the statistical interpretation on characteristic surfaces is not against the spirit of quantum theory but seems a natural choice minimizing the interpretational difficulties.

Returning to the general situation in Fig. 5.3 one obtains a new picture of a statistical theory. In it, the initial *pure states* are assigned to a certain null surface. The aspect of *mobility* is represented by a variety of the space-time leaves and the corresponding class of the evolution operations. The *elementary observables* define the statistical interpretation on the null surfaces. The general observables are measured by sending the system to one of the possible space-time leaves (an operation $a \in G$) and then by measuring new elementary observables on a new null surface there. Hence, the full class of observables F is defined and the convex set S is determined by the construction of Section 3. The resulting scheme possesses some general features resembling quantum mechanics but is less restrictive. In particular, the functional observables and the mobility aspect seem to permit the statistics of non-linear quanta (Haag and Bannier 1978; Bialynicki-Birula 1976; Davies 1979*a*; Mielnik 1980). A variant of the scheme assuming still less (pure states and functional observables, no mobility) has recently been developed to formulate a non-linear quantum field theory (Kibble 1978, 1979). The question arises, whether this helps in gravity quantization? What certainly follows from the philosophy of Fig. 5.3 is that the non-linear graviton (if it exists) might be out of the formalism of the operator algebras. More specific predictions would be premature. What the whole scheme can offer at the moment is to increase our freedom on the 'quantum side of the abyss' be relaxing the linearity. However, what if the non-linear quanta do not exist and the micro-cosmos is linear? Personally, I share the intuitive disbelief of Penrose (1975) and Kibble (1978, 1979). But even if it is so (that nature is so inflexible), it is still good to think that it could be otherwise: it seems that the structure of our world can be much better understood if we compare it with a variety of fictitious universes which God has forgotten to create.

REFERENCES

Alfsen, E. and Schultz, W. (1976). *Mem. Am. Math. Soc.* **172**.

—— —— (1978). *Acta Math.* **140**, 155.

Araki, H. (1979). On Characterization of the State Space of Quantum Mechanics. Preprint.

Bialynicki-Birula, I. and Mycielski, J. (1976). *Ann. Phys.* **100**, 62.

Birkhoff, G. and von Neumann, J. (1936). *Ann. Math.* **37**, 823.

Davies, E.B. (1979*a*). *Commun. Math. Phys.* **64**, 191.

—— (1979*b*). Private discussion, Lausanne.

—— and Lewis, J.T. (1970). *Commun. Math. Phys.* **17**, 239.

Finkelstein, D. (1962). Private discussion, Warsaw.

—— (1963). The Logic of Quantum Physics. *Trans. Y.Y. Acad. Sci.*

Foulis, D.J. and Randall, C.H. (1979). In *Physical theory as logico-operational structure*. (ed. C.A. Hooker). Reidel, Dordrecht.

Gudder, S. (1973). *Commun. Math. Phys.* **29**, 249.

Haag, R. (1980). Private discussion, Warsaw.

—— and Bannier, U. (1978). *Commun. Math. Phys.* **60**, 1.

—— and Kastler, D. (1964). *J. math. Phys.* **5**, 848.

Kibble, T.W.B. (1978). *Commun. Math. Phys.* **64**, 73.

—— (1979). *Commun. Math. Phys.* **65**, 189.

Kijowski, J. (1974). *Rep. Math. Phys.* **6**, 361.

Lamb, W.E. Jr. (1969). *Phys. Today.* **22**, 23.

Lindbladt, G. (1979). Private discussion, Geneva.

Lubkin, E. (1974). *J. math. Phys.* **15**, 673.

Ludwig, G. (1967). *Z. Naturforsch.* **22a**, 1303; **22a**, 1324.

Mielnik, B. (1969). *Commun. Math. Phys.* **15**, 1.

—— (1974). *Commun. Math. Phys.* **37**, 221.

—— (1976). In *Quantum Mechanics, Determinism, Causality and Particles.* (ed. Flato *et al.*). Reidel, Dordrecht.

—— (1977). *Rep. Math. Phys.* **12**, 331.

—— (1979). Motion and Form. In *Workshop on Quantum Logic.* Erice, Majorana Centre.

—— (1980). *J. math. Phys.* **21**, 44.

Misner, C.W., Thorne, K.S., and Wheeler, J.A. (1973). *Gravitation.* Freeman and Co, San Fransisco.

Misra, B. and Sudarshan, E.C.G. (1977). *J. math. Phys.* **18**, 756.

Penrose, R. (1975). *Non-linear Graviton.* First Award Gravity Foundation Essay.

—— (1976). *Gen. Rel. Grav.* 7, 31.

Piron, C. (1964). *Helv. phys. Acta.* **37**, 439.

—— (1976). *Foundations of quantum physics.* W.A. Benjamin, Inc., Menlo Park, CA.

—— (1978). *C.R. Acad. Sc. Paris.* **t.286**,A, 713.

—— (1979). Private discussion, Geneva.

Plebanski, J.F. (1970). *Lectures on Non-linear Electrodynamics.* NORDITA, Copenhagen.

—— (1975). *J. math. Phys.* **16**, 2395.

—— and Robinson, I. (1976). *Phys. Rev. Lett.* **37**, 493.

—— —— (1977). In *Group theoretical methods in physics.* (ed. R.T. Sharp and B. Kolman). Academic Press, N.Y.

Schwinger, J. (1953). *Phys. Rev.* **91**, 713.

Waniewski, J. (1980). *Commun. Math. Phys.* **76**, 27.

29

THE QUANTIZED RESPONSE OF A STABLE PARTICLE OF TRAPPED ENERGY AND ITS BEHAVIOUR IN A GRAVITATIONAL FIELD

R.C. Jennison

University of Kent at Canterbury

1. INTRODUCTION

It is usually assumed that quantization is an intrinsic property of the communicating fields which relate particles of matter. The photon is the classical example of the quantization of the electromagnetic field and the graviton has been suggested for the gravitational field. It is then assumed that the particles of matter respond in a quantized manner because of the stimulus from these quantized intercommunication fields and it has not been usual to consider that the particles themselves may have an in-built transfer function which requires them always to respond in a quantized manner irrespective of the continuity of the input field. Recent research into the problems of relativistic rigidity has shown that the conservation of proper length and proper time exhibited by a relativistically rigid body such as a proton, requires that the particles have an intrinsically quantized transfer function between the input stimulus and the resulting action. This gives rise to a range of properties, many of which are long established but now have an alternative and common explanation, dispensing with many previous postulates, whilst others, such as that discussed in this paper, are entirely new.

Consider an arbitrary particle of matter. Let this particle have inertial rest mass and a very long lifetime such that the internal energy in the proper frame of the particle is invariant over the time scale of interest, Furthermore let this proper internal energy and the proper length of the particle be the same when the proper measurements are repeated after the particle has been accelerated to another inertial frame. Since no particle can have absolute rigidity, it is clear that the internal energy distribution must be perturbed when it is moved but it is also clear that the particle cannot exhibit this internal perturbation as excess proper energy after it has come to rest in the second inertial frame, although it will have gained in total energy relative to the frame in which it was originally at rest by the acquisition of overall kinetic energy. The invariance of the proper energy and proper length imply that the redistribution of energy must be such that the internal perturbations are internally reflected or refracted such that they are self-cancelling in the proper frame after the removal of the perturbing agent and any excess must be re-radiated. Particles which rigorously satisfy all these requirements have been analyzed under the general title of phase-locked cavities. The expression 'phase-locking' in this context refers to the perfect continuity in the recycling of internal energy so that the dimensions of the particle adjust automatically to fit the internal fundamental wave system carrying the internal energy; thus the proper length is controlled entirely by the wavelength of this wave and is related to it by simple geometry.

The finite dimensions of these particles depend upon the auto-correlation of the internal wave system, continuously comparing the present state with that at an earlier epoch. This is maintained by the 100 per cent feedback of the internal wave system. It gives the particles the property of relativistic rigidity and ensures that they act as proper units of length and time (since the internal period = $1/\nu_0 = \lambda_0/c$). They are thus central to the whole of the theory of relativity and transform accordingly. This is an important and fundamental point. The mechanism is such that the proper units are transferred by the particle to any new state that it may acquire and the relative measurements to an external observer then follow as a necessary outcome of the geometry and the measurement procedure. It has been shown that the delayed response of the particles accounts very simply for Newton's first and second laws but the analysis shows that the laws are quantized and that the particle does not lock into the new state of motion until the feedback cycle is complete (Jennison and Drinkwater 1977). This phenomenon is extremely precise and it accounts quite rigorously for the energies and momenta in the Compton effect (Jennison 1978). It has been shown very recently that phase-locked particles have spin ½, which further identifies their properties with those of everyday particles of matter (Jennison 1980). At the opposite extreme, a variety of macroscopic phase-locked particles have been constructed in the laboratory and all of their major properties have been demonstrated.

One aspect of the general properties of phase-locked particles is worthy of further comment in this account. A phase-locked particle derives its inertial reactive force entirely from the closed feedback circuit of its own internal field systems. It comes about simply from the change in the flux of the internal fields when the boundaries are moved. It is quantitatively and qualitatively precise and in the local (proper) frame it is independent of the influence of the distant masses of the universe except insofar as these, and any other large masses, may influence the direction and curvature of the local null geodesics sampled within the particles. This does not change the magnitude of the quantities in the proper frame because the trapped and free wave systems are equally affected so that the inertial rest mass retains its value and the reactive force remains inviolate even if the mass content of the universe is drastically reduced or if a super massive system is in the vicinity. The effect of such a super massive system is simply to redefine a 'straight line' in the system of the particle by changing the geometry of the local space and it will, of course, affect the relative magnitude of quantities measured at a distance but the particle preserves its own proper units as these are protected by the basic memory of the system.

It may be noted that the effect of the gravitational potential on electromagnetic waves, and on particles formed from or converted into these waves (by pair production and annihilation), is such that the local relationships are conserved. For example the annihilation wavelength, or Compton wavelength, of an electron is always the same at the relevant electron and all electrons formed in any region are identical to those which have arrived there from quite different parts of space. If this were not so then the physical properties of matter would become so randomized that the laws of physics could no longer hold and spectra, in particular, would be meaningless. Thus the local or proper parameters which determine the inertial rest mass and the associated inertial reactive force of a phase-locked particle, always retain the same values in proper terms and are not affected by the changes in gravitational potential, albeit they are *relatively* different to those at another potential or another velocity. There is therefore a very different situation to that in which Mach's principle is called upon to account for the origin and magnitude of the reactive force of a particle and, given the fact that one can produce macroscopic particles on the phase-locked principle, one cannot also have a second, Machian, agent to produce the inertial phenomena; one criminal is not only sufficient but mandatory, two would produce chaos with Newton's law varying at will.

It is, however, interesting in the context of this conference on quantum gravity, to note that if one accepts the Machian gravitational origin of the inertial reaction force, then the whole universe must be filled with quantized gravity, for all actions are known to be quantized and it follows inexorably that all reactions must also be quantized, yet the total energy involved is enormous. This reactive quantization must be gravitational on the Machian hypothesis, for no other connection between local and distant matter has been proposed and even if one endeavours to construct a physics in which there is no delay between the action and the reaction from the limits of the universe, there remains a very real problem in accommodating the sudden quantum changes of the gravitational reactive component without propagation of the correspondingly energetic gravitational quanta. If, on the other hand, one accepts that the particles of matter are finite relativistically rigid bodies then the application of the principles proposed by the author is appropriate and the reactive quantization is simply an internal phenomenon within the finite particles themselves.

2. THE NODAL FREQUENCY

In this paper, it will be shown that some categories of phase-locked particles may possess a unique property associated with a frequency that is related to the force acting upon them.

From Jennison and Drinkwater (1977), the relationship between the internal incremental velocity, v, of a phase-locked particle during the feedback time, δt, and the resulting mean acceleration, a, measured at the node of the cavity, is given by

$$a = vc/L \ , \tag{2.1}$$

where $2L = c\delta t$ corresponds to the feedback transmission path length of the particle in its proper frame, L depends upon the internal wavelength and is related to the physical size or radius of the particle by a factor K which depends also upon the cavity mode of the particle. At high values of a the length L may be replaced by $L' = L(1 - v^2/c^2)^{\frac{1}{2}}$.

For a symmetrical, centrally noded particle in these circumstances, there is a difference of frequency between the forward and trailing components at the node, corresponding to the difference in Doppler shift imposed within the particle by the acceleration. Jennison (1978) shows that this couples to an external frequency ν' at the node:

$$\nu' = \nu_0 \frac{v/c}{(1 - v^2/c^2)^{\frac{1}{2}}} \tag{2.2}$$

where ν_0 is the internal frequency of the particle.

From (2.1) and (2.2)

$$\frac{\nu'}{\nu_0}(1 - v^2/c^2)^{\frac{1}{2}} = aL(1 - v^2/c^2)^{\frac{1}{2}}/c^2$$

$$\nu' = \frac{aK\lambda_0\nu_0}{c^2}$$

$$= \frac{aK}{c} \ . \tag{2.3}$$

Note that this is independent of the scale or rest mass of the phase-locked particle but depends only upon its modal geometry or its 'form factor' K. For this particular (push-pull) phase-locked particle, ν' represents a cyclic pulsation of the magnetic field and of the effective inertial mass if the internal radiation is plane polarized, or a rotational precession about a vector parallel to the axis of acceleration if the internal radiation is rotational.

Equation (2.3) shows that when a force acts upon this type of particle it must always be accompanied by, or give rise to, a slow pulsation or rotation. Such a rotation may be considered as a redistribution of the original magnetic moment and angular momentum of the system (cf. the precession of a gyroscope) and need not imply non-conservation. Whether or not it may couple to an external mechanical or magnetic system has not yet been established, and it is of interest to compute the order of magnitude of the phenomenon in various circumstances. The frequency ν' is very low in the majority of man-made circumstances, for example if a = 10.0 ms^{-2} then v' is of the order of K cycles per annum. In order to assess whether or not such a phenomenon would apply when a particle is simply held against the earth's gravity at the surface of the earth, it is necessary to compare the behaviour of a phase-locked particle of this type in curved space with that of a similar kinematically accelerated particle in flat space.

3. THE PHASE-LOCKED CAVITY IN GRAVITATIONALLY DISTORTED SPACE

It is first necessary to see if the phase-locked particle exhibits the property of gravitational rest mass. In order to investigate this problem it is useful to commence by determining the change of symmetry associated with the Doppler shift within a minute phase-locked particle when it is kinematically accelerated on a linear path in flat space. Insertion of the appropriate values into the equation given in Jennison 1980, shows that for a particle comparable to the proton, the blue shifted and red shifted wavelengths which are proportional to the compressed front and rarefied rear parts of the particle are $\lambda\ (1 \pm 1.088 \times 10^{-31})$ m for an acceleration of $g = 9.7802\ \text{ms}^{-2}$. The difference in the positive and negative values gives the effective imbalance at the node under this acceleration. It is remarkable that this simple process is so sensitive to such a small imbalance and we shall now see if a similar mechanism may affect a phase-locked particle subjected only to the gravitational distortion of space.

Einstein, in Section 13 of his paper on General Relativity

(1916) *Equations of motion of a material point in the gravitational field*, clearly associated the gravitational rest mass of a material particle with a point and the theory beautifully describes the nature of space in these terms. There is no question as to the correctness of the General Theory but nevertheless it is well known that there is no explanation of how a point mass, deprived of the concept of a force acting over a macroscopic distance, is able to assess the gradient of the metric in order to follow a preferred path when it is in a state of freedom, or to exert a specific force on a fixed support when it is held captive in curved space. In the absence of action at a distance it does not follow that a point mass can assess the relative potential of neighbouring space, or the relative metric of that space, in order to know in which direction to fall or to exert a pressure according to the relative potential. In order to possess this property it is necessary for the test mass to have finite physical extent and to act as a differential operator which may assess the local gradient of the metric or gravitational potential in the immediate environment, and transfer this into a local force. This property is possessed by a phase-locked particle.

Where a phase-locked particle is imbedded in a region where the spatial metric is uniform but at a high gravitational potential, it is clear that, though the system will be balanced and in equilibrium, its relative dimensions will differ from those of a particle in locally flat space at a different potential. Since in each case the local invariance of the velocity of light determines that $c = \nu\lambda$, the internal frequency in each case, constituting the mechanism of its role as a proper clock, will also be in accord with general relativity.

Thus in these circumstances, measurements of lengths and time intervals between frames at different gravitational potentials will be scaled in accordance with general relativity, but no differential stress appears across a particle in either uniform region and the particles have no tendency to fall. Now consider a phase-locked particle on the surface of a massive body such as the earth, where the metric is no longer uniform. The only way in which this can exert a force at the node is for there to be a difference in the flux from the upper and lower halves.

There will be a very small difference in the velocity of light, c_N, at the central node of the cavity relative to the mean velocity of light c_1 in the upper- and c_2 in the lower feedback loops of the internal radiation.

From the General Theory of Relativity the radial component of the velocity of light, in the vicinity of a mass M, is given by

$$c_R = c\left(1 - \frac{2MG}{c^2R}\right) \tag{3.1}$$

in which c is the velocity of light in distance flat space. Thus, to a high degree of accuracy in the present context:

$$\frac{c_1 - c_2}{c_N} = \frac{2GM}{c^2R^2_N}(R_1 - R_2) \tag{3.2}$$

where R_N, R_1 and R_2 are the radii from the centre of the earth to the central node and to shells passing mid-way through the upper and lower feedback loops of the particle, respectively. Thus $(R_1 - R_2)$ is roughly equal in length to the radius of the particular particle.

Expressing eqn. (3.2) in terms of the incremental velocities relative to the value c_N at the central node, noting that, by the principle of relativity, c_N will be interpreted as c by an observer at the node and substituting the nominal radius of a phase-locked particle corresponding to the proton, we obtain, at the equator of the earth,

$$1 \pm v_p/c = 1 \pm 1.091 \times 10^{-31} \tag{3.3}$$

The difference in the positive and negative values gives the effective imbalance at the node but in this case the particle is at rest and the imbalance arises from the distortion of space.

The calculation illustrates how the phase-locked particle is a natural sensor of the gradient of the gravitational potential and thereby the instrument by which the resulting force is manifest. It shows that the principle is complementary and not an identity imposed by gravitational influences on both sides of the equation. A particle which is restrained in a gravitational field suffers a force entirely from the asymmetry imposed by the local metric, whereas a particle which is kinematically accelerated suffers a similar but complementary force from the complementary asymmetry imposed by the motion of the node relative to its own trapped radiation. If the support is removed from a particle which was previously restrained in a gravitational field, it will move in such a way as to endeavour to achieve symmetry in the local metric, for otherwise it would still be stressed; thus it must adopt an internal incremental velocity and thence an ultimate acceleration that the gravitational and kinematic distortions perfectly cancel.

The principle of the phase-locked particle is the only mechanism that has yet been proposed for translating fields of force or potential gradients into the local prime-moving force on particles of matter, for it is capable of assessing the presence and precise magnitude of the gradient and translating this into either a local force or an acceleration according to whether it is constrained or free. It is therefore proper to suggest that this same mechanism is applicable to the action of all forces on particulate matter, with consequent relevance to the nuclear forces.

The author suggested recently that all 'rigid' particles with invariant rest energy (J_o particles) may possess this property together with inertia and quantization. We limit the analysis to the case of push-pull phase-locked particles. These on the earth's surface should suffer a continuous internal differential frequency giving rise to a rotation (or vibration for a simple man-made particle) given by ν' in eqn. (2.3) where $a = g$ is now the equivalent acceleration due to gravity, i.e. the acceleration that the particle would have if the support were to be removed. For a particle on the earth's surface, the period $1/\nu' = 0.97/K$ years. It is not yet clear how this might be detected or whether it is likely to couple with either the magnetic or mechanical spin frequency of the earth. It could possibly be associated with the twist of biological molecules irrespective of latitude on the earth's surface. If one uses a value of K corresponding to the ratio of the classical electron radius to the Compton wavelength, then the period is similar to that of the difference of spin frequencies of the earth's mechanical and magnetic systems.

The origins of the earth are too fraught with uncertainty to be able to trace this frequency from its formative years but this may not be so in the case of stellar bodies. Consider the mechanics of the self gravitation of a cloud of interstellar gas. It is known that the collapse of the cloud may proceed until an embryonic self supporting system forms out of the gas. The rotation of the new star is assumed to come from the initial shear and turbulent eddies in the interstellar gas. The condensing stars of a particular spectral type and mass have a narrow range of rotational periods, and although, for many stars, the environment appears to provide too rich a source of angular momentum, it is of interest to enquire whether or not an intrinsic coupling with the gravitational field may be responsible for a part of the initial angular momentum or magnetic moment especially in the presence of an aligning process such as a magnetic field. It is a simple matter to compute ν' for a newly formed star (perhaps the symbol ν^* is more apt in this instance) and it shows a remarkable order of agreement with observed parameters. Even in the case of an established main sequence G type star such as the sun, $1/\nu^*$ is $14.3/K$ days per cycle in the surface layers and progressively longer at greater depths. It can be shown that K may be almost unity for a proton and, as hydrogen predominates in the solar interior, the period is not dissimilar to the rotation period. One model of the electron gives a value of $K = 1/861$ resulting in a period of the same order as that of the sunspot cycle. This is of interest in view of the slipping frequency between the earth's mechanical and magnetic spin frequency, mentioned earlier.

In the later stages of the life of a star, the conservation of angular momentum on the one hand and the shedding of angular momentum with the material of outer layers on the other, renders the extrapolation more difficult except insofar as the principle could still be operative in even the most highly condensed matter, provided that particles are still present. For a typical white dwarf star, the period $1/\nu^*$ is of the order of $1/K$ minutes and at the surface of a neutron star

$\nu^* \gtrsim K \times 10$ Hz (giving a nodal frequency $\gtrsim 1$ Hz for an electron with $K = 1/861$): these figures do not allow for the reduction of the effective g due to centrifugal acceleration, the effect of which will be to lengthen the periods, and they assume a homogeneous distribution of mass.

4. STATIONARY STATES OF AN ACCELERATED PARTICLE

Charged or magnetized particles which are accelerated radiate electromagnetic waves under all normal circumstances. The Bohr orbits are the exception to the rule and lack of knowledge of the physical reason for this exception has prevented the practical exploitation of the principle to particles in non-atomic accelerated systems. It is unlikely that gravitational radiation is emitted by an accelerated mass in a stationary orbit for this would cause the ultimate decay of the orbit, albeit in a very long time scale.

The present analysis shows that the acceleration of some phase-locked particles is accompanied by a characteristic frequency. If a perturbation can be applied to the particle such that it induces a precisely equal and opposite effect, then it would seem probable that the particle will not experience the effect of acceleration and accordingly may not radiate.

Consider the case of centrifugal acceleration in a central field of electric force. Let the angular frequency be ω and the angular acceleration $a = \omega^2 r$. The nodal frequency will be

$$\nu' = \frac{aK}{c} = \frac{\omega^2 r}{c} K \, . \tag{4.1}$$

We seek a natural phenomenon which may resonate with this frequency but we are hampered by lack of knowledge of K. It may be that ν synchronises with an asymmetry in the spinning structure of the proton or the electron. Nevertheless two natural frequencies are immediately available for consideration, one of these is the orbital frequency (or spin if the orbital radius is the particle radius) and the other is the Somerfield precession resulting from the relativistic change of mass.

The first case gives:

$$\frac{\omega^2 r}{c} K = \frac{\omega}{2\pi}$$

$$1 > K = \frac{c}{2\pi\omega r} > \frac{1}{2\pi} . \qquad (4.2)$$

If we consider the specific case of an electron then $1/2\pi$ is certainly a possible value for K, corresponding to the tentative model discussed at the end of Jennison (1978) but it may alternatively apply to the proton. Using this value of K for an orbiting electron, it is found that synchronism should occur after every 137 $(1/\alpha)$ Bohr orbits. Another value of K may be obtained using the ratio of the classical electron radius, e^2/mc^2 to the annihilation wavelength of the electron. This ratio is $\alpha/2\pi$ where α is the fine structure constant, $(2\pi e^2)/(hc) \approx 1/137$, giving $K_e = e^2/hc = 1/861 = 1.162 \times 10^{-3}$.

Turning now to the precession caused by the relativistic variation of mass of an orbiting particle, this gives rise to a term in v^2/c^2 times the orbiting frequency. In order to assess if this precession may synchronise with the nodal frequency of the orbiting particle, we put

$$\nu' = \frac{v^2}{c^2} \cdot \frac{\omega}{2\pi}$$

whence $$\frac{\omega^2 r}{c} K = \frac{\omega^2 r^2}{c^2} \cdot \frac{\omega}{2\pi}$$

$$K = \frac{\omega r}{2\pi c} .$$

Now putting $K_e = \alpha/2\pi$, we obtain

$$\frac{\alpha}{2\pi} = \frac{\omega r}{2\pi c}$$

and for harmonic synchronisation of the pth harmonic

$$\omega r/c = \alpha/p \qquad (4.3)$$

which is precisely the condition for the Bohr orbits. This result may, however, be fortuitous for the solution is not unique.

5. CONCLUSION

One of the more interesting predictions of this analysis is that it may be possible to induce artificially a synchronous perturbation which may neutralize the local effects of acceleration on a fundamental particle, thereby producing a state analogous to the free fall conditions in a gravitational field in which the particle itself is not stressed. Whether or not this condition may also be non-radiative will require further investigation.

The remarkably near coincidence of the predicted frequency with that for the drift of the earth's magnetic field system, the sunspot cycle, pulsar frequencies etc. is intriguing but may be purely accidental and it should be noted that, for simple models of the particles, there is usually a factor of two between the magnetic periodicity and that of the complex Poynting vector for a linearly polarised wave. If the internal wave is circularly polarised the force is constant but the field rotates at the wave frequency.

A definitive statement of the phenomenon referred to in this paper is still required but it may possibly be that, when a body with a magnetic field is subject to acceleration or held against a gravitational field, the magnetic field associated with the particles of that body will slowly precess. Thus if the overall magnetic field aligns the particles, it too may show a modulation or relative rotation at the precessional frequency. However the analysis does not apply to many other models satisfying the conditions for J_0 particles.

REFERENCES

Einstein, A. (1916). *Ann. Physik*. **49**.
Jennison, R.C. (1978). *J. Phys*. **A11**, 1525.
—— (1979). *Wireless Wld*. **85**, 1522.
—— (1980). *J. Phys*. **A13**, 2247.
—— and Drinkwater, A.J. (1977). *J. Phys*. A10, 167.